Protein-Nucleic Acid Interactions
Structural Biology

RSC Biomolecular Sciences

Editorial Board:

Professor Stephen Neidle (Chairman), *The School of Pharmacy, University of London, UK*
Dr Simon F. Campbell CBE, FRS
Dr Marius Clore, *National Institutes of Health, USA*
Professor David M.J. Lilley FRS, *University of Dundee, UK*

This Series is devoted to coverage of the interface between the chemical and biological sciences, especially structural biology, chemical biology, bio- and chemo-informatics, drug discovery and development, chemical enzymology and biophysical chemistry. Ideal as reference and state-of-the-art guides at the graduate and post-graduate level.

Titles in the Series:

Biophysical and Structural Aspects of Bioenergetics
Edited by Mårten Wikström, *University of Helsinki, Finland*
Computational and Structural Approaches to Drug Discovery: Ligand-Protein Interactions
Edited by Robert M. Stroud and Janet Finer-Moore, *University of California in San Francisco, San Francisco, CA, USA*
Exploiting Chemical Diversity for Drug Discovery
Edited by Paul A. Bartlett, *Department of Chemistry, University of California, Berkeley* and Michael Entzeroth, *S*Bio Pte Ltd, Singapore*
Metabolomics, Metabonomics and Metabolite Profiling
Edited by William J. Griffiths, *University of London, The School of Pharmacy, University of London, London, UK*
Protein–Carbohydrate Interactions in Infectious Disease
Edited by Carole A. Bewley, *National Institutes of Health, Bethesda, Maryland, USA*
Protein-Nucleic Acid Interactions: Structural Biology
Edited by Phoebe A. Rice, *Department of Biochemistry & Molecular Biology, The University of Chicago, Chicago IL, USA* and Carl C. Correll, *Dept of Biochemistry and Molecular Biology, Rosalind Franklin University, North Chicago, IL, USA*
Quadruplex Nucleic Acids
Edited by Stephen Neidle, *The School of Pharmacy, University of London, London, UK* and Shankar Balasubramanian, *Department of Chemistry, University of Cambridge, Cambridge, UK*
Ribozymes and RNA Catalysis
Edited by David M.J. Lilley FRS, *University of Dundee, Dundee, UK* and Fritz Eckstein, *Max-Planck-Institut for Experimental Medicine, Goettingen, Germany*
Sequence-specific DNA Binding Agents
Edited by Michael Waring, *Department of Pharmacology, University of Cambridge, Cambridge, UK*
Structural Biology of Membrane Proteins
Edited by Reinhard Grisshammer and Susan K. Buchanan, *Laboratory of Molecular Biology, National Institutes of Health, Bethesda, Maryland, USA*
Structure-based Drug Discovery: An Overview
Edited by Roderick E. Hubbard, *University of York, UK and Vernalis (R&D) Ltd, Cambridge, UK*

Visit our website on www.rsc.org/biomolecularsciences

For further information please contact:
Sales and Customer Care, Royal Society of Chemistry, Thomas Graham House, Science Park, Milton Road, Cambridge, CB4 0WF, UK
Telephone: +44 (0)1223 432360, Fax: +44 (0)1223 426017, Email: sales@rsc.org

Protein-Nucleic Acid Interactions
Structural Biology

Edited by

Phoebe A. Rice
Department of Biochemistry & Molecular Biology, The University of Chicago, Chicago, IL, USA

Carl C. Correll
Department of Biochemistry and Molecular Biology, Rosalind Franklin University of Medicine and Science, North Chicago, IL, USA

RSCPublishing

ISBN: 978-0-85404-272-2

A catalogue record for this book is available from the British Library

© Royal Society of Chemistry 2008

All rights reserved

Apart from fair dealing for the purposes of research for non-commercial purposes or for private study, criticism or review, as permitted under the Copyright, Designs and Patents Act 1988 and the Copyright and Related Rights Regulations 2003, this publication may not be reproduced, stored or transmitted, in any form or by any means, without the prior permission in writing of The Royal Society of Chemistry or the copyright owner, or in the case of reproduction in accordance with the terms of licences issued by the Copyright Licensing Agency in the UK, or in accordance with the terms of the licences issued by the appropriate Reproduction Rights Organization outside the UK. Enquiries concerning reproduction outside the terms stated here should be sent to The Royal Society of Chemistry at the address printed on this page.

Published by The Royal Society of Chemistry,
Thomas Graham House, Science Park, Milton Road,
Cambridge CB4 0WF, UK

Registered Charity Number 207890

For further information see our web site at www.rsc.org

Preface

The structural biology of protein-nucleic acid interactions is in some ways a mature field and in others in its infancy. High-resolution structures of protein-DNA complexes have been studied since the mid 1980s. A vast array of such structures has now been determined, but surprising and novel structures still appear quite frequently.

High-resolution structures of protein-RNA complexes were relatively rare until the last decade. Propelled by advances in technology as well as a blossoming realization of RNA's importance to biology, the number of example structures has ballooned in recent years. As with many other fields, the deeper one digs, the more questions surface. New insights are now being gained from comparative studies only recently made possible due to the size of the database, as well as from careful biochemical and biophysical studies.

The field has, in some ways, been a victim of its own success: it is no longer possible to write a comprehensive review. Instead, current review articles tend to focus on particular subtopics of interest. This makes it difficult for newcomers to the field to attain a solid understanding of the basics. One goal of this book is therefore to provide in-depth discussions of the fundamental principles of protein-nucleic acid interactions as well as to illustrate those fundamentals with up-to-date and fascinating examples for those who already possess some familiarity with the field.

This book also aims to bridge the gap between the DNA- and the RNA-centric views of nucleic acid–protein recognition, which are often treated as separate fields. However, this is a false dichotomy because protein–DNA and protein–RNA interactions share many general principles. This book therefore includes relevant examples from both sides, and frames discussions of the fundamentals in terms that are relevant to both. History supports this approach: despite the amazing conformational versatility discovered for RNA, many of the lessons learned from early studies of protein-DNA complexes

RSC Biomolecular Sciences
Protein-Nucleic Acid Interactions: Structural Biology
Edited by Phoebe A. Rice and Carl C. Correll
© Royal Society of Chemistry 2008

could be applied directly to understanding newer protein-RNA complexes. Conversely, recent structures of proteins bound to noncanonical DNA structures reveal recognition strategies more commonly associated with RNA-binding proteins.

We have assembled a team of experts to write the individual chapters of this book. The beginning chapters (1–8) focus on more fundamental aspects of protein-nucleic acid interactions, such as thermodynamics and recognition strategies, while later chapters (9–14) highlight more specialized topics. Since it is impossible to cover all aspects of this rapidly expanding field, we have chosen to highlight a few topics that are fascinating in their own right while also providing a broad range of examples to underscore the basic principles laid out earlier. We hope that readers at all levels will find this an interesting guide and a useful reference.

Phoebe A. Rice
Carl C. Correll

Contents

Chapter 1 Introduction
Carl C. Correll and Phoebe A. Rice

1.1	Overview	1
1.2	Fundamentals of DNA and RNA Structure	1
	1.2.1 Stabilizing Forces	1
	1.2.2 Chemical Differences between DNA and RNA	3
	1.2.3 Canonical A- and B-form Helices	4
	1.2.4 Deviation is the Norm	6
	1.2.5 Bending and Supercoiling DNA	6
	1.2.6 Folded RNA and Noncanonical DNA	7
1.3	Principles of Recognition	7
	1.3.1 Forces that Contribute to Complex Formation	8
	1.3.2 Site Recognition Overview	8
	1.3.3 Recognizing Duplex DNA *via* Direct and Indirect Readout	8
	1.3.4 Recognizing Single-stranded Nucleic Acids	9
	1.3.5 Recognizing Folded RNAs	9
	1.3.6 Recognizing Noncanonical DNA Structures	10
	1.3.7 Conformational Rearrangements	10
1.4	Future Directions	11
	References	11

RSC Biomolecular Sciences
Protein-Nucleic Acid Interactions: Structural Biology
Edited by Phoebe A. Rice and Carl C. Correll
© Royal Society of Chemistry 2008

Chapter 2 Role of Water and Effects of Small Ions in Site-specific Protein-DNA Interactions
Linda Jen-Jacobson and Lewis A. Jacobson

2.1	Introduction	13
2.2	Affinity and Specificity	14
2.3	Macromolecular Hydration Influences $\Delta H°$, $\Delta S°$ and $\Delta C°_P$	15
2.4	Water Release Attending Protein-DNA Association	17
2.5	Retained Water Molecules Contribute to Affinity and Specificity	26
2.6	Thermodynamic Effects of Retained Water	27
2.7	Overview of Small Ion Effects on Protein-DNA Interactions	28
2.8	Multiple Physical Phenomena Associated with Salt Dependence	28
2.9	Cation Release Favors Protein-DNA Association	30
2.10	Selective Effects of Anions on Protein-DNA Binding	32
2.11	Divalent Cation Binding at Active Sites Relieves Electrostatic Strain	34
2.12	Ion Effects and Cosolute Effects are Mechanistically Independent	36
2.13	Comparison with Nonspecific Binding: How Water and Ions Affect Specificity	36
2.14	Conclusions	39
Acknowledgements		40
References		40

Chapter 3 Structural Basis for Sequence-specific DNA Recognition by Transcription Factors and their Complexes
Manqing Hong and Ronen Marmorstein

3.1	Introduction	47
3.2	Transcriptional Regulators that Bind Core DNA Elements	48
	3.2.1 Helix-turn-helix and Winged Helix-turn-helix	48
	3.2.2 Basic Leucine-zipper and Basic Helix-loop-helix	50
	3.2.3 Zinc-binding Domains that Bind as Monomeric Units	52
	3.2.4 DNA Recognition by β-Ribbons	53
	3.2.5 Immunoglobulin Fold	53
	3.2.6 HMG Domain	54

Contents

	3.3	Transcriptional Regulators that Bind as Dimers to two DNA Half Sites with Different Spacing and Polarity	54
		3.3.1 Zn$_2$Cys$_6$ Binuclear Cluster	54
		3.3.2 Nuclear Receptors	57
	3.4	Transcription Regulatory Complexes that use a Combination of Different DNA-binding Motifs	59
		3.4.1 Combinatorial DNA Interactions	59
		3.4.2 ETS Family Ternary Complexes	60
		3.4.3 NFAT/Fos-Jun/DNA Quaternary Complex	62
	3.5	Conclusions	62
	References		63

Chapter 4 Indirect Readout of DNA Sequence by Proteins
Catherine L. Lawson and Helen M. Berman

	4.1	Introduction	66
		4.1.1 DNA Sequence Recognition: A Historical Perspective	66
	4.2	Indirect Readout	68
		4.2.1 Direct *vs.* Indirect Readout	68
		4.2.2 Language of Indirect Readout: DNA Geometry	69
		4.2.3 Sequence-dependent Polymorphisms of B-DNA	69
		4.2.3.1 Base Stacking	70
		4.2.3.2 Hydrogen Bonding	72
		4.2.3.3 Steric Repulsion	72
		4.2.3.4 DNA Bending	72
		4.2.4 Indirect Readout: A Universal Feature of Protein-DNA Interactions	72
	4.3	DNA Sequence Recognition by CAP	73
		4.3.1 Direct Readout by CAP	76
		4.3.2 Indirect Readout by CAP	79
		4.3.2.1 Conformation and Flexibility of the DNA Site for CAP	79
		4.3.2.2 Indirect Readout at Positions 1-2	80
		4.3.2.3 Indirect Readout at Position 6	80
		4.3.2.4 Comparison with Other Protein-induced Positive Roll Deformations	81
		4.3.3 DNA Bending *vs.* DNA Kinking – A Dynamic Duo?	83
	4.4	Conclusions	86
	Acknowledgements		86
	References		86

Chapter 5 Single-stranded Nucleic Acid (SSNA)-binding Proteins
Martin P. Horvath

5.1	Introduction		91
5.2	Basic Elements		93
	5.2.1	Interaction Types	93
		5.2.1.1 Salt Bridges and Electrostatics	93
		5.2.1.2 Stacking Interactions	95
		5.2.1.3 Steric Packing and van der Waals Interactions	95
		5.2.1.4 Hydrogen Bonding	96
	5.2.2	Folds, Evolution and Function	98
		5.2.2.1 OB-fold	98
		5.2.2.2 Sm-fold	103
		5.2.2.3 RRM	106
		5.2.2.4 KH	106
		5.2.2.5 Others: Pumilio, TRAP and Whirly	108
5.3	Emergent Properties		110
	5.3.1	Molecular Recognition: Specificity, Adaptability and Degeneracy	111
		5.3.1.1 Specific yet Adaptable Recognition by Modular Puf Proteins	112
		5.3.1.2 A "Hot-spot" for Recognition of Telomere DNA by Cdc13	113
		5.3.1.3 "Nucleotide Shuffling" and TEBP-α/β	113
		5.3.1.4 Degeneracy in Splicing Branch Site Identification	114
	5.3.2	Cooperativity	115
		5.3.2.1 SSB and Multiple Cooperativity Modes	115
		5.3.2.2 Anti-cooperativity and TEBP-α	116
		5.3.2.3 Positive Heterotypic Cooperativity at Telomere Ends	117
	5.3.3	Allostery	117
		5.3.3.1 Small Molecule Effectors and SSNA-binding	118
		5.3.3.2 Proteins as Allosteric Effectors for Binding and Release of SSNA	119
5.4	Conclusion and Perspective		120
	Acknowledgements		120
	References		121

Contents

Chapter 6 DNA Junctions and their Interaction with Resolving Enzymes
David M.J. Lilley

6.1	The Four-way Junction in Genetic Recombination	129
6.2	Structure and Dynamics of DNA Junctions	129
	6.2.1 Dynamics of the Four-way Junction	131
	6.2.2 Metal Ions and the Electrostatics of the Four-way Junction	131
	6.2.3 Branch Migration	132
	6.2.4 Comparison with Four-way RNA Junctions	134
6.3	Proteins that Interact with DNA Junctions	134
6.4	Junction-resolving Enzymes	134
	6.4.1 Occurrence of the Junction-resolving Enzymes	135
	6.4.2 Phylogeny	135
	6.4.3 Junction-resolving Enzymes are Dimeric	135
	6.4.4 Structures of the Junction-resolving Enzymes	135
6.5	Molecular Recognition and Distortion of the Structure of DNA Junctions by Resolving Enzymes	137
	6.5.1 Sequence Specificity of the Junction-resolving Enzymes	138
	6.5.2 Structural Distortion of DNA Junctions by the Junction-resolving Enzymes	138
	6.5.3 Coordination of the Resolution Process	139
6.6	T7 Endonuclease I	139
	6.6.1 Biochemistry of Endonuclease I	139
	6.6.2 Structure of Endonuclease I	140
	6.6.3 The Active Site	141
	6.6.4 Catalysis of Phosphodiester Bond Hydrolysis	141
	6.6.5 Interaction between Endonuclease I and DNA Junctions	142
6.7	In Conclusion	144
Acknowledgements		145
References		145

Chapter 7 RNA-protein Interactions in Ribonucleoprotein Particles and Ribonucleases
Hong Li

7.1	Introduction	150
7.2	Experimental Methods used to Determine RNA-protein Complex Structures	151

7.3	RNA-protein Interactions in Ribonucleoprotein Particles		152
	7.3.1	Ribosome	153
	7.3.2	RNAi Complexes	156
	7.3.3	Signal Recognition Particle	159
	7.3.4	s(no)RNPs	160
	7.3.5	RNA Editing Complexes	164
7.4	RNA-protein Interactions in Ribonucleases		165
	7.4.1	RNase E	165
	7.4.2	RNase II	166
	7.4.3	RNase III	167
	7.4.4	Restrictocin	168
	7.4.5	RNA Splicing Endonucleases	169
	7.4.6	tRNase Z	170
7.5	Concluding Remarks		170
Acknowledgements			171
References			171

Chapter 8 Bending and Compaction of DNA by Proteins
Reid C. Johnson, Stefano Stella and John K. Heiss

8.1	Introduction		176
8.2	Forces Controlling DNA Rigidity		178
	8.2.1	DNA Elasticity and the Influence of DNA Sequence	178
	8.2.2	Base Stacking Primarily Controls Helix Rigidity	179
	8.2.3	Electrostatic Forces Modulate DNA Bending	180
8.3	Bending of DNA at High Resolution		183
	8.3.1	Helix Parameters Controlling DNA Structure	184
		8.3.1.1 Roll and Tilt	184
		8.3.1.2 Twist	185
		8.3.1.3 Propeller Twist, Slide, and Shift	186
		8.3.1.4 Changes in DNA Groove Width	186
	8.3.2	Influence of Exocyclic Groups on Base Stacking	188
	8.3.3	Flexibility of Dinucleotide Steps	189
		8.3.3.1 Pyrimidine-purine (Y-R) Steps	189
		8.3.3.2 Purine-purine (R-R) or Pyrimidine-pyrimidine (Y-Y) Steps	189
		8.3.3.3 Purine-pyrimidine (R-Y) Steps	190

	8.4	Examples of DNA Bending Proteins	191
		8.4.1 Histone Binding to DNA	191
		8.4.2 Phage λ Xis Protein	194
		8.4.3 Papillomavirus E2 Protein	194
		8.4.4 *Escherichia coli* Fis Protein	197
		8.4.4.1 Long-range DNA Condensation by Fis	198
		8.4.5 *Escherichia coli* CAP Protein	201
		8.4.6 Prokaryotic HU/IHF Protein Family	203
		8.4.6.1 Single-DNA Molecule Analysis of HU/IHF Protein Binding	207
		8.4.7 HMGB Protein Family	208
		8.4.7.1 Single DNA Molecule Analyses of HMGB Protein Binding	211
		8.4.7.2 DNA Binding by HMGB Shares Features with TBP	212
	8.5	Concluding Remarks	212
	References		213

Chapter 9 Mode of Action of Proteins with RNA Chaperone Activity
Sabine Stampfl, Lukas Rajkowitsch, Katharina Semrad and Renée Schroeder

	9.1	Introduction	221
		9.1.1 RNA Folding	221
		9.1.2 Proteins with RNA Chaperone Activity (RCA)	222
		9.1.3 Measuring RCA	223
	9.2	Mode of Action of Proteins with RCA	223
		9.2.1 RNA Annealing Activity	223
		9.2.1.1 Annealing of Protein-bound Guide RNAs with Target RNAs	226
		9.2.2 Nucleic Acid Melting Activity	227
	9.3	RNA Binding and Restructuring	228
		9.3.1 Proteins with RCA Interact with RNA only Weakly	228
		9.3.2 Proteins with Specific RNA-binding Affinity	229
		9.3.3 Protein Structure and RNA Chaperone Activity	230
	Acknowledgements		231
	References		231

Chapter 10 Structure and Function of DNA Topoisomerases
Ken C. Dong and James M. Berger

	10.1	Introduction	234
	10.2	Type IA Topoisomerases	238

		10.2.1	Overview	238
		10.2.2	Structures and Mechanism	239
		10.2.3	Type IA Topoisomerase Paralogs	242
			10.2.3.1 Topoisomerase III	243
			10.2.3.2 Reverse Gyrase	243
	10.3	Type IB Topoisomerases		244
		10.3.1	Overview	244
		10.3.2	General Architecture	245
		10.3.3	DNA Recognition and Cleavage	246
		10.3.4	Mechanism	248
	10.4	Topoisomerase V – The Defining Member of the Type IC Topoisomerases?		249
	10.5	Type IIA Topoisomerases		249
		10.5.1	Overview	249
		10.5.2	Structural Organization	253
			10.5.2.1 ATPase Domain	253
			10.5.2.2 DNA Breakage/Reunion Domain and the DNA Binding/Cleavage Core	254
		10.5.3	Duplex DNA Transport Mechanism	255
			10.5.3.1 Type IIA Topoisomerase Paralogs: Role of the C-terminal Domain in Modulating Duplex Transport	256
		10.5.4	Physiological Specialization of Type IIA Topoisomerases	258
	10.6	Type IIB Topoisomerases		259
		10.6.1	Overview	259
		10.6.2	Structure	259
		10.6.3	Mechanism	260
	10.7	Conclusions		261
	Acknowledgements			261
	References			261

Chapter 11 DNA Transposases
Fred Dyda and Alison Burgess Hickman

	11.1	Introduction		270
		11.1.1	Nomenclature, Classification, and Overview of Transposition Systems	271
			11.1.1.1 Prokaryotic Elements	271
			11.1.1.2 Eukaryotic Elements	273
	11.2	Transposases and DNA		274
		11.2.1	Ends of Transposons	274
		11.2.2	Chemistry of DNA Transposition	276

	11.2.3	DDE Transposases	276
	11.2.4	Cut-and-paste, Copy-in, and Copy-out Transposition	277
		11.2.4.1 Mechanism of DNA Cleavage and Strand Transfer	277
		11.2.4.2 Second Strand Cleavage by Cut-and-paste Transposases	280
		11.2.4.3 Replicative Transposition	281
	11.2.5	Transposition Happens in Context	281
11.3	Tn5 Transposase: The Minimum Necessary	282
	11.3.1	Overview of the Tn5 Transposition Pathway	283
	11.3.2	Structure of a Tn5 Transposase Dimer Bound to DNA	283
11.4	Bacteriophage Mu	284
	11.4.1	Organization of the Mu Genome Ends	285
	11.4.2	Domain Structure of MuA	285
	11.4.3	Putting Mu Ends and MuA together: The Mu Transpososome	287
11.5	Tc1/*mariner* Transposases	288
	11.5.1	Tc3 N-terminal Domains and DNA	289
	11.5.2	*Mos1* Transposase	290
	11.5.3	What is the Active Assembly?	290
11.6	*hAT* Elements: A First Glimpse	290
	11.6.1	Domain Organization of Hermes	292
11.7	Transposases that Form Covalent Phosphotyrosine or Phosphoserine Intermediates	293
	11.7.1	Y and S Transposases	293
	11.7.2	Y2 Transposases	294
	11.7.3	Y1 Transposases	296
		11.7.3.1 ISHp608 TnpA	296
Acknowledgements	299
References	299

Chapter 12 Site-specific Recombinases
Gregory D. Van Duyne

12.1	Introduction	303
12.2	Tyrosine Recombinases	307
	12.2.1 Cre and Flp Recombinases	309
	12.2.2 λ-Integrase and XerCD Recombinases	313
	12.2.3 Integron Integrases	316
	12.2.4 Other Tyrosine Recombinases	318
12.3	Serine Recombinases	318
	12.3.1 Tn3 and γδ-Resolvases	321
	12.3.2 Hin and Gin Recombinases	323

	12.3.3	Regulation by Accessory Sites	324
	12.3.4	Large Serine Recombinases	326
12.4	Summary		327
References			328

Chapter 13 DNA Nucleases
Nancy C. Horton

13.1	Introduction		333
13.2	Summaries by Fold		334
	13.2.1	Restriction Endonuclease-like Fold	334
	13.2.2	RNaseH-like Fold	341
	13.2.3	Homing Endonuclease-like Fold	343
	13.2.4	His-Me Finger Endonucleases	345
	13.2.5	SAM Domain-like/PIN Domain-like Fold	348
	13.2.6	DNase I-like Fold	350
	13.2.7	Phospholipase C/P1 Nuclease Fold	352
	13.2.8	Phospholipase D/Nuclease Fold	354
	13.2.9	TIM beta/alpha Barrel Fold	355
	13.2.10	DHH Phosphoesterases Fold	357
	13.2.11	GIY-YIG Endonuclease Fold	358
	13.2.12	Metallo-dependent Phosphatases Fold	360
	13.2.13	*Bacillus chorismate* Mutase-like Fold	361
13.3	Conclusion		361
References			363

Chapter 14 RNA-modifying Enzymes
Adrian R. Ferré-D'amaré

14.1	Introduction: Scope of RNA Modification		367
14.2	The tRNA Adenosine Deaminase TadA		369
	14.2.1	RNA Recognition by Loop Eversion	369
	14.2.2	Hydrolytic Deamination by a Zinc-activated Water	370
14.3	The tRNA Pseudouridine Synthase RluA		372
	14.3.1	RNA Recognition through Protein-induced Base-pairing	374
	14.3.2	In-line Displacement or Michael Addition?	374
14.4	The tRNA Archaeosine Transglycosylase ArcTGT		376
	14.4.1	RNA Recognition by Tertiary Structure Rearrangement	376

14.4.2	Transglycosylation Using Two Aspartate Residues	378
14.5	Conclusions	379
Acknowledgements		379
References		380

Subject Index   382

CHAPTER 1
Introduction

CARL C. CORRELL[a] AND PHOEBE A. RICE[b]

[a] Department of Biochemistry and Molecular Biology, Rosalind Franklin University of Medicine and Science, 3333 Green Bay Road, North Chicago, IL 60064, USA; [b] Department of Biochemistry and Molecular Biology, The University of Chicago, 929 E. 57th St., Chicago, IL 60637, USA

1.1 Overview

Nucleic acids are the information storehouse of life and in many cases serve as the regulators and construction workers as well. Indeed, self-replicating RNAs may have been the beginning of life itself, predating evolution of the first protein. Modern organisms, however, depend on a complex interplay between nucleic acids and proteins. This chapter reviews the basic features of the structures that nucleic acids can adopt and highlights the ways that nucleic acids and their cognate proteins interact with one another.

1.2 Fundamentals of DNA and RNA Structure

1.2.1 Stabilizing Forces

The fundamental forces that stabilize nucleic acid and protein structure are the same. The hydrophobic effect drives folding of these molecules as they attempt to simultaneously satisfy multiple goals: minimizing the exposure of hydrophobic surfaces to water, satisfying all the hydrogen bond donors and acceptors that become buried from solvent, maximizing van der Waals interactions, and ensuring that all charges are either solvated or neutralized with opposing charges. Nucleic acids differ from proteins in that the hydrogen bonds that are key to secondary structure formation are between the variable moieties (the

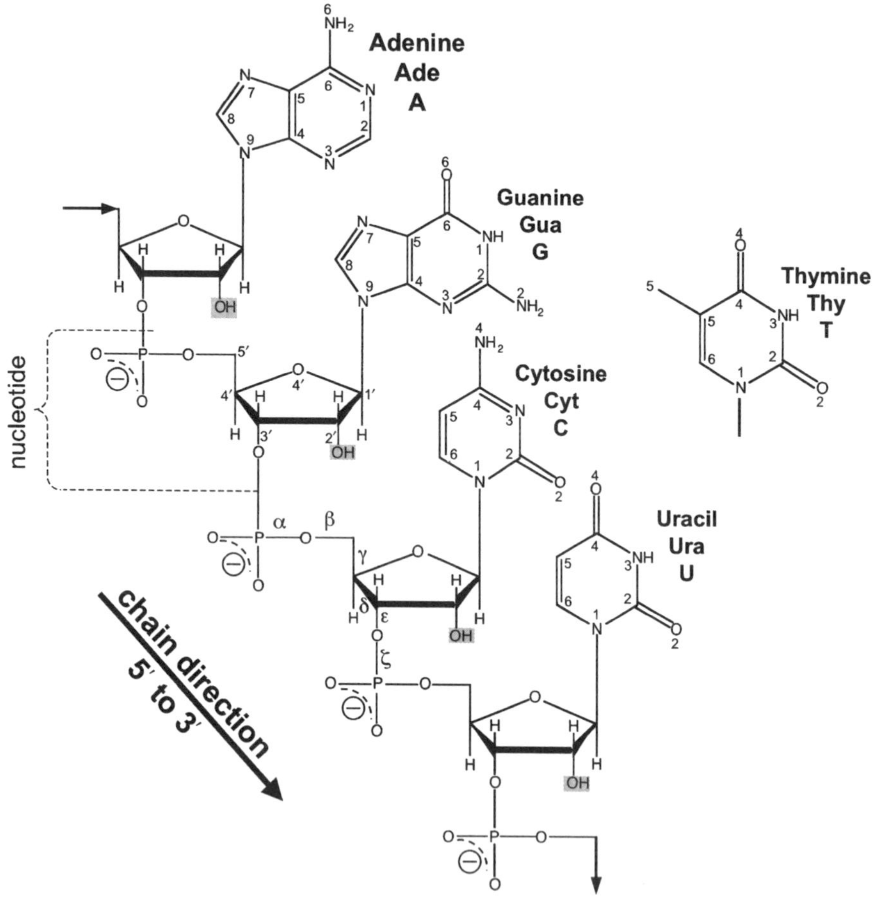

Figure 1.1 Nucleic acid primary structure. A fragment of RNA is shown, with the sequence 5′-AGCU-3′. Replacing the 2′-hydroxyl groups (shaded gray squares) with 2′ hydrogen atoms creates DNA, which uses thymine (inset) instead of uracil nucleobases. Standard atom numbers for the sugar are primed (labeled on the guanine) and for the bases are unprimed (shown for each base). The backbone torsion angles are denoted with Greek letters (labeled on the cytosine).

bases) rather than the constant ones (the backbones). The backbone of nucleic acids is more flexible than that of proteins, with six variable torsion angles rather than two (Figure 1.1). Nucleic acid backbones also differ from proteins in that their backbones are uniformly negatively charged. To form large tertiary structures with buried backbones nucleic acids must rely on external sources of positive charge such as solvent cations or helper proteins. Relative to tertiary structure, secondary structure is more stable in nucleic acids than proteins. As a consequence, nucleic acid folding is generally less cooperative than protein folding, with the formation of tertiary structure following that of secondary structure.

1.2.2 Chemical Differences between DNA and RNA

The defining difference between RNA and DNA is the presence of the 2'-hydroxyl group on the pentose ring of RNA. This group is distinctive in two fundamental ways. First, it is an Achilles heel that renders the RNA chain more susceptible to cleavage than the DNA chain (Figure 1.2). Apparently, susceptibility to cleavage is the reason why the 2'-hydroxyl is removed, at considerable metabolic expense, to make a more stable molecule (DNA) for information storage. Second, the 2'-hydroxyl is the glue permitting RNA to readily fold: it is the only group on the entire phosphodiester backbone that can donate as well as accept hydrogen bonds (Figure 1.1). This hydrogen-bonding capacity plays an important role in stabilizing the large variety of structures adopted by RNA molecules.

DNA also differs from RNA in that the pyrimidine base with two keto groups is thymine rather than uracil. Chemically, the difference is minor: thymine is merely 5-methyluracil, and the additional methyl group does not change the overall structure, but does provide a recognition opportunity. However, like the removal of the sugar ring's 2'-hydroxyl group, the addition of this methyl group requires considerable metabolic expense. Mother Nature's presumed logic in this case is slightly more convoluted. Because cytosine is disturbingly readily deaminated to form uracil, the methyl group added to thymine allows repair enzymes to discriminate between pyrimidines that were intended to have two keto groups (*i.e.*, thymine) and those that are the products of cytosine deamination and need to be removed (*i.e.*, uracil).

The chemical repertoire of nucleic acids can be greatly expanded by modifications introduced after replication or transcription. Particularly for functional RNAs, species from all kingdoms of life have evolved a vast number of enzymes, comprising up to approximately 10% of coding genomes, that modify nucleobases

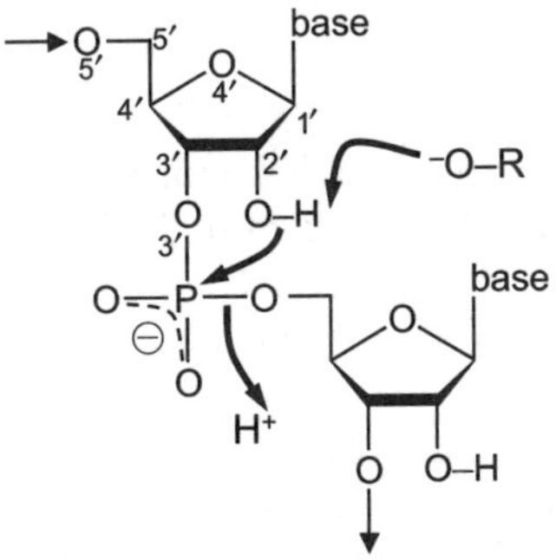

Figure 1.2 The susceptibility of RNA to hydrolysis arises from the ability of the 2' hydroxyl group to attack the 3'-adjacent phosphate group. This severs the phosphodiester backbone, generating a 2',3'-cyclic phosphate on one side of the break and a 5'-hydroxyl on the other. The reaction can be catalyzed by a base. It also requires a 180° angle between the attacking 2'O, P, and leaving 5'O atoms, which is not possible in a canonical A-form double helix.

after transcription (Chapter 14). At last count, about 100 different modified nucleosides have been identified in RNA. It is becoming clear that base modifications, in particular, are functionally significant: they are required for pre-mRNA splicing, they improve translational fidelity, and they increase RNA stability.

1.2.3 Canonical A- and B-form Helices

Despite the great variety of folds that RNA can adopt and the variation observed in DNA structure, the double helix remains the most common element of nucleic acid structure. The base pairing scheme first suggested by Watson and Crick is special in that the distances and angles between glycosidic bonds are constant, creating a regular structure that is independent of sequence, to a first approximation (Figure 1.3). Thus, nucleic acids with Watson–Crick base pairs adopt a deceptively simple structure: the double helix. When viewed in more detail however, the double helix is really a family of related conformations, by far the most common of which are the two forms termed A and B. Detailed descriptions of their anatomy can be found in many texts;[1] only a basic reminder is included here.

The backbone conformations of A- and B-form helices differ primarily in the puckers of their sugar rings: $C2'$-*endo* for B-form, and $C3'$-*endo* for A-form (Figure 1.4). The pucker makes relatively little difference to the placement of the atoms within the sugar ring itself. However, the pucker determines the relative placement of the substituents, namely the base and the flanking phosphates, which are critical to the overall conformation of the duplex.

Duplex RNA is largely limited to the A-form, for two reasons: canonical B-form helices are sterically incompatible with the protruding $2'$-hydroxyl groups of RNA, and the intrinsic sugar pucker preferences are affected by the $2'$-substituent. Both A- and B-forms are readily accessible to DNA, although the B-form predominates under physiological conditions. Local B-to-A conformational transitions can be triggered by DNA binding proteins, and are often associated with DNA bending (Chapter 8).

Overall, A- and B-form double helices differ most dramatically in the relative sizes and shapes of their grooves, with important consequences for their interactions with proteins (Figure 1.3). In B-form duplexes, the major groove is wider than the minor groove and both are readily accessible for protein recognition. In A-form, the major groove is deeper and narrower and thus less accessible to probing proteins; the shallower and wider minor groove is accessible but offers limited opportunity for sequence specific recognition (Section 1.3.3). Another difference between A- and B-form duplexes is the position of the sugar's $C2'$ atom. In the A-form, the $C2'$ atom positions the $2'$ hydroxyl groups to line the outside rim of the minor groove, whereas in the B-form, it protrudes toward the major groove. The minor groove of A-form RNA duplexes is thus lined with the hydrogen bond donor and acceptor moieties of the $2'$-hydroxyl groups. When DNA adopts A-form geometry, the resulting sugar puckers present more hydrophobic surface area to the minor groove than do B-form ones, a feature sometimes exploited by DNA bending proteins.

Introduction

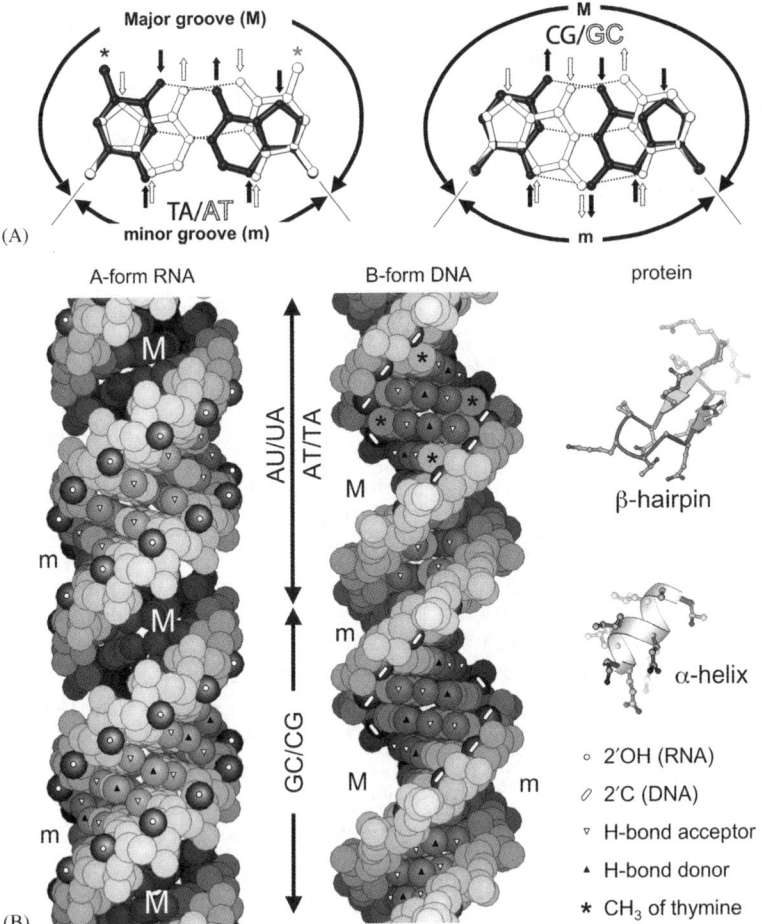

Figure 1.3 Watson–Crick base pairs generate canonical A-form RNA and B-form DNA helices. (*A*) On the left is a superposition of a TA (unfilled lines) and an AT (filled lines) base pair and on the right a CG and a GC pair. Note the similar distances and angles between the glycosidic bonds in all four cases. The functional groups that present recognition opportunities to the major (M) and minor (m) grooves are highlighted: hydrogen bond donors (arrows pointing away from the functional group), acceptors (arrows pointing toward the functional group), and methyl groups (asterisks). (*B*) Space filling representation of an A-form RNA helix (left), a B-form DNA helix (middle) and ribbon drawings of two common protein fragments used for recognition (right): a β-hairpin and α-helix (all on the same scale). The drawing illustrates differences between the major (M) and minor (m) grooves of DNA and RNA, including the placement of hydrogen bond donor (filled triangles) and acceptor (open triangles) groups, the 2′C of DNA (ellipses), the 2′OH of RNA (open circles) and the methyl groups of thymine (asterisk). The sequence is the same for both RNA and DNA: Ten alternating CG and GC base pairs followed by ten alternating UA (TA) and AU (AT) base pairs.

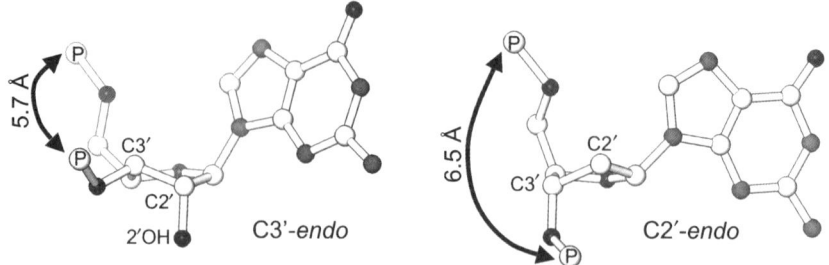

Figure 1.4 G nucleotides with the C3′-*endo* (left) and C2′-*endo* (right) sugar puckers that are found in A- and B-form helices, respectively. The distances between neighboring phosphorus atoms are indicated.

1.2.4 Deviation is the Norm

When nucleic acid structures are examined in detail, it becomes clear that very few are strictly canonical. As described in Chapters 4 and 8, even B-form DNA is quite flexible, with wide variations in parameters such as twist and groove width, even in the absence of proteins. The degree of flexibility and the conformational preferences are highly sequence-dependent, and these features are often important in site recognition by proteins. DNA damage can also change the structure and/or the stability of the double helix. Recognizing such damage is crucial for genetic integrity. In fact, the types of possible damage and the recognition strategies used by repair enzymes are so numerous that entire books have been dedicated to them[2,3] and thus they are not covered here. However, the repertoire of structures seen for DNA is still relatively limited compared to the tremendous variety seen for folded RNA (Section 1.2.6).

1.2.5 Bending and Supercoiling DNA

The double helix is a stiff but not inflexible structure: in the absence of proteins, B-form DNA has a persistence length of ∼150 base pairs (Chapter 8). Thus, the probability of two ends of a long DNA molecule meeting peaks at ∼450–500 bp. The main features of the double helix that resist bending are the stacking of the aromatic bases, and the mutual repulsion of the negatively charged phosphate groups. Proteins that induce large bends in DNA have evolved mechanisms to counteract one or both of these forces (Chapters 4 and 8).

Long DNA segments can become torsionally strained (supercoiled) when the ends are restrained such that one strand of the duplex cannot rotate freely about the other. This is the case not only for circular DNA molecules such as bacterial plasmids but also for chromosomal segments that are restrained by bacterial nucleoid-associated proteins or eukaryotic chromatin. Chapter 10 describes the many families of enzymes that modulate DNA supercoiling.

1.2.6 Folded RNA and Noncanonical DNA

Not all nucleic acids are double-stranded, and the two base pairs proposed by Watson and Crick are by no means the only possible base–base hydrogen bonding schemes. In fact, long stretches of fully Watson–Crick base paired duplex are generally only found in the genetic storage material – and thus, for most organisms, only in DNA (a few viruses use double-stranded RNA as their genetic material). In folded RNA, all types of non-Watson–Crick base pairings have been observed (*e.g.*, A can pair with A, C, G or U).

In contrast with DNA, all RNAs are synthesized as single strands that often then fold into far more elaborate and idiosyncratic structures than the simple A- and B-forms described above. Folding creates surfaces that can form catalytic sites, thereby transforming this polymer from merely a carrier of information into one that also catalyzes chemical reactions (peptide bond formation, pre-mRNA splicing and RNA processing to name a few).

To create a global structure, the RNA backbone twists and turns, permitting it to fold back upon itself.[4,5] As stated above, 2′-hydroxyl groups play key roles in stabilizing RNA secondary, tertiary and quaternary structure. RNA secondary structures are built of reoccurring modular motifs reminiscent of "lego" pieces. These pieces include one Watson–Crick element (A-form helices) and many loop elements that are stabilized by non-Watson–Crick interactions, including bulges, turns, linker regions, and multi-way junctions. RNA tertiary structure is formed by interactions among these elements, often involving the formation of base pairs, triples and/or quadruples. Interestingly, adenine, the most hydrophobic of the four bases, is highly over-represented in such interactions. Perhaps the most common recurring feature of RNA tertiary structure is the "A minor" motif where two adjacent adenosines dock into the minor groove of an RNA helical structure, often resulting in tandem base triples. Base stacking also plays a role by stabilizing docking of co-axial helices and interactions with side-by-side A base pairs (designated as A-platform/receptor interactions).

Although DNA has a limited structural repertoire compared to RNA, DNA is not always found as a canonical duplex. For example, during replication, the strands must be separated so that the precious information within can be copied, and repair and recombination also often involve unusual DNA structures such as hairpins and Holliday junctions (Section 1.3.6).

1.3 Principles of Recognition

Despite the differences between RNA and DNA, proteins use similar strategies to recognize them. Nucleic-acid binding proteins discriminate among potential binding sites based on their sequence, structure, or a combination of both. Although the details of the strategies used by proteins to recognize their cognate sites vary widely, the general principles remain the same.

1.3.1 Forces that Contribute to Complex Formation

As with other macromolecular interactions, the forces involved in noncovalent protein-nucleic acid complex formation are all relatively weak, and overall affinity results from the sum of many interactions, some favorable and others not. Perhaps the most obvious force among these is electrostatics: because nucleic acids are polyanions, most proteins that bind to them are rich in the positively charged amino acids lysine and arginine. As described in Chapter 2, screening by solvent ions modulates the strength of these interactions. However, hydrophobic interactions are surprisingly prevalent: even for proteins that bind B-form DNA, an average of nearly 50% of the surface area buried at the interface is nonpolar.[6] Many complexes involving less canonical nucleic acid structures are surprisingly resistant to high salt, often reflecting strong hydrophobic stacking interactions between the protein and nucleic acid bases (Chapter 6). Polar interactions involving direct and water-mediated hydrogen bonds are also widespread, and often of particular importance to sequence specificity. The above interactions are revealed upon inspection of the structure of a protein-nucleic acid complex. However, the initial states of the partners are also important, since complex formation generally involves desolvation of the interface and changes in the conformation and flexibility of one or both partners (Chapter 2).

1.3.2 Site Recognition Overview

Many proteins display high specificity for a particular nucleic acid sequence or structure, often binding their cognate site several orders of magnitude more tightly than random ones. The challenges presented by this recognition problem vary with the type of nucleic acid to be recognized, and are perhaps the most formidable for Watson–Crick paired duplexes (usually DNA) due to their structural homogeneity (Figure 1.3). Relatively unstructured single-stranded nucleic acids are the most flexible, allowing easy access to discriminating functional groups. In contrast, protein-RNA recognition is in many ways more similar to protein-protein recognition because of the great variety of surfaces that folded RNA can create; both types of recognition involve interacting surfaces that have each evolved to complement the other's shape and electrostatic properties.

1.3.3 Recognizing Duplex DNA *via* Direct and Indirect Readout

The sequence of a duplex DNA molecule can be read by examining the pattern of unique functional groups exposed in the major groove (Figure 1.3). The minor groove is less interesting: all four bases display hydrogen bond acceptors in similar locations. The protruding 2-amino group of G distinguishes G:C pairs from A:T ones, but is centrally located such that it is still hard for a protein to determine which base is the G and which is the C.

The simplest way for a protein to recognize a specific sequence in DNA is thus to bind in the major groove, and "interrogate" the unique features of the bases that are exposed there by making favorable contacts with the correct sequence and by making unfavorable contacts with incorrect sequences. As proposed by Zubay and Doty in 1959, an alpha helix fits nicely into the major groove of B-form DNA.[7] All combinations of DNA grooves and protein secondary structure elements have now been seen, but a "recognition" helix that inserts into the major groove is still the most common feature of sequence-specific DNA binding proteins (Chapter 3). The major groove of A-form helices is narrower and deeper than that of B-form helices, and thus less accessible to proteins (Figure 1.3). This could be a problem for RNA-protein recognition if RNAs were constrained to forming only A-form helices. However, as described below, the folding of RNA molecules generally produces other features that can guide protein recognition.

Comparison of specific and nonspecific protein-DNA complexes highlights the importance of dehydration in this process: formation of nonspecific complexes generally displaces fewer water molecules than formation of specific ones, and the resulting interfaces are less complementary (Chapter 2).

In contrast to the simple, direct readout described above, proteins can also recognize DNA sequences through sequence-dependent variations in flexibility and structural parameters such as the groove width and the twist between base pairs, a strategy referred to as "indirect" readout. Many proteins use both direct and indirect readout to identify their preferred binding sites. However, as described in Chapters 4 and 8, proteins that bend DNA and those that primarily contact the minor groove generally rely more heavily on indirect readout.

1.3.4 Recognizing Single-stranded Nucleic Acids

Chapter 5 focuses on proteins that bind single-stranded DNA and RNA. Although in aqueous solution base-stacking interactions still occur in single-stranded substrates, they are readily disrupted and proteins have easy access to the entire surface of each base. This simplifies the problem of sequence recognition, although it may actually complicate life for non-sequence-specific single-stranded binding proteins. For example, proteins involved in replication need to bind all sequences essentially equally, yet still protect the critical Watson–Crick faces of the bases from chemical damage until they are paired with a new partner. Specific and nonspecific interactions with single-stranded nucleic acids also usually involve extensive stacking interactions between the bases and aromatic side chains. In addition, proteins can induce portions of RNA structures to extrude out of their fold such that they adopt single-stranded character with its corresponding recognition opportunities (Chapters 7 and 14).

1.3.5 Recognizing Folded RNAs

RNA folds into a larger variety of structures than DNA due to its synthesis as a single strand and the stabilizing contacts made by 2'-hydroxyl groups. The structure of RNA shares features with protein and DNA. Like proteins, its folds create various surfaces that can evolve to bind to an unexpectedly large number of substrates, both large and small. Like DNA, each nucleotide of RNA is negatively charged and the polymer is sometimes single stranded, double helical or involved in multi-way helical junctions. Also like DNA, the most common structural element in a folded RNA structure is the regular or distorted double helix.

Folded RNA presents a virtually limitless number of surfaces that can be recognized by evolving a protein surface to complement its shape and electrostatic properties (Chapter 7). Minor groove recognition is greatly facilitated by non-Watson–Crick base pairs that present distinctive functional groups into this otherwise uninformative groove. Bulges or other non-Watson–Crick interactions can widen the major groove for recognition. In contrast, recognition of the regular geometry of A-form helices is important in RNAi.

1.3.6 Recognizing Noncanonical DNA Structures

Unusual DNA structures are generally intermediates in replication, repair, and recombination. For example, DNA strand exchange *via* homologous recombination can lead to the formation of four-armed branched structures known as Holliday junctions. As described in Chapter 6, these junctions are specifically recognized by the enzymes that process them, partly through their structure and flexibility. Holliday junctions are also formed as the intermediates of some site-specific recombination reactions, and in these cases are stabilized by the recombinases (Chapter 12).

There are a few cases in which the two strands of a broken DNA duplex become covalently joined to form a hairpinned DNA end. Such hairpins occur as enzyme-stabilized intermediates in some transposition reactions (Chapter 11). Similar hairpins are the final product of another recombinase family, the hairpin telomere resolvases (Chapter 12).

Although formation of stable secondary structure in single-stranded nucleic acids is usually associated with RNA, there are examples from the DNA world as well. These are sometimes found in connection with mobile DNA elements, many of which carry genes that humans would rather bacteria did not share with one another (such as those encoding drug resistance). In the few cases so far studied in detail, the ssDNA folds into duplexes with characteristic distortions such as bulged-out bases that are specifically recognized by the cognate recombinases (Chapters 11 and 12). Recognizing such noncanonical structural features is a strategy more commonly associated with RNA-binding proteins.

1.3.7 Conformational Rearrangements

Protein-nucleic acid interactions often involve induced fit – that is, conformational changes in one or both partners upon binding. In some cases, inducing or stabilizing a particular DNA or RNA conformation is the protein's *raison d'etre:* for example, the DNA bending proteins discussed in Chapter 8. Interestingly, conformational changes in both the protein and the nucleic acid can be coupled to site specificity.

Introduction

Induced fit is of particular interest with regard to nucleic-acid modifying enzymes (Chapter 14). Here it can add a second level of specificity, beyond merely binding or not binding. For example, some restriction enzymes show surprisingly little sequence discrimination at the binding step, but cleave with exquisite specificity because only the cognate sequence can readily undergo the distortion required to position the scissile phosphate in the active site (Chapters 2 and 13). Other enzymes actually flip one or more bases (or entire nucleotides) completely out of the ground state structure, be it a double helix or some other motif. Base-modifying enzymes of both the DNA and RNA worlds often use this strategy to gain easy access to the substrate base (Chapter 14 and refs. 8 and 9). Alternatively, ribonucleases (Chapter 7) can use this strategy to move the RNA from an inactive form to an active form, in which the nucleophilic 2′-hydroxyl group is positioned in-line with the scissile phosphate bond, and thereby poised for attack.

RNAs, like proteins, can misfold, and are perhaps even more susceptible to such problems. RNA chaperones have evolved to alleviate this problem. These proteins can accelerate RNA annealing, strand displacement and/or helix destabilizing activities to guide the RNA into a functional form (Chapter 9). DNA and RNA helicases also rearrange nucleic acid structure, but unlike chaperones they hydrolyze ATP to drive duplex unwinding and/or to dislodge bound proteins (refs. 10–12 are excellent recent reviews).

1.4 Future Directions

As this book illustrates, the study of protein-nucleic acid complex structures has shed much light on the basis of action of these fundamental cellular assemblies. So what future directions remain un- or underexplored? Of course, the most exciting discoveries are often in areas where we did not even realize we were ignorant. However, some areas where even more data are needed include multi-component protein-nucleic acid complexes in which the whole is often different than the sum of the parts, and the role of induced fit in recognition and catalysis. The opportunity to compare matching bound and unbound structures is crucial for the latter, but such pairs are often lacking. Other areas where a better understanding is needed include (but are by no means limited to) the solution dynamics of these systems, the mechanisms by which proteins chaperone the folding of RNA and rearrange the structure of DNA, and the development of better methods for predicting and designing specificity in nucleic-acid binding proteins. Based on the recent explosion of information, the next decade promises to be exciting indeed.

References

1. W. Saenger, *Principles of Nucleic Acid Structure*, Springer-Verlag Inc., New York, 1984.
2. E.C. Friedberg, G.C. Walker, W. Siede, R.D. Wood, R.A. Schultz and T. Ellenberger, *DNA Repair and Mutagenesis*, 2nd edn., ASM Press, Washington D.C., 2005.

3. *DNA Damage Recognition*, ed. W. Siede, Y.W. Kow and P.W. Doetsch, CRC Press, Boca Raton, FL, 2005.
4. D.K. Hendrix, S.E. Brenner and S.R. Holbrook, *Q. Rev. Biophys.*, 2005, **38**, 221.
5. N.B. Leontis, A. Lescoute and E. Westhof, *Curr. Opin. Struct. Biol.*, 2006, **16**, 279.
6. K. Nadassy, S.J. Wodak and J. Janin, *Biochemistry*, 1999, **38**, 1999.
7. G. Zubay and P. Doty, *J. Mol. Biol.*, 1959, **1**, 1.
8. X. Cheng and R.J. Roberts, *Nucleic Acids Res.*, 2001, **29**, 3784.
9. J.T. Stivers, *Prog. Nucleic Acid Res. Mol. Biol.*, 2004, **77**, 37.
10. M.R. Singleton, M.S. Dillingham and D.B. Wigley, *Annu. Rev. Biochem.*, 2007, **76**, 23.
11. E. Jankowsky and M.E. Fairman, *Curr. Opin. Struct. Biol.*, 2007, **17**, 316.
12. F. Bleichert and S.J. Baserga, *Mol. Cell*, 2007, **27**, 339.

CHAPTER 2
Role of Water and Effects of Small Ions in Site-specific Protein-DNA Interactions

LINDA JEN-JACOBSON AND LEWIS A. JACOBSON

Department of Biological Sciences, University of Pittsburgh, Pittsburgh PA 15260, USA

2.1 Introduction

It has long been understood that the differential expression of particular genes with temporal, cellular or developmental specificity ultimately depends on the dynamic interaction of site-specific DNA binding proteins with regulatory regions of genomic DNA. This perception has in turn provided a strong motivation for studying the structures of site-specific protein-DNA complexes, and two decades of such structural studies, primarily by X-ray crystallography and NMR, have provided us with many invaluable insights into the intermolecular contacts that confer specificity in protein-DNA interactions.

At the same time, the finely detailed, esthetically pleasing views of the structures of specific protein-DNA complexes have perhaps distracted us from appreciating the extent to which site-specific intermolecular association is driven by "invisible" thermodynamic factors. By this we mean bulk processes such as the changes in surface hydration, redistribution of macromolecule-associated cations and anions, and interaction with other solution components (*e.g.*, osmolytes). These bulk processes are not directly reflected in the structures of protein-DNA complexes, yet they have profound, or even dominant, effects on both the overall thermodynamics of site-specific protein-DNA binding and, perhaps surprisingly given their apparent "bulk" nature, on the

RSC Biomolecular Sciences
Protein-Nucleic Acid Interactions: Structural Biology
Edited by Phoebe A. Rice and Carl C. Correll
© Royal Society of Chemistry 2008

specificity of such binding. These processes also form a principal fraction of the basis by which environmental variables (salt, osmolytes, *etc.*) modulate both the affinity and the specificity of protein-DNA interactions.

The focus of this chapter is on these "invisible" contributions to protein-DNA interactions and how they are affected by environmental parameters. Rather than attempting an exhaustive review of the literature, we use a limited number of examples to illustrate the physicochemical principles that govern the influence of these factors. Illustrative data are drawn from work in the Jen-Jacobson laboratory on DNA recognition by the EcoRI, BamHI and EcoRV restriction endonucleases.

2.2 Affinity and Specificity

It is a fundamental principle of intermolecular interactions that affinity and specificity are not equivalent, although they are related. This is most easily appreciated by considering the binding of a site-specific protein (P) to either DNA containing the specific recognition site (D_S) (Reaction 2.1) or to so-called "non-specific" DNA lacking such a site (D_{NS}) (Reaction 2.2). (See Jen-Jacobson[1] for discussion of the conceptual issues involved in precisely defining "non-specific" DNA, and the experimental issues in selecting a "non-specific" DNA.)

$$P + D_S \xrightleftharpoons{K_S} P \cdot D_S \quad (2.1)$$

$$P + D_{NS} \xrightleftharpoons{K_{NS}} P \cdot D_{NS} \quad (2.2)$$

The *affinity* for either DNA can be measured by the equilibrium association constant K_i or by the standard free energy change $\Delta G°_i = -RT \ln K_i$ ($i = $ S or NS) for the association reaction. The *specificity*, by contrast, is measured by the relative preference for specific over nonspecific DNA binding, and can be defined, according to preference, as the specificity ratio K_S/K_{NS} or as the difference in standard free energy changes $\Delta \Delta G°_{bind} = RT \ln(K_S/K_{NS})$.

To assess how any environmental change X affects the *affinity*, consider a thermodynamic cycle (Figure 2.1), in which the horizontal arrows represent observable specific binding as in Reaction 2.1, and the vertical arrows represent the (often non-observable) effects of environmental variable X on the free macromolecules (left-hand vertical arrow) or the protein-DNA complex (right-hand vertical arrow). Since the sum of free energy changes around a cycle must be zero, it follows that:

$$\Delta G°_{bind,0} + \Delta G°_{X,complex} = \Delta G°_{X,free} + \Delta G°_{bind,X} \quad (2.1)$$

and:

$$\Delta \Delta G°_{bind,X} = \Delta G°_{bind,X} - \Delta G°_{bind,0} = \Delta G°_{X,complex} - \Delta G°_{X,free} \quad (2.2)$$

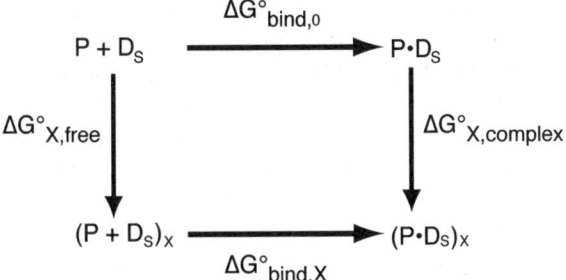

Figure 2.1 Thermodynamic cycle for the influence of environmental variable X on specific protein-DNA binding. The subscript "0" refers to the absence of X.

This implies that any environmental variable X has an observable effect on *affinity* (binding) if and only if it has a differential free energy effect on the free and complexed macromolecules.

Considering together Reactions 2.1 and 2.2 and Figure 2.1, we can ask what requirements must be met for any environmental variable X to affect *specificity*. The crucial point is that in comparing specific to nonspecific binding, the free proteins in all reactions (Reactions 2.1 and 2.2 with and without X) are the same, and provided that the "non-specific DNA" is correctly designed (*i.e.*, the nonspecific DNA has the same base composition as the specific DNA but scrambled nucleotide sequence in the recognition site, and identical nucleotide sequence in the surrounding region), then we can with near-certainty exclude differential effects of the environmental variable X on the molecules on the left-hand side of the pseudocycle (Figure 2.1) for specific binding and the analogous pseudocycle for nonspecific binding; that is, $\Delta\Delta G°_{X,free} = (\Delta G°_{S,X,free} - \Delta G°_{NS,X,free}) = 0$. In consequence, any environmental variable X can affect specificity if and only if it differentially affects the specific and non-specific protein-DNA *complexes* such that $\Delta\Delta G°_{X,complex} = (\Delta G°_{S,X,complex} - \Delta G°_{NS,X,complex}) \neq 0$. We will return to the question of specificity modulation after discussing how water and small ions affect the formation of site-specific protein-DNA complexes.

2.3 Macromolecular Hydration Influences $\Delta H°$, $\Delta S°$ and $\Delta C°_P$

When a protein binds to its specific DNA recognition site, the high surface complementarity of the protein-DNA interface causes much of the water that solvated these regions of both protein and DNA to be squeezed out. Thus, water release from the macromolecular surfaces has a major influence on the overall thermodynamics of the association. However, the interfacial region is not fully dehydrated; X-ray structures of co-crystalline complexes show that the tightly apposed surfaces are "damp" rather than dry.[2] The retained water molecules that bridge between protein and DNA often enhance binding affinity,[3] sometimes contribute to specificity determination, and also have important consequences for the thermodynamics of association (see below).

Water release from both polar and nonpolar surfaces is entropically favorable ($\Delta S° > 0$). However, release from polar and nonpolar surfaces has different effects on enthalpy change $\Delta H°$ and opposing effects on heat capacity change $\Delta C°_P$. For macromolecular association reactions, $\Delta C°_P$ reflects to a large extent the changes in thermally accessible vibrational motions. These may derive from the macromolecules themselves or from the associated water molecules. The dependence of the heat capacity on the energies of such motions is not monotonic, however, because it is "determined by a balance between fluctuations ... between different states ... and the energy changes occurring upon transition between these states".[4] The interactions of water molecules with polar surfaces of macromolecules are much stronger than water-water interactions in bulk water, such that fluctuation between states is improbable, reducing the heat capacity.[5] Consequently, the release of water from such polar surface during protein-DNA binding increases $\Delta C°_P$. On the other hand, water molecules associated with nonpolar surfaces also interact with each other more strongly than they would in bulk water,[6] yet the energy difference between states is small enough to yield a high probability of thermal fluctuation, thus contributing a greater heat capacity.[5] Consequently, the release of water from such nonpolar surfaces during protein-DNA binding contributes to a negative $\Delta C°_P$.

Based upon transfer studies with small-molecule model compounds, water release from a nonpolar surface (the "hydrophobic effect") has little effect on $\Delta H°$ near room temperature,[7-9] and results in loss of heat capacity such that $\Delta C°_P$ is made more negative.[9-14] By contrast, water release from a polar surface is enthalpically unfavorable and increases heat capacity such that $\Delta C°_P$ is made less negative.[15] These effects of polar desolvation are usually compensated by the formation of direct protein-DNA contacts (e.g., hydrogen bonds with bases and phosphates), which are enthalpically favorable but entropically unfavorable. Experimental measurements of the effects of base-analogue[1] or phosphate analogue[16,17] substitutions that were carefully chosen to disrupt single interactions indicate that the net $\Delta G°$ of forming each such contact (including the desolvation of the contacting groups) is slightly favorable.

Various semi-empirical equations[9,14,15,18,19] have been derived that relate $\Delta C°_P$ for protein folding/unfolding to changes in accessible surface area (ΔASA), subdivided to account for the burial of nonpolar surfaces (ΔASA_{NP}) and polar surfaces (ΔASA_{POL}) or, alternatively, to the total ΔASA or even the total number of residues involved. It appears[20] that differences in the numerical values of the coefficients reflect the underlying reality that data on various sets of proteins can be decently well fit by widely varying pairs of coefficients. Furthermore, because ΔASA_{POL} and ΔASA_{NP} tend to be correlated, treating them separately for proteins may not be much better than simply relating $\Delta C°_P$ to total ΔASA.[20] There has been some effort to derive a separate set of coefficients for DNA surfaces,[21] and it has been suggested[22] that water release from the major and minor grooves of DNA has somewhat different effects on $\Delta C°_P$.

There is another, more fundamental, objection to surface-area parameterization of $\Delta C°_P$ for protein-DNA interactions. These empirical relationships have been studied almost exclusively for protein folding, where water release is likely the dominant influence on $\Delta C°_P$. By contrast, for protein-DNA association there are also very substantial contributions to a negative $\Delta C°_P$ from vibrational constraints on the macromolecules,[23-28] vibrational restrictions on interfacial water molecules,[24,29] and coupled equilibria such as protonation,[30-32] ion binding[33] or conformational changes.[31,34] For some DNA-binding proteins there are countervailing contributions from electrostatic repulsion (strain) in unoccupied metal-binding sites that make $\Delta C°_P$ less negative. Consequently, empirical equations based on protein folding or model compounds give only the water-release contribution to $\Delta C°_P$ but this may account for a minor fraction[24,35,36] of the observed negative $\Delta C°_P$ (perhaps less than 25%[27]). It is therefore incorrect to combine the empirical relationships with experimental values of $\Delta C°_P$ to calculate ΔASA, water stoichiometry, or the hydrophobic effect for protein-DNA association.

2.4 Water Release Attending Protein-DNA Association

Macromolecular surfaces are almost certainly solvated by water that is more densely packed than in bulk solvent. There are many subtle and difficult issues that arise in deriving densities of solvation shells from experimental data[37] and/or from simulations,[38] and the solvent density may vary across chemically different regions of protein surface.[39] Makarov et al.[38] infer that "the density value averaged over the first hydration shell . . . is higher than that of the bulk solvent . . . (by) at most 10%". However, it is not known whether this average differs for different proteins, depending on the proportions and/or arrangement of nonpolar, polar uncharged and charged surfaces. For DNA, the first hydration shell deviates more pronouncedly from bulk solvent, showing a density from 2 to 6 times greater than that of bulk water, with excess density extending out to second or third hydration shells.[38,40]

Quantifying the release of bound water from the macromolecular surfaces upon protein-DNA association is not straightforward. There are predominantly computational approaches that use some experimental data, and predominantly experimental approaches that use some computational information; both have their strengths and weaknesses.

Where a high-resolution structure of the complex is available, one can compute[41,42] the solvent accessible surface area (ASA) of the complex and subdivide this surface into polar and nonpolar fractions. If the structures of the free protein and the free DNA are also available, it is in principle possible to make separate computations of ASA and derive the change ($\Delta ASA < 0$) upon association. In practice, there are often obstacles to simplistic application of this approach: All three structures may not be available, or one or more of them may be affected by crystal lattice constraints such that the structure differs from that in solution, and/or one or more may contain "disordered" regions that are invisible in the crystal structures. Where a structure for one or both free molecules is unavailable, a common expedient is to "undock" the protein and DNA from the complex as rigid bodies, then to calculate their ASA values. This probably gives a fairly good approximation for the DNA, unless the DNA in the complex is strongly distorted, and a rough but useful approximation for the protein, unless there is substantial folding of disordered protein regions coupled to DNA binding. The issue is nearly impossible to avoid, because it appears to be the case that DNA distortion and coupled protein folding have a somewhat reciprocal relationship for many protein-DNA pairs,[26] such that proteins that most distort their DNA binding sites have little coupled folding, whereas those proteins that bind with nearly undistorted DNA tend to have more extensive disorder-to-order transitions. Some protein-DNA pairs show both DNA distortion and coupled folding. In cases where the structure of the free protein shows missing or invisible regions, it is generally difficult to know from crystallography if such regions are distributed among a few predominant conformations or are in fact highly mobile. Such invisible regions can be modeled onto the structure for purposes of computation, and it appears that they are best modeled as more compact structures[43,44] rather than as extended chains. The degree to which these considerations influence the final calculation of ΔASA depends on the particulars of each protein-DNA pair. In the calculations in Table 2.1, we account as best we can for conformational changes in both protein and DNA.

The interpretation of ΔASA in terms of hydration changes upon binding is also not entirely straightforward. It is tempting to calculate from ΔASA a stoichiometric number of water molecules released, assuming that only a single hydration layer is perturbed. This simplifying assumption is adequate for nonpolar surfaces, but is likely a poor approximation for polar or ionized surfaces.

To estimate the contribution of the hydrophobic effect ($\Delta G°_{HE}$) to $\Delta G°_{bind}$, ΔASA$_{NP}$ can be used with a "contact free energy" value derived from transfer studies with small-molecule model compounds.[42,45–49] This value relates the free energy of transfer to the ΔASA$_{NP}$, independent of assumptions about how many solvation layers are affected by macromolecular binding, or the structure or density of solvent in those layers. Numerical conversion factors that relate $\Delta G°_{HE}$ to ΔASA$_{NP}$ range from about 20 to 47 cal mol^{-1} Å^2,[9,42,45–53] with the variation arising in several ways (see discussion in Brem et al.[53]). We use the contact free energy value estimated by Dill and co-workers[53] at 40 cal mol^{-1} Å^2. Taken with computed ΔASA values that account as best we can for conformational changes in protein and DNA, this yields estimates of the $\Delta G°_{HE}$ (kcal mol^{-1}) as follows (Table 2.1): EcoRI, −80; EcoRV, −88; BamHI, −80. This should be compared with experimental net $\Delta G°_{bind}$ values for these proteins in the range of −11 to −15 kcal mol^{-1}.[27,54,55] Purely computational studies[56] of 40 protein-DNA systems suggest that the average contribution of the hydrophobic effect may be as much as twice that large. Despite the numerical uncertainties, the release of water from nonpolar surfaces is almost certainly the largest single favorable contribution to the free energy of protein-DNA association.[57]

As an alternative to these computational/empirical methods, it would be highly desirable to have a direct experimental determination of water release during macromolecular association. One approach is based on the idea that the perturbation of association equilibrium by small-molecule osmolytes (or "cosolutes") can be used to measure the stoichiometric release of water

Table 2.1 Water stoichiometries from accessible surface areas and from cosolute experiments.[a]

EcoRI			EcoRV			BamHI		
ΔASA (Å^2) Nonpolar, polar	ΔN_w	$\Delta G°_{HE}$ (kcal mol^{-1})[b]	ΔASA (Å^2) Nonpolar, polar	ΔN_w	$\Delta G°_{HE}$ (kcal mol^{-1})	ΔASA (Å^2) Nonpolar, polar	ΔN_w	$\Delta G°_{HE}$ (kcal mol^{-1})
−1992, −2340	−433 (−383)[c]	−80 −80	−2188, −2508	−470 (−414)[c]	−88 −100	−2006, −2050	−407 (−357)[c]	−80 −88
$\partial \ln K_A / \partial \ln a_w$	−350		$\partial \ln K_A / \partial \ln a_w$	−430		$\partial \ln K_A / \partial \ln a_w$	−356	

[a] Solvent accessible surface areas (ASA) were calculated using the program NACCESS[165] (version July, 1996), which had been modified to include van der Waals radii and atom types for DNA. A water probe of radius 1.4 Å was used with z-slice of 0.01 Å. The program GRASP[166] gave similar ASA values. Modeling takes into account conformational changes of the protein upon complex formation. Models for specific complexes and free DNA were as follows: EcoRI complex, a highly refined version of PDB ID 1CKQ,[167] CGCGAATTCGCG, 355D; BamHI complex 2BAM, 10-mer ATGGATCCAT built with QUANTA software (Accelerys, Inc.); EcoRV complex 1B94, 11mer AAAGATATCTT built with QUANTA. Free protein models were as follows: apo-EcoRI,[167] missing residues modeled according to Creamer and Rose;[43,44] BamHI, ASA of residues 79–91 of undocked protein from nonspecific complex (1ESG) substituted for same region on undocked protein from specific complex; apo-EcoRV (1RVE). The net change in ASA on complex formation was calculated as $\Delta ASA = ASA_{complex} - (ASA_{protein} + ASA_{DNA})$. Cosolute dependences $\partial \ln K_A / \ln a_w$ are TEG data from Figure 2.3.
[b] The free energy contribution of the hydrophobic effect $\Delta G°_{HE}$ was calculated as described in the text.
[c] Changes upon binding in the number of macromolecule-associated waters (ΔN_w) from ΔASA computation use surface occupancy of 10 Å^2 per H$_2$O and are corrected for retained interfacial H$_2$O: EcoRI and BamHI (50 waters); EcoRV (56 waters).

during macromolecular association by means of "osmotic stress" analysis.[58,59] Because many such cosolutes bind to protein and/or DNA surfaces in preference to water, and this would clearly oppose and obscure the effect of the cosolute in decreasing water activity, the "osmotic stress" measurement of hydration changes has been stated to require "inert" or "noninteracting" cosolutes. This requirement has been cogently criticized[60,61] on the grounds that exclusion of the cosolute from the macromolecular surface itself entails an energetic cost that is not properly taken into account in the "osmotic stress" analysis.

Alternative treatments of the effect of cosolutes on protein-DNA binding have been based either on thermodynamic descriptions (*e.g.*, "preferential interactions") that require no molecular picture of the phenomena[61-70] or on various statistical-mechanical treatments that incorporate a molecular picture of steric exclusion.[71-74] The most comprehensive views encompass both perspectives.[75,76]

When macromolecules coexist in solution with a small-molecule cosolute, there is mutual exclusion by virtue of steric repulsion ("molecular crowding"). This mutual steric exclusion exacts an energetic cost (ΔG_{SE}) that depends on the molecular volumes of the macromolecules and cosolute, but not on their chemical properties. Because the excluded volume of the protein-DNA complex is smaller than the sum of excluded volumes of the free macromolecules, the energetic cost of steric exclusion of the cosolute is less for the complex than for the free macromolecules (*i.e.*, $\Delta\Delta G_{SE} < 0$). On this basis alone, the presence of cosolute promotes protein-DNA association. In addition to steric exclusion, there is a free energy term ΔG_{int} for the interaction of the cosolute with the macromolecules; this term is negative ($\Delta G_{int} < 0$) if the cosolute interacts in preference to water, and is positive ($\Delta G_{int} > 0$) if the cosolute is preferentially excluded (*i.e.*, the macromolecular surfaces are preferentially hydrated). The observed effect of the cosolute on binding is thus:

$$\Delta\Delta G_{obs,cosolute} = \Delta\Delta G_{SE} + \Delta\Delta G_{int} \quad (2.3)$$

Because the complex offers a smaller interaction surface than the sum of free protein and DNA, for interactive cosolutes $\Delta\Delta G_{int} > 0$, so $\Delta\Delta G_{obs,cosolute}$ is less favorable than that predicted from steric exclusion alone.

In practical terms, these effects are most clearly analyzed with a cosolute that is neither preferentially bound to, nor preferentially excluded from, the macromolecular surfaces, such that $\Delta\Delta G_{int} \approx 0$, but there is no purely experimental criterion by which to identify such a cosolute. We have taken the expedient of comparing the thermodynamic nonideality with that predicted from steric exclusion (mutually excluded volume) alone. We measure experimentally the dependence on cosolute molar concentration c_3 of the thermodynamic nonideality factor:

$$\Gamma = \frac{K_{cosolute}}{K_o} = \frac{\gamma_{PD}}{\gamma_P \gamma_D} \quad (2.4)$$

where $K_{cosolute}$ and K_o are the equilibrium association constants with and without cosolute, respectively, and γ_i are the molar activity coefficients in the presence of cosolute. A molar scale is used here because this allows the thermodynamic nonideality factor to be directly compared with the statistical-mechanical quantity of covolume change;[71] see discussion in Davis-Searles et al.[74] Specifically, the slopes $\partial \ln \Gamma/\partial c_3$ approximate the difference in second virial coefficients ΔB_{ij} between the protein-DNA complex and the free molecules.

The second virial coefficient B_{ij} appears in the expansion:[77]

$$\ln \gamma_i = \sum_j B_{ij} C_j + \sum_j \sum_k B_{ijk} C_j C_k + \ldots \quad (2.5)$$

where successive terms refer to two-body interactions between species i and j, three-body interactions among species i, j, k and so forth. Because macromolecule concentrations are low and much lower than the concentration of cosolute, one can neglect interactions between macromolecules and also neglect the higher-order (three-body) terms. If there is no contribution to thermodynamic nonideality from cosolute binding or chemical exclusion from the macromolecular vicinity (i.e., ΔG_{int} is zero), then the second virial coefficient B_{ij} can be related to the statistical-mechanical concept of mutually excluded volume (covolume) between cosolute and macromolecules.[71–74,77]

When $\Delta \Delta G_{int} \approx 0$, it should be true that $\Delta B_{ij} = (\partial \ln \Gamma/\partial c_3)_{T,P,salt} = \Delta U$, where ΔU is the computed covolume change calculated from crystal structures.[78] (The calculation of ΔU is subject to some uncertainties due to conformational changes upon binding, etc.) Figure 2.2(A) shows such data for EcoRI endonuclease and various cosolutes. Figure 2.2(B) shows that for triethylene glycol (TEG) and polyethylene glycol 200 (PEG200) the experimental values of $\partial \ln \Gamma/\partial c_3$ (middle bars of each group) are slightly larger than the computed covolume changes ΔU (right-hand bars of each group), showing that TEG and PEG200 are not preferentially interacting with the macromolecular surfaces. The experimental values are also smaller than the covolumes calculated with one hydration layer around the macromolecules (left-hand bars of each group), implying that TEG and PEG200 are not excluded from the first hydration layer. This implies that TEG and PEG200 are present in the region near the macromolecules at concentrations that closely approach their concentrations in the bulk solution, and thereby fulfill the criterion for $\Delta \Delta G_{int} \approx 0$. Thus, for site-specific DNA binding of *Eco*RI, *Bam*HI and *Eco*RV endonucleases, the thermodynamic nonideality for TEG and PEG200 is nearly perfectly accounted for by steric exclusion.

These cosolute effects are independent of other influences on the thermodynamics that do not affect the steric exclusion. For example, varying the nucleotide sequence flanking the recognition site can cause the value of K_o for EcoRI-DNA binding to vary by a factor of 850, and the $\Delta C°_p$ to vary between -1.2 and -2.5 kcal mol^{-1} K^{-1}.[27] Nevertheless, the dependence of binding $\partial \ln \Gamma/\partial c_3$ on TEG concentration is the same for all flanking context variants.[79]

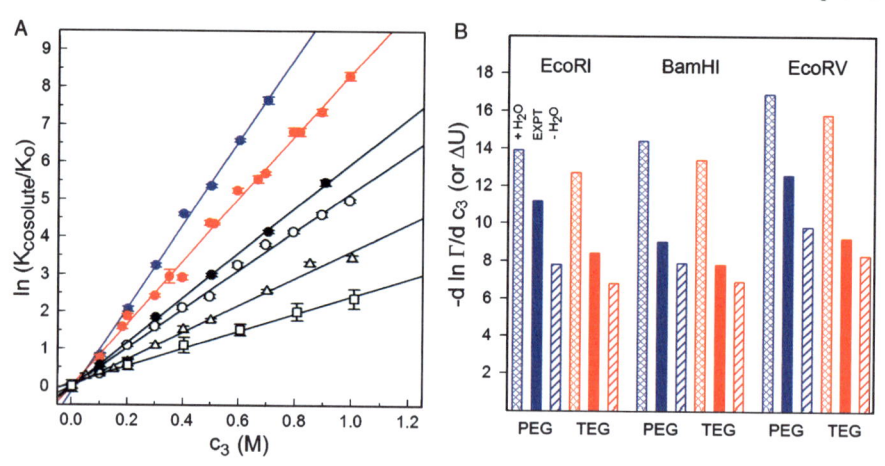

Figure 2.2 (A) Cosolutes promote different degrees of thermodynamic nonideality in specific binding of EcoRI endonuclease to DNA. The nonideality factor is $\Gamma = K_{cosolute}/K_o$ where K is the equilibrium association constant and the subscripts denote the presence or absence of cosolutes as follows: PEG200 (●), TEG (●), trehalose (●), betaine (○), sucrose (△), glycerol (□). Binding reactions at 21°C were in 10 mM bis-tris-propane, pH 7.3; 0.22 M NaCl; 1 mM EDTA; 50 μM dithiothreitol; 100 μg mL^{-1} bovine serum albumin. Binding was measured as described in ref. 168. The sequence of the double-stranded oligonucleotide was GGGCGGGAGCgaattcGCTGGCGC. Activity of Na$^+$ for each cosolute concentration was measured with a Na$^+$-selective electrode and all K_A values corrected (see text) to the Na$^+$ activity in the absence of cosolute. Data points are means±std. dev. of ≥3 determinations. (B) Experimental slopes (solid bars) from panel (A) are compared with covolume changes ΔU calculated without (horizontal striped bars) or with (diagonal striped bars) a single hydration layer. The free proteins were approximated by undocking from the structures of the protein-DNA complexes. For the free EcoRI DNA we used the crystal structure[169] of the free dodecamer CGCGAATTCGCG; for EcoRV (11 bp) and BamHI (10 bp) we used QUANTA software (Accelerys, Inc.) to build a B-form DNA containing the appropriate recognition site. Covolumes were calculated using GRASP.[166] The computed ΔU does not change if we use longer oligonucleotides, since the proteins contact only about 10 bp of DNA. (Unpublished data of D.F. Cao, M. Kurpiewski, T. Niu and L. Jen-Jacobson.)

Other cosolutes that are commonly used to stabilize proteins (*e.g.*, glycerol, sucrose, trehalose), by contrast, show evidence of varying degrees of chemical interaction with the macromolecules (Figures 2.2A and 2.2B) such that $\Delta\Delta G_{int} > 0$. For example, for trehalose, which has a larger radius (4 Å) than PEG200 (3.7 Å) but a lower slope (Figure 2.2A), the experimental value of $\partial \ln \Gamma / \partial c_3$ is less than ΔU as a result of the chemical interaction. We emphasize that deconvoluting the effects of $\Delta\Delta G_{int}$ and $\Delta\Delta G_{SE}$ for such cosolutes is nontrivial, but essential if their effects are to be understood. It is also noteworthy that, in a homologous series, ethylene glycol and diethylene glycol interact with the macromolecules such that $\Delta\Delta G_{int} > 0$, TEG and PEG200 are neither preferentially interacting nor preferentially excluded, and PEGs larger than 200 are actively excluded such that $\Delta\Delta G_{int} < 0$ (M. Kurpiewski, T. Niu and L. Jen-Jacobson, unpublished). It remains to be determined if TEG and PEG200 have the same favorable properties for other macromolecular systems.

A rigorous statistical-mechanical treatment of cosolute effects based on the Kirkwood–Buff theory of solutions[80] has been elaborated by Schurr et al.[76] and also (with minor variations) by others[81–85] following the work of Ben-Naim.[86] (We use the notation of Schurr et al.[76] except that we renamed the chemical interaction term that they called ΔS as ΔI to avoid confusion with entropy change.) This treatment predicts that K_A, the experimentally observed equilibrium association constant for protein and DNA, varies as cosolute alters a_W, the activity of water in the solution (determined experimentally, e.g., by vapor-pressure osmometry), according to:

$$\frac{\partial \ln K_A}{\partial \ln a_W} = \Delta X + \Delta I \qquad (2.6)$$

where ΔI is a term that measures the change in interaction or exclusion of the cosolute with the macromolecules and $\Delta X = X_{PD} - (X_P + X_D)$ where the subscripts P, D and PD denote the protein, DNA and their complex, respectively, and:

$$X_i = \frac{\Delta V_i^{ac}}{\bar{V}_1} \qquad (2.7)$$

$$\Delta X = \frac{\Delta \Delta V_i^{ac}}{\bar{V}_1} \qquad (2.8)$$

Here ΔV_i^{ac} is the difference in macromolecular volumes excluded to cosolute and to water[76] and $\bar{V}_1$ is the partial molar volume of water. In other words, ΔX is a steric exclusion term that measures for the protein-DNA association reaction the change in volume excluded to cosolute, corrected for the change in volume excluded to solvent. We believe it is appropriate to use a value of $\bar{V}_1$ that pertains in the hydration shell(s) of the macromolecules; we use Gerstein and Chothia's value of 24.5 Å³/H$_2$O for proteins,[39] which seems a reasonable average, rather than the value of 29.9 Å³/H$_2$O that applies to bulk water. (The water density is higher around the DNA,[38] but a DNA oligonucleotide contributes only a small fraction of the total ΔV_i^{ac}, so this may not be a major issue here.) Note that X_i and ΔX have units of numbers of waters, but since cosolute molecules are larger than water molecules, the volume excluded to cosolute is larger than the volume excluded to water, i.e., $\Delta V_i^{ac} > 0$. As a result, X_i and ΔX relate to the numbers of water molecules (or the change in this number upon protein-DNA binding) in a volume that extends beyond the first hydration layer.

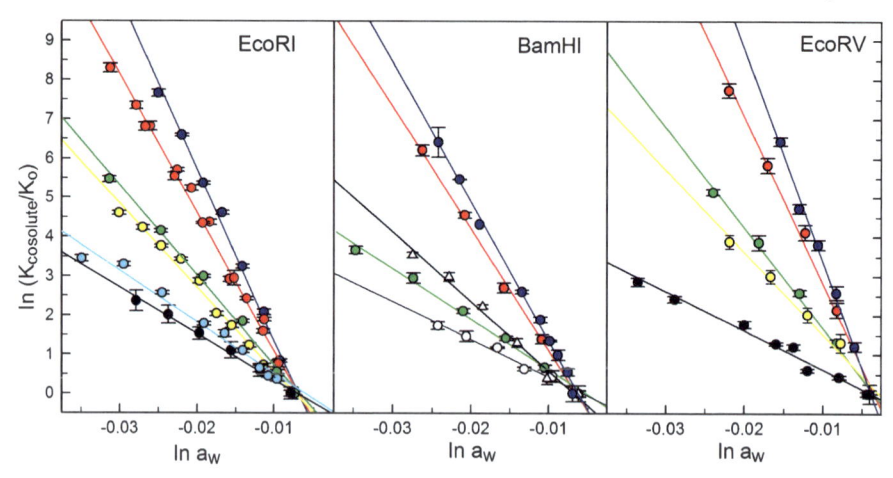

Figure 2.3 Effect of water activity on EcoRI, BamHI, and EcoRV specific binding. K_o is the equilibrium association constant in the absence of cosolute. $K_{cosolute}$ is the equilibrium association constant in the presence of PEG200 (●), TEG (●), trehalose (●), betaine (●), sucrose (●), glycerol (●), diethylene glycol (△), ethylene glycol (○). Buffer conditions were as given in the legend to Figure 2.2, except salt concentrations were as follows: EcoRI, 0.24 M NaCl; BamHI, 0.18 M Na acetate; EcoRV, 0.12 M NaCl. Data points are means±SD of at least three determinations. Water activity was determined by vapor pressure osmometry (mOsm ≡ $-(10^6 \ln a_w)/18$). Duplex DNA substrates with recognition sites in lower-case are: EcoRI (GGGCGGGAGCgaattcGCTGGCGC); BamHI (GGGATGGGTGggatccCACCCAC) and EcoRV (GTGTTGTAGgatatcCTACTGG). (Unpublished data of D.F. Cao, M. Kurpiewski, T. Niu and L. Jen-Jacobson.)

Figure 2.3 shows the dependence of ln K_A on ln a_w for various cosolutes in three endonuclease-DNA systems. It is immediately evident that the slope is different for each cosolute, thus completely excluding any conceptual model (e.g., the simplest form of "osmotic stress"[58–59,87–89]) in which the cosolute exerts its effect exclusively by modifying water activity. Furthermore, the hierarchical effects of various cosolutes do not correlate with their effects on the viscosity or dielectric constant of the solution.[79] The neutral cosolutes alter the salt activity in the solution, but this potentially confounding effect can be avoided by direct measurement of salt activity, e.g., with a Na^+-selective electrode, so that the data are represented at constant Na^+ activity.

The argument made from Figure 2.2 implies that for TEG and PEG200 we expect $\Delta I \approx 0$, and indeed Figure 2.4 shows that this is the case for EcoRI and EcoRV endonucleases. The excellent agreement between the experimental $\partial \ln K_A / \partial \ln a_w$ and the computed ΔX for these two cosolutes shows that the Kirkwood–Buff treatment of Schurr et al.[76] correctly predicts the experimental nonideality for EcoRI and EcoRV within about 1–3%. Some biochemical experiments[54] imply that the asymmetry seen in the crystal structure of the BamHI complex[90] is a result of crystal packing and does not reflect the solution structure; we believe this would cause our calculated ΔX to be too negative.

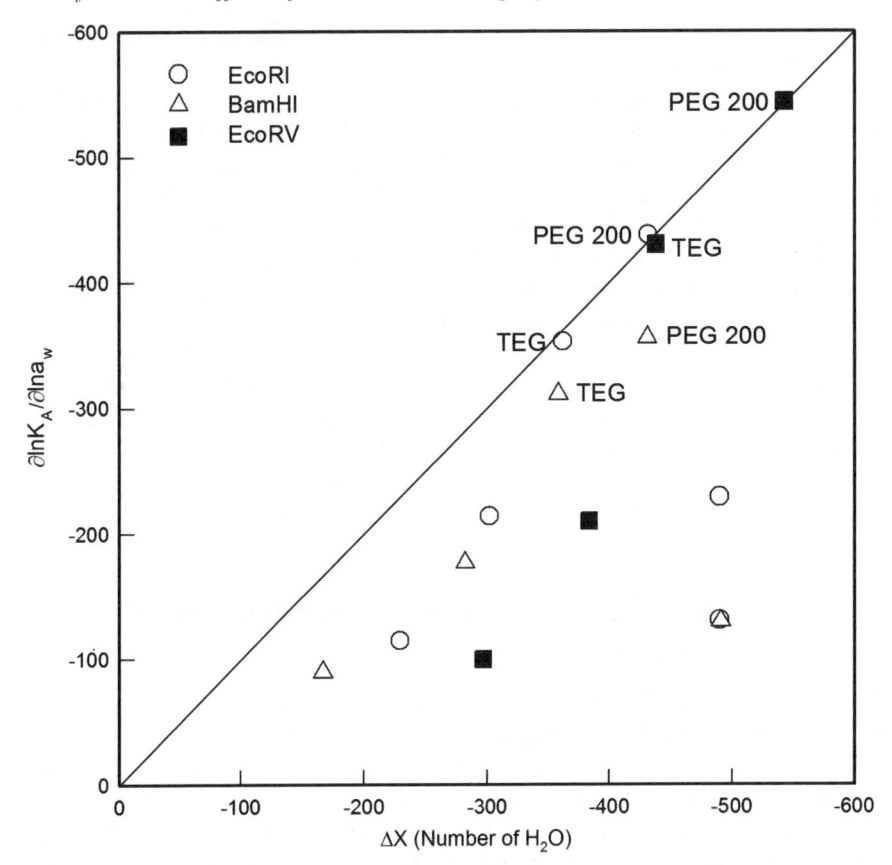

Role of Water and Effects of Small Ions in Site-specific Protein-DNA Interactions 25

Figure 2.4 Experimental data for $\partial \ln K_A / \partial \ln a_w$ (slopes of plots in Figure 2.3) plotted against ΔX according to (Equation 2.6). Deviations from the 1 : 1 line represent values of ΔI, which measures interaction with (below the line) or exclusion from (above the line) the macromolecular surfaces. (Unpublished data of D.F. Cao, M. Kurpiewski, T. Niu and L. Jen-Jacobson.)

Since the values of $\Delta X \approx \partial \ln K_A / \partial \ln a_w$ have the units of numbers of waters, it is instructive to compare these experimental values with those derived from ΔASA calculations, noting that the computational method assumes release of a single hydration layer, whereas the experimental (cosolute) measurements make no such assumption. As shown in Table 2.1, the values from the cosolute experiments with TEG are very close to the computed values for a single hydration layer.

What is the "correct" water value to consider as we try to understand the thermodynamics? It is unlikely to be a single hydration layer because, as noted above, the solvent domain influenced by macromolecules, defined by some criterion such as density deviation from bulk solvent, is likely greater than a single layer.[38] On the other hand, the domain of macromolecular hydration influenced by cosolutes depends on the diameter of the cosolute[76,91] and there is no simple and rigorous numerical relationship of water release from this domain to the "stoichiometric" water release upon protein-DNA binding in the absence of the cosolute. Nevertheless, the average rugosity (presence of crevices and grooves) of the protein surfaces turns out to be such that there are regions accessible to water but not to cosolute, and for cosolutes the size of TEG (calculated[92] van der Waals radius $r = 3.4$ Å) or PEG200 ($r = 3.74$ Å) this factor compensates to nearly the right extent for the greater diameter of the cosolute molecule. (A more detailed discussion of accessibility issues may be found in Schurr et al.[76]) We can alternatively calculate the contribution of the hydrophobic effect as follows: We use the experimental water release values ($\partial \ln K_A / \partial \ln a_w$) from the TEG effects, partitioning these according to the fractions of nonpolar and polar surfaces derived computationally (Table 2.1). Because water release from nonpolar surfaces has negligible effect on $\Delta H°$ near room temperature,[7-9,48] we can then use an estimated value ($+1.7$ cal mol^{-1} K^{-1} per H$_2$O) of the entropic contribution of the hydrophobic effect[93] to calculate the free energy component $-T\Delta S°$. Such calculations give values for 294 K as follows: EcoRI, -80 kcal mol^{-1}; EcoRV, -100 kcal mol^{-1}; BamHI, -88 kcal mol^{-1}. These are within about 10% of the estimates from ΔASA and the contact free energy.

2.5 Retained Water Molecules Contribute to Affinity and Specificity

Nearly all protein-DNA complexes whose structures have been determined at sufficient resolution show multiple "trapped" water molecules that bridge between protein and DNA, although the TATA-box binding protein TBP is a notable exception.[94] Crystallography cannot distinguish between site occupancy and residence time, but our experience with many molecular dynamics simulations of wild-type, mutant and modified-DNA versions of the EcoRI endonuclease-DNA complex[17] (P.J. Sapienza and L. Jen-Jacobson, unpublished) and the EcoRV-DNA complex (S. Hancock and L. Jen-Jacobson, unpublished) suggests that even for water sites that have essentially complete occupancy there is a range of exchange rates with bulk solvent. Some water molecules that bridge protein and DNA, but make only two hydrogen bonds, have residence times of only 20–50 ps, but range up to 200 ps.[95] Water molecules that make three or four hydrogen bonds with surrounding functional groups, and thus potentially contribute to specificity, have longer residence times and appear to fall into two groups: those with residence times of 100–300 ps, and those that are sufficiently immobilized such that no exchange occurs in a 5 ns simulation. For the AntP homeodomain-DNA complex, NMR data suggested a boundary of residence times from nanoseconds to milliseconds,[96] but in computational simulations such waters have residence times in the range 400–1000 ps.[97,98] Thus, while it is well not to construct too static a mental picture of the water molecules that bridge between protein and DNA, these waters are still very much immobilized relative to bulk liquid water, where water-water interactions have lifetimes less than a picosecond,[99] and also relative to the hydration shell of uncomplexed protein.[100]

A survey of 129 crystal structures of protein-DNA complexes[101] indicated that water-mediated protein-DNA interactions are nearly as common as direct protein-base hydrogen bonds, constituting about 15% of all contacts. Only about 30% of the water-mediated interactions involved DNA bases, and only about 20% of these latter waters formed three or more hydrogen bonds. Water molecules that mediate protein-DNA recognition often occupy the same positions that were hydrated in the free DNA,[102] so that sequence-specific proteins exploit complementarity with the hydrated DNA surface. (Even at direct protein-DNA contacts not mediated by water, the protein atoms involved often occupy the same positions that were occupied by water in the free DNA.[103]) It has been suggested that in the majority of the water-mediated protein-DNA interactions the water acts merely as "space fillers" to increase overall complementarity of the surfaces and/or to improve the electrostatics. Electrostatic calculations[104,105] support the idea that most retained waters act to screen electrostatic repulsions between electronegative atoms in protein and DNA. The exchangeability of some of these interfacial waters with bulk solvent may not only reduce the entropic disadvantage but may provide dynamic flexibility to the interface.[106]

Nevertheless, it is unquestioned that water mediated interactions can contribute to specificity in individual cases. In the *trp* repressor-operator complex[107] water-mediated hydrogen bonds provide the only contacts between protein and the DNA bases, and contribute to sequence specificity.[108] In the EcoRI-DNA complex,[109] the initial G of the GAATTC site is recognized by a single water molecule, of residence time > 5 ns in simulations, that donates hydrogen bonds to both N7 and O6 of the G while receiving hydrogen bonds from the guanidino side-chains of two arginines. Deletion of the N7 group by replacing G with 7-deaza-G causes a binding penalty[110] $\Delta\Delta G°_{bind} = +1.1$ kcal mol^{-1}, very similar to the value (+1.2 to +1.4 kcal mol^{-1}) for direct protein-base hydrogen bonds.[1,57] A similar energetic equivalence was observed for mutational disruption of direct and water-mediated interactions of papillomavirus E2C protein with DNA.[111] The formation of a water-mediated protein-DNA interaction is likely to contribute favorable $\Delta H°$ and unfavorable $\Delta S°$, but these values for a single water-mediated interaction are probably too small to be measured accurately by current technology.

2.6 Thermodynamic Effects of Retained Water

When water molecules become sandwiched between protein and DNA in the complex, the accessible motions of these waters become more restricted, decreasing the heat capacity.[24,36] The extent to which this is a major contribution to the negative $\Delta C°_P$ for any given protein-DNA association varies from system to system, depending not only on the number of interfacial waters but also on the particular environments in which they dwell. The $\Delta C°_P$ for incorporation of water from bulk solvent to the interface of a protein-ligand complex has been estimated at -8.6 cal K^{-1} per mol of water.[112] Thus, for the EcoRI-DNA and BamHI-DNA complexes, each of which retains about 50 water molecules between protein and DNA,[27] the total contribution can be estimated as $\Delta C°_P \approx -0.4$ kcal mol^{-1} K^{-1}. Compared with total observed $\Delta C°_P$ values of -1.2 to -2.5 kcal mol^{-1} K^{-1} for these systems[27], the interfacial water contribution is significant but not dominant. In addition to this, the release of water from protein and DNA surfaces probably makes a smaller contribution to the negative $\Delta C°_P$, so that all hydration terms together account for less than half the observed $\Delta C°_P$. By contrast, in studies of the interaction of an archaeal TBP with DNA,[36] large changes in $\Delta C°_P$ were observed upon mutational perturbation of a network of five to seven water molecules lying in a highly charged cleft. The authors inferred that immobilization in such a circumstance could produce not only a greater heat capacity loss ($\Delta C°_P \approx -17.9$ cal K^{-1} per mol of water) for the networked waters but also a reduction of heat capacity in a second solvent layer *via* hydrogen bonding.

2.7 Overview of Small Ion Effects on Protein-DNA Interactions

The equilibrium affinity of proteins for DNA can be very strongly influenced by salt concentration. In the range of moderate monovalent salt concentration (~ 0.15 to 0.4 M), K_A for various DNA-binding proteins typically decreases by factors of 10^5 to $> 10^{11}$ for a ten-fold increase in salt concentration.[65,113,114] Experimentally, there is little or no distinction between the effects of Na$^+$ and K$^+$ on binding.[115] Elegant theoretical treatments[116–119] of this phenomenon have focused on the polyanionic nature of DNA and the thermodynamically favorable displacement of cations from DNA phosphates upon protein binding. Although terms for anion effects appear explicitly in the equations,[65] the role of anions is often underemphasized. In part because of anion effects, it is incorrect to relate the experimental salt dependence ($\partial \log K_A / \partial \log[MX])_{T,P,pH}$ (where MX is a monovalent salt M$^+$X$^-$) only to the stoichiometry of cation release (neglecting anions) or to use this dependence to enumerate "ion pairs" (*i.e.*, charge-charge interactions) formed between protein and DNA phosphates. Furthermore, hindsight and newer experimental data have made it clear that small ions may have additional effects on binding equilibria beyond those on the direct electrostatic interactions between protein and DNA, and that the magnitudes of these effects may depend on peculiarities of individual proteins.

Figure 2.5 shows the effect of salt concentration on site-specific equilibrium binding of BamHI endonuclease to DNA, using the sodium salts of chloride, fluoride, acetate and glutamate. The theoretical treatments of the "polyelectrolyte effect", focusing on the cations, propose a linear dependence of log K_A on log [M$^+$]. It is evident in Figure 2.5 that (a) the nature of the anion markedly influences affinity; (b) there are significant deviations from linearity at low salt concentrations; (c) the occurrence and direction of deviations from linearity at high salt depend on the identity of the anion. We discuss below the various factors that contribute to these observations.

2.8 Multiple Physical Phenomena Associated with Salt Dependence

The various theoretical treatments of salt effects on binding do not predict the marked departures from linearity in plots such as that of Figure 2.5. Close

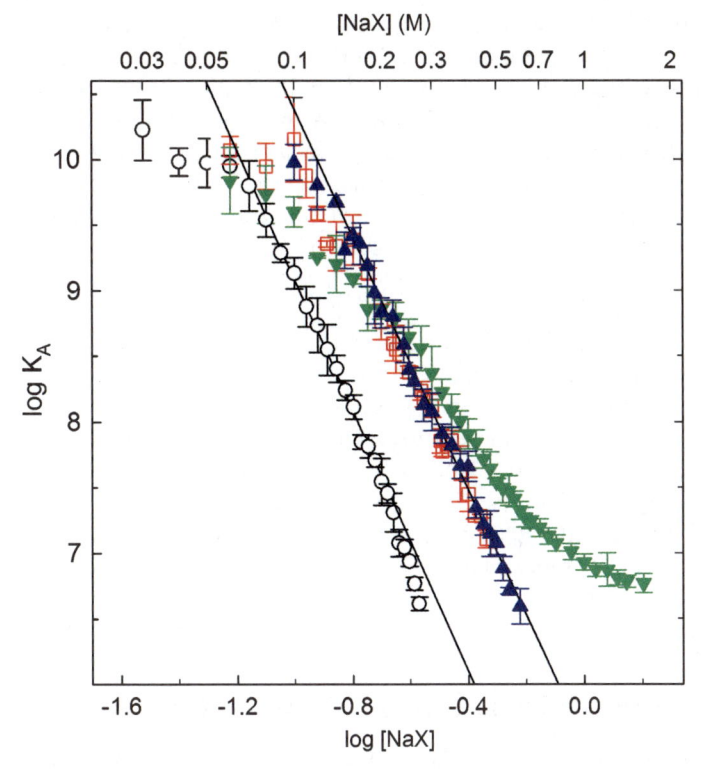

Figure 2.5 Complex dependence of specific BamHI-DNA binding on monovalent salt concentration. Inhibition of binding at increasing [NaX] is dominated by Na^+ cations in the mid-range. The onset of nonlinearity at low salt occurs at higher [NaX] for excluded (Glu) or non-interactive (F^-, acetate) anions, and at lower [NaX] for the interactive Cl^-. Deviations from linearity at high salt for Cl^- and glutamate are the result of binding or chemical exclusion, respectively (see text). Conditions for binding to CGCGGGCGGCggatccGGGCGGGC were as given in the legend to Figure 2.2, except for the indicated concentrations of the Na^+ salts of glutamate (▼), F^- (▲), acetate (□) or Cl^- (O). Data points are means ± std. dev. of > 3 determinations.[170]

examination of published datasets on salt dependence often reveals the decrease in slope at low [salt].[115,120–123] If the data points are few, and/or if an insensitive assay for protein-DNA binding restricts measurements to the low-salt range, imposing a linear fit on the data across this region of changing slope will underestimate the salt dependence significantly. When binding is measured with large DNA fragments, the slope decrease at low salt (<0.1 M) is attributable to effects on the facilitated diffusion ("sliding") of the protein along nonspecific DNA sequences.[124] In the case of *Eco*RI endonuclease binding to small (≤24 bp) oligonucleotides, where such sliding is not an issue, it was shown that the decrease in slope at low salt is a result of a competing process of protein-protein interaction (*e.g.*, a shift from dimers to tetramers) leading in the limit to protein precipitation.[125] Few published studies take account of this possibility. In our experience with *Eco*RI, *Bam*HI and *Eco*RV endonucleases (M. Kurpiewski and L. Jen-Jacobson, unpublished observations), the low-salt threshold for older enzyme stock solutions shifts over months to progressively higher salt, which we interpret in terms of nucleation of higher-order aggregates, since the threshold can be again lowered by re-purification. Note also that the low salt threshold for nonlinearity depends on the anion, such that the interactive anion Cl^- moves the threshold to lower salt and the excluded anion glutamate moves it to higher salt. We believe this reflects inhibition of protein-protein interaction by protein-bound Cl^-, and promotion of protein-protein interaction by the excluded anion glutamate by virtue of a steric exclusion effect as discussed above for other cosolutes.

Figure 2.5 also shows that at high salt concentration the curve remains linear when using NaF, but curves downward in NaCl and upward in sodium glutamate. The NaF curve shows that the underlying inhibition of protein-DNA binding by cations (see below) remains linear at higher salt. However, the cation effect is counteracted by the steric exclusion of glutamate, which at high concentration promotes protein-DNA binding, and reinforced by binding of Cl^-, which inhibits protein-DNA binding. A similar upward curvature at high glutamate concentration was observed for *lac* repressor-operator binding.[115]

2.9 Cation Release Favors Protein-DNA Association

In the mid-range of Figure 2.5 at moderate salt concentrations, the linear dependence fits the common expectation of the theoretical frameworks, which variously treat cation effects in terms of stoichiometric release of cations from DNA phosphates,[116–119] or in terms of non-stoichiometric reorganization of cations around the macromolecules, including cation–solvent interactions.[126] We have discussed elsewhere[57] the likelihood that cations can be displaced not only by protein-DNA ion pairs but also by protein-phosphate hydrogen bonds. Consequently, any attempt to draw numerical correspondences between the linear slopes of salt-dependence curves and the number of protein-phosphate contacts that can be counted in a structure is necessarily hazardous when there are multiple protein groups interacting with a single DNA phosphoryl oxygen. Furthermore, cations bound to the phosphates of free DNA, and released upon protein binding, may not be the only cations that affect binding. Monovalent cations may have significant thermodynamic consequences by affecting intra-protein salt bridges[127] and/or by Debye screening of electrostatically unfavorable negative charge clusters in enzyme active sites[128] and/or by reducing inter-phosphate repulsion to affect DNA distortability. For papillomavirus E2 proteins[123] an increase of binding with increasing salt concentrations has been interpreted as showing cation *uptake* by DNA at <0.35 M salt. This stimulates protein-DNA binding because cations bound in the DNA minor groove stabilize bent DNA conformations found in the protein-DNA complex.

Figure 2.6 shows that if we use the noninteractive anion F^- [129,130] then increasing the salt concentration has only very slight effects on $\Delta H°_0$ and

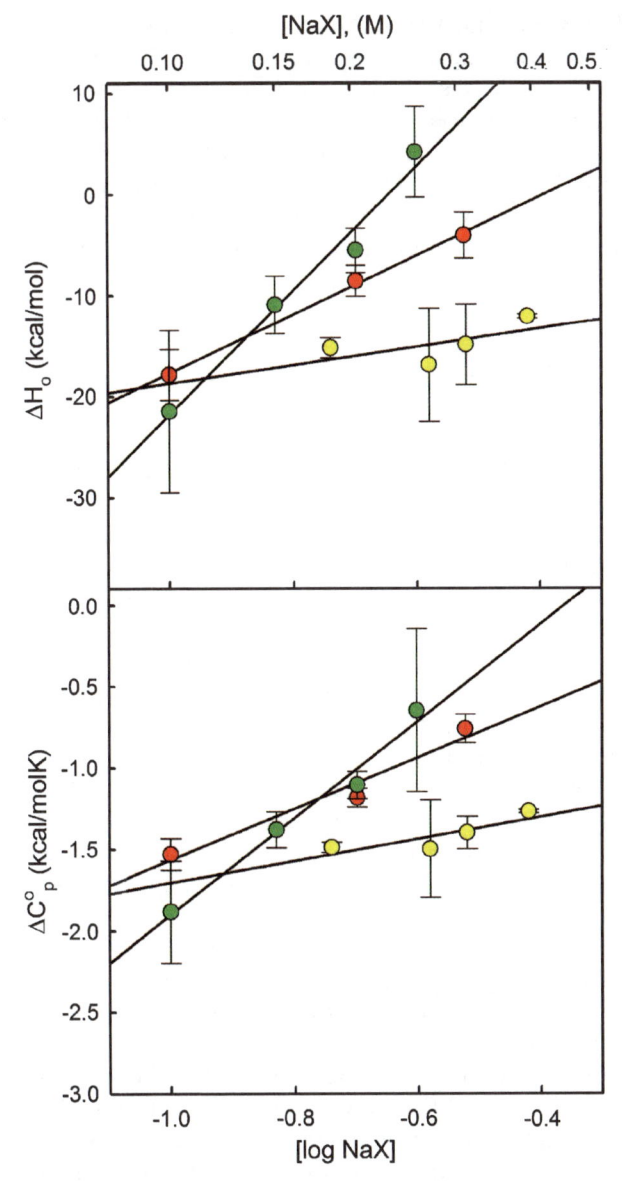

Role of Water and Effects of Small Ions in Site-specific Protein-DNA Interactions 31

Figure 2.6 $\Delta H°_0$ and $\Delta C°_P$ for specific BamHI binding depend on salt concentration and anion identity. Isothermal titration calorimetry (ITC) measurements of heats of binding (to specific DNA sequence GGGATGGGTGggatcc-CACCCAC) were performed as described[27] over temperatures from 2 to 15°C. Above 15°C, the stoichiometry of binding (DNA:protein dimer) goes down markedly as a result of protein aggregation. Measured heats of reaction $\Delta H°_{obs}$ were corrected for buffer heats of ionization to obtain $\Delta H°_0$.[32] Note that, for BamHI, $\Delta H°_0$ closely approximates the intrinsic binding enthalpy change $\Delta H°_{intrinsic}$;[30,32] BamHI binding is coupled to carboxylate protonation (see Section 2.11), but $\Delta H_{protonation}$ for carboxylates is very small. Salts at the indicated concentrations were NaF (●), NaCl (●), or NaBr (●). (L.E. Engler and L. Jen-Jacobson, unpublished data.)

$\Delta C°_p$ for BamHI endonuclease-DNA binding, so that cation release can be considered as contributing primarily or exclusively favorable $\Delta S°$.[131,132] (The notation $\Delta H°_0$ denotes the measured heat of reaction $\Delta H°_{obs}$ corrected for buffer heat of ionization. See legend to Figure 2.6. We henceforth refer to this corrected value as $\Delta H°$.) Similar observations have been made for the non-sequence-specific binding of SSB protein to single-stranded DNA[130] and for sequence-specific binding of homeodomain proteins to their recognition sites.[133] There is some disagreement about whether this is entirely attributable to entropy of mixing.[113,126,134]

2.10 Selective Effects of Anions on Protein-DNA Binding

The effects of monovalent anions on protein-DNA association follow the Hofmeister series[135] $Br^- > Cl^- > F^-$, with acetate about equivalent to F^-.[33,130] (The differential interaction of anions with macromolecules depends on the relative free energies of macromolecule-bound anion and hydrated free anion in solution; for example, F^- is more strongly hydrated than Cl^-,[136] and therefore F^- interacts little or not at all with protein.) The physical basis for the behavior of various anions can be explained in various ways.[136–140] At the level of $\Delta G°_{bind}$, the effect of anions in the mid-range of salt concentrations (~ 0.1 to 0.3 M) appears to be simple (Figure 2.5). The slopes of the curves for F^-, Cl^- and acetate are the same within experimental error, so that if Cl^- interacts with the protein, while F^- and acetate do not, the presence of Cl^- imposes an approximately constant free-energy penalty $\Delta\Delta G°_{anion} \approx +1.2 \text{ kcal mol}^{-1}$, increasing only slightly at higher salt. Similar data for *lac* repressor-operator binding (comparing Cl^- to acetate) gave $\Delta\Delta G°_{anion} = +1.7 \text{ kcal mol}^{-1}$ (Na^+ salts, Barkley *et al.*;[141] K^+ salts, Ha *et al.*[115]).

However, calorimetric dissection into $\Delta H°$ and $\Delta S°$ reveals more complex phenomena. $\Delta H°$ is only very slightly dependent on NaF concentration (Figure 2.6); because F^- is non-interactive, this implies that $\Delta H°$ does not depend significantly on cation concentration.

Similar observations have also been made for SSB-DNA binding.[33] By contrast, interactive anions ($Br^- > Cl^-$) have concentration-dependent effects on $\Delta H°$, which implies that anion interactions also affect $\Delta S°$.

The anion effect on $\Delta H°$ appears to vanish at salt concentrations below about 0.1 M (Figure 2.6). Data cannot be acquired for this system at lower salt because of the coupled protein-protein equilibria, in which formation of higher-order protein aggregates competes with protein-DNA binding. However, data on SSB binding to DNA suggest that the $\Delta H°$ values in various anions converge at or near 0.05 M salt.[130] This convergence, together with the fact that an anion penalty on $\Delta\Delta G°_{bind}$ already exists at low salt (Figure 2.5), implies that interactive anions like Cl^- exert an unfavorable entropic influence on protein-DNA binding at low salt concentrations. The inference that bound anions inhibit aggregation of free protein at low salt (see above) would also be consistent with an anion influence on protein-protein interaction sites of the free protein. However, there is no direct evidence to show such a high-affinity anion interaction with any free DNA-binding protein. Like others[115,141] who have considered the anion dependence of $\Delta G°_{bind}$, we have not yet been able to devise a molecular explanation for these thermodynamic observations that is entirely satisfactory while maintaining conceptual consistency with the thermodynamic effects of anions at higher concentrations, where $\Delta H°$ is also affected (see below).

As the concentrations of interactive anions increase above ~ 0.1 M, $\Delta H°$ and $\Delta C°_P$ become progressively less negative (Figure 2.6), as has also been observed for SSB binding to DNA.[130] The same directions of trend in $\Delta H°$ and $\Delta C°_P$ for specific DNA binding of IHF[127] was attributed to salt destabilization of intramolecular salt bridges between cationic side-chains and carboxylates in the unbound protein to free cationic groups to interact with DNA. After careful examination of the free and DNA-bound BamHI structures, we found that this explanation does not pertain for BamHI endonuclease. It, therefore, appears that the effects on $\Delta H°$ and $\Delta C°_P$ are the symptoms of low-affinity interactions that occur for $Br^- > Cl^-$ but not (or very weakly) with F^-.

Low-affinity anion binding sites should cause effects that become evident only at high salt, and thus introduce apparent nonlinearity into the salt dependences. This can be seen as a nonlinear effect on $\Delta G°_{bind}$ at high NaCl concentration (Figure 2.5), but the calorimetric data (Figure 2.6) are too sparse to detect such nonlinearity. (Calorimetric data at higher salt concentrations are difficult to obtain for most proteins because protein-DNA binding becomes too weak.) It is not simple to relate such sites to structure, because low-affinity binding sites could plausibly exist in the free protein, in the protein-DNA complex or in both. Given the anionic nature of DNA, it is often posited that anions bound in low stoichiometry to the free protein are displaced upon DNA binding,[33,115] but in the absence of definitive evidence it remains possible that both of these mechanisms may contribute (perhaps to unequal extents) to the observed low-affinity effects of interactive anions such as Cl^-. That is, anion binding could lower the free energy of the free protein, destabilize the protein-DNA complex, or both. The alternative rationales are as follows:

(a) Anions such as Cl^- associate weakly with regions of the free protein that have positive electrostatic potential. This makes $\Delta H°$ more positive because there is less enthalpic benefit for binding the protein to the negative surface of the DNA. The replacement of protein-water interactions with stronger protein-anion interactions also lowers the heat capacity of the free protein, thus making $\Delta C°_P$ less negative. Anion binding has been portrayed as competition[115] between small anion and DNA phosphates for the protein regions of positive potential, but at the molecular level there is no compelling reason to limit this argument to the regions of the protein that directly interact with DNA, since not all regions of positive potential are engaged in such direct interaction.

(b) Anions such as Cl^- might also bind to the protein-DNA complex, at distal regions of positive potential that do not interact directly with DNA. This would make the electrostatics of the protein-DNA complex less favorable, thus raising the heat capacity and making $\Delta H°$ and $\Delta C°_P$ less negative.

We have presented this analysis to illustrate how even the seemingly simple and approximately constant inhibition of binding by interactive anions (Figure 2.5) may veil more complex phenomena (Figure 2.6). Molecular interpretations should suggest themselves more readily when the component $\Delta H°$ and $\Delta S°$ and the $\Delta C°_P$ have been determined, but ambiguity may yet remain because it is not always clear which side of the binding equilibrium is being affected. Recent technical advances in biocalorimetry will likely lead to a more rapid accumulation of data such as these for diverse protein-DNA systems, which will perhaps provide a firmer base from which to seek general thermodynamic explanations. We also note that the interactive anions are generally found at low intracellular concentrations (*e.g.*, 3 mM Cl^- [136]) *in vivo*.

2.11 Divalent Cation Binding at Active Sites Relieves Electrostatic Strain

Some DNA-binding proteins, such as polymerases and restriction nucleases, are also enzymes; these usually require divalent metal ion(s) for catalytic activity but not for DNA binding. The metal-ion chelating sites are formed (in some cases contingent upon site-specific DNA binding[17]) by the close apposition of several carboxylate side-chains to each other and to DNA phosphate.[90,142] Such a negative charge cluster produces an unfavorable electrostatic repulsion in the absence of the divalent metal ion.[54] This can be directly shown because mutation of one or more of the asp/glu residues to remove a charge from the cluster stimulates binding while abolishing catalysis.[54,143] The same stimulation occurs upon deletion of the relevant DNA phosphate.[54] It is a widespread observation that binding of Ca^{2+} to such sites, which does not support catalysis, nevertheless strongly stimulates protein-DNA association,[17,54,144-146] but Ca^{2+} stimulation is not observed in the mutants where one or more active-site carboxylates is removed (L.E. Engler and L. Jen-Jacobson, unpublished) or where the DNA phosphate is replaced by an uncharged analogue.[17] The affinities for Ca^{2+} are in the low millimolar to sub-millimolar range,[17,54] so these phenomena can be observed at Ca^{2+} concentrations that avoid confounding with the polyelectrolyte effect.

It is also of interest that these electrostatically repulsive active-site clusters, when present, may be the source of much of the pH-dependence of protein-DNA binding. A carboxylate in a negative cluster has its pK_a raised above that of an isolated carboxylate. DNA binding is much stronger (in the absence of metal) when these residues are protonated below pH 7, and the pH dependence in the neutral range nearly disappears when a key active-site carboxylate is removed by mutation, or the cluster neutralized in the presence of Ca^{2+} (Figure 2.7).

In terms of catalysis, the electrostatic repulsion (a form of strain) in the active sites can be considered as a stratagem for destabilizing the metal-free ground state of the enzyme-DNA complex, such that there is less energy required to

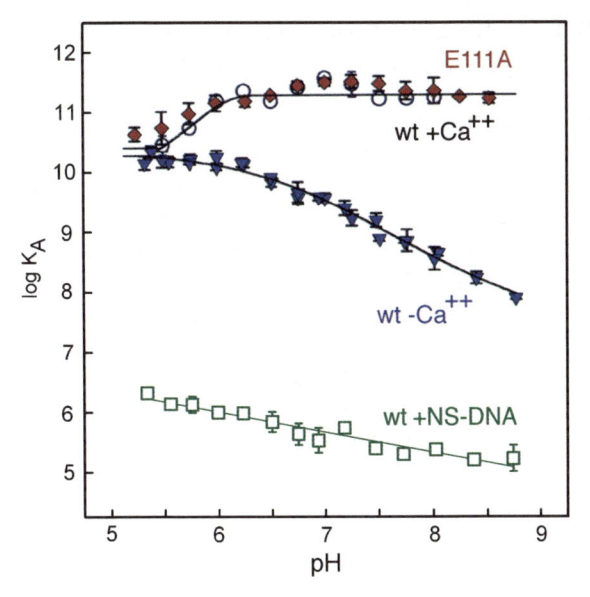

Figure 2.7 Role of active-site electrostatic repulsion in the pH-dependence of BamHI binding. Wild-type binding to the specific ggatcc site (▼) increases strongly from pH 8.8 to ∼6, but neutralization of the active-site charge cluster by mutant E111A[54] (◆) or with 10 mM Ca^{2+} (○) nearly abolishes pH dependence above pH 6. Binding of wild-type enzyme to nonspecific (NS) DNA (□) is much more weakly pH dependent, because the active-site cluster does not assemble in the nonspecific complex.[155] Binding reactions were conducted in the buffer given in the legend of Figure 2.2, except that citrate was used as the buffer at lower pH and the salt was 0.18 M sodium acetate. Specific DNA sequence given in the legend of Figure 2.3. Nonspecific DNA sequence with inverted site in lowercase was CGCGGGCGGCcctaggGGGCGGGC. Data points are means ± std. dev. of >3 determinations. (L.E. Engler, O. Pham, G. Bosco and L. Jen-Jacobson, unpublished.)

reach the transition state[17,54,57] and consequent acceleration of catalysis. Electrostatic strain, expressed in the form of perturbed pK_a values for ionizable side-chains, is a reasonably common feature of enzyme active sites.[147,148]

Unpublished studies in the Jen-Jacobson laboratory (L. Engler and L. Jen-Jacobson, unpublished) have shown that active-site binding of Ca^{2+}, or equivalent neutralization by mutational removal of an active-site carboxylate, cause the $\Delta C°_P$ to be dramatically more negative. This implies that electrostatic repulsion in the absence of divalent cation contributes positive $\Delta C°_P$. Inasmuch as such electrostatic strain and binding-coupled protonation may be considered reciprocal aspects of a single phenomenon, it is important to understand that the protonation equilibrium *per se* should contribute negative $\Delta C°_P$, so its abolition by mutation should make $\Delta C°_P$ less negative. The fact that the observed effect is the opposite of this emphasizes the importance of electrostatic strain to the heat capacity.

2.12 Ion Effects and Cosolute Effects are Mechanistically Independent

Given that the stimulation of protein-DNA binding by cosolutes derives primarily from excluded-volume (steric) effects, whereas the inhibition of binding by salts derives also from the purely entropic (cations) or electrostatic (anions) effects of small-ion release from or binding to the macromolecules, it is of interest to inquire whether the effects are independent. (Of course, salts are also "cosolutes" that produce excluded-volume effects at sufficient concentrations.) The slope of the log K_A vs. log[NaCl] plot in the range of moderate salt does not change upon addition of various cosolutes at various concentrations, although the absolute values of K_A are dramatically affected (M.R. Kurpiewski and L. Jen-Jacobson, unpublished). Our data show that this is true for cosolutes (like TEG and PEG200) that are neither interactive nor preferentially excluded, and is also true for cosolutes (like trehalose) that preferentially interact with the macromolecules. Thus, the influence of two important environmental variables on protein-DNA association can be regarded as mechanistically independent, although operationally in a real solution they influence each other because the activities of all molecular species are thermodynamically coupled. This implies, as proposed by Record and co-workers,[149] that protein-DNA affinities could be maintained *in vivo* in the face of varying intracellular concentration of monovalent cation (primarily K^+) by varying the concentration of neutral osmolyte (cosolute) in the same direction to compensate, such that protein-DNA binding affinity remains approximately invariant. We consider below how such compensatory action might affect the specificity of binding.

2.13 Comparison with Nonspecific Binding: How Water and Ions Affect Specificity

All known site-specific DNA binding proteins also bind weakly to nonspecific DNA, although the definition of what constitutes "nonspecific" binding varies somewhat from case to case.[1,27] This presents a substantial challenge for the ability of site-specific binding proteins to function at their specific recognition sites *in vivo*, because although the binding affinity for specific site(s) is orders of magnitude greater than that for nonspecific DNA sites, the nonspecific sites are present in very large molar excess over recognition site. It thus seems likely that site-specific proteins must spend the large majority of their time interacting with nonspecific DNA,[110,150] which nevertheless lowers kinetic barriers to association with the specific site(s) by means of "facilitated diffusion".[124,151]

Nonspecific binding involves the formation of protein complexes that differ significantly from the corresponding site-specific complex in several properties, including crystal structure,[152–155] the degree to which motion is restricted in the complexes,[25,156] and the absence of a localized footprint on the DNA.[55,110] Table 2.2 shows how the various component factors that contribute to the thermodynamics of site-specific protein-DNA association differ between specific and nonspecific complexes.

Role of Water and Effects of Small Ions in Site-specific Protein-DNA Interactions 37

Table 2.2 Energetic components of protein-DNA interaction.

Component in site-specific complex	Difference in nonspecific complex
Favorable	
Protein-base H-bonds[a]	Absent
Protein-phosphate contacts[b]	More favorable
Nonpolar desolvation (hydrophobic effect)	Less favorable
Unfavorable	
Translational-rotational entropy loss	Same
Vibrational restriction[c]	Less unfavorable
DNA distortion[d]	Absent
Coupled protein folding[e]	Absent
Active-site electrostatic repulsion[f]	Absent
Polar desolvation[g] (excluding base and phosphate contacts)	Less unfavorable?

[a] Includes desolvation of the polar groups involved.
[b] Includes desolvation of polar and charged groups and cation release from DNA phosphates.
[c] Includes both the macromolecules and any water molecules trapped in the interface.
[d] Not always present. Primarily effect of base destacking on $\Delta H°$.[26] Some particular DNA distortions may be affected by hydration or cation neutralization of phosphates.
[e] Not always present. Includes unfavorable configurational $\Delta S°$ counterbalanced by hydrophobic effect.
[f] Also subsumes the effects of coupled protonation equilibria (see text).
[g] Desolvation of polar groups that do not contact DNA bases or phosphates.

One signature property that distinguishes specific from nonspecific binding is that site-specific binding is always accompanied by a strongly negative $\Delta C°_P$, whereas for nonspecific binding the measured values of $\Delta C°_P$ are near zero.[24,27,157–160] This difference reflects the fact that the nonspecific complexes lack nearly all the factors that contribute to the negative $\Delta C°_P$ for formation of the specific complexes: Nonspecific binding involves much less water release from nonpolar surfaces, vibrational restriction of the macromolecules[25] and presumably of interfacial waters is lessened, and in some cases the coupled protonation equilibria are absent (Figure 2.7). At the same time, the electrostatic repulsions that for metal-binding enzymes contribute a positive increment to $\Delta C°_P$ in specific binding do not do so for nonspecific binding, because the active-site charge clusters remain unassembled in the nonspecific complex.[155] In consequence, nonspecific binding is not stimulated by Ca^{2+} (P. Sapienza, L. Engler and L. Jen-Jacobson, unpublished). These differences in basic properties all fundamentally derive from the fact that nonspecific complexes lack recognition interactions with the DNA base sequence, so that the protein and DNA surfaces are less intimately apposed.

The differences in hydration can be shown experimentally by the significantly reduced sensitivity of nonspecific binding to the presence of cosolutes (Figure 2.8). This is qualitatively consistent with the inference from the few crystal structures of nonspecific complexes,[155] where the protein and DNA surfaces appear much less closely apposed than they are in the specific

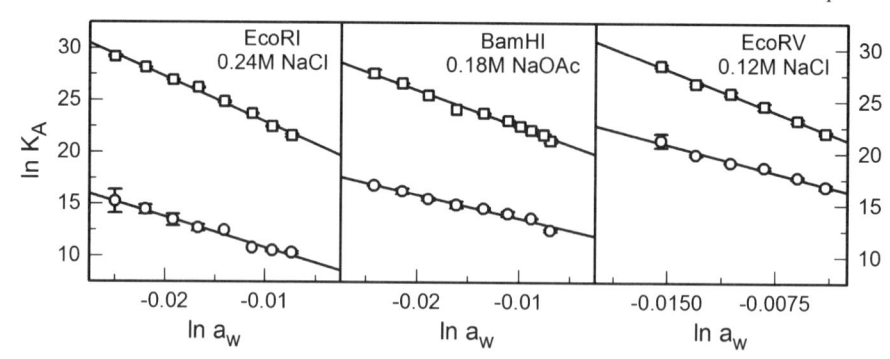

Figure 2.8 Comparison of PEG200 effects on specific and nonspecific binding. Procedures were as given in the legends to Figures 2.2 and 2.3, except that nonspecific binding was measured by equilibrium competition.[168] Specific DNA sequences are given in legend to Figure 2.3; nonspecific DNA sequences with inverted sites in lowercase were: EcoRI (GTGCcttaagCGCG); BamHI (GGGATGGGTGcctaggCACCCAC); EcoRV (GGCTCctatagCTATG). (M. Kurpiewski and L. Jen-Jacobson, unpublished.)

complexes, leaving room for retention of substantial amounts of water between them. The retention of this interfacial water means that the nonspecific complexes are less compacted than specific complexes, i.e., $\Delta\Delta V^{ac}$ in (Equation 2.8) is smaller for nonspecific binding. Quantitatively, the roughly two-fold lower slopes for nonspecific binding in Figure 2.8 suggest that as much as half the water that is released during specific protein-DNA binding may be retained in the nonspecific complex. Because the nonspecific complexes remain mobile and traverse the DNA by "facilitated diffusion", it seems unlikely that the additional waters in the nonspecific complex are much restricted in motion, or that the interfacial waters of the specific complex are similarly immobilized in the nonspecific complex. Regardless of detailed interpretation, experiments such as those in Figure 2.8 provide a reliable empirical criterion for distinguishing specific from nonspecific binding, and are far less technically demanding than accurate measurement of $\Delta C°_P$.

The hydration (or compaction) differences between specific and nonspecific complexes (Figure 2.8) imply that the cosolute effect fulfills the criterion for an influence of an environmental variable on specificity, namely a differential effect on the complexes (see Section 2.2). Because site-specific binding affinity is more sensitive to cosolute concentration than is nonspecific binding affinity, it follows that an increased concentration of cosolute enhances specificity; that is, the equilibrium preference for the specific DNA binding site over nonspecific sites increases.

The effects of salt concentration on specific and nonspecific binding are quite similar, but the slope dlog K_A/dlog[NaCl] is generally greater for nonspecific than for specific binding.[25,54,55,110,115,161,162] This presumably reflects the fact (Table 2.2) that protein-phosphate interactions are proportionally more important to nonspecific than to specific complex formation. For example, for EcoRV endonuclease this difference is such that an increase in NaCl from 0.1 to 0.2 M *in vitro* increases the specificity ratio K_S/K_{NS} by about a factor of 3. The internal concentration of K^+ in *E. coli* can vary over a wider range than this in response to varying external conditions,[149] so the salt effect on binding specificity may be considered physiologically significant. There are insufficient published data to assess with confidence whether the specific and nonspecific binding processes are differentially sensitive to anion identity, so we cannot evaluate the significance of using unphysiological Cl^- concentrations for these comparisons from *in vitro* experiments. Studies by the Lohman group[33,130,163] on the non-sequence-specific binding of SSB protein indicate that the general trends of anion effects on nonspecific binding will be the same.

Considering the cosolute (osmolyte) effects and the salt effects together, we can see an extension to the idea[149,161,164] that cells compensate for increasing intracellular K^+ by raising the internal concentration of osmolytes so as to maintain the affinity of protein DNA interactions. (The stimulatory effect of cosolute is mitigated but not eliminated if the cosolute is one like trehalose that interacts preferentially with the macromolecular surfaces, rather than one like trimethylamine oxide that interacts little with the macromolecules.) Since increasing the concentration of both salt and cosolute raises the specificity ratio K_S/K_{NS}, it follows that the effect of such adjustment of the internal milieu is also to increase the equilibrium specificity of protein-DNA interactions. It is important to bear in mind that these environmental variables also affect the kinetics of the protein-DNA interaction, but as far as we know there are no sufficiently complete data now available to allow an assessment of how cosolute and salt concentrations interact *in vivo* to affect the kinetics of association and dissociation.

2.14 Conclusions

It appears clear that water and small ions play important or even preeminent roles in determining the specificity and thermodynamic properties of protein-DNA interactions. These influences are exerted not only by water or ionic ligands that are released from the macromolecules upon protein-DNA binding but also by water molecules, and perhaps small ions, that remain bound to the protein-DNA complexes. The release of water from nonpolar surfaces to bulk solvent (hydrophobic effect) is likely the largest single energetic factor that drives site-specific protein-DNA binding. This bulk solvent-release process cannot itself be a determinant of sequence specificity, yet paradoxically enhances specificity because it does not occur unless sequence-specific interactions bring complementary protein and DNA surfaces into sufficiently intimate apposition to expel water from the interface.

Our knowledge of water and small ion effects rests somewhat uneasily at present upon an intersection between experiment and computational simulation. Experiments achieve much greater quantitative precision than current computational analyses, but may relate to the process of interest only indirectly or in a model-dependent fashion. Analysis of experiments may often divide an overall process like protein-DNA association into a different set of elementary subprocesses than does a purely computational treatment, since the latter is to some extent unconstrained by physical "reality". For example, a computational dissection might calculate the energy for transferring a macromolecule from vacuum to water, a "process" that would not be measured by experiment. The difference in how processes are subdivided remains an obstacle to direct comparison between experimental and computational results. Both computational simulations and model-dependent analysis of experiments often make the assumption that changes in thermodynamic parameters for component subprocesses can be treated as additive. This assumption is usually difficult to test rigorously.

If we are to aspire to a quantitative rather than qualitative understanding of the roles of water and small ions in macromolecular processes, we will ultimately be compelled to seek deeper knowledge of fundamental physicochemical issues, such as the detailed structure and dynamics of water surrounding macromolecules, the hydration of small ions, and the dynamic behavior of proteins, DNA and their complexes in solution.

Acknowledgements

We thank present and former members of the Jen-Jacobson laboratory for discussions and use of unpublished information: M. Kurpiewski, D.F. Cao, T. Niu, P. Sapienza, O. Pham, P. Mehta, L. Engler, S. Hancock, and G. Bosco. We especially thank Mike Kurpiewski for help in preparing figures. This work was supported by NIH MERIT Award 5R37GM029207 and allocations of advanced computing resources supported by NSF (MCB050027P and MCB060029P) to L.J.J.

References

1. L. Jen-Jacobson, *Methods Enzymol.*, 1995, **259**, 305.
2. J. Janin, *Structure*, 1999, **7**, R277.
3. F. Spyrakis, P. Cozzini, C. Bertoli, A. Marabotti, G.E. Kellogg and A. Mozzarelli, *BMC Struct. Biol.*, 2007, **7**, 4.
4. N.V. Prabhu and K.A. Sharp, *Annu. Rev. Phys. Chem.*, 2005, **56**, 521.
5. K.A. Sharp and B. Madan, *J. Phys. Chem. B*, 1997, **101**, 4343.
6. T.M. Raschke, *Curr. Opin. Struct. Biol.*, 2006, **16**, 152.
7. B. Lee, *Biopolymers*, 1991, **31**, 993.
8. W. Blokzijl and B.F.N. Engberts, *Angew. Chem., Int. Ed. Engl.*, 1993, **32**, 1545.
9. G.I. Makhatadze and P.L. Privalov, *Adv. Protein Chem.*, 1995, **47**, 307.
10. J.T. Edsall, *J. Am. Chem. Soc.*, 1935, **57**, 1506.
11. S.J. Gill, S.F. Dec, G. Olofsson and I. Wadso, *J. Phys. Chem.*, 1985, **89**, 3758.

12. R.L. Baldwin, *Proc. Natl. Acad. Sci. U.S.A.*, 1986, **83**, 8069.
13. K.P. Murphy and S.J. Gill, *J. Mol. Biol.*, 1991, **222**, 699.
14. J.R. Livingstone, R.S. Spolar and M.T. Record Jr, *Biochemistry*, 1991, **30**, 4237.
15. R.S. Spolar, J.R. Livingstone and M.T. Record Jr, *Biochemistry*, 1992, **31**, 3947.
16. M.R. Kurpiewski, M. Koziolkiewicz, A. Wilk, W.J. Stec and L. Jen-Jacobson, *Biochemistry*, 1996, **35**, 8846.
17. M.R. Kurpiewski, L.E. Engler, L.A. Wozniak, A. Kobylanska, M. Koziolkiewicz, W.J. Stec and L. Jen-Jacobson, *Structure (Cambridge)*, 2004, **12**, 1775.
18. J. Gomez, V.J. Hilser, D. Xie and E. Freire, *Proteins*, 1995, **22**, 404.
19. J.K. Myers, C.N. Pace and J.M. Scholtz, *Protein Sci.*, 1995, **4**, 2138.
20. A.D. Robertson and K.P. Murphy, *Chem. Rev.*, 1997, **97**, 1251.
21. B. Madan and K.A. Sharp, *Biophys. J.*, 2001, **81**, 1881.
22. P.L. Privalov, A.I. Dragan, C. Crane-Robinson, K.J. Breslauer, D.P. Remeta and C.A. Minetti, *J. Mol. Biol.*, 2007, **365**, 1.
23. J.M. Sturtevant, *Proc. Natl. Acad. Sci. U.S.A.*, 1977, **74**, 2236.
24. J.E. Ladbury, J.G. Wright, J.M. Sturtevant and P.B. Sigler, *J. Mol. Biol.*, 1994, **218**, 669.
25. C.G. Kalodimos, N. Biris, A.M. Bonvin, M.M. Levandoski, M. Guennuegues, R. Boelens and R. Kaptein, *Science*, 2004, **305**, 386.
26. L. Jen-Jacobson, L.E. Engler and L.A. Jacobson, *Struct. Fold. Des.*, 2000, **8**, 1015.
27. L. Jen-Jacobson, L.E. Engler, J.T. Ames, M.R. Kurpiewski and A. Grigorescu, *Supramol. Chem.*, 2000, **12**, 143.
28. P.L. Privalov and A.I. Dragan, *Biophys. Chem.*, 2007, **126**, 16.
29. C.J. Morton and J.E. Ladbury, *Protein Sci.*, 1996, **5**, 2115.
30. A.G. Kozlov and T.M. Lohman, *Proteins*, 2000, Suppl 4, 8.
31. M.R. Eftink, A.C. Anusiem and R.L. Biltonen, *Biochemistry*, 1983, **22**, 3884.
32. B.M. Baker and K.P. Murphy, *Biophys. J.*, 1996, **71**, 2049.
33. A.G. Kozlov and T.M. Lohman, *Biochemistry*, 2006, **45**, 5190.
34. T. Lundback, J.F. Chang, K. Phillips, B. Luisi and J.E. Ladbury, *Biochemistry*, 2000, **39**, 7570.
35. V. Petri, M. Hsieh and M. Brenowitz, *Biochemistry*, 1995, **34**, 9977.
36. S. Bergqvist, M.A. Williams, R. O'Brien and J.E. Ladbury, *J. Mol. Biol.*, 2004, **336**, 829.
37. D.I. Svergun, S. Richard, M.H. Koch, Z. Sayers, S. Kuprin and G. Zaccai, *Proc. Natl. Acad. Sci. U.S.A.*, 1998, **95**, 2267.
38. V. Makarov, B.M. Pettitt and M. Feig, *Acc. Chem. Res.*, 2002, **35**, 376.
39. M. Gerstein and C. Chothia, *Proc. Natl. Acad. Sci. U.S.A.*, 1996, **93**, 10167.
40. T.V. Chalikian, J. Volker, A.R. Srinivasan, W.K. Olson and K.J. Breslauer, *Biopolymers*, 1999, **50**, 459.
41. B. Lee and F.M. Richards, *J. Mol. Biol.*, 1971, **55**, 379.

42. F.M. Richards, *Annu. Rev. Biophys. Bioeng.*, 1977, **6**, 151.
43. T.P. Creamer, R. Srinivasan and G.D. Rose, *Biochemistry*, 1997, **36**, 2832.
44. T.P. Creamer, R. Srinivasan and G.D. Rose, *Biochemistry*, 1995, **34**, 16245.
45. C. Chothia, *Nature*, 1974, **248**, 338.
46. J.A. Reynolds, D.B. Gilbert and C. Tanford, *Proc. Natl. Acad. Sci. U.S.A.*, 1974, **71**, 2925.
47. R.B. Hermann, *Proc. Natl. Acad. Sci. U.S.A.*, 1977, **74**, 4144.
48. G.I. Makhatadze and P.L. Privalov, *J. Mol. Biol.*, 1993, **232**, 639.
49. P.A. Karplus, *Protein Sci.*, 1997, **6**, 1302.
50. K.A. Sharp, A. Nicholls, R.F. Fine and B. Honig, *Science*, 1991, **252**, 106.
51. H.S. Chan and K.A. Dill, *J. Chem. Phys.*, 1994, **101**, 7007.
52. H.S. Chan and K.A. Dill, *Annu. Rev. Biophys. Biomol. Struct.*, 1997, **26**, 425.
53. R. Brem, H.S. Chan and K.A. Dill, *J. Phys. Chem. B*, 2000, **104**, 7471.
54. L.E. Engler, P. Sapienza, L.F. Dorner, R. Kucera, I. Schildkraut and L. Jen-Jacobson, *J. Mol. Biol.*, 2001, **307**, 619.
55. L.E. Engler, K.K. Welch and L. Jen-Jacobson, *J. Mol. Biol.*, 1997, **269**, 82.
56. B. Jayaram, K. McConnell, S.B. Dixit, A. Das and D.L. Beveridge, *J. Comput. Chem.*, 2002, **23**, 1.
57. L. Jen-Jacobson, *Biopolymers*, 1997, **44**, 153.
58. V.A. Parsegian, R.P. Rand and D.C. Rau, *Methods Enzymol.*, 1995, **259**, 43.
59. V.A. Parsegian, R.P. Rand and D.C. Rau, *Proc. Natl. Acad. Sci. U.S.A.*, 2000, **97**, 3987.
60. S.N. Timasheff, *Biophys. Chem.*, 2002, **101–102**, 99.
61. S.N. Timasheff, *Proc. Natl. Acad. Sci. U.S.A.*, 1998, **95**, 7363.
62. J.A. Schellman, *Biopolymers*, 1987, **26**, 549.
63. J.A. Schellman, *Biophys. Chem.*, 1990, **37**, 121.
64. S.N. Timasheff, *Adv. Protein Chem.*, 1998, **51**, 355.
65. M.T. Record Jr, W. Zhang and C.F. Anderson, *Adv. Protein Chem.*, 1998, **51**, 281.
66. E.S. Courtenay, M.W. Capp, R.M. Saeckerc and M.T. Record Jr, *Proteins*, 2000, Suppl 4, 72.
67. E.S. Courtenay, M.W. Capp and M.T. Record Jr, *Protein Sci.*, 2001, **10**, 2485.
68. S.N. Timasheff, *Biochemistry*, 2002, **41**, 13473.
69. S.N. Timasheff, *Proc. Natl. Acad. Sci. U.S.A.*, 2002, **99**, 9721.
70. C.F. Anderson, D.J. Felitsky, J. Hong and M.T. Record, *Biophys. Chem.*, 2002, **101–102**, 497.
71. W.G. McMillan Jr and J.E. Mayer, *J. Chem. Phys.*, 1945, **13**, 276.
72. D.J. Winzor and P.R. Wills, Thermodynamic nonideality and protein solvation, In *Protein-Solvation Interactions*, ed. R.B. Gregory, Marcel Dekker, New York, 1995, p. 483.
73. A.P. Minton, *Methods Enzymol.*, 1998, **295**, 127.

74. P.R. Davis-Searles, A.J. Saunders, D.A. Erie, D.J. Winzor and G.J. Pielak, *Annu. Rev. Biophys. Biomol. Struct.*, 2001, **30**, 271.
75. J. A. Schellman, *Biophys. J.*, 2003, **85**, 108.
76. J.M. Schurr, D.P. Rangel and S.R. Aragon, *Biophys. J.*, 2005, **89**, 2258.
77. A.P. Minton, *Mol. Cell. Biochem.*, 1983, **55**, 119.
78. B.L. Neal, D. Asthagiri and A.M. Lenhoff, *Biophys. J.*, 1998, **75**, 2469.
79. D.F. Cao, Ph.D. thesis, University of Pittsburgh, PA, 2002.
80. J.G. Kirkwood and F.P. Buff, *J. Chem. Phys.*, 1951, **19**, 774.
81. J.A. Schellman, *Q. Rev. Biophys.*, 2005, **38**, 351.
82. S. Shimizu and C.L. Boon, *J. Chem. Phys.*, 2004, **121**, 9147.
83. S. Shimizu and N. Matubayashi, *Chem. Phys. Lett.*, 2006, **420**, 518.
84. P.E. Smith, *Biophys. J.*, 2006, **91**, 849.
85. J. Rosgen, B.M. Pettitt and D.W. Bolen, *Protein Sci.*, 2007, **16**, 733.
86. A. Ben-Naim, *Statistical Thermodynamics for Chemists and Biochemists*, Plenum Press, New York, 1992.
87. N.Y. Sidorova and D.C. Rau, *Proc. Natl. Acad. Sci. U.S.A.*, 1996, **93**, 12272.
88. M.G. Fried, D.F. Stickle, K.V. Smirnakis, C. Adams, D. MacDonald and P. Lu, *J. Biol. Chem.*, 2002, **277**, 50676.
89. N.Y. Sidorova, S. Muradymov and D.C. Rau, *J. Biol. Chem.*, 2006, **281**, 35656.
90. M. Newman, T. Strzelecka, L.F. Dorner, I. Schildkraut and A.K. Aggarwal, *Science*, 1995, **269**, 656.
91. K.E. Tang and V.A. Bloomfield, *Biophys. J.*, 2002, **82**, 2876.
92. J.T. Edward, *J. Chem. Educ.*, 1970, **47**, 261.
93. J.D. Dunitz, *Chem. Biol.*, 1995, **2**, 709.
94. D.B. Nikolov, H. Chen, E.D. Halay, A. Hoffman, R.G. Roeder and S.K. Burley, *Proc. Natl. Acad. Sci. U.S.A.*, 1996, **93**, 4862.
95. S. Sen and L. Nilsson, *Biophys. J.*, 1999, **77**, 1782.
96. Y.Q. Qian, G. Otting, M. Billeter, M. Muller, W. Gehring and K. Wuthrich, *J. Mol. Biol.*, 1993, **234**, 1070.
97. K. Wuthrich, M. Billeter, P. Guntert, P. Luginbuhl, R. Riek and G. Wider, *Faraday Discuss.*, 1996, **103**, 245.
98. M. Billeter, P. Guntert, P. Luginbuhl and K. Wuthrich, *Cell*, 1996, **85**, 1057.
99. M.F. Kropman and H.J. Bakker, *Science*, 2001, **291**, 2118.
100. B. Halle, *Philos. Trans. R. Soc. London, Ser. B: Biol. Sci.*, 2004, **359**, 1207.
101. N.M. Luscombe, R.A. Laskowski and J.M. Thornton, *Nucleic Acids Res.*, 2001, **29**, 2860.
102. Z. Shakked, G. Guzikevich-Guerstein, F. Frolow, D. Rabinovich, A. Joachimiak and P.B. Sigler, *Nature*, 1994, **368**, 469.
103. J. Woda, B. Schneider, K. Patel, K. Mistry and H.M. Berman, *Biophys. J.*, 1998, **75**, 2170.
104. C.K. Reddy, A. Das and B. Jayaram, *J. Mol. Biol.*, 2001, **314**, 619.
105. B. Jayaram and T. Jain, *Annu. Rev. Biophys. Biomol. Struct.*, 2004, **33**, 343.
106. Y. Levy and J.N. Onuchich, *Annu. Rev. Biophys. Biomol. Struct.*, 2006 **35**, 389.

107. Z. Otwinowski, R.W. Schevitz, R.G. Zhang, C.L. Lawson, A. Joachimiak, R.Q. Marmorstein, B.F. Luisi and P.B. Sigler, *Nature*, 1988, **335**, 321.
108. A. Joachimiak, T.E. Haran and P.B. Sigler, *EMBO J.*, 1994, **13**, 367.
109. J.M. Rosenberg, *Curr. Opin. Struct. Biol.*, 1991, **1**, 104.
110. D.R. Lesser, M.R. Kurpiewski and L. Jen-Jacobson, *Science*, 1990, **250**, 776.
111. D.U. Ferreiro, M. Dellarole, A.D. Nadra and G. de Prat-Gay, *J. Biol. Chem.*, 2005, **280**, 32480.
112. P.R. Connelly, The cost of releasing site-specific, bound water molecules from proteins: Toward a quantitative guide for structure-based drug design, in *Structure Based Drug Design Thermodynamics, Modeling and strategy*, ed. J.E. Ladbury and P.R. Connelly, Springer-Verlag, New York, 1997, ch. 5, p. 143.
113. M.T. Record Jr, C.F. Anderson and T.M. Lohman, *Q. Rev. Biophys.*, 1978, **11**, 103.
114. M.T. Record Jr, J.H. Ha and M.A. Fisher, *Methods Enzymol.*, 1991 **208**, 291.
115. J.H. Ha, M.W. Capp, M.D. Hohenwalter, M. Baskerville and M.T. Record Jr, *J. Mol. Biol.*, 1992, **228**, 252.
116. G.S. Manning, *Q. Rev. Biophys.*, 1978, **11**, 179.
117. G.S. Manning, *Biophys. Chem.*, 1978, **9**, 65.
118. C.F. Anderson and M.T. Record Jr, *Annu. Rev. Biophys. Biophys. Chem.*, 1990, **19**, 423.
119. C.F. Anderson and M.T. Record Jr, *Annu. Rev. Phys. Chem.*, 1995, **46**, 657.
120. A.K. Vershon, S.M. Liao, W.R. McClure and R.T. Sauer, *J. Mol. Biol.*, 1987, **195**, 311.
121. M.G. Fried and D.F. Stickle, *Eur. J. Biochem.*, 1993, **218**, 469.
122. R.B. Winter and P.H. von Hippel, *Biochemistry*, 1981, **20**, 6948.
123. D.M. Blakaj, C. Kattamuri, S. Khrapunov, R.S. Hegde and M. Brenowitz, *J. Mol. Biol.*, 2006, **358**, 224.
124. R.B. Winter, O.G. Berg and P.H. von Hippel, *Biochemistry*, 1981, **20**, 6961.
125. L. Jen-Jacobson, M. Kurpiewski, D. Lesser, J. Grable, H.W. Boyer, J.M. Rosenberg and P.J. Greene, *J. Biol. Chem.*, 1983, **258**, 14638.
126. K.A. Sharp, R.A. Friedman, V. Misra, J. Hecht and B. Honig, *Biopolymers*, 1995, **36**, 245.
127. J.A. Holbrook, O.V. Tsodikov, R.M. Saecker and M.T. Record, Jr., *J. Mol. Biol.*, 2001, **310**, 379.
128. E.B. Garcia-Moreno and C.A. Fitch, *Methods Enzymol.*, 2004, **380**, 20.
129. P.H. von Hippel and T. Schleich, *Acc. Chem. Res.*, 1969, **2**, 257.
130. A.G. Kozlov and T.M. Lohman, *J. Mol. Biol.*, 1998, **278**, 999.
131. T.M. Lohman, P.L. deHaseth and M.T. Record Jr, *Biochemistry*, 1980, **19**, 3522.
132. D.P. Mascotti and T.M. Lohman, *Proc. Natl. Acad. Sci. U.S.A.*, 1990, **87**, 3142.
133. A.I. Dragan, Z. Li, E.N. Makeyeva, E.I. Milgotina, Y. Liu, C. Crane-Robinson and P.L. Privalov, *Biochemistry*, 2006, **45**, 141.

134. K.A. Sharp, *Biopolymers*, 1995, **36**, 227.
135. F. Hofmeister, *Arch. Exp. Pathol. Pharmakol.*, 1888, **24**, 247.
136. K.D. Collins, *Biophys. J.*, 1997, **72**, 65.
137. P.H. von Hippel, V. Peticolas, L. Schack and L. Karlson, *Biochemistry*, 1973, **12**, 1256.
138. K.D. Collins and M.W. Washabaugh, *Q. Rev. Biophys.*, 1985, **18**, 323.
139. R.L. Baldwin, *Biophys. J.*, 1996, **71**, 2056.
140. K.A. Dill, T.M. Truskett, V. Vlachy and B. Hribar-Lee, *Annu. Rev. Biophys. Biomol. Struct.*, 2005, **34**, 173.
141. M.D. Barkley, P.A. Lewis and G.E. Sullivan, *Biochemistry*, 1981, **20**, 3842.
142. H. Viadiu and A.K. Aggarwal, *Nat. Struct. Biol.*, 1998, **5**, 910.
143. S.Y. Xu and I. Schildkraut, *J. Biol. Chem.*, 1991, **266**, 4425.
144. I.B. Vipond and S.E. Halford, *Biochemistry*, 1995, **34**, 1113.
145. A.M. Martin, N.C. Horton, S. Lusetti, N.O. Reich and J.J. Perona, *Biochemistry*, 1999, **38**, 8430.
146. L.H. Conlan and C.M. Dupureur, *Biochemistry*, 2002, **41**, 14848.
147. M.J. Ondrechen, J.G. Clifton and D. Ringe, *Proc. Natl. Acad. Sci. U.S.A.*, 2001, **98**, 12473.
148. T.K. Harris and G.J. Turner, *IUBMB Life*, 2002, **53**, 85.
149. M.T. Record Jr, E.S. Courtenay, S. Cayley and H.J. Guttman, *Trends Biochem. Sci.*, 1998, **23**, 190.
150. P.J. Sapienza, C.A. Dela Torre, W.H. McCoy, S.V. Jana and L. Jen-Jacobson, *J. Mol. Biol.*, 2005, **348**, 307.
151. O.G. Berg, R.B. Winter and P.H. von Hippel, *Biochemistry*, 1981, **20**, 6929.
152. B. Luisi, W.X. Xu, Z. Otwinowski, L.P. Freedman, K.R. Yamamoto and P.B. Sigler, *Nature*, 1991, **352**, 497.
153. R.A. Albright, M.C. Mossing and B.W. Matthews, *Protein Sci.*, 1998, **7**, 1485.
154. F.K. Winkler, D.W. Banner, C. Oefner, D. Tsernoglou, R.S. Brown, S.P. Heathman, R.K. Bryan, P.D. Martin, K. Petratos and K.S. Wilson, *EMBO J.*, 1993, **12**, 1781.
155. H. Viadiu and A.K. Aggarwal, *Mol. Cell*, 2000, **5**, 889.
156. Y. Duan, P. Wilkosz and J.M. Rosenberg, *J. Mol. Biol.*, 1996, **264**, 546.
157. E. Merabet and G.K. Ackers, *Biochemistry*, 1995, **34**, 8554.
158. D.E. Frank, R.M. Saecker, J.P. Bond, M.W. Capp, O.V. Tsodikov, S.E. Melcher, M.M. Levandoski and M.T. Record Jr, *J. Mol. Biol.*, 1997, **267**, 1186.
159. Y. Takeda, P.D. Ross and C.P. Mudd, *Proc. Natl. Acad. Sci. U.S.A.*, 1992, **89**, 8180.
160. C. Berger, I. Jelesarov and H.R. Bosshard, *Biochemistry*, 1996, **35**, 14984.
161. M.T. Record, C.F. Anderson, P. Mills, M.C. Mossing and J.-H. Roe, *Adv. Biophys.*, 1985, **20**, 109.
162. M. T. Record and R. S. Spolar, *Some thermodynamic principles of nonspecific and site-specific DNA interactions*, in *Nonspecific DNA-Protein Interactions*, ed. A. Revzin, CRC Press, Boca Raton, Florida, 1990, p. 33.

163. T.M. Lohman, L.B. Overman, M.E. Ferrari and A.G. Kozlov, *Biochemistry*, 1996, **35**, 5272.
164. M.T. Record Jr, E.S. Courtenay, D.S. Cayley and H.J. Guttman, *Trends Biochem. Sci.*, 1998, **23**, 143.
165. S.J. Hubbard and J.M. Thornton, NACCESS, University College, London, 1993.
166. A. Nicholls, K.A. Sharp and B. Honig, *Proteins*, 1991, **11**, 281.
167. A. Grigorescu, Ph.D. thesis, University of Pittsburgh, Pittsburgh, PA, 2003.
168. L. Jen-Jacobson, D. Lesser and M. Kurpiewski, *Cell*, 1986, **45**, 619.
169. X. Shui, L. McFail-Isom, G.G. Hu and L.D. Williams, *Biochemistry*, 1998, **37**, 8341.
170. L.E. Engler, Ph.D. thesis, University of Pittsburgh, Pittsburgh, PA, 1998.

CHAPTER 3

Structural Basis for Sequence-specific DNA Recognition by Transcription Factors and their Complexes

MANQING HONG AND RONEN MARMORSTEIN

The Wistar Institute, Department of Chemistry and Department of Biochemistry and Molecular Biophysics, University of Pennsylvania, Philadelphia, Pennsylvania 19104, USA

3.1 Introduction

Sequence-specific DNA-binding proteins play key roles in several different DNA regulatory activities, including replication, recombination, repair and transcription. One of the largest and most diverse classes of DNA-binding proteins are the transcriptional regulators that bind DNA to initiate mRNA transcription.[1] Transcriptional regulators control numerous cell processes, including development, growth, and differentiation; and the aberrant activities of transcriptional regulators often lead to disease states. For example, the p53 DNA-binding transcription factor and tumor suppressor protein is mutated in most human cancers.[2] Transcription factors that bind DNA do so by binding to specific DNA sequences located upstream of the gene promoter region and promoting either gene activation or repression through interactions with RNA polymerase II, general transcription factors, and/or other cofactors bound at or near the gene promoter.[3] Many transcriptional regulators contain an autonomous sequence-specific DNA-binding domain as well as other domains

---

RSC Biomolecular Sciences
Protein-Nucleic Acid Interactions: Structural Biology
Edited by Phoebe A. Rice and Carl C. Correll
© Royal Society of Chemistry 2008

associated with other protein interactions to modulate the transcriptional state. Moreover, many of these proteins are grouped into families based on sequence and structural homologies within their DNA-binding domains.

The mode of sequence-specific recognition of DNA by transcriptional regulators has been studied using genetic and biochemical tools with molecular insights coming from structural studies of DNA complexes with transcriptional regulators and their associated complexes using both X-ray crystallography and NMR techniques. Correlations between available functional and structural data on DNA binding by transcription factors have established that although there are certain amino acid base-pair preferences there is not an amino acid base-pair code that underlies protein-DNA binding specificity.[4–6] Instead, protein-DNA interactions appear to be governed by many of the same stabilizing forces that promote stable protein-protein interactions. In addition to the stabilizing effects of amino acid base-pair interactions, van der Waals interactions, water-mediated protein-DNA hydrogen bonds, and conformational transitions of local protein and/or DNA regions play key roles in stabilizing specific protein-DNA complexes.[7] Many protein-DNA complexes also exploit the polarity of DNA half sites and the base-pair separation between two contacted DNA half-sites for contributions to binding specificity. Another paradigm that has emerged regarding protein-DNA recognition, especially in higher eukaryotes, is that the DNA-binding properties of many transcriptional activators are modulated by the binding of another DNA-binding protein nearby. This cooperative effect of nearby DNA-binding proteins, often referred to as combinatorial regulation, is a hallmark of gene regulation in eukaryotes.

The transcriptional regulators that bind DNA can be grouped into three broad classes. These classes include (1) proteins that bind as monomeric or dimeric units to a core DNA sequence or, in the case of some dimers, to two DNA half sequences of fixed separation and polarity; (2) proteins that bind as dimers to DNA targets containing related DNA half site sequences but different half site separation and/or polarity; and (3) combinatorial protein interactions with DNA by different proteins from class (1). In this chapter, we survey the structural framework for DNA recognition by transcriptional regulators from these three different DNA-binding classes. Readers interested in a more chemical understanding of protein-DNA recognition by the DNA binding proteins described here are referred to several excellent review articles that focus on particular DNA-binding structural scaffolds.

3.2 Transcriptional Regulators that Bind Core DNA Elements

3.2.1 Helix-turn-helix and Winged Helix-turn-helix

The helix-turn-helix (HTH) is a common DNA recognition element found in prokaryotes and bacteriophage. The bacteriophage λ-repressor,[8] 434 repressor,[9] and bacterial *Trp* repressor[10] were among the first DNA binding proteins characterized at the biochemical and structural levels. In eukaryotic transcription factors, the homeodomain and Myb domain contain eukaryotic versions of HTH motifs. While the HTH transcription factors from bacteriophage and bacteria are generally homodimers, the eukaryotic versions are monomeric. The basic HTH fold consists of three core helices that form a right handed helical bundle with a tight turn between helices 2 and 3. All HTH motifs contain a hydrophobic core at the interface of the three helices serving to stabilize the motif and to present helix 3 for sequence-specific DNA recognition. Helix 3 of the HTH motif, or so-called DNA-recognition helix, inserts into the major groove of the target DNA duplex, with its side chains specifically contacting nucleotide bases in the major groove and forming both specific base and sugar-phosphate backbone contacts. Helix 1 and the turn between helices 2 and 3 also contribute additional DNA contacts although sequence specificity is largely dictated by the residues of helix 3 that make base specific contacts. A comparison of DNA complexes with HTH proteins shows that the orientation of the DNA-recognition helix with the DNA major groove is variable and thus different regions of the DNA recognition helix are employed for sequence-specific interaction.

Figure 3.1(a) shows an example of a HTH domain protein bound to DNA. In this example, the HTH containing bacterial *Trp* repressor forms an intertwined dimer in which the DNA-recognition helices of each subunit sit in successive major grooves along one face of the DNA.[10] In this case, the binding of the amino acid, L-tryptophan, induces DNA binding by appropriately separating the two DNA recognition helices of the *Trp* repressor dimer such that they can engage two successive major grooves of B-form DNA. Interestingly, for the *Trp* repressor, the vast majority of major grove hydrogen bonds are water-mediated, highlighting the use of water-mediated interactions in protein-DNA recognition (Chapter 2).

Winged helix-turn-helix (wHTH) domains, also called fork head domains due to their identification within the *Drosophila* homeotic fork head protein,[11] are variants of the HTH domains. In addition to the three helix HTH bundle, these domains contain an additional antiparallel β-sheet that sits adjacent to the HTH motif and over the DNA minor groove to make additional DNA backbone contacts. Many of the proteins of this family also contain a "second wing", which sometimes consists of the turn between helices 2 and 3, such as the case with the Ets subfamily of wHTH domains, but more often by N- or C-terminal extensions to the wHTH domain. This second wing often makes DNA minor groove backbone contacts on the side opposite the HTH to the first wing.

The prototypical wHTH domain is HNF-3 (hepatocyte nuclear factor 3).[12] The structure of the HNF-3γ/DNA complex shows that helix 3 sits within the DNA major groove, mediating sequence specific DNA contacts, while wings 1 (the loops between S1 and S2) and 2 (S3 to the C-terminal end of the DNA binding domain) mediate backbone contacts to the flanking DNA minor grooves (Figure 3.1b). There are several variants of the wHTH domain proteins including the ETS-domain transcription factors. Wing 2 within these proteins derives from the longer turn between helices 2 and 3 of the HTH. Interestingly, ETS-domain transcription factors bind very related DNA core sites with

50   *Chapter 3*

sequence-specificity being augmented by subtle sequence-dependant DNA conformations outside the core sequence[13,14] and association with other DNA binding proteins through combinatorial regulation.

3.2.2 Basic Leucine-zipper and Basic Helix-loop-helix

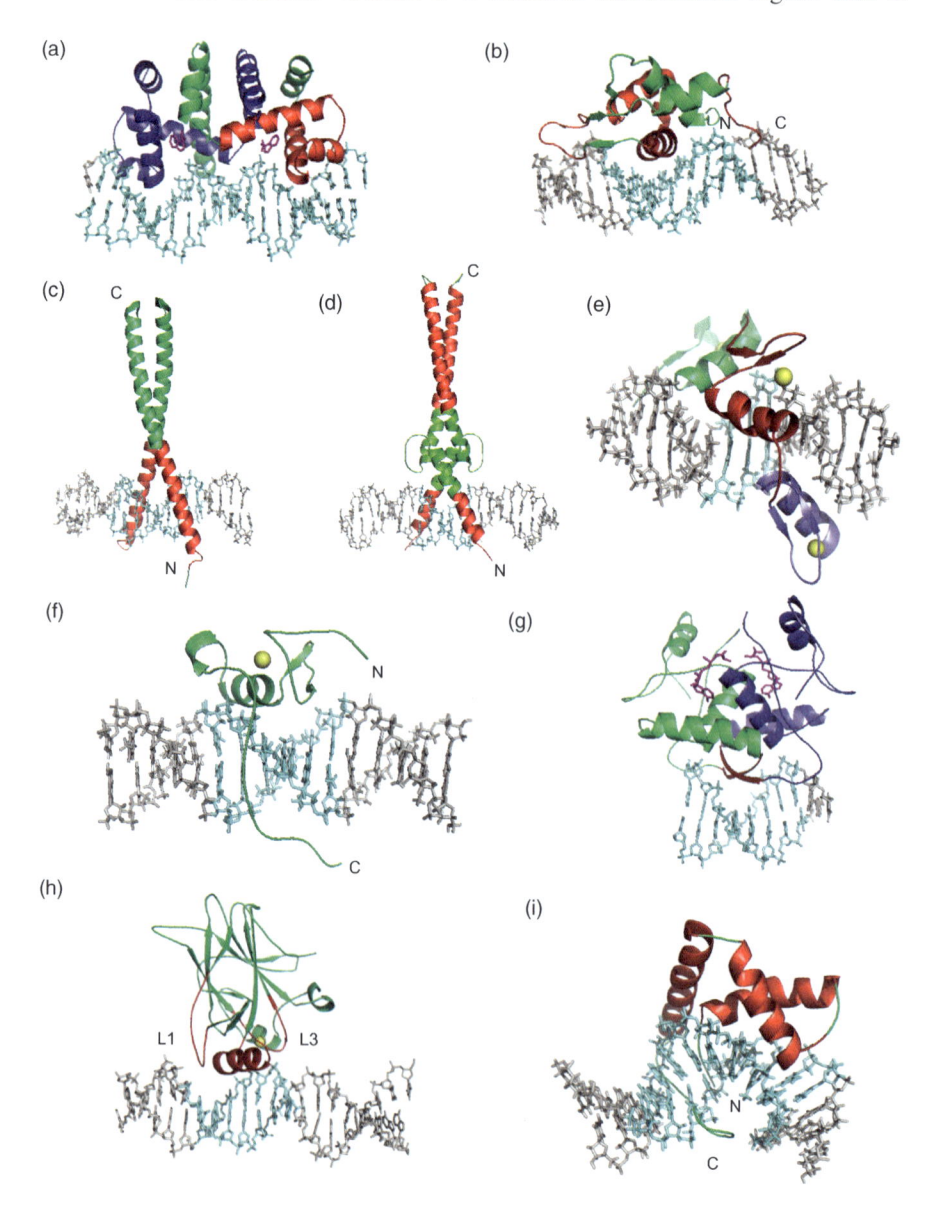

The basic leucine zipper (bZIP) DNA-binding motif is a dimer of extended α-helices. These α-helices contain a C-terminal dimerization region that is stabilized by a heptad repeat of hydrophobic residues (usually leucine or isoleucine residues) along one face of the helices that facilitates a "zipper-like" coiled-coil. The N-terminal end of the helices contains more basic and polar DNA-binding segments that insert into opposite sides of the same DNA major groove. The two helices look like chopsticks that grab the DNA through the major groove. Within the basic segment, one side of the helices contacts the DNA base functional groups while neighboring residues contact the phosphodiester backbone. bZIP proteins are only found in eukaryotes and can form homo- or hetero- dimers. Interestingly, the basic DNA binding regions adopt secondary structure only upon DNA binding. This highlights the DNA-dependent allosteric transition that many transcriptional regulators use for DNA recognition.

The yeast GCN4 protein is the prototypical bZIP protein. In the structure of GCN4,[15] the C-terminal leucine zipper forms a left-handed coiled-coil dimerization domain and the helices transition into the basic region that mediates both direct and water-mediated sequence-specific contacts to the DNA functional groups within the major groove and the flanking phosphodiester backbone (Figure 3.1c). Since nearly all the sequence-specific contacts are mediated by the basic region, the formation of both homo- and heterodimers increases the repertoire of DNA sequences that can be specified by bZIP proteins.

The basic helix-loop-helix (bHLH) domain shares many similarities to the bZIP domains. Like the bZIP domains, the bHLH contain a C-terminal coiled-coil dimerization region and an N-terminal basic region that contacts the DNA major groove. Also like the bZIP proteins, the bHLH proteins also form either homo- or heterodimers to increase the spectrum of DNA specific sequences that can be bound by these proteins. The bHLH domains differ from the bZIP domains in that the N- and C-terminal segments are not part of the same helices but are two separate helices that are connected by a loop and helical segments

---

Figure 3.1 Transcriptional regulators that bind core DNA elements. (*a*) The *Trp* repressor/DNA complex. The subunits of the protein dimer are shown in green and blue, with the helix-turn-helix (HTH) domain of the green subunit highlighted in red. The core DNA site is in aqua with the activating L-tryptophan corepressor in magenta. (*b*) The HNF3/DNA complex. The winged-helix-turn-helix (wHTH) domain as well as the "second wing" is highlighted in red. The core DNA site is in aqua. (*c*) GCN4/DNA complex. The basic-zipper (bZip) region of the dimer highlights the coiled-coil zipper and basic regions in green and red, respectively. (*d*) Max-homodimer/DNA complex. The basic and coiled-coil regions of the basic-helix-loop-helix (bHLH) are highlighted in red. (*e*) TFIIIA/DNA complex. The three zinc-finger regions are in green, red and blue and the zinc ions are in yellow. (*f*) GATA-1/DNA complex. A single zinc binding domain of the Cys_2-Cys_2 family of DNA binding proteins is shown in green. (*g*) Met repressor/DNA complex. The subunits of the dimer are blue and green with the β-sheet of the b-ribbon motif in red. The S-adenosyl-methionine (SAM) cofactors are magenta. (*h*) p53 monomer/DNA complex. The immunoglobulin-like fold is shown, highlighting the L1 and L3 loops and H2 helix that interact with the DNA in red. (*i*) LEF-1/DNA complex. The HMG domain is highlighted in red.

of the dimer form a four-helix bundle at the interface between the dimerization and DNA recognition segments. The loop region can also participate in DNA backbone contacts. Figure 3.1(d) shows the structure of the homodimeric bHLH domain from the Max transcription factor bound to DNA.[16]

3.2.3 Zinc-binding Domains that Bind as Monomeric Units

DNA-binding domains that employ one or more structural zinc ions are the most common DNA binding domains in eukaryotes and about 3% of the human genome encode such domains. There are at least ten different subfamilies of these domains that are defined by the type (cysteine and/or histidine) and arrangement of amino acids that mediate tetrahedral zinc coordination and thus the overall fold of the domain. What is common in each case is that the zinc ion(s) are used as a structural element to mediate the folding of a relatively short polypeptide chain that would otherwise not be able to properly fold due the absence of a sufficient number of hydrophobic residues to stabilize the core of the domain. In addition to DNA binding, many such domains also serve as protein-protein interaction regions. Of the zinc binding domains that recognize core DNA elements, the classical Cys_2His_2 zinc fingers and Cys_2Cys_2 GATA-1 like domains have been the most extensively studied.

Classical Cys_2His_2 zinc fingers are small motifs often found in multiple copies in "zinc finger proteins". Such proteins comprise the largest subgroup of zinc-containing DNA-binding proteins and are present in both prokaryotes and eukaryotes. The first zinc finger protein characterized was TFIIIA, a transcriptional regulator from *Xenopus* oocytes that contains nine zinc fingers.[17] These zinc finger proteins generally have between 3 and 15 copies of this motif, but some have many more. In such multi-finger proteins, some but not all of the zinc fingers generally contact DNA. The typical sequence motif for a single classical zinc finger is X2-Cys-X2-4-Cys-X12-His-X2-8-His, which coordinates a single zinc ion. Each $\sim$30-residue repeat forms a two-stranded antiparallel β-sheet followed by an α-helix, which are held together by a small conserved hydrophobic core and the zinc ion. On binding DNA, the β sheet contacts the backbone and the α-helix inserts into the major groove, making sequence-specific side chain-to-base contacts (Figure 3.1e). Figure 3.1(e) shows three of the zinc-fingers from the transcription factor TFIIIA bound to cognate DNA.[17] Each zinc finger contacts a three-base-pair sequence, which is typical for zinc-finger DNA interactions.

The Cys_2-Cys_2 GATA-like family zinc-containing DNA-binding domain ligates the zinc ion through four cysteine residues. Unlike the Cys_2-His_2 zinc finger proteins, the Cys_2-Cys_2 motif occurs only twice per polypeptide chain, with each motif binding to DNA. Each Cys_2-Cys_2 core motif is composed of two irregular antiparallel β-strands and an α-helix, followed by a long carboxyl-terminal loop. This loop and the helix interact with the major groove of the DNA, while a carboxyl-terminal tail makes additional base-specific contacts with the minor groove. Figure 3.1(f) shows one Cys_2-Cys_2 core motif of the GATA-1 transcription factor bound to its six bases DNA consensus sequence.[18]

3.2.4 DNA Recognition by β-Ribbons

The β-ribbon motif is also a common DNA-binding domain found in prokaryotic and bacteriophagic transcription factors and is found in the prokaryotic integration host factor (IHF),[19] arc repressor of bacteriophage P22[20] and the TATA box-binding protein (TBP),[21] among others. Generally, β-ribbon transcription factors bind to DNA as dimers with each monomer contributing the same number of β-strands to interact with DNA. The β-strands of the ribbon insert into either the major or minor groove of its target DNA site, with its side chains making sequence-specific contacts with base and/or phosphate backbone atoms of the DNA. Structural elements outside the β-sheet region often contribute to protein dimer contacts and/or to additional contacts to DNA backbone atoms.

The *E. coli Met* repressor was the first DNA binding protein shown to not use a helix in the DNA major groove for sequence-specific DNA binding[22] (Figure 3.1g). The *Met* repressor recognizes a two-fold symmetric eight-base conserved DNA sequence. Two copies of the Met repressor interact, forming a tight symmetrical dimer with an antiparallel two-stranded β-ribbon motif fitting snugly into the DNA major groove, making direct side-chain base pair hydrogen bonds. In addition, there are two α-helices C-terminal to the β-ribbon motif and a loop N-terminal to the strands of the β-ribbon that mediate DNA phosphate backbone and protein dimer contacts (Figure 3.1g). Interestingly, binding of the corepressor SAM to the Met repressor enhances its DNA binding affinity although the mechanisms for this are not known.[22]

3.2.5 Immunoglobulin Fold

Surprisingly, a protein fold found in immunoglobulins is employed for DNA recognition by a subset of higher eukaryotic transcriptional regulatory proteins. Like immunoglobulins, DNA binding domains with this fold contain a core β-sandwich with two β-sheets and loops of variable size connecting the β-strands. These proteins employ a subset of these β-strand connecting loops as important determinants for sequence specific DNA binding, while the β-sandwich provides the scaffold for loop presentation to the DNA. DNA binding transcription factors with the immunoglobulin-like domain have low sequence conservation and include proteins such as NF-κB, STAT, NFAT and p53.[23–25]

The p53 tumor suppressor protein is one of the most well-known proteins that bind DNA through an immunoglobulin-like domain. The p53 DNA-binding core domain contains a sandwich of two antiparallel β-sheets with four and five strands respectively, and a loop-helix motif that packs against one edge of the β sandwich for DNA interaction. Specifically, the L1 loop and H2 helix from this motif interacts with the DNA major groove and an additional L3 loop interacts with the phosphate backbone of an adjacent DNA minor groove (Figure 3.1h). Correlating with the importance of the L3 loop for DNA recognition by p53, it is a hot-spot for mutations associated with p53-mediated cancers.[26]

3.2.6 HMG Domain

The high mobility group (HMG) family of transcriptional regulatory proteins is found from yeast to human. These proteins can be further divided further into two subclasses, depending on whether the protein binds specifically or non-specifically to DNA. Generally, proteins containing a single HMG domain bind DNA in a sequence-specific fashion while proteins with multiple HMG domains bind to DNA non-specifically. All HMG proteins have been shown to bind DNA primarily in the minor groove. A typical HMG domain is about 80 amino acids long and contains three α-helices that associate through an extensive hydrophobic core with a characteristic L-shaped fold, in which two of the short α-helices forms one arm of the L while the third longer α-helix forms the long arm of the L. The L-shape of the HMG domain adopts a concave surface that accommodates a highly bent DNA duplex along the minor groove.

Lymphoid enhancer factor (LEF-1) is one of the best-characterized sequence-specific HMG domain containing transcription factors.[27] The L-shaped arrangement of three LEF-1 α-helices along with an extended N-terminal coil region binds extensively and continuously to a widened minor groove of DNA. An additional stretch of highly basic residues C-terminal to helix 3 binds across the narrowed major groove, leading to a greatly improved DNA binding affinity and specificity and a sharp DNA bending towards the major groove by approximately 100° (Figure 3.1i). Both the N- and C-terminal regions of LEF-1 were shown to be ordered only when bound to DNA, which implies a DNA-dependent conformational transition of the protein upon DNA binding. The bending of DNA by LEF-1 also facilitates interactions between transcription factors bound on either side of LEF-1 and promotes the interactions of other enhancer-binding proteins, which is an important factor for assembling a higher-order nucleoprotein complex by bringing nonadjacent transcription factor binding sites into close proximity. Therefore, LEF-1 is also referred to as an architectural transcription factor.

3.3 Transcriptional Regulators that Bind as Dimers to two DNA Half Sites with Different Spacing and Polarity

3.3.1 Zn$_2$Cys$_6$ Binuclear Cluster

With over 80 members identified to date, proteins that contain a Zn$_2$Cys$_6$ binuclear cluster domain form a subfamily of DNA-binding domains that are found exclusively in fungi such as yeast.[28–32] As inferred from the name, the domain is folded around two zinc ions that are each tetrahedrally coordinated by four cysteine residues, with two of the cysteine residues shared by the two zinc ions. The folded domain contains two short helices, each followed by an extended strand. The first and fourth residue of each helix and third residue of each strand provide the cysteine residues for zinc ligation. Many proteins in this family contain a linker region that connects the Zn_2Cys_6 domain to a C-terminal coiled-coil domain that mediates homodimerization. These protein dimers generally bind to DNA sites containing CG-rich DNA half sites, and most often CGG, that are contacted by the Zn_2Cys_6 domain through major groove and phosphate backbone interactions.

While proteins that contain a Zn_2Cys_6 domain bind the same or highly related CGG half sites, specificity for the half-site separation and polarity is dictated by the distinct configurations of the linker and coiled-coil dimerization regions of these proteins (Figure 3.2). Based on the difference of polarity, most proteins in this family can be grouped into three types, which recognize inverted, everted and directed half-sites, respectively. Within each subgroup, different proteins specify different base-pair separations (Figure 3.2a).

A subgroup of the Zn_2Cys_6 domain proteins recognize inverted CGG half-site repeats with different half site separation. This includes the prototypical Zn_2Cys_6 domain protein, Gal4, as well as Ppr1 and Put3, which recognize inter base pair separations of 11, 6 and 10 base-pairs, respectively.[33–35] In the case of Gal4, the base of the coiled-coil and the extended linker region make complementary electrostatic interactions with the DNA minor groove to space the Zn_2Cys_6 domains of the dimer 11 base pair apart (Figure 3.2b).[33] In contrast, the linker and coiled-coil regions of Ppr1 do not make DNA contacts but form a more compact and asymmetric structure that is largely stabilized by van der Waals interactions that specifies a six base pair separation between CGG half-sites (Figure 3.2c).[34] The Put3 coiled-coil and linker region also forms an asymmetric structure that involves partial intercalation of the linker region into the DNA minor groove that concomitantly bends the DNA by about 40° to specify a ten base pair inter half-site separation (Figure 3.2d).[35]

Leu3 and Pdr1 recognize everted CGG half-sites (opposite orientation to inverted sites) with separations of 4 and 0 base pairs, respectively,[10,36,37] and Hap1 binds to direct CGG half-site repeat with a six base-pair separation.[38] The structure of the Leu3/DNA complex reveals striking similarity to the Gal4/DNA complex except for the fact that the two Zn_2Cys_6 domains adopt a flipped orientation relative to the domains of Gal4, causing Leu3 to recognize CGG sites in everted orientation and with a four base pair separation (Figure 3.2f).[36] The Hap1/DNA complex shows that multiple hydrophobic residues, within the Zn_2Cys_6 domain of one subunit, form an extensive hydrophobic network with residues from the dimerization region, linker, and Zn_2Cys_6 domain of the other subunit and that other residues in the N-termini of the Zn_2Cys_6 domains form additional interactions with the DNA.[39] All of these interactions cooperate to help configure the Zn_2Cys_6 domains in a head-to-tail fashion, so that they are able to interact with CGG half-sites in direct orientation separated by six inter half-site base-pairs (Figure 3.2e).

Taken together, a comparison of all of the Zn_2Cys_6 domains in complex with their respective cognate DNA half-sites reveals a diverse set of ways in which the linker, dimerization domains, and in some cases non-conserved positions in the Zn_2Cys_6 domains, interact to modulate sequence-specific DNA recognition that exploits the base-pair separation and polarity of common DNA half-sites.

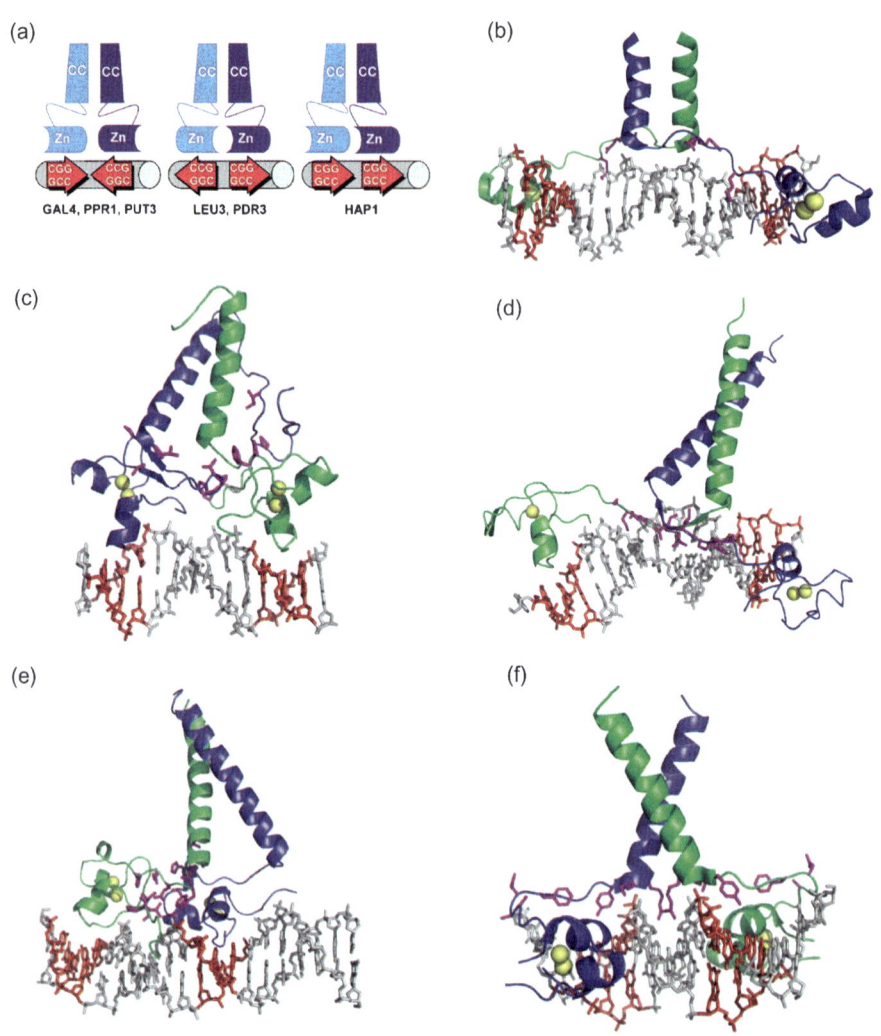

Figure 3.2 Zn$_2$Cys$_6$ binuclear cluster family. (*a*) Schematic of DNA recognition by proteins that contain a Zn$_2$Cys$_6$ binuclear cluster. The three subfamilies that recognize inverted, everted and direct CGG DNA half sites are indicated, highlighting the participation of the Zn$_2$Cys$_6$ domain (Zn), coiled-coil dimerization domain (CC) and linker region connecting these domains in specifying the recognition of DNA half-site polarity. (*b*) Gal4/DNA complex. The two subunits of the dimer are shown in blue and green, respectively, with the DNA half sites in red. The zinc ions are in yellow and side chains within the linker and coiled-coil regions that specify the polarity and base-pair separation of the half-sites are shown as purple side chains. (*c*) Ppr1/DNA complex. (*d*) Put3/DNA complex. (*e*) Leu3/DNA complex. (*f*) Hap-1/DNA complex.

3.3.2 Nuclear Receptors

Nuclear receptors comprise the largest known family of eukaryotic transcription factors, with at least 150 members involved in numerous processes such as development, growth, cell differentiation, proliferation and the maintenance of homeostasis.[40-44] The nuclear receptors contain a highly conserved DNA-binding domain (DBD), a connecting hinge region and a ligand binding domain (LBD), which binds to a cognate ligand and stimulates DNA binding of the dimeric DBD to cognate hormone response elements (HRE).[40-41,45] A minority of the receptors have also been shown to bind DNA as monomers. However, the discussion here will focus on the dimeric receptors that exploit half-site polarity and inter-half-site separation for sequence specific DNA binding. In the absence of the LBD domain, the homo- and heterodimeric DBDs bind constitutively to their cognate HREs and dimerization of these domains is DNA-dependant. The structures of the DBDs from several different receptors have been determined in complex with specific DNA targets and reveal a structurally conserved DBD. This conserved domain contains two bound zinc ions and a DNA recognition helix that makes base-specific contacts to highly homologous half-sites, which contain the sequence AGAACA or AGGTCA.[41,45] Although some DNA target discrimination is achieved through subtle differences in half-site sequences, most of the discrimination is achieved by differences in how homo- and heterodimers recognize half-site base-pairs that differ in polarity and/or separation. This discrimination is dictated in large part by different dimerization interfaces between receptors on DNA. The DBD of the same protein can change its conformation or dimerization interface in response to different dimerization partners. Generally, the zinc-II module and C-terminal extension (CTE) of DBDs play an important role in forming specific dimerization interfaces. The interacting residues between the second zinc-module of one subunit and the first zinc-module of the other subunit vary in response to distinct dimerization partners. Moreover, the CTEs are flexible and vary widely in sequence for different receptors, which enables the DBD to adopt variable conformations for distinct dimerization interfaces without steric clash. In a sense, CTEs function as "molecular rulers" that prevent co-occupancy of receptors on the wrong response elements. Therefore, the dimerization interfaces specify their cognate DNA sequences, which in turn reinforce dimer interactions.

The structures of the glucocorticoid receptor (GR)[46] and estrogen receptor (ER)[47] homodimers, bound to their respective cognate DNA sites, demonstrate that the dimerization interface between the two subunits specifies a cognate three base-pair separation in the DNA half-site pair. Both structures bind as symmetrical homodimers to sites containing inverted half-sites. Specifically, the structures of both homodimers reveal two independent zinc binding sites; one helps orient N-terminal secondary structural elements of the domain for DNA-recognition, the other provides a structural scaffold for the C-terminal region of the domain for homodimeric interactions (Figures 3.3a and b). These homodimeric interactions are largely mediated by 7–8 residues within the

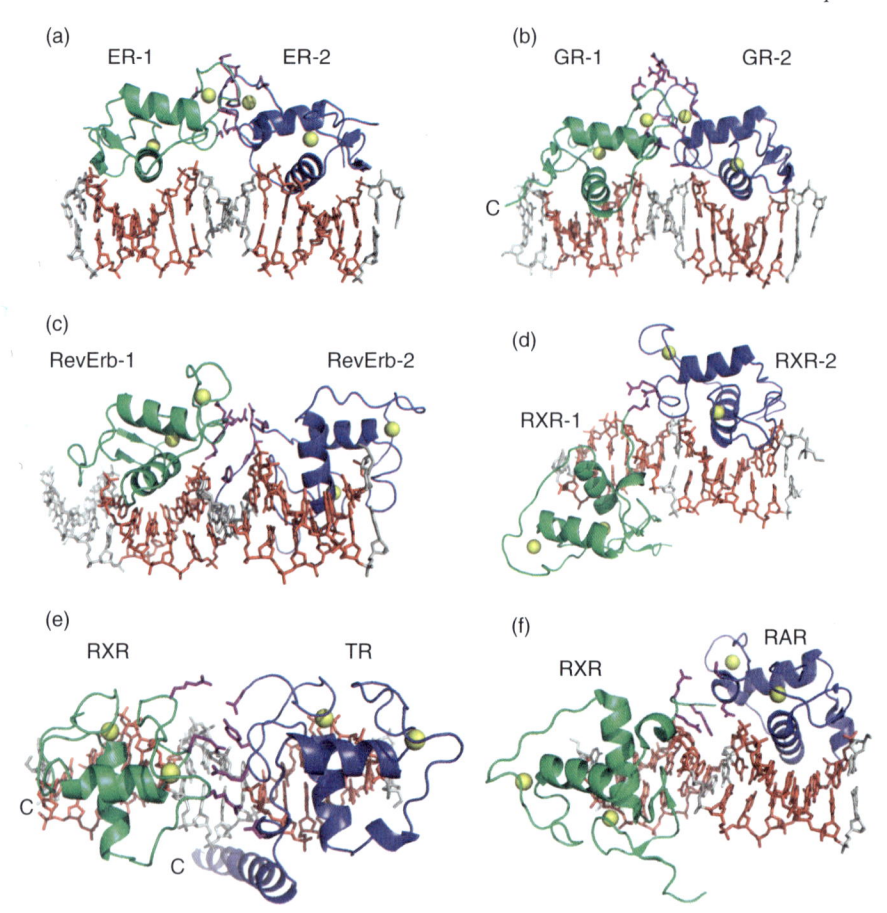

Figure 3.3 The nuclear receptor family. (*a*) ER-homodimer/DNA complex. The two subunits of the dimer are shown in blue and green, respectively, with the DNA half sites in red. The zinc ions are in yellow and side chains that mediate dimer contacts and that specify the polarity and base-pair separation of the half-sites are shown in purple. (*b*) GR-homodimer/DNA complex. (*c*) RevErb1-homodimer/DNA complex. (*d*) RXR-homodimer/DNA complex. (*e*) RXR-TR/DNA complex. (*f*) RXR-RAR/DNA complex.

second zinc module of both receptors. The dimerization interface of GR is further stabilized with DNA phosphate backbone interactions. This is consistent with the observation that dimerization of the DBD of these proteins is DNA dependant and argues that the dimerization interface is a particularly important component of sequence-specific binding by these proteins.

In contrast to GR and ER, the orphan receptor RevErb binds to half sites in direct orientation with a minimum base-pair separation of 2. The structure of the RevErb homodimer bound to DNA reveals how dimer contacts restrict the minimum base pair separation to two and also shows novel protein-DNA interactions between a C-terminal extension (CTE) of the RevErb DBD and minor groove sequence 5' to the half sites (Figure 3.3c).[48] These CTE-minor groove interactions are as extensive as the RevErb DBD half-site interactions.

The retinoid X receptor (RXR) binds to direct repeats of AGGTCA, both as a homodimer and a heterodimer with several other receptors. DNA sequence discrimination by these RXR dimers is dictated by their inter half-site separation. Structures of the RXR homodimer,[49] the RXR-retinoic acid receptor (RAR) heterodimer,[50] and the RXR-thyroid receptor (TR) heterodimer,[51] bound to their respective cognate DNA sites with half-site spacing of 1, 1 and 4, respectively, have been determined. These structures show that RXR's DBD changes its dimerization contacts to accommodate the half-site separation required for DNA binding with its corresponding partner protein. Specifically, the RXR/TR heterodimer bound to a consensus site, containing a four base-pair half-site separation, reveals that several arginine residues, within the second zinc module of RXR, reach over to hydrogen bond to side chain and backbone carbonyl oxygens in the N-terminal and first zinc module of TR (Figure 3.3d). TR also contributes an arginine that contacts a backbone carbonyl within the second zinc module of RXR. The RXR homodimer and the heterodimer with RAR are both bound to cognate sites with a one base-pair separation and show less extensive dimer contacts than the RXR/TR heterodimer (Figure 3.3e and 3f). The differences in contacts between RXR/RAR and RXR homodimer are both compensated with an observed DNA deformation and DNA minor groove-dimer interactions that appear to play a more important role in DNA binding for these dimers. In the RXR/RXR and RXR/RAR dimer interfaces, the second metal domain of the upstream subunit, which binds to the upstream half-site, also makes dimer contacts with non-conserved positions outside the C-terminal extension.

Together, the receptor structures reveal that the second zinc-binding module of one subunit acts cooperatively with DNA binding to form a dimerization interface, which specifically recognizes the inter half-site separation. In general, different residues mediate the dimer contacts, revealing that each receptor combination adapts a unique way to dimerize on its cognate DNA site.

3.4 Transcription Regulatory Complexes that use a Combination of Different DNA-binding Motifs

3.4.1 Combinatorial DNA Interactions

There have now been several reports of complexes in which two or more transcriptional regulators bind cooperatively to DNA. In the cases that have been directly visualized, protein-protein interactions occur between both DNA-binding domains or one of the DNA-binding domains contacts a short polypeptide attached to the second DNA-binding domain. In most cases, these interactions facilitate cooperative DNA binding by both proteins but do not alter the DNA-binding specificity of the partner protein. Examples of this type of cooperativity are seen in the ternary DNA complexes with

Mata1/Matα2,[56] MCM1/Matα2[57] and GABPα/β.[58] In some cases, however, protein-protein contacts on DNA serve to modify the way in which one or both DNA-binding proteins contact DNA. Examples of this type of cooperativity are seen in the ternary DNA complexes containing the Ets proteins Ets-1[54] and SAP-1/Elk-1[52,53] and the quaternary NFAT/Fos-June,[55] which we will use as illustrative examples below.

3.4.2 ETS Family Ternary Complexes

Ets proteins contain a highly conserved ~85 amino acid winged helix-turn-helix ETS DNA-binding domain that mediates binding to DNA sites containing a GGA core sequence. The base-pairs flanking the GGA core provide further specificities for different proteins within the ETS family. Additionally, Ets proteins can form heterodimers with other DNA-binding transcriptional regulators to augment their sequence-specific DNA binding activity in what is referred to as combinatorial regulation. The Ets-proteins contact their respective heteromeric partners either through direct interaction within the ETS-domain or through regions outside the ETS-domain. Two examples of combinatorial regulation with Ets proteins are the complexes of the SAP-1 and Elk-1 Ets proteins with the MADS-box serum response factor (SRF)[52,53] and the Ets-1 protein with the helix-turn-helix containing paired domain protein, Pax5.[54]

SAP-1 and Elk-1 are highly related members of the ETS family that cooperatively bind to the *c-fos* promoter with SRF. Part of this cooperativity stems from an interaction between a 20-residue B-box, which is C-terminal to the ETS domains of SAP-1 and Elk-1 and required for interaction with SRF, and the DNA binding domain of SRF.[13,14] In addition, direct interactions between the DNA binding domains of SAP-1/Elk-1 with the SRF MADS-box also effects DNA recognition by the complex. In particular, although SAP-1 and Elk-1 contain identical amino acid sequences within their DNA recognition helices, their isolated DNA-binding domains exhibit very different DNA-binding properties.[59] While SAP-1 binds a *c-fos* site efficiently in the absence of SRF, Elk-1 does not, suggesting that SRF may directly modulate the DNA-binding properties of Elk-1. The structure of a ternary SAP-1/SRF/DNA complex reveals that the DNA recognition helices of SRF and SAP-1 directly interact[53] (Figure 3.4a). The most extensive interaction is mediated by a tyrosine residue (Y 65) that is conserved in nearly all ETS-domains and is also contacted by Pax5 in the Ets-1/Pax5/DNA complex. Specifically, SRF orients the tyrosine of SAP-1 to make optimal contacts to the GGA core sequence. Comparison of the structure of SAP-1/SRF/DNA complex with that of an isolated Elk-1/DNA complex with an optimal consensus site reveals that this tyrosine is oriented in a way that prevents interactions with the GGA DNA sequence,[14] thus suggesting why the nascent Elk-1 cannot bind a *c-fos* site without SRF present. Modeling studies of a SRF/Elk-1 complex on a *c-fos* promoter suggests that interactions between SRF and Elk-1 would reorient this tyrosine residue to make "SAP-1 like" interactions, which would allow Elk-1 to bind to an otherwise low affinity site in the presence of SRF.[52,53]

Structural Basis for Sequence-specific DNA Recognition 61

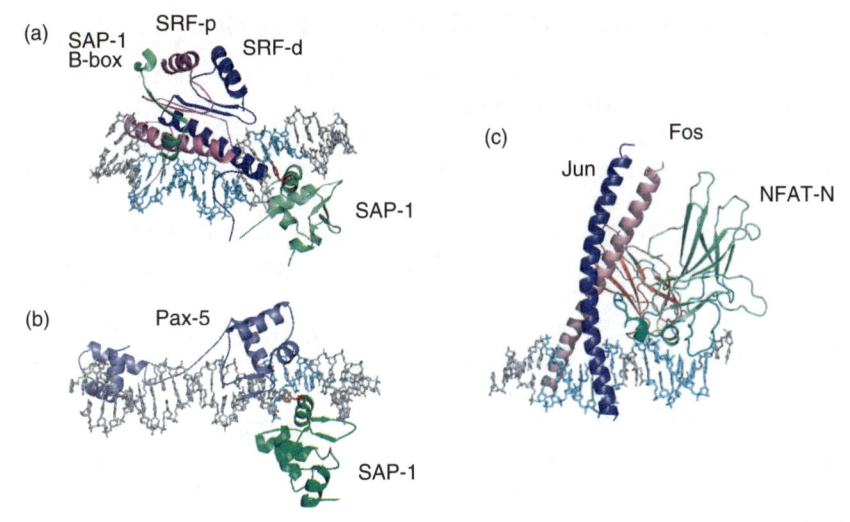

Figure 3.4 Combinatorial transcriptional regulatory complexes. (*a*) SAP-1/SRF/DNA complex. The wHTH ETS-domain of Sap-1 is shown in green and the MADS-box DNA binding domains of SRF are in blue and purple. The Tyr65 side chain of SAP-1 that is reoriented for optimal DNA interactions upon SRF binding is highlighted as a red side chain. The core DNA binding site is highlighted in aqua. The linker between the B-box and ETS-domain of Sap-1 is not visible in the crystal structure here because of flexibility of this region. (*b*) Ets-1/Pax5/DNA complex. The wHTH ETS-domain of Ets-1 is shown in green and Pax5 is in blue. The Tyr395 side chain of Ets-1 that is reoriented for optimal DNA interactions upon Pax5 binding is highlighted as a red side chain. (*c*) NFAT/Fos-Jun/DNA complex. The N- and C-terminal immunoglobulin-like folds of NFAT are in green and red, respectively. The bZIP proteins Fos and Jun are in violet and blue, respectively.

A similar strategy for combinatorial regulation is seen in the ternary Ets1/Pax5/DNA complex. Pax-5 can selectively recruit Ets-1 to the *mb-1* promoter while Ets1 has poor affinity for this promoter in the absence of Pax5.[54] A comparison of the Ets1/Pax5/DNA complex with Ets1 bound to its high and low affinity DNA binding sequence reveals that side chain interactions at the Ets1/Pax5 interface alters the DNA contacts through reorientation of the same conserved tyrosine residue of SAP-1/Elk-1 (Tyr395 in Ets-1) within the DNA recognition helix of Ets-1 for optimal DNA contacts (Figure 3.4b). This tyrosine residue adopts different rotamers on a low affinity Ets-1 binding site or in the absence of Pax5. Taken together, the ternary structures of Ets proteins in complex with DNA and their heteromeric partners reveal how these partners can reorient a conserved tyrosine residue to make optimal DNA contacts to an otherwise suboptimal DNA recognition site. While other ternary DNA complexes may involve the reorientation of other amino acid residues, the general principle of one DNA binding domain reorienting residues for more optimal DNA binding by the other DNA binding domain is likely to be a conserved feature of such complexes.

3.4.3 NFAT/Fos-Jun/DNA Quaternary Complex

The NFAT (nuclear factor of activated T cell) transcription factors contain an immunoglobulin-fold DNA-binding domain that shows weak sequence conservation with the Rel homology region (RHR) of proteins such as NF-κB. There are DNA binding sites for the Fos-Jun bZip-containing heterodimers immediately downstream of most NFAT sites and including the IL-2 promoter and activation from such promoters requires the binding of both NFAT and Fos-Jun. Indeed, NFAT or Fos-Jun alone have been shown to bind weakly to such DNA sites.[55,60]

The structure of the NFAT/Fos-Jun/DNA quaternary complex reveals two RHR domains with the C-terminal RHR domain contacting DNA mostly through major groove interactions that are similar to DNA interactions that are mediated by the NF-κB p50 subunit.[23] Similarly, the Fos-Jun interactions in the quaternary complex show overall similarity to the Fos-Jun interactions that are mediated by the ternary Fos-Jun DNA complex.[61] However, the structure of the quaternary complex shows an extensive interaction surface between NFAT and Fos-Jun that allows the two DNA binding elements to cooperatively bind to DNA[62] (Figure 3.4c). Specifically, the cleft between the N- and C-terminal RHR domains of NFAT interact with an edge of the Fos-Jun heterodimer causing the coiled-coil of Fos-Jun to incline about 15° towards NFAT, resulting in a 20° DNA bend at the center of the AP-1 (Fos-Jun) site and assuring that the Jun subunit binds to the half of the AP-1 site closest to NFAT. Together the NFAT/Fos-Jun complex forms a composite continuous DNA binding surface that interacts specifically with 15 base-pairs of DNA by spanning two major grooves and the intervening minor groove through extensive hydrogen bond and van der Waals interactions.

3.5 Conclusions

In this chapter we have surveyed several ways in which transcriptional regulators bind to DNA in a sequence-specific fashion. Together, we have described how nature has done this through the use of different structural scaffolds to recognize core DNA elements, either alone or in combination, or DNA half-sites with defined polarity and half-site separation. Amazingly, even within a particular structural scaffold, such as the helix-turn-helix, an extremely broad array of DNA sites can be specified and, indeed, an entire chapter can be written of similar length to the present one describing the subtitles of sequence-specific DNA recognition within a single structural scaffold. Despite this seemingly overwhelming diversity in the way transcriptional regulators can bind DNA in a sequence specific fashion, several themes emerge. Transcriptional regulators make sequence specific interactions with the DNA, predominantly through the DNA major groove and mostly by using a protein helix. The minor groves flanking the contacted major groove are also often exploited, *via* phosphate backbone contacts, to enhance protein affinity for DNA. Transcriptional regulators that bind DNA as dimers also exploit the half-site polarity and inter half-site base pairs for sequence specific recognition. Finally, combinatorial regulation allows for further modulation of protein-DNA contacts to facilitate promoter-specific binding. Through these themes, protein-DNA interactions all appear to involve many of the stabilizing forces that influence protein-protein interactions, including direct and water-mediated hydrogen bonds, van der Waals contacts and in some cases conformational changes within the protein and/or DNA. As much as we understand a bout DNA recognition by transcriptional regulators, continued studies in this area will undoubtedly lead to additional new insights and surprises that are likely to have broader implications for understanding molecular recognition in general.

References

1. H.C. Nelson, Structure and function of DNA binding proteins, *Curr. Opin. Genet. Dev.*, 1995, **5**, 180–189.
2. G.J. Semenza, *Transcription Factors and Human Disease*, Oxford University Press, New York, 1998.
3. R. Tijan and T. Maniatis, Transcriptional activation: A complex puzzle with few easy pieces, *Cell*, 1994, **77**, 5–8.
4. C.O. Pabo and R.T. Sauer, *Annu. Rev. Biochem.*, 1984, **53**, 293–321.
5. N.C. Seeman, J.M. Rosenberg and A. Rich, *Proc. Natl. Acad. Sci. U.S.A.*, 1976, **73**, 804–808.
6. B.W. Matthews, *Nature*, 1988, **335**, 294–295.
7. R.J. Reece and M. Ptashne, *Science*, 1993, **261**, 909–911.
8. L.J. Beamer and C.O. Pabo, *J. Mol. Biol.*, 1992, **227**, 177–196.
9. A.K. Aggarwal, D.W. Rodgers, M. Drottar, M. Ptashne and S.C. Harrison, *Science*, 1988, **242**, 899–907.
10. Z. Otwinowski, R.W. Schevitz, R.G. Zhang, C.L. Lawson, A. Joachimiak, R.Q. Marmorstein, B.F. Luisi and P.B. Sigler, *Nature*, 1988, **335**, 321–329.
11. B. Kuzin, S. Tillib, Y. Sedkov, L. Mizrokhi and A. Mazo, *Genes Dev.*, 1994, **8**(20), 2478–2490.
12. C. Jin, L. Marsden, X. Chen and X. Liao, *J. Mol. Biol.*, 1999, **289**, 683–690.
13. Y. Mo, B. Vaessen, K. Johnston and R. Marmorstein, *Mol. Cell*, 1998, **2**(2), 201–212.
14. Y. Mo, B. Vaessen, K. Johnston and R. Marmorstein, *Nat. Struct. Biol.*, 2000, **7**(4), 292–297.
15. T.E. Ellenberger, C.J. Brandl, K. Struhl and S.C. Harrison, *Cell*, 1992, **71**, 1223–1237.
16. P. Brownlie, T.A. Ceska, M. Lamers, C. Romier, G. Stier, H. Teo and D. Suck, *Structure*, 1997, **5**, 509–520.
17. M.P. Foster, D.S. Wuttke, I. Radhakrishnan, D.A. Case, J.M. Gottesfeld and P.E. Wright, *Nat. Struct. Biol.*, 1997, **4**, 605–608.
18. N. Tjandra, J.G. Omichinshi, A.M. Gronenborn, G.M. Clore and A. Bax, *Nat. Struct. Biol.*, 1997, **4**, 732–738.

19. T.W. Lynch, E.K. Read, A.N. Mattis, J.F. Gardner and P.A. Rice, *J. Mol. Biol.*, 2003, **330**, 493–502.
20. B.E. Raumann, M.A. Rould, C.O. Pabo and R.T. Sauer, *Nature*, 1994, **367**, 754–757.
21. Y. Kim, J.H. Geiger, S. Hahn and P.B. Sigler, *Nature*, 1993, **365**, 512–520.
22. W. Somers and S.E.V. Phillips, *Nature*, 1992, **359**, 387–393.
23. G. Ghosh, G.V. Duyne, S. Ghosh and P.B. Sigler, *Nature*, 1995, **373**, 303–310.
24. S. Becker, B. Groner and C.W. Muller, *Nature*, 1998, **394**, 145–151.
25. P. Zhou, L. Sun, V. Dotsch, G. Wagner and G.L. Verdine, *Cell*, 1998, **92**, 687–696.
26. Y. Cho, S. Gorina, P.D. Jeffrey and N.P. Pavletich, *Science*, 1994, **265**, 346–355.
27. J.J. Love, X. Li, D.A. Case, K. Giese, R. Grosschedl and P.E. Wright, *Nature*, 1995, **376**, 791–795.
28. J. Strauss, M.I. Muro-Pastor and C. Sazzocchio, *Mol. Cell. Biol.*, 1998, **18**(3), 1339–1348.
29. R. Cerdan, B. Cahuzac, B. Felenbok and E. Guittet, *J. Mol. Biol.*, 2000, **295**, 729–736.
30. B. Cahuzac, R. Cerdan, B. Felenbok and E. Guitter, *Structure*, 2001, **9**, 827–836.
31. Y.M. Mamnun, R. Pandjaitan, Y. Mahé, A. Delahodde and K. Kuchler, *Mol. Microbiol.*, 2002, **46**(5), 1429–14406.
32. R.B. Todd and A. Andrianopoulos, *Fungal Genetics Biol.*, 1997, **21**, 388–405.
33. R. Marmorstein, M. Carey, M. Ptashne and S.C. Harrison, *Nature*, 1992, **356**, 408–414.
34. R. Marmorstein and S.C. Harrison, *Genes Dev.*, 1994, **8**, 2504–2512.
35. K. Swaminathan, P. Flynn, R.J. Reece and R. Marmorstein, *Nat. Struct. Biol.*, 1997, **4**(9), 751–759.
36. M.X. Fitzgerald, J.R. Rojas, J.M. Kim, G.B. Kohlhaw and R. Marmorstein, *Structure*, 2006, **14**, 725–735.
37. K. Hellauer, M.H. Rochon and B. Turcotte, *Mol. Cell. Biol.*, 1996, **16**(11), 6096–6102.
38. K. Pfeifer, T. Prezant and L. Guarente, *Cell*, 1987, **49**, 19–27.
39. D.A. King, L. Zhang, L. Guarente and R. Marmorstein, *Nat. Struct. Biol.*, 1999, **6**(1), 64–71.
40. J.P. Renaud and D. Moras, *Cell. Mol. Life Sci.*, 2000, **57**, 1748–1769.
41. S. Khorasanizadeh and F. Rastinejad, *Trends Biochem. Sci.*, 2001, **26**(6), 384–390.
42. R. Marmorstein and M.X. Fitzgerald, *Gene*, 2003, **304**, 1–12.
43. F. Claessens and D.T. Gewirth, *Essays Biochem.*, 2004, **40**, 59–72.
44. J. Bastien and C.R. Egly, *Gene*, 2004, **328**, 1–16.
45. F. Rastinejad, *Curr. Opin. Struct. Biol.*, 2001, **11**, 33–38.
46. B.F. Luisi, W.X. Xu, Z. Otwinowski, L.P. Freedman, K.R. Yamamoto and P.B. Sigler, *Nature*, 1991, **352**, 497–505.

47. J.W. Schwabe, L. Chapman, J.T. Finch and D. Rhodes, *Cell*, 1993, **75**, 567–578.
48. Q. Zhao, S. Khorasanizadeh, Y. Miyoshi, M.A. Lazar and F. Rastinejad, *Mol. Cell.*, 1998, **1**, 849–861.
49. Q. Zhao, S.A. Chasse, S.D.M.L. Sierk, B. Ahvazi and F. Rastinejad, *J. Mol. Biol.*, 2000, **296**, 509–520.
50. F. Rastinejad, T. Wagner, Q. Zhao and S. Khorasanizadeh, *EMBO J.*, 2000, **19**, 1045–1054.
51. F. Rastinejad, T. Perlmann, R.M. Evans and P.B. Sigler, *Nature*, 1995, **375**, 203–211.
52. M. Hassler and T.J. Richmond, *EMBO J.*, 2001, **20**(12), 3018–3028.
53. Y. Mo, W. Ho, K. Johnston and R. Marmorstein, *J. Mol. Biol.*, 2001, **314**, 495–506.
54. C.W. Garvie, J. Hagman and C. Wolberger, *Mol. Cell*, 2001, **8**, 1267–1276.
55. L. Chen, J.N.M. Glover, P.G. Hogan, A. Rao and S.C. Harrison, *Nature*, 1998, **392**, 42–48.
56. T. Li, M.R. Stark, A.D. Johnson and C. Wolberger, *Science*, 1995, **270**, 262–269.
57. S. Tan and T.J. Richmond, *Nature*, 1998, **391**, 660–666.
58. A.H. Batchelor, D.E. Piper, F.C. de la Brousse, S.L. McKnight and C. Wolberger, *Science*, 1998, **279**(5353), 1037–1041.
59. P. Shore and D.A. Sharrocks, *Nucleic Acids Res.*, 1995, **23**(22), 4698–4706.
60. S.A. Wolfe, P. Zhou, V. Dotsch, L. Chen, A. You, S.N. Ho, G.R. Crabtree, G. Wagner and G.L. Verdine, *Nature*, 1997, **385**, 172–176.
61. J.N.M. Glover and S.C. Harrison, *Nature*, 1995, **373**, 257–261.
62. P. Berk, L. Holm and C.J. Brandl, *J. Mol. Biol.*, 1994, **242**, 309–320.

CHAPTER 4
Indirect Readout of DNA Sequence by Proteins

CATHERINE L. LAWSON AND HELEN M. BERMAN

Department of Chemistry and Chemical Biology, Rutgers University, 610 Taylor Road, Piscataway, NJ 08854, USA

4.1 Introduction

Interactions between proteins and DNA play a key role in many vital processes, including regulation of gene expression, DNA replication and repair, and packaging. The remarkable specificity with which proteins recognize target DNA sequences is of considerable theoretical and practical importance. In this chapter, we examine a key concept in recognition of DNA by proteins called "indirect readout", which, briefly stated, is the recognition of DNA sequence through conformational and dynamic properties of the DNA. We begin with a short history that explains how this term came into use. We then provide a description of indirect readout and what we currently understand about physicochemical properties of DNA that give rise to sequence-dependent effects on conformation. We next review the role of indirect readout in DNA sequence recognition by a protein that has been extensively studied, the catabolite activator protein (CAP), and conclude with a summary of important features of indirect readout gleaned from analysis of the CAP-DNA interaction.

4.1.1 DNA Sequence Recognition: A Historical Perspective

A fundamental goal since the discovery of the DNA double helix in 1953[1] has been to identify how DNA sequence preference is achieved. Initially, only fiber models of double helical structures of DNA were available and the only known structural variability was the B to A transition between high and low humidity forms.[2,3] These low-resolution models reinforced the idea of a simple, uniform structure for DNA capable of accommodating all possible combinations of base pairs within the common framework of a helical ladder.

In the 1970s crystallographic studies of fragments of the double helix began to yield additional information about DNA structure. A key prediction from these studies was that sequence-specific recognition could occur through a series of direct hydrogen-bonding interactions between specific amino acid side chains and base pairs.[4] For example, a pair of hydrogen bond donors from an arginine guanidinium might form bonds with similarly spaced hydrogen bond acceptors from a guanine base exposed in the major groove. At the same time, differences in base stacking geometries depending on base composition challenged the picture of DNA as a uniform ladder.[5]

The 1970s also brought the elucidation of the first structure of tRNA, revealing the anti-codon loop as the structural basis for the genetic code.[6] The early structural data provided strong evidence that the translation machinery could read the sequence encoded in the messenger RNA through direct hydrogen-bonding. Some scientists at that time thought that recognition of DNA sequence by proteins could involve a similarly simple code.

The arrival of the first structures of DNA duplex fragments, and in particular the structure of a B form dodecamer DNA duplex in 1981,[7] confirmed the main features of the Watson-Crick model, but also revealed local deviations produced by base-sequence. Analysis of this model led to the first articulation of rules describing sequence-dependent structural non-uniformities of DNA and the recognition that sequence-dependent structural and dynamic properties could play an important role in recognition by proteins.[8]

The stage was set for examination of crystal structures of DNA-protein complexes. In 1984, electron density maps of the EcoRI restriction enzyme bound to its cognate DNA binding site provided the first glimpse of a protein-DNA interaction.[9] Subsequent revision of some structural details of this complex[10,11] did not alter two fundamental observations from the first density maps: (1) the major determinant of sequence specificity was a tight complementary interface between the enzyme and the major groove of the DNA and (2) the protein-bound DNA contained departures from canonical B conformation. Remarkably, nearly all subsequently determined structures of sequence-specific protein-DNA complexes have yielded essentially this same basic result.

By the late 1980s several structures of helix-turn-helix motif phage and bacterial repressors bound to cognate DNA sequences had also been elucidated,[12-16] and the hypothesis that sequence-recognition occurs largely through multidentate, hydrogen-bonded and van der Waals interactions between protein side-chains and DNA base-pairs was confirmed.[17-19] But it was also clear that the extensive complementarity of protein-DNA interfaces that characterizes sequence-specific recognition involves deviations in DNA structure, including kinks and bends. With only a handful of structures available it was already clear that recognition of DNA sequence was complex, involving both "direct" and "indirect" readout modes, and that it would not be possible to articulate a simple code to define DNA sequence recognition, even within a single class of DNA-binding proteins.[20]

4.2 Indirect Readout

4.2.1 Direct *vs.* Indirect Readout

The term "direct readout" has a specific stereochemical meaning: it is the preference for a particular pattern of hydrogen bond donors, hydrogen bond acceptors, and non-polar groups of the base-pairs presented in the major and/or minor grooves of the DNA. The term direct readout also applies in situations where sequence preference arises from tight hydrogen-bonded networks involving water molecules embedded within the protein-DNA interface.[21]

Technically speaking, "indirect readout" is the opposite of direct readout. It is sometimes described as a catchall term for mechanisms of sequence-specific recognition that do not involve direct readout, and it is typically invoked when the DNA sequence-dependence of a protein-DNA interaction that has been established by biochemical and molecular biology methods cannot be satisfactorily explained by direct protein-base pair interactions. However, the term is most appropriately used to describe situations where direct interactions are confirmed to be lacking between the protein and one or more mutationally-sensitive DNA base pairs, and where it is also possible to identify either sequence-dependent effects on DNA conformation or sequence-dependent effects on susceptibility to DNA deformation.

Direct and indirect readout are the ying and yang of protein-DNA interactions. Their dual roles are required for DNA sequence recognition by proteins with a diverse range of regulatory functions. Major classes include (with selected examples) (a) transcriptional repressors (434 repressor,[22] papillomavirus E2,[23] *met* repressor,[24] and *trp* repressor[12]), (b) transcriptional activators (TATA box binding protein[25] and CAP[26]), (c) restriction enzymes (EcoRI[27] and EcoRV[28]), and (d) DNA packaging proteins (bacterial integration host factor[29] and yeast nucleosome[30]).

When establishing the role of indirect readout in a particular system, it is important to establish that the reconstituted complex is not missing additional factors that could contribute to sequence preference. The sequence sensitivities of central base-pairs in the *trp* repressor-operator DNA interaction were initially attributed entirely to indirect readout,[12] but it was subsequently realized that flanking tandem molecules also make contacts with the central base-pairs.[31]

4.2.2 Language of Indirect Readout: DNA Geometry

Indirect readout is described either in terms of deviations in DNA geometry from ideal structure or in terms of changes in geometry between sets of observed structures. To enable complete descriptions of DNA geometry, a common reference frame based on idealized base-pair coordinates has been established.[32] Figure 4.1 illustrates parameters commonly used to describe DNA geometry based on such a reference frame. These parameters fall into three classes:

- The six *base-pair parameters* – shear, buckle, stretch, propeller, stagger, and opening (Figure 4.1, upper left) – together provide a complete definition of the positions and orientations of the two bases relative to ideal base-pair geometry.
- The six *base-pair step parameters* – shift, tilt, slide, roll, rise, and twist (Figure 4.1, upper right) – together provide a complete definition of the relative positions and orientations of the two consecutive base-pairs.
- The six *helical parameters* – x-displacement, y-displacement, inclination, tip (Figure 4.1, lower right), helical rise and helical twist (not shown) – describe the geometry with respect to the helical axis.

In cases where the overall path of the DNA helical axis is significantly bent, as is often true for indirect readout mechanisms, base-pair step parameters are generally more useful for comparative purposes than are overall helical parameters. One set of parameters can be deduced from the other.[33] Table 4.1 summarizes average parameters and standard deviations for B-DNA.

4.2.3 Sequence-dependent Polymorphisms of B-DNA

Our general understanding of how a particular DNA sequence influences conformation and flexibility is at best limited. The main difficulty is that DNA structure is highly dynamic and a single DNA sequence can be compatible with a wide range of structures, even crossing between the A- B- and/or Z-DNA helical families. The B-DNA helical family, which most closely represents the range of conformations recognized by proteins, is thought to be the most flexible of all families because of the wide range of conformations that can be adopted by the deoxyribose sugars in the backbone.

Some rules that appear to govern sequence-dependence of B-DNA conformation and flexibility are outlined here. By far the most dominant factor is base stacking. The sugar-phosphate backbone plays a relatively passive role, yielding wider or narrower grooves that reflect the base-stacking interactions. Environment also plays a role: variables such as cation concentration,[34–36] bound drugs, bound proteins, and even crystal lattice environment[37] can influence the structure of a DNA fragment.

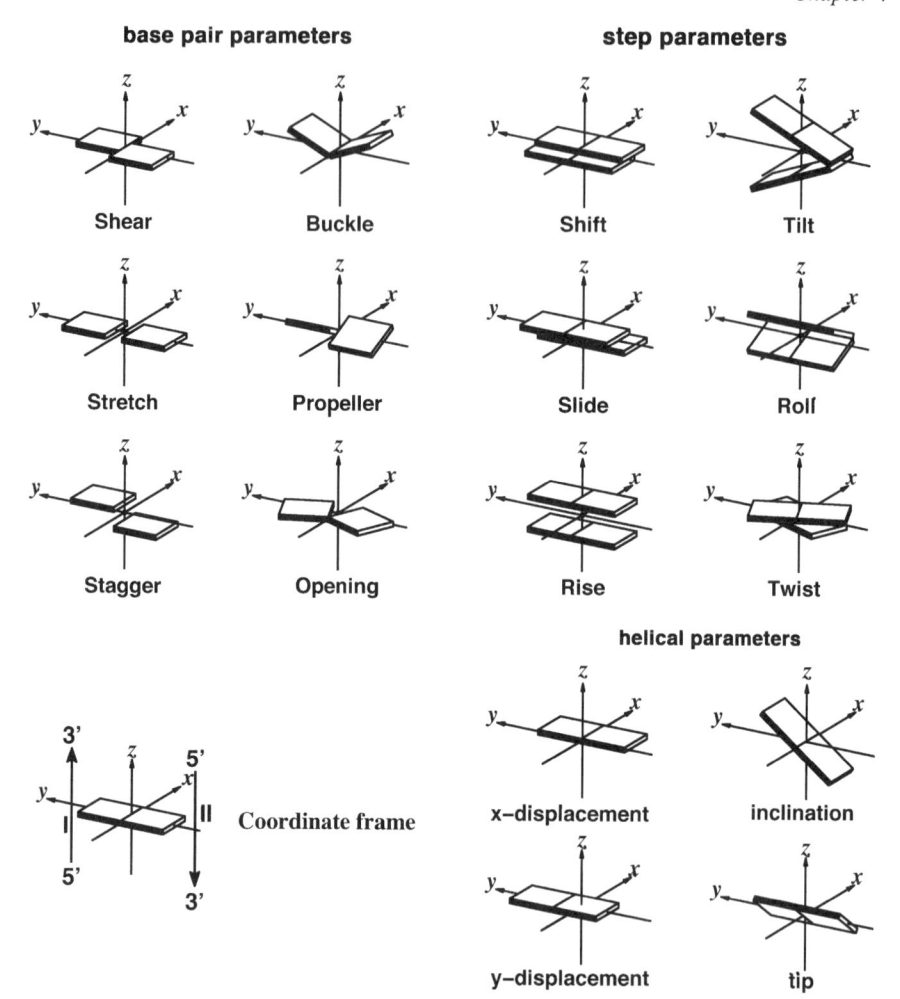

Figure 4.1 Illustration of parameters describing the geometry of DNA.[32] If a base or base pair is taken as a rigid block, six parameters are required to rigorously describe the position and orientation of one block relative to another. The six *base-pair parameters* describe the relative positions of the two bases in a pair. The six *step parameters* describe the local stacking geometry between neighboring base pairs. Six *helical parameters* describe the position and orientation of a base pair with respect to the helical axis (helical rise and helical twist are not shown). The step and helical parameter sets are interrelated: from either set, the other can be deduced. The values of local *vs.* helical rise and twist can be quite different in structures that deviate significantly from B-DNA. Diagrams were drawn using the program 3DNA.[33] Positive displacements of each parameter are shown.

4.2.3.1 Base Stacking

Whether free in solution or linked in a polymer through a sugar-phosphate backbone, nucleic acid bases tend to pile up like coins in a roll. They stack

Table 4.1 Average geometric parameters for B-DNA.

Parameter	Value[a]
Roll (°)	0.6 (5.2)
Twist (°)	36.0 (6.8)
Tilt (°)	0.6 (5.2)
Slide (Å)	0.23 (0.81)
Rise (Å)	3.32 (0.19)
Shift (Å)	−0.02 (0.45)
Buckle (°)	0.5 (6.7)
Propeller (°)	−11.4 (5.3)
Opening (°)	0.6 (3.1)
Shear (Å)	0.00 (0.21)
Stretch (Å)	−0.15 (0.12)
Stagger (Å)	0.09 (0.19)

[a] Values and standard deviations in parenthesis are from Olson et al.[32]

Table 4.2 Stacking energies of base-pair steps.

Base-pair step[a]	Type	Energy[b]
GC	RY	−15.5
CG	YR	−12.3
GG=CC	RR=YY	−11.7
AC=GT	RY	−11.3
TC=GA	RR=YY	−10.6
CT=AG	RR=YY	−10.4
TG=CA	YR	−9.8
AT	RY	−7.4
AA=TT	RR=YY	−6.7
TA	YR	−6.5

[a] Base-pair steps are denoted in shorthand. GC corresponds to the self-complementary base-pair step (5'-GC-3')$_2$; AC and GT correspond to the same base-pair step, (5'-AC-3')(5'-GT-3'), and are therefore denoted as AC=GT.
[b] Energies are from Ornstein et al.[38] Units are kcal mol^{-1} base-pair dimer.

snugly with their planes separated by the minimal possible distance, 3.4 Å. Base-stacking is driven by the hydrophobic effect and by dipolar and London dispersion forces. Stacking minimizes exposed hydrophobic surface area and therefore reduces the number of structured water molecules that must interact with the bases. Permanent dipoles and fluctuating electronic structure together strongly influence the relative orientations of the bases within a stack.

The formation of a DNA duplex joins the base stacks of two individual strands. Stacking strength varies with both composition and sequence; Table 4.2 displays calculated energies for the ten possible base-pair steps.[38] In general, the stacking interactions of base-pair steps involving G:C base pairs are more stable than those involving A:T base pairs. Another important trend is that the three possible base-pair steps with purine-pyrimidine sequences (RY: GC, GT=AC, and AT) are more stable than the corresponding pyrimidine-purine sequence (YR: CG, TG=CA, and TA). The reason is that the overlap between adjacent base pairs is much greater for RY steps than for YR steps. In particular, the polar groups are more precisely centered within the π-electron cloud of adjacent bases in RY compared to YR steps. For this reason, YR steps are more flexible than RY steps; TG=CA and TA in particular facilitate bending of DNA into either the major or minor groove.

4.2.3.2 Hydrogen Bonding

A:T base pairs have only two hydrogen bonds instead of three, and hence offer less resistance to twisting than G:C base pairs. As a consequence, A:T base pairs exhibit greater variability in propeller twist than do G:C pairs. "A-tracts" are extended runs of four or more A:T base pairs on one strand without a disruptive TA step. A-tracts tend to form a rigid stack with high propeller twist and a narrowed major groove with an ordered zigzag spine of hydration,[37] but can open up their minor groove to accommodate drugs.[39]

4.2.3.3 Steric Repulsion

Propeller twist improves the strength of base-pair stacking interactions but also leads to sequence-dependent steric repulsion effects. This applies mainly to purines from opposite strands in YR and RY steps; for continuous runs of purines or pyrimidines, steric repulsion is negligible. The most severe steric repulsion effects involve guanine N_2,N_3/adenine N_3 atoms in the minor groove of YR steps. Clashes were observed to be avoided in the B-DNA dodecamer crystal structure[7] by combinations of four different maneuvers, known as Calladine's Rules:[8] (1) reducing propeller twist, (2) opening base-pair roll angle (YR: expand at minor groove; RY: expand at major groove), (3) increasing base-pair step slide (which leads to alterations in sugar pucker), and (4) decreasing base-pair step twist.

4.2.3.4 DNA Bending

In a phenomenon called "helical phasing", particular DNA sequences that are phased by an integral number of turns give rise to anomalously high curvatures observed by techniques such as cyclization and electrophoretic mobility shift assays.[40] The prototype and most widely studied case of helical phasing is phased A-tracts. The physical basis for bending in DNA sequences with A-tracts is not well understood, but appears to mainly involve bending outside of rigid A-tract blocks, particularly at YR steps.[41]

4.2.4 Indirect Readout: A Universal Feature of Protein-DNA Interactions

In the past decade, the number of structures of protein-DNA complexes archived in the Nucleic Acid Database[42] and Protein Data Bank[43] has grown from about 200 to over 1000. The wealth of data available in this period has

Indirect Readout of DNA Sequence by Proteins

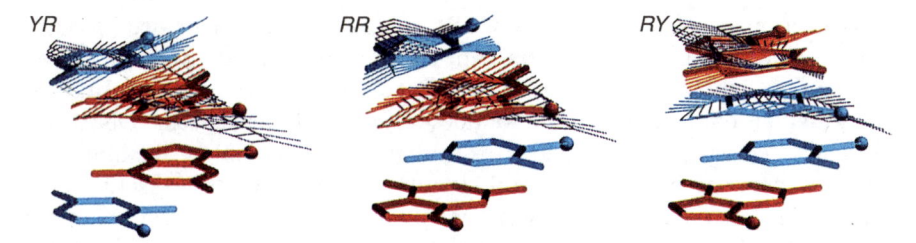

Figure 4.2 Sequence-dependent positions and motions of dimer steps found in protein-DNA structures: pyrimidine-purine (YR), purine-purine (RR), and purine-pyrimidine (RY) steps. Motions are illustrated for the 3′ (upper) base pair relative to the 5′ (lower) base pair. Note the decreasing range of motions from left to right. (Reproduced with permission from Olson et al.[44] Copyright (1998), National Academy of Sciences, U.S.A.)

enabled researchers to begin to examine the roles of direct and indirect readout in sequence specific protein-DNA interactions in terms of broad surveys and statistical analyses.[44–48] Every protein-DNA interaction is unique, but clear trends have emerged.

An important result of these analyses has been the development of sequence-dependent empirical energy functions to describe DNA deformability by proteins. Each of the ten possible base pair steps has a unique fingerprint based on average value and standard deviation for each of the six step parameters, plus volume of conformation space occupied, in the presence and absence of bound protein.[44] Figure 4.2 shows illustrations of occupied conformation space in protein-bound DNA, averaged for YR, RR, and RY base-pair step types. The statistics confirm that the three YR steps are by far the most deformable, while RR=YY steps are intermediate and RY steps are the least deformable.

These empirical DNA deformability energy functions have been employed in the context of a threading algorithm that varies the DNA sequence within a protein-DNA complex structure to yield a computational estimate of the contribution of indirect readout. When combined with complementary energy functions to measure direct interactions between protein and DNA, a profile of the direct *vs.* indirect readout contributions is obtained for the protein-DNA complex structure.[49] A survey of profiles obtained for 62 non-redundant protein-DNA complex structures showed that the relative contributions of direct and indirect readout can vary substantially, but both components are present in every interaction.[47] Thus, we can state with confidence that indirect readout is a universal mechanism of protein-DNA recognition.

4.3 DNA Sequence Recognition by CAP

The catabolite gene activator protein (CAP, also known as the cyclic AMP receptor protein, or CRP) is a regulatory protein that enables *E. coli* to survive in situations where glucose is not readily available as an energy source. When glucose levels drop, the cell-surface enzyme adenyl cyclase generates the second messenger molecule cyclic AMP (cAMP) from ATP. When activated by binding of cAMP ligand, CAP alters transcription at more than one hundred different promoters, mobilizing enzymes needed to utilize other sources of energy, and repressing others.[50] CAP does this by binding to specific DNA sites in or near target promoters and either enhancing or repressing the ability of RNA polymerase holoenzyme (RNAP) to bind and initiate transcription.[51] Activation of transcription by CAP involves formation of specific contacts between CAP and the α and/or σ subunits of RNA polymerase.[26,51] DNA-binding and transcription activation by CAP continues to be the subject of extensive biophysical, biochemical, and genetic investigations.[26]

CAP is a dimer of two identical subunits 209 residues in length. Each subunit is composed of an N-terminal cAMP ligand binding domain[52] and a C-terminal DNA binding-domain with a winged helix-turn-helix motif (see Chapter 3). CAP subunits can adopt a range of conformations that differ in relative orientations of the N- and C-terminal subunits.[53–56] The precise mechanism of activation by cAMP is not known because there are no crystal structures of unliganded CAP. Careful analysis of available structures has led to the suggestion that cAMP activates CAP by reorganizing the dimer interface, which in turn alters the relative positions of the two DNA-binding domains in the dimer.[53]

The DNA binding site for CAP (Figure 4.3) is a 22 base pair twofold-symmetric site with preferred sequence 5′-AAATGTGATCTAGATCACA-TTT-3′.[57] The interaction between CAP and its DNA binding site has been

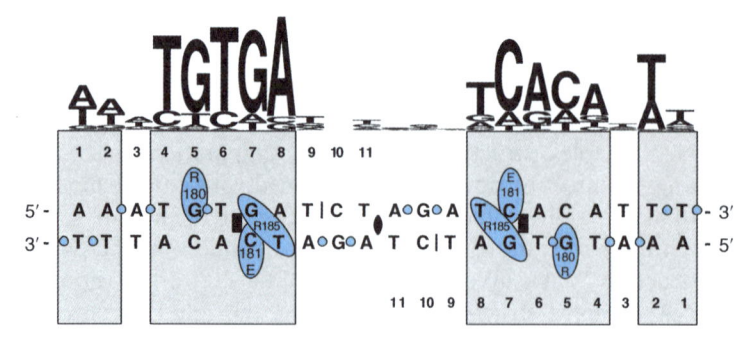

Figure 4.3 DNA sequence preference of CAP and summary of CAP-DNA interactions. The sequence logo (top) represents the consensus and variability of 58 known DNA binding sites for CAP.[70,93] The schematic diagram (bottom) represents the consensus of CAP-DNA interactions in 22 independent crystallographic observations of CAP-DNA half-complexes.[26,64] Shaded boxes indicate positions at which CAP exhibits strong sequence preferences.[57,67–69] The central black oval and black rectangles indicate, respectively, the twofold-symmetry axis of the complex and the primary kink sites. The two black vertical lines indicate positions of single-phosphate gaps present in crystallization DNA fragments. The ovals and circles indicate, respectively, amino acid-base contacts and amino acid-phosphate contacts.

studied extensively. Eleven crystal structures of CAP bound to DNA and three structures of CAP mutants bound to DNA have been determined to date.[58–64]

The crystal structure of a ternary complex of CAP, the C-terminal domain of the RNA polymerase α subunit (α-CTD), and DNA containing consensus sites for CAP and α-CTD has also been determined.[65] All CAP-DNA complex structures are either perfectly twofold symmetric or nearly so; each subunit of CAP interacts with one half of the DNA site (Figure 4.4). In each of the structures, CAP bends the DNA by approximately 80°, wrapping the DNA toward and around the sides of the CAP dimer. The same degree of DNA bending is observed in CAP-DNA complexes in solution.[66] Bending usually includes distinct kinks between base-pair positions 6 and 7 within each half-complex, and has a large out-of-plane component (Figure 4.5).

CAP exhibits strong sequence preferences at seven positions within each DNA half-site: positions 1, 2, and 4–8 (boxed positions in Figure 4.3),[57,67–69] and experimentally verified CAP binding sites closely follow this preference (sequence logo in Figure 4.3).[70] At positions 1 and 2, CAP faces the DNA

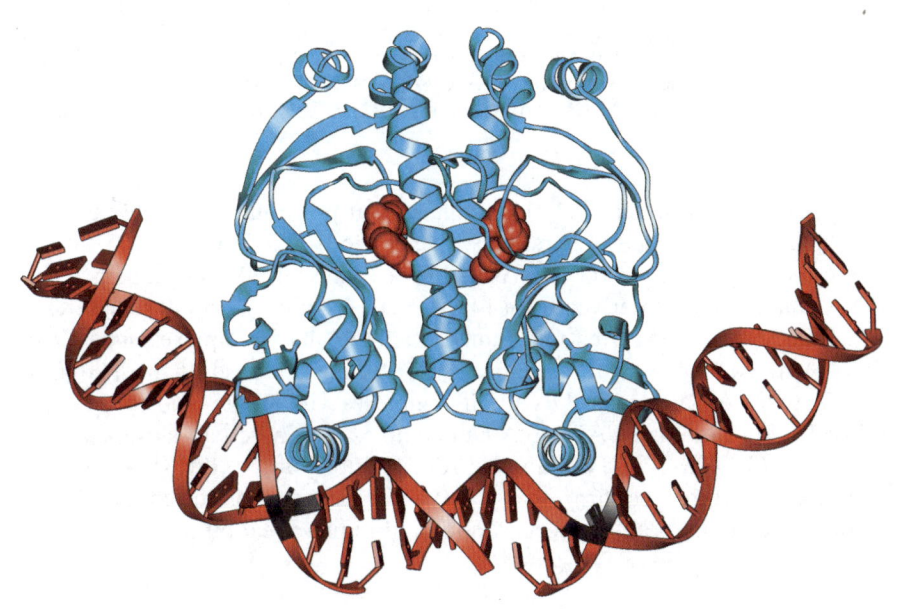

Figure 4.4 Schematic representation of a CAP-DNA complex. The two cyclic AMP ligands that activate CAP to bind DNA are shown in spacefill. DNA base-pairs are shown as blocks and the base-pairs at the two symmetry-related positions 6 are shaded black. In most independent observations of CAP-DNA half-complexes (18/22), there is a pronounced kink between base-pairs 6 and 7 with high roll angle, reduced twist, and increase in rise. In the example shown here (PDB entry 1ZRC[64]), a pronounced kink is visible within the half-complex on the left, but the DNA is smoothly bent in the half-complex on the right. The presence of both kinked and smoothly bent DNA conformations within a single crystal structure suggests that deformation of DNA by CAP is a dynamic process with at least two states.

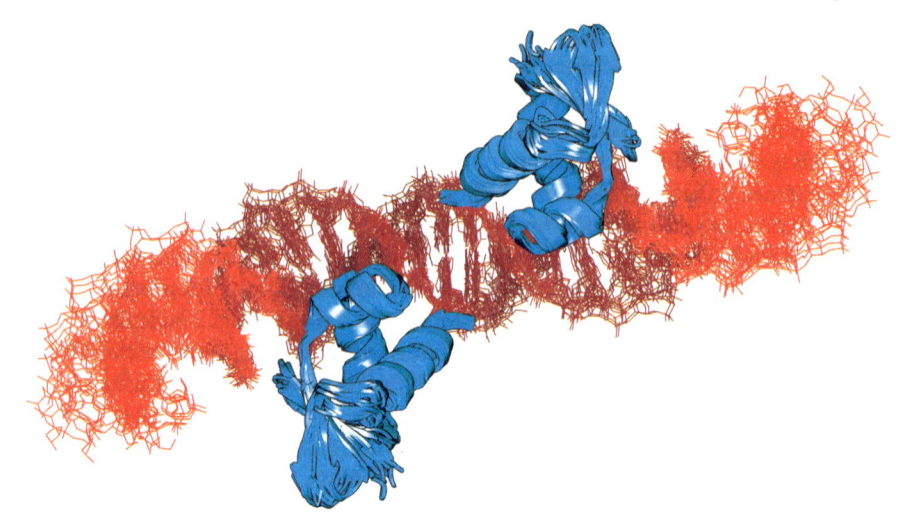

Figure 4.5 Comparison of CAP-DNA crystal structures. View is down the approximate molecular two-fold symmetry axis. The 14 crystal structures are shown superimposed; structures with two non-identical half-complexes are represented in both orientations. In this view the out-of-plane component of bending of the DNA is readily seen. For simplicity the N-terminal cAMP binding domains of the CAP dimers have been omitted.

minor groove and prefers A,T rich sequences; at positions 4–8, CAP faces the DNA major groove and prefers the sequence motif 5′-TGTGA-3′.

We recently surveyed the DNA-binding features of 22 independent CAP-DNA half-complexes derived from 14 CAP-DNA complex crystal structures.[64] The set of 22 consists of ten CAP-consensus DNA half complexes, nine CAP-DNA half complexes with base-pair substitutions at position 6 (TGNGA, N≠T), two CAP-consensus DNA half complexes with amino-acid substitution Glu181→Phe, and one CAP-DNA half complex with base-pair substitution at position 6 (TGCGA) and amino-acid residue substitution Glu181→Asp. The collected structures confirm that CAP recognizes its preferred DNA site through a combination of direct readout and indirect readout.

4.3.1 Direct Readout by CAP

Direct contacts between CAP and DNA base-pairs are made only within the major groove of the TGTGA motif, and involve interaction of CAP recognition helix residues Arg-180, Glu-181, and Arg-185.[58–64,71–73] The hydrogen-bonded interactions between CAP and the TGTGA motif are shown schematically in Figure 4.3 and are illustrated in Figure 4.6. The sequence specificities measured by Gunasekera *et al.*[57] and interactions to TGTGA motif base-pairs observed in the 22 CAP-DNA half complexes by Napoli *et al.*[64] are summarized below, ordered by base-pair position:

Indirect Readout of DNA Sequence by Proteins

Base-pair position 4 (consensus = T:A): Sequence preference at this position is modest (T:A > C:G > G:C ≈ A:T). The structural data show that the side chain of Arg-180 either stacks against the thymine base, making contact to the thymine C5 methyl group (14/22) (Figure 4.6B), or does not make direct contact to the thymine base but is in very close proximity (8/22). Direct readout at position 4 involves only contact to the C5 methyl group of the thymine.

Base-pair position 5 (consensus = G:C): Sequence preference at this position is strong (G:C ≫ T:A > A:T > C:G). In all but one half-complex structure, the side chain guanidinium of Arg-180 forms a pair of hydrogen bonds with the guanine O6 and N7 atoms of the consensus G:C base pair at position 5 (Figure 4.6B). In the only exception, the guanidinium is relocated 5 Å away to a position between the carboxylate of Glu-181 and a backbone phosphate.[62] Even though this exception demonstrates the flexibility of Arg-180, model building indicates that it cannot reach guanine of a nonconsensus C:G. Sequence preference at this base pair is, therefore, fully explained by the direct readout requirement for an adjacent pair of hydrogen bond acceptors.

Base-pair position 6 (consensus = T:A): Sequence preference at this position is modest (T:A > A:T ≈ C:G ≈ G:C). However, no hydrogen bonds are made to this base pair, except when modified from the consensus. When the base-pair is substituted with C:G or G:C, the carboxylate of Glu-181 usually forms a hydrogen-bond with the N4 atom of the cytidine (not shown).[61,62,64] In a few structures either Glu-181 or Arg-185 make van der Waals contact with the thymine C5 methyl group, but this is not a consistent feature. There is no direct readout of this base pair by CAP.

Base-pair position 7 (consensus = G:C): Sequence preference at this position is very strong (G:C ≫ T:A > A:T ≈ C:G). Without exception, the side chain of residue 181 forms a hydrogen bond with the cytosine N4 atom (Figure 4.6C). When Glu-181 is substituted with Phe, an aromatic hydrogen bond is made.[59] In 16/22 half-complexes, the side chain guanidinium of Arg-185 forms at least one hydrogen bond and sometimes two with the guanine O6 and/or N7 atoms (Figure 4.6C). These two interactions constitute a direct recognition pattern for this base-pair position that can be satisfied only by the consensus, G:C.

Base-pair position 8 (consensus A:T): Sequence preference at this position is moderately strong (A:T > C:G > G:C ≈ T:A). Without exception, a tight van der Waals contact is made between the residue 181 side chain and the thymine C5 methyl group of the consensus A:T base pair at position 8 (Figure 4.6C). In 13/22 half-complexes, the side chain guanidinium of Arg-185 also forms a hydrogen bond with the thymine O4 atom of the consensus A:T base pair at position 8 (Figure 4.6C). These two interactions constitute a direct recognition pattern for this base-pair position that can be satisfied only by the consensus, A:T.

In summary, the CAP-DNA half-site interaction specifies a unique pattern of hydrogen-bonding and van der Waals interactions that varies slightly from

78 Chapter 4

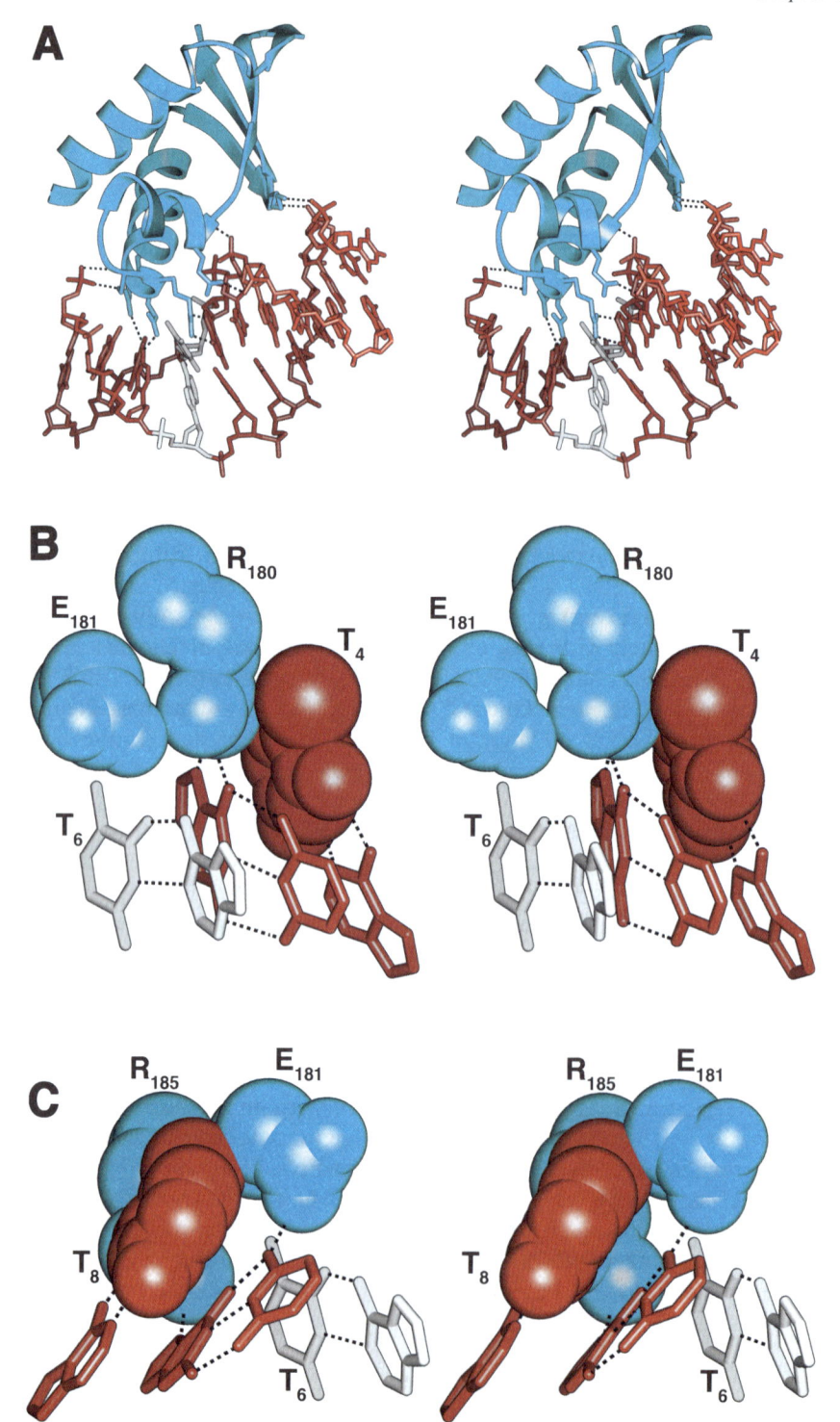

Indirect Readout of DNA Sequence by Proteins

structure to structure but is uniquely satisfied by the sequence TGNGA. In contrast, sequence preference at positions 1, 2, and 6 occurs in the absence of "directional" amino acid-base contacts and must therefore involve indirect readout.[58,60]

4.3.2 Indirect Readout by CAP

In the previous section we reviewed each of the direct interactions between CAP and DNA base-pairs in CAP-DNA complexes. The astute reader will notice that we did so without a single reference to DNA geometry. The briefest glance at the bent DNA conformations illustrated in Figures 4.4 and 4.5 makes it obvious that we have not yet revealed the whole story of how CAP recognizes its DNA binding site.

4.3.2.1 Conformation and Flexibility of the DNA Site for CAP

To what extent does the conformation of the DNA site for CAP, when free in solution, resemble its conformation in the protein-DNA complex? Remarkably, the consensus DNA site contains an intrinsic bend in the absence of CAP.[74–76] Elements of the consensus sequence that likely contribute to bending in the absence of CAP include the pairs of consecutive TG=CA steps within each half site, phased one helical turn apart, and the short A,T-rich region at the ends of the consensus site (positions 1, 2), phased two helical turns apart.

Estimates of the DNA bend angle in the absence of CAP vary from 15°, obtained in cyclization assays employing tandem arrays of DNA sites,[74] to the much higher estimate of 52°, obtained in electrophoretic mobility shift phasing assays.[76] Time-resolved luminescence resonance energy transfer experiments indicate that the upper limit of the DNA bend angle in the absence of CAP is 40–50°.[75] Thus it is clear that in bending the DNA to 80°, CAP actively deforms its DNA binding site.

How does CAP achieve such a high overall bend in the DNA? A key factor appears to be formation of electrostatic interactions between positively charged residues on the sides of the CAP dimer and negatively charged DNA phosphates.

Figure 4.6 CAP-DNA interactions (stereo pairs). In each panel the position of the sixth base pair is shown in light gray; position numbers increase from right to left. (*A*) CAP-DNA half-complex. Hydrogen bonds (thin dotted lines) are shown between CAP residues Arg-180, Glu-181, and Arg-185 and DNA base pairs at positions 5 and 7, and between CAP residues Gln-170, Ser-179, and Lys-201 and DNA backbone phosphate oxygens. The view is midway between the right half-complexes shown in Figures 4.4 and 4.5. Note the compression of the DNA minor groove on the right-hand side at DNA base-pair positions 1 and 2. Coordinates are from PDB entry 1ZRC.[64] (*B*) and (*C*) illustrate consensus contacts of CAP-DNA complex crystal structures to the TGTGA motif. Coordinates are from the two independent half-complexes of PDB entry 1RUN.[60]

Electrostatic interactions are proposed to contribute to DNA bending both by stabilizing the bent state through amino acid-phosphate and helix dipole-phosphate interactions,[58,66,67,75] and by destabilizing the unbent state through asymmetric phosphate neutralization.[76,77] Substitution of positively charged residues on the sides of the CAP dimer (Lys-22, Lys-26, Lys-44, and Lys-166) with neutral residues reduces the mean DNA bend angle in the CAP-DNA complex in solution by $\sim 5°$ per residue per half-complex.[75]

The conformation and flexibility of the DNA targets of several other proteins that bend DNA have also been investigated, most notably the targets of papillomavirus E2 proteins,[23] *trp* repressor,[78] 434 repressor[22] and TATA binding protein.[79] In general, a positive correlation exists between DNA conformational flexibility and protein-DNA binding affinity, as well as a positive correlation between protein-DNA binding affinity and favorable pre-bending of the DNA target, although the behavior of E2 proteins is more complex.[23] Chapter 8 also includes a nice discussion of the role of indirect readout in site selection by DNA bending proteins.

4.3.2.2 Indirect Readout at Positions 1-2

CAP strongly prefers A:T or T:A base pairs at DNA positions 1 and 2 where the protein faces the minor groove.[67,69] However, in CAP-DNA complex structures there are no direct interactions to these base pairs. The sequence preference in this region results from a pattern of hydrogen bonding between CAP residues and backbone phosphates that is best accommodated by a severely narrowed minor groove in the DNA (Figure 4.6A). A,T-rich DNA sequences are uniquely able to deform to the extent required (see Section 4.2.3.2).

4.3.2.3 Indirect Readout at Position 6

The consensus T:A base-pair at position 6 sits at the center of the major groove binding motif, TGTGA. In most CAP-DNA half-complex structures (17/22), this base pair is the locus for an extraordinarily large local deformation in the DNA structure. The local deformation, when present, is a kink with high positive roll angle, reduced twist, and increased rise between positions 6 and 7. The local deformation is so severe that it largely breaks the base-pair stacking interaction at this step.

The substitution of any other base pair at position 6 other than T:A results in a loss of free energy of ~ 1.4 kcal mol^{-1} half-site for the CAP-DNA interaction.[57] The preference for T:A cannot arise from direct readout because no direct contacts are made between the protein and DNA to the consensus base-pair. It has, therefore, been proposed that specificity for T:A at position 6 is a consequence of formation of the kink between DNA base-pair positions 6 and 7, and of effects of the consensus TG=CA step on the geometry of DNA kinking, the energetics of DNA kinking, or both.[60]

To investigate how sequence information is indirectly sensed at this site of large local DNA deformation, we have obtained and analyzed a series of CAP-DNA

Indirect Readout of DNA Sequence by Proteins 81

structures with all possible base pairs at position 6.[62,64] The collected structures show that CAP can introduce a kink at positions 6–7 with any nucleotide at position 6 (Figure 4.7). The kink is equally sharp with the consensus YR step TG and the nonconsensus YR step CG (roll angles $\sim 42°$, twist angles $\sim 16°$, rise ~ 4.8 Å; Figure 4.7, compare panels A and B).[62,64] In several of the structures with YG steps, one or two ordered water molecules are observed to sit within the wedge formed by the two base-pairs at the minor groove.

A kink is also present but is much less sharp with the nonconsensus RR=YY steps AG and GG (roll angles $\sim 20°$, twist angles $\sim 17°$, rise ~ 3.4 Å; Figure 4.7C).[64] The kinked RG steps, unlike YG steps, do not open up enough for water to fit. Although of reduced magnitude relative to the YG steps, the kinked RG steps still represent extraordinarily large local deformations, as can be seen in a roll *vs.* twist plot (Figure 4.8).

We infer from these studies that CAP discriminates between consensus and non-consensus YR steps at positions 6–7 based solely on differences in the energetics of DNA deformation (compare CG *vs.* TG base-pair step strengths in Table 4.2). In contrast, CAP discriminates between the consensus YR step and non-consensus RR=YY steps at positions 6–7 based on differences in the energetics of DNA deformation and qualitative differences in the deformability of DNA at these base pair steps. The results of these studies confirm that the main determinant of local DNA geometry in this system is protein-DNA interaction, and not DNA sequence.

4.3.2.4 Comparison with Other Protein-induced Positive Roll Deformations

In the CAP-DNA system, a large positive roll deformation (kink) at position 6–7 plays an important role in DNA sequence recognition. Positive roll deformations are frequently observed in sequence-specific protein-DNA interactions (Table 4.3). The decrease in width of the major groove and increase in width of the minor groove that result from positive roll deformation improves overall shape complementarity to many protein recognition elements, whether major or minor groove binders.

Positive roll deformations induced by proteins range from modest to extreme (Table 4.3). *trp* Repressor deforms the central TA step of its recognition half-site by just 15°,[12] while integration host factor deforms an AA step in its preferred binding site by 65°.[29] In general, proteins that contact the major groove tend to induce gentle deformations, while proteins that contact the minor groove tend to induce extreme kinks with one or two hydrophobic protein side chains intercalating into the minor groove. In every instance, positive role deformations involve either YR or RR steps, but never RY steps. The protein-induced kink at position 6–7 YG steps in the CAP-DNA interaction is the most extreme example of positive roll for a major groove binder, and interestingly the only such case where water is sometimes observed to

82   Chapter 4

intercalate in the base-pair step. Water is also seen to intercalate into the minor groove of kinked base-pair steps in EcoRV-DNA complexes,[28] but in these cases the protein fully envelops the DNA with contacts to both major and minor grooves.

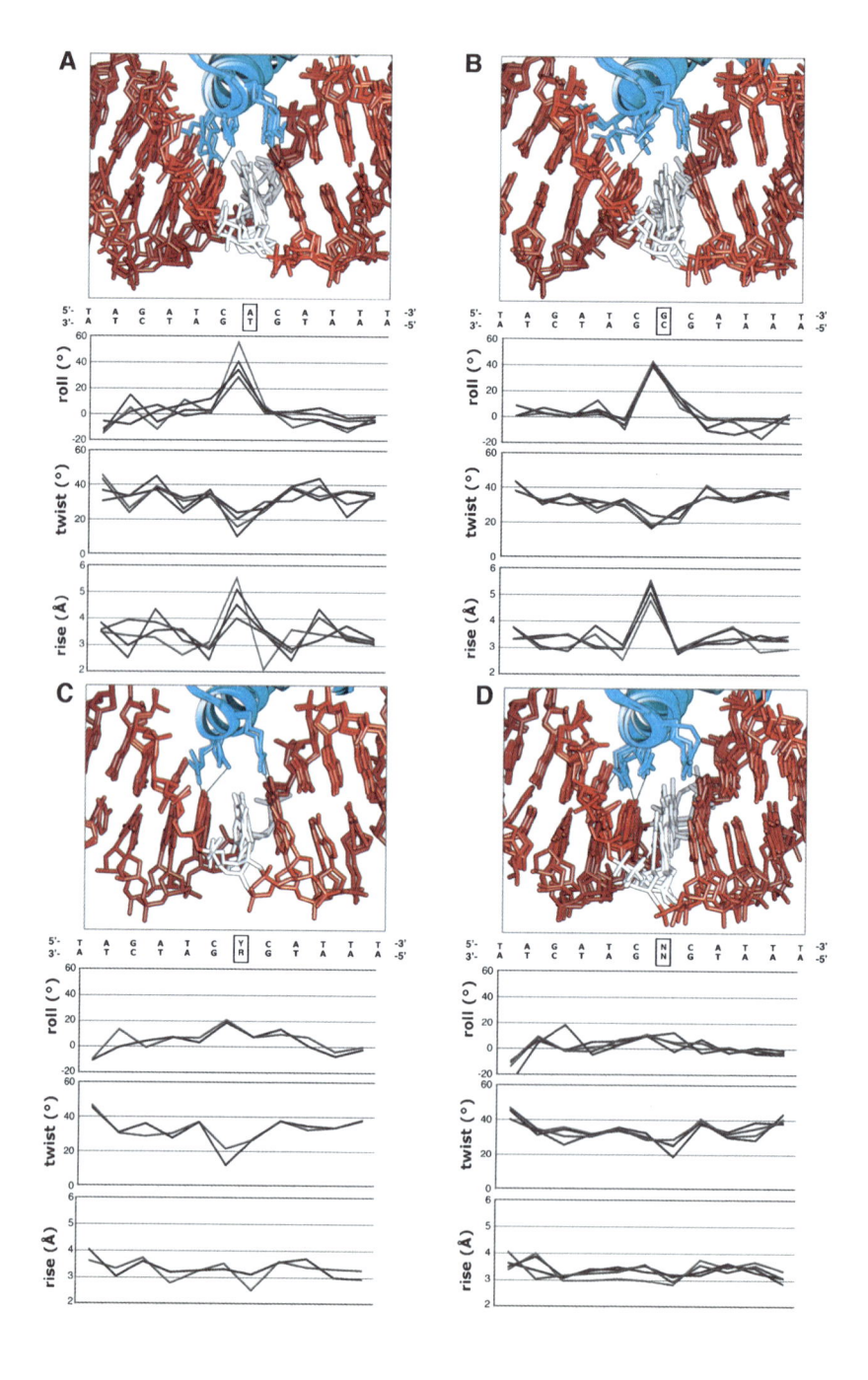

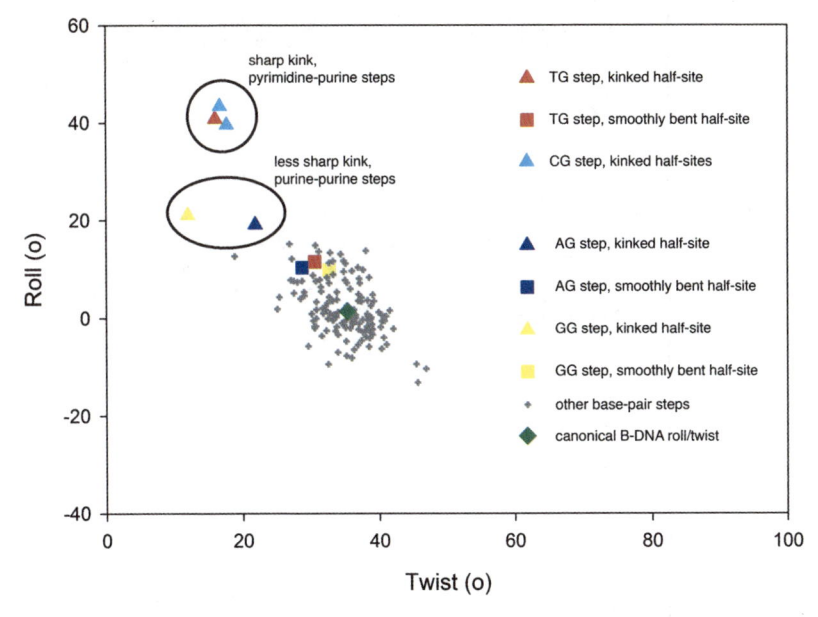

Indirect Readout of DNA Sequence by Proteins 83

Figure 4.8 DNA helical parameters of CAP-DNA complexes with four different base-pairs at position 6 of the primary-kink site. The figure presents a plot of DNA roll angle versus DNA twist angle for each base pair step of each half-complex. The eight base pair 6–7 dinucleotide steps are represented as colored triangles and squares. All other base pair steps are represented as grey crosses. Modified from Napoli et al.[64]

4.3.3 DNA Bending *vs.* DNA Kinking – A Dynamic Duo?

We have learned from Section 4.3.2.3 that CAP discriminates among all possible NG base-pair steps at positions 6–7 both by the energy required to

Figure 4.7 DNA conformations in CAP-DNA structures. Most CAP-DNA structures with YG=CR base pair steps at positions 6–7 display a pronounced kink [TG=CA, panel (*A*); CG, panel (*B*)]. CAP-DNA structures with RG=CY base-pair steps [AG=CT or GG=CC, panel (*C*)] are only modestly deformed by CAP. CAP-DNA structures with any base pair at position 6 (NG=CN) can instead display smoothly bent DNA (*D*). Each panel shows overlays of 2–4 example structures in the same orientation; representative hydrogen bonds are shown as thin lines. The recognition helix of CAP is displayed with side chains Arg-180, Glu-181 and Arg-185 (left, middle, right). The DNA strand closest to the viewer corresponds to the top strand in the label, with the 5′ end at top left and the 3′ end at bottom right. Position 6 base pairs are shaded light grey. Base-pair step parameters roll, twist, and rise are plotted below each set of representative structures. The parameter profiles of CAP-bound DNA with kinked TG=CA or CG position 6–7 steps form a distinctive pattern that is muted when RG=CY steps are substituted, or when the DNA adopts a smoothly bent conformation.

Table 4.3 Positive roll deformations in protein-DNA complexes.

DNA binding protein	Roll (°)	Step	Step type	Groove contact[a] major	Groove contact[a] minor	Minor groove intercalator	Representative NDB id	Ref.
trp Repressor	15	TA	YR	x			PDR009	12
Excisionase XIS	20	TG	YR	x			PD0872	81
E2 protein	21	CG	YR	x			PD0717	23
TN5 transposase	22	TA	YR	x			PD0340	82
Put3 binding protein	23	GG	RR		x	Val	PDT044	83
MecI Repressor	26	TG	YR	x			PD0718	84
Endonuclease PI-SceI	32	GG	RR	x			PD0319	85
Endonuclease I-HmuI	37	TG	YR		x	Val	PD0575	86
γ δ resolvase	40	TA	YR		x	Thr	PDE0115	87
DNA Pol κ	40	AA	RR	x			PD0962	88
CAP	44	TG	YR	x		Water[b]	PD0306	64
TATA binding protein	53	TA	YR		x	Phe	PD0156	89
MutM repair enzyme	54	AG	RR		x	Phe	PD0751	90
Sac7D protein	61	CG	YR		x	Val + Met	PDR047	91
lac Repressor	61	CG	YR		x	Leu + Leu	PD0118	92
EcoRV endonuclease	65	TA	YR	x	x	Water	PD0532	28
Integration host factor	65	AA	RR		x	Pro	PD0903	29

[a] DNA groove contact by protein in the vicinity of the positive roll deformation.
[b] Observed in some CAP/DNA complex structures.

deform the step and by the intrinsic deformability of the step. But CAP does not always kink DNA: in 5 of 22 CAP-DNA half-complexes the DNA is smoothly bent rather than kinked, with small deformations distributed throughout the TGTGA motif rather than a single central kink at positions 6–7. The structural data indicate two different factors that can lead to this difference in local DNA geometry: local environment or a mutation in CAP.

When smooth bending was first observed in a CAP-DNA crystal structure,[63] it was not possible to discern what factors might be responsible for the altered local geometry. In the recent study by Napoli et al.,[64] three CAP-DNA structures are described in a novel crystal form in which the DNA is kinked in one half-complex and smoothly bent in the other, as illustrated in Figure 4.4. Thus even within the environment of a single crystal, CAP can bend DNA to essentially the same overall DNA bend angle ($\sim 47°$ per half-complex) through a kink or through a smooth bend (Figure 4.7, compare panels A, B, and C to D). Based on these new structures we infer that kinked and smoothly bent DNA correspond to two states of a dynamic protein-DNA complex. The two states are close enough in energy that external environment variables can influence which state is preferred.

Smooth bending is also observed in the crystal structure of the CAP mutant Glu181→Asp bound to DNA.[61] The amino acid substitution eliminates specificity for T:A at position 6. Because a DNA kink is observed in the isomorphous crystal structure of wild-type CAP bound to the same DNA duplex,[62] the local change in DNA geometry can be directly attributed to the mutation.

When the different sets of CAP-DNA half complex structures in Figure 4.7 are superimposed, a striking difference that emerges is that the DNA in structures with YG step kinks (panels A and B) appear "stretched" relative to the DNA in structures with RG step kinks (panel C) or DNA that is smoothly bent (panel D). The stretching is localized to the kinked base-pair step, where the disruption of base-pair stacking yields an increase in rise parameter of up to ~ 1.5 Å (see Figure 4.7 parameter plots).

In the full CAP-DNA dimer, the ~ 3 Å increase in helical path produced by two kinks is accommodated by small positional shifts in the DNA between the kinks, while the conformation and position of DNA directly flanking the TGTGA motif (i.e., base-pair positions 1–2, where CAP faces the minor groove) remain essentially unchanged. Even so, we speculate that conformation and dynamics beyond the CAP site could be influenced by conformational changes within the CAP site from kink to bend or bend to kink (e.g., with positions 1–2 acting as a pivot point). During the exercise of building a theoretical model of a Class II CAP transcription activation complex based on combined structural, genetic and biochemical data, it was not possible to satisfy the full set of refinement constraints unless the kink initially present in the downstream half of the DNA site for CAP was replaced with a smooth bend.[26] It is likely that both kinked and smoothly bent conformational states of DNA observed in CAP-DNA complex crystals play important roles in recognition of DNA and in activation of transcription by CAP.

4.4 Conclusions

The wealth of high-resolution structural data now available for protein-DNA interactions confirms that indirect readout plays a key role in recognition of DNA sequence. An important result of the data analysis is that the sequence–structure relationships of protein-bound DNA are statistical and probabilistic, not rigidly one-to-one cause and effect.[44–46,48]

Distinct deformability profiles for the ten possible base-pair steps in protein-DNA complexes demonstrate that our genomic material is a heterogeneous polymer with varying physicochemical properties along its length.[44] When stressed by external forces such as bending induced in a protein-DNA interaction, deformation of DNA is not evenly distributed but instead localizes to weaker base-pair steps, generally TA and TG=CA, that are situated where roll can relieve strain. Deformation of roll angle can be negative, bending into the minor groove, or positive, bending into the major groove; either roll angle deformation is easier to accomplish than the orthogonal deformations of tilt angle.

In the CAP-DNA interaction, the two serial TG steps per half-complex each lend flexibility to the DNA, and both TG steps play crucial roles in sequence recognition, but only one is deformed to an extraordinary extent by CAP. The pair of TG steps that are kinked by CAP (positions 6–7) are in phase, precisely one helical turn (10 bp) apart, and are therefore ideally positioned to promote wrapping of the DNA around the protein. In contrast, the flanking TG steps (positions 4–5) that are only modestly deformed by CAP are out of phase and so cannot contribute as extensively to DNA bending.

Although indirect readout is generally described in terms of one or more local sequence-dependent geometric deviations of DNA, it is important to recognize that indirect readout is not a local phenomenon but is a global consequence of the whole protein-DNA interaction. One isolated CAP recognition helix or DNA binding domain cannot by itself discriminate among all possible TGNGA sequences; it is the collected set of protein-DNA interactions in the full complex producing the overall DNA bend that enables discrimination of DNA deformability. Future efforts to generate complete predictive models of protein-DNA interactions will ultimately have to take these longer range effects into account.

Acknowledgements

We thank our colleagues Wilma Olson and Richard Ebright for many informative discussions. This work was supported the National Institutes of Health grant GM21589 to C.L.L. and H.M.B. All molecular graphics were produced using UCSF Chimera.[80]

References

1. J.D. Watson and F.H. Crick, *Nature*, 1953, **171**, 737–738.
2. S. Arnott and D.W. Hukins, *J. Mol. Biol.*, 1973, **81**, 93–105.

3. S. Arnott, D.W. Hukins, S.D. Dover, W. Fuller and A.R. Hodgson, *J. Mol. Biol.*, 1973, **81**, 107–122.
4. N.C. Seeman, J.M. Rosenberg and A. Rich, *Proc. Natl. Acad. Sci. U.S.A.*, 1976, **73**, 804–808.
5. C.E. Bugg, J.M. Thomas, M. Sundaralingam and S.T. Rao, *Biopolymers*, 1971, **10**, 175–219.
6. S.H. Kim, F.L. Suddath, G.J. Quigley, A. McPherson, J.L. Sussman, A.H. Wang, N.C. Seeman and A. Rich, *Science*, 1974, **185**, 435–440.
7. R.E. Dickerson and H.R. Drew, *J. Mol. Biol.*, 1981, **149**, 761–786.
8. C.R. Calladine, *J. Mol. Biol.*, 1982, **161**, 343–352.
9. C.A. Frederick, J. Grable, M. Melia, C. Samudzi, L. Jen-Jacobson, B.C. Wang, P. Greene, H.W. Boyer and J.M. Rosenberg, *Nature*, 1984, **309**, 327–331.
10. J.A. McClarin, C.A. Frederick, B.C. Wang, P. Greene, H.W. Boyer, J. Grable and J.M. Rosenberg, *Science*, 1986, **234**, 1526–1541.
11. Y.C. Kim, J.C. Grable, R. Love, P.J. Greene and J.M. Rosenberg, *Science*, 1990, **249**, 1307–1309.
12. Z. Otwinowski, R.W. Schevitz, R.G. Zhang, C.L. Lawson, A. Joachimiak, R.Q. Marmorstein, B.F. Luisi and P.B. Sigler, *Nature*, 1988, **335**, 321–329.
13. A.K. Aggarwal, D.W. Rodgers, M. Drottar, M. Ptashne and S.C. Harrison, *Science*, 1988, **242**, 899–907.
14. C. Wolberger, Y.C. Dong, M. Ptashne and S.C. Harrison, *Nature*, 1988, **335**, 789–795.
15. S.R. Jordan and C.O. Pabo, *Science*, 1988, **242**, 893–899.
16. J.E. Anderson, M. Ptashne and S.C. Harrison, *Nature*, 1987, **326**, 846–852.
17. P.H. von Hippel and O.G. Berg, *Proc. Natl. Acad. Sci. U.S.A.*, 1986, **83**, 1608–1612.
18. R.G. Brennan and B.W. Matthews, *Trends Biochem. Sci.*, 1989, **14**, 286–290.
19. C.O. Pabo and R.T. Sauer, *Annu. Rev. Biochem.*, 1992, **61**, 1053–1095.
20. B.W. Matthews, *Nature*, 1988, **335**, 294–295.
21. A. Joachimiak, T.E. Haran and P.B. Sigler, *EMBO J.*, 1994, **13**, 367–372.
22. G.B. Koudelka, S.A. Mauro and M. Ciubotaru, *Prog. Nucleic Acid Res. Mol. Biol.*, 2006, **81**, 143–177.
23. R.S. Hegde, *Annu. Rev. Biophys. Biomol. Struct.*, 2002, **31**, 343–360.
24. C.W. Garvie and S.E. Phillips, *Structure Fold Des.*, 2000, **8**, 905–914.
25. A. Bareket-Samish, I. Cohen and T.E. Haran, *J. Mol. Biol.*, 2000, **299**, 965–977.
26. C.L. Lawson, D. Swigon, K.S. Murakami, S.A. Darst, H.M. Berman and R.H. Ebright, *Curr. Opin. Struct. Biol.*, 2004, **14**, 10–20.
27. D.R. Lesser, M.R. Kurpiewski and L. Jen-Jacobson, *Science*, 1990, **250**, 776–786.
28. A.M. Martin, M.D. Sam, N.O. Reich and J.J. Perona, *Nat. Struct. Biol.*, 1999, **6**, 269–277.
29. T.W. Lynch, E.K. Read, A.N. Mattis, J.F. Gardner and P.A. Rice, *J. Mol. Biol.*, 2003, **330**, 493–502.

30. T.J. Richmond and C.A. Davey, *Nature*, 2003, **423**, 145–150.
31. C.L. Lawson and J. Carey, *Nature*, 1993, **366**, 178–182.
32. W.K. Olson, M. Bansal, S.K. Burley, R.E. Dickerson, M. Gerstein, S.C. Harvey, U. Heinemann, X.J. Lu, S. Neidle, Z. Shakked, H. Sklenar, M. Suzuki, C.S. Tung, E. Westhof, C. Wolberger and H.M. Berman, *J. Mol. Biol.*, 2001, **313**, 229–237.
33. X.J. Lu and W.K. Olson, *Nucleic Acids Res.*, 2003, **31**, 5108–5121.
34. T.K. Chiu and R.E. Dickerson, *J. Mol. Biol.*, 2000, **301**, 915–945.
35. K.K. Woods, L. McFail-Isom, C.C. Sines, S.B. Howerton, R.K. Stephens and L.D. Williams, *J. Am. Chem. Soc.*, 2000, **122**, 1546–1547.
36. N.V. Hud and M. Polak, *Curr. Opin. Struct. Biol.*, 2001, **11**, 293–301.
37. R.E. Dickerson, D.S. Goodsell and S. Neidle, *Proc. Natl. Acad. Sci. U.S.A.*, 1994, **91**, 3579–3583.
38. R.L. Ornstein, R. Rein, D.L. Breen and R.D. MacElroy, *Biopolymers*, 1978, **17**, 2341–2360.
39. J.G. Pelton and D.E. Wemmer, *Proc. Natl. Acad. Sci. U.S.A.*, 1989, **86**, 5723–5727.
40. P.J. Hagerman, *Annu. Rev. Biochem.*, 1990, **59**, 755–781.
41. D.L. Beveridge, S.B. Dixit, G. Barreiro and K.M. Thayer, *Biopolymers*, 2004, **73**, 380–403.
42. H.M. Berman, W.K. Olson, D.L. Beveridge, J. Westbrook, A. Gelbin, T. Demeny, S.H. Hsieh, A.R. Srinivasan and B. Schneider, *Biophys. J.*, 1992, **63**, 751–759.
43. H.M. Berman, J. Westbrook, Z. Feng, G. Gilliland, T.N. Bhat, H. Weissig, I.N. Shindyalov and P.E. Bourne, *Nucleic Acids Res.*, 2000, **28**, 235–242.
44. W.K. Olson, A.A. Gorin, X.J. Lu, L.M. Hock and V.B. Zhurkin, *Proc. Natl. Acad. Sci. U.S.A.*, 1998, **95**, 11163–11168.
45. N.M. Luscombe, S.E. Austin, H.M. Berman and J.M. Thornton, *Genome Biol.*, 2000, **1**, 1–37.
46. S. Jones, D.T. Daley, N.M. Luscombe, H.M. Berman and J.M. Thornton, *Nucleic Acids Res.*, 2001, **29**, 943–954.
47. M. Gromiha, J.G. Siebers, S. Selvaraj, H. Kono and A. Sarai, *J. Mol. Biol.*, 2004, **337**, 285–294.
48. R.E. Dickerson, in *Oxford Handbook of Nucleic Acid Structure*, ed. S. Neidle, Oxford University Press, New York, 1999, pp. 145–197.
49. A. Sarai, J. Siebers, S. Selvaraj, M.M. Gromiha and H. Kono, *J. Bioinform. Comput. Biol.*, 2005, **3**, 169–183.
50. G. Gosset, Z. Zhang, S. Nayyar, W.A. Cuevas and M.H. Saier, Jr., *J. Bacteriol.*, 2004, **186**, 3516–3524.
51. S. Busby and R.H. Ebright, *J. Mol. Biol.*, 1999, **293**, 199–213.
52. H.M. Berman, L.F. Ten Eyck, D.S. Goodsell, N.M. Haste, A. Kornev and S.S. Taylor, *Proc. Natl. Acad. Sci. U.S.A.*, 2005, **102**, 45–50.
53. J.M. Passner, S.C. Schultz and T.A. Steitz, *J. Mol. Biol.*, 2000, **304**, 847–859.
54. I.T. Weber and T.A. Steitz, *J. Mol. Biol.*, 1987, **198**, 311–326.

55. S.Y. Chu, M. Tordova, G.L. Gilliland, I. Gorshkova, Y. Shi, S. Wang and F.P. Schwarz, *J. Biol. Chem.*, 2001, **276**, 11230–11236.
56. M.C. Vaney, G.L. Gilliland, J.G. Harman, A. Peterkofsky and I.T. Weber, *Biochemistry*, 1989, **28**, 4568–4574.
57. A. Gunasekera, Y.W. Ebright and R.H. Ebright, *J. Biol. Chem.*, 1992, **267**, 14713–14720.
58. S.C. Schultz, G.C. Shields and T.A. Steitz, *Science*, 1991, **253**, 1001–1007.
59. G. Parkinson, A. Gunasekera, J. Vojtechovsky, X. Zhang, T.A. Kunkel, H. Berman and R.H. Ebright, *Nat. Struct. Biol.*, 1996, **3**, 837–841.
60. G. Parkinson, C. Wilson, A. Gunasekera, Y.W. Ebright, R.E. Ebright and H.M. Berman, *J. Mol. Biol.*, 1996, **260**, 395–408.
61. S. Chen, A. Gunasekera, X. Zhang, T.A. Kunkel, R.H. Ebright and H.M. Berman, *J. Mol. Biol.*, 2001, **314**, 75–82.
62. S. Chen, J. Vojtechovsky, G.N. Parkinson, R.H. Ebright and H.M. Berman, *J. Mol. Biol.*, 2001, **314**, 63–74.
63. J.M. Passner and T.A. Steitz, *Proc. Natl. Acad. Sci. U.S.A.*, 1997, **94**, 2843–2847.
64. A.A. Napoli, C.L. Lawson, R.H. Ebright and H.M. Berman, *J. Mol. Biol.*, 2006, **357**, 173–183.
65. B. Benoff, H. Yang, C.L. Lawson, G. Parkinson, J. Liu, E. Blatter, Y.W. Ebright, H.M. Berman and R.H. Ebright, *Science*, 2002, **297**, 1562–1566.
66. A.N. Kapanidis, Y.W. Ebright, R.D. Ludescher, S. Chan and R.H. Ebright, *J. Mol. Biol.*, 2001, **312**, 453–468.
67. M.R. Gartenberg and D.M. Crothers, *Nature*, 1988, **333**, 824–829.
68. R.H. Ebright, Y.W. Ebright and A. Gunasekera, *Nucleic Acids Res.*, 1989, **17**, 10295–10305.
69. D.D. Dalma-Weiszhausz, M.R. Gartenberg and D.M. Crothers, *Nucleic Acids Res.*, 1991, **19**, 611–616.
70. K. Robison, A.M. McGuire and G.M. Church, *J. Mol. Biol.*, 1998, **284**, 241–254.
71. R.H. Ebright, P. Cossart, B. Gicquel-Sanzey and J. Beckwith, *Nature*, 1984, **311**, 232–235.
72. R.H. Ebright, A. Kolb, H. Buc, T.A. Kunkel, J.S. Krakow and J. Beckwith, *Proc. Natl. Acad. Sci. U.S.A.*, 1987, **84**, 6083–6087.
73. X.P. Zhang and R.H. Ebright, *Proc. Natl. Acad. Sci. U.S.A.*, 1990, **87**, 4717–4721.
74. J.D. Kahn and D.M. Crothers, *J. Mol. Biol.*, 1998, **276**, 287–309.
75. A.N. Kapanidis, Ph.D. Thesis, Rutgers University, 1999.
76. P.R. Hardwidge, J.M. Zimmerman and L.J. Maher, 3rd, *Nucleic Acids Res.*, 2002, **30**, 1879–1885.
77. A.D. Mirzabekov and A. Rich, *Proc. Natl. Acad. Sci. U.S.A.*, 1979, **76**, 1118–1121.
78. Z. Shakked, G. Guzikevich-Guerstein, F. Frolow, D. Rabinovich, A. Joachimiak and P.B. Sigler, *Nature*, 1994, **368**, 469–473.

79. J.D. Parvin, R.J. McCormick, P.A. Sharp and D.E. Fisher, *Nature*, 1995, **373**, 724–727.
80. E.F. Petterson, T.D. Goddard, C.C. Huang, G.S. Couch, D.M. Greenblatt, E.C. Meng and T.E. Ferrin, *J. Comput. Chem.*, 2004, **25**, 1605–1612.
81. M.A. Abbani, C.V. Papagiannis, M.D. Sam, D. Cascio, R.C. Johnson and R.T. Clubb, *Proc. Natl. Acad. Sci. U.S.A.*, 2007, **104**, 2109–2114.
82. M. Steiniger-White, A. Bhasin, S. Lovell, I. Rayment and W.S. Reznikoff, *J. Mol. Biol.*, 2002, **322**, 971–982.
83. K. Swaminathan, P. Flynn, R.J. Reece and R. Marmorstein, *Nat. Struct. Biol.*, 1997, **4**, 751–759.
84. M.K. Safo, T.P. Ko, F.N. Musayev, Q. Zhao, A.H. Wang and G.L. Archer, *Acta Crystallogr. Sect. F: Struct. Biol. Cryst. Commun.*, 2006, **62**, 320–324.
85. C.M. Moure, F.S. Gimble and F.A. Quiocho, *J. Mol. Biol.*, 2003, **334**, 685–695.
86. B.W. Shen, M. Landthaler, D.A. Shub and B.L. Stoddard, *J. Mol. Biol.*, 2004, **342**, 43–56.
87. W. Yang and T.A. Steitz, *Cell*, 1995, **82**, 193–207.
88. S. Lone, S.A. Townson, S.N. Uljon, R.E. Johnson, A. Brahma, D.T. Nair, S. Prakash, L. Prakash and A.K. Aggarwal, *Mol. Cell*, 2007, **25**, 601–614.
89. G.A. Patikoglou, J.L. Kim, L. Sun, S.H. Yang, T. Kodadek and S.K. Burley, *Genes Dev.*, 1999, **13**, 3217–3230.
90. J.C. Fromme, A. Banerjee and G.L. Verdine, *Curr. Opin. Struct. Biol.*, 2004, **14**, 43–49.
91. H. Robinson, Y.G. Gao, B.S. McCrary, S.P. Edmondson, J.W. Shriver and A.H. Wang, *Nature*, 1998, **392**, 202–205.
92. M.A. Schumacher, K.Y. Choi, H. Zalkin and R.G. Brennan, *Science*, 1994, **266**, 763–770.
93. T.D. Schneider and R.M. Stephens, *Nucleic Acids Res.*, 1990, **18**, 6097–6100.

CHAPTER 5
Single-stranded Nucleic Acid (SSNA)-binding Proteins

MARTIN P. HORVATH

Biology Department, University of Utah, 257 S 1400 E, Salt Lake City UT 84112-0840, USA

5.1 Introduction[†]

Single-stranded nucleic acid is encountered at critical points in DNA and RNA metabolism, including DNA replication, DNA repair, transcription, mRNA maturation, translation, and nucleoprotein complex assembly. Many of these processes require access to the information stored within base sequences, yet the single-stranded form that provides such access is liable to unproductive and deleterious side-reactions resulting from nuclease attack, chemical modification such as depurination and deamination, inappropriate recombination, formation of intra-strand secondary structures or premature reannealing with complementary strands. Proteins that bind and stabilize the single-stranded form of DNA and RNA have evolved to ensure that these side-reactions are avoided and that DNA and RNA processes stay on-track. These single-stranded nucleic acid (SSNA)-binding factors thus act as essential chaperones without which catalytic events related to information processing would be impossible or fraught with risk to the genome and proper cell function.

SSNA-binding proteins have been studied from the very beginning of the molecular biology revolution using genetic and biochemical methods. More

---

[†] The following abbreviations are used in this chapter: SSNA, single-stranded nucleic acid; SSB, single-stranded DNA-binding protein; RPA, replication protein A; TEBP, telomere end-binding protein; Pot, protection of telomere.

recently, structure determination by X-ray crystallography and NMR has provided detailed molecular pictures for SSNA-binding proteins and nucleoprotein complexes formed with DNA and RNA. These structures speak to diverse mechanisms for nucleic acid recognition and often reveal underlying evolutionary relationships not previously apparent. Portions of this field have been reviewed as separate, more focused topics.[1-7] This chapter provides an introduction and broad overview to the field of SSNA-binding protein structural biology so as to compare recognition strategies for both DNA and RNA-specific proteins.

At first glance, the numerous details particular to each single-stranded RNA or DNA-protein structure give the impression that every example is unique and specialized. Common themes do emerge, however. For example, constellations of atomic interactions at nucleic acid-protein interfaces can be deconstructed as a combinatorial application of hydrogen bonds, aromatic stacks, van der Waals contacts, and electrostatic interactions. These interaction types are commonly encountered in all macromolecular interaction systems, including double-stranded nucleic acid-protein complexes; however, the extended open conformations adopted by single-stranded RNA and DNA allow for substantially richer interfaces with highly intricate interaction networks. Another theme by which to approach SSNA-binding proteins relates to protein structure architecture. A few folding motifs are represented in various forms among these diverse proteins, and for some classes of SSNA-binding proteins, a protein fold defines the function and reflects the evolutionary history of family members. By way of overview, the chapter begins with a description of these basic elements, interaction types and protein folds, for recognition of single-stranded nucleic acid.

Real-life examples illustrating these basic elements can be divided into two general groups, which I will call "modular" and "eclectic". The modular proteins are sequence-specific, and the eclectic proteins come in both sequence-specific and sequence-non-specific varieties. Members of the modular group, which include Pumilio and Sm proteins, use repeated structural motifs that each recognize individual nucleotides or short repeated sequences in a stereotypical manner. Members of the eclectic group, which include telomere end-binding proteins (TEBPs) and hnRNP K-homology (KH) proteins, also make use of multiple motifs; however, in these cases the motifs are not symmetrically arranged or necessarily equivalent and segments of nucleic acid are recognized in a seemingly haphazard manner without consistently repeating patterns. The repeated forms seen for modular structures provide encouragement for anyone hoping to design novel RNA and DNA-binding proteins with particular sequence-specificities, and such examples are emerging.[8-11] The eclectic structures, by contrast, make it clear that evolution can also put together proteins with complicated recognition properties in ways that would be difficult to rationally re-design.

Following this overview of basic elements, complex emergent properties are discussed. These include properties readily apparent from structural models, such as recognition specificity, and less obvious traits, such as degeneracy, cooperativity and allostery. To develop accurate mechanistic models for how

SSNA-binding proteins work inside cells, we will need to understand how molecular structure relates to these emergent properties that likely drive the biological functions of SSNA-binding proteins.

5.2 Basic Elements

5.2.1 Interaction Types

Reviews of nucleic acid-protein interactions consistently emphasize the importance of electrostatics and hydrogen-bonding networks for understanding molecular recognition. Positive electrostatic potential is generally associated with single-stranded DNA and RNA-binding surfaces. SSNA-binding proteins substantially augment the interaction repertoire with stacking and steric packing interactions. The more extensive interaction networks typical for SSNA-protein interfaces are possible (and likely required) because of the less constrained nature of single-stranded nucleic acid. Whereas double-stranded helical DNA generally retains base-pairing and base-stacking structure, thermodynamic characterizations of SSNA-protein association reactions indicate complex formation is coupled with base unstacking and "melting" of secondary structure.[12–16] The enthalpy cost for breaking these intramolecular interactions is offset by the formation of new intermolecular interactions. The resulting interaction surfaces typically bury 1000–2000 Å^2, and involve hydrophobic, aromatic, charged and polar groups. The following section reviews the four main interaction types seen at SSNA-protein interfaces: (1) salt bridges, (2) stacking of aromatic groups, (3) steric packing, and (4) hydrogen bonds.

The prominence of particular interaction types correlates with whether a protein recognizes single-stranded nucleic acid in a sequence-specific manner. Sequence-non-specific binding proteins rely heavily on electrostatics and stacking interactions that recognize charged phosphate groups and planar heteroaromatic bases common to all sequences. Sequence-specific binding proteins construct interfaces with a high degree of steric complementarity, including rich networks of hydrogen bonds to the exclusion of close-range phosphate interactions, in some cases.

5.2.1.1 Salt Bridges and Electrostatics

Phosphodiester linkages carry a formal negative charge at physiological pH and many SSNA-interaction surfaces provide positively charged lysine and arginine side chains to encourage association. The degree to which electrostatics drive DNA or RNA-binding can be assessed by measuring the binding constant ($K_{A\text{-DNA}}$) over a range of salt concentrations since salt ions shield Coulombic interactions yet have a less pronounced effect on other interaction types. For *E. coli* single-stranded DNA-binding (SSB) protein, much of the binding free energy is electrostatically generated,[17,18] and complexes undergo substantial transitions and eventually dissociate as a function of salt concentration,[19] consistent with the observation of lysine-phosphate and arginine-phosphate salt bridges in the co-crystal structure.[20] Care must be taken, however, in interpretation of salt effects on K_A and correlation with structural information. Coulombic forces extend over considerable distances so that electrostatic contributions can arise even for groups not in direct contact with each other. Detailed analysis of salt type and concentration has shown that single-stranded DNA-SSB association is coupled to dissociation of ions, especially anions, at specific sites on the protein,[17,21,22] meaning that, in addition to general screening effects, binding linkage cycles also apply to salt-dependent SSNA-binding behaviour.[21,23]

Sequence-specific proteins, such as those that bind with single-stranded DNA found at the ends of telomeres, rely less on electrostatic interactions. The first telomere end-binding protein (TEBP) to be isolated and characterized was initially fractionated and purified from the ciliated protozoan *Sterkiella nova* (formerly *Oxytricha nova*) on the basis of forming a salt-resistant DNA complex that remained intact even at 1 M sodium chloride.[24–26] A crystal structure of this protein in complex with telomere-derived single-stranded DNA showed a few lysine and arginine-mediated salt bridges, but most phosphate groups were unpaired and solvent exposed.[27,28] Co-crystal structures of the single-stranded DNA-binding domains of TEBP homologues from human and fission yeast support the idea that evolution has applied positive selective pressure to minimize salt bridges for adaptation of SSB-like proteins for sequence-specific recognition of telomere DNA.[29,30] Salt-dependence of $K_{A\text{-DNA}}$ for DNA-binding portions of the fission yeast and protist proteins attribute about 15% or less of binding free energy to electrostatic sources.[16,31,32] Scanning alanine mutagenesis applied to Cdc13, the single-stranded telomere DNA-binding protein from budding yeast, showed that many of the lysine or arginine residues could be removed with little effect on DNA-binding *in vitro* or cellular performance *in vivo*,[33,34] consistent with the idea that sequence-specific recognition of single-stranded DNA does not strongly depend on direct phosphate interactions with any one lysine or arginine residue.

Similar trends have been noted for single-stranded RNA-binding proteins. The complex formed by RNA-binding proteins MRP1 and MRP2 with guide RNA (gRNA) is highly salt-sensitive and dissociates at concentrations of sodium chloride greater than 0.2 M.[35] The crystal co-complex structure shows use of arginine/lysine-phosphate salt bridges and only a few stacking or hydrogen-bonding interactions with bases of the gRNA,[35] suggesting that this protein functions to facilitate downstream annealing reactions, but lacks any sequence recognition ability. By contrast, some highly sequence-specific RNA-protein systems show that lysine and arginine side chains are employed to construct base-specific hydrogen-bonding networks or base-stacking interactions, not salt bridges, as is the case for the tryptophan and RNA-binding attenuation protein (TRAP) and its complex with mRNA.[36] The RNA-binding Puf domain derived from human Pumilio 1 also recognizes mRNA in a sequence-specific manner, makes no salt bridges, and reserves arginine residues for stacking between bases of the mRNA.[8] Most single-stranded RNA-binding proteins fall somewhere between these extremes and complement a few arginine or lysine-mediated phosphate-salt bridges with an ensemble of other interaction types.

5.2.1.2 Stacking Interactions

Stacking interactions formed between aromatic amino-acid residues and heteroaromatic bases are especially prominent in single-stranded nucleic acid-protein complexes. First evident in tRNA anticodon loop recognition by amino-acyl tRNA synthetases,[37] face-to-face contacts between aromatic amino-acid residues and nucleotide bases are now generally expected for recognition of extended single-stranded nucleic acids in both sequence-non-specific and sequence-specific contexts. Human replication protein A (RPA) and *E. coli* SSB, two proteins with little sequence preference, employ phenylalanine, tyrosine and tryptophan residues to make several such base-stacking interactions.[20,38] The single-stranded telomere DNA-binding factors, which are sequence-specific, also employ stacking interactions.[27,29,30,39] Stacking of planar (sp^2), resonance-stabilized chemical groups with nucleotide bases often involves non-canonical groups such as main-chain peptide linkages and the side chains of asparagine and arginine residues.[40] Arginine-base stacking is a type of cation–π interaction,[41–43] whereby favorable van der Waals contacts between the planar groups are additionally supported by favorable electrostatic contributions between the positively charged guanidinium group of arginine and the partial negative charge associated with the face of heteroaromatic bases.

Arginine residues are frequently numerous at SSNA-protein interfaces,[44] probably because these residues have several chemical characters suitable for interaction with single-stranded DNA. The planar nature of this group makes it able to stack with planar base groups. The positively charged guanidinium group establishes a favorable electrostatic field for attracting negatively charged polyanions and can form a hydrogen-bonded salt bridge when placed in close proximity with phosphate groups. The multivalent hydrogen-bonding capacity of arginine residues allows for hydrogen bonds either with base groups or with other protein moieties. In certain settings, all of these chemical characters can be operational at the same time such that the arginine side-chain simultaneously makes salt-bridging, base-stacking, and hydrogen-bonding interactions.

5.2.1.3 Steric Packing and van der Waals Interactions

"Stacking" has been applied rather generously in the field as a description for face-to-face approximation of planar aromatic groups. The degree of overlap, inclination angle and distance between such groups varies considerably and does not always correspond with completely overlapping π-orbital systems. It may be appropriate to think of these interactions as favored by close van der Waals contact and the hydrophobic effect. Hydrophobic residues such as leucine, isoleucine and valine, as well as the aliphatic portions of otherwise polar residues such as lysine, arginine and glutamine play a similar space filling or steric packing role. For example, the splicing factor SF1 uses leucine, valine, and isoleucine to line hydrophobic pockets surrounding the bases of intron branch site RNA sequences.[45] The penultimate nucleotide of telomere single-stranded DNA fits into a similarly constructed pocket close to the surface of the telomere end-binding protein (TEBP)-α subunit.[27] The prevalence of hydrophobic interactions for recognition of telomere DNA makes comparison with the hydrophobic cores of folded proteins useful.[7] As is widely appreciated for protein folding and protein-protein association reactions,[46,47] removal of solvent from these hydrophobic groups contributes a large driving force favoring single-stranded nucleic acid-protein association.

Close packing of hydrophobic groups plays both direct and supportive roles in establishing recognition specificity. In its direct role, steric packing can exclude groups such as the 2′ hydroxyl for deoxyribose specific recognition or the C5 methyl group for uridine *versus* thymidine base discrimination. Steric packing interactions can also play a supportive role in molecular recognition by exacting a penalty for mismatched hydrogen-bonding groups. Enforcement of specificity by close packing is familiar in the context of DNA polymerase-catalyzed nucleotidyl transfer reactions. The Watson–Crick base pairs are nearly isoenergetic with mismatched alternatives. Higher than expected fidelity is nonetheless achievable since correct pairing is examined by close packing interactions at the polymerase active site that do not permit catalysis if steric clashes are encountered.[48,49] Analogous enforcement by steric packing interactions likely contributes to recognition of single-stranded nucleic acid by hydrogen bonding interactions.

5.2.1.4 Hydrogen Bonding

Hydrogen bonds depend strongly on geometry and require a particular match of donor and acceptor groups. As a consequence, hydrogen bonds are vector-like and establish more exacting recognition specificities than the other three interaction types. SSNA-binding proteins incorporate a rich and varied assortment of hydrogen bonds for recognition of phosphate, sugar and base groups. The sequence-specific proteins use polar residues, especially asparagine, aspartate, glutamine, glutamate, and arginine residues, to closely examine the Watson–Crick edges of bases in single-stranded nucleic acid ligands. Many of these interactions resemble authentic Watson–Crick base pairs to some degree (Figure 5.1). Carboxylate-bearing glutamate and aspartate residues mimic cytosine by making two hydrogen bonds with N1 and N2 hydrogen-bond donor groups of G bases. Glutamate-G pairing was seen in the modular recognition of mRNA by Pumilio's Puf domain,[50] and both aspartate-G and glutamate-G pairs were seen in a TEBP-telomere DNA complex (Figure 5.1a).[27] Asparagine and glutamine residues are G, A, C or U/T mimics, depending on amide orientation. Through collaborating interactions with other hydrogen-bonding groups one rotamer can be locked into place, enabling specific recognition of C, U/T, G or A with a pair of hydrogen bonds to matching hydrogen-bond donor and acceptor groups (Figure 5.1b). Glutamine-U and asparagine-U pairs were seen in eubacterial Hfq-RNA[51] and archaeal Sm protein-RNA complexes,[52] respectively. An asparagine-G pair was seen for the DNA-binding domain of human Pot1 (protection of telomere-1 protein) in complex with oligos derived from human telomere repeats.[30] The repeating pattern of glutamate-G and asparagine-U pairs

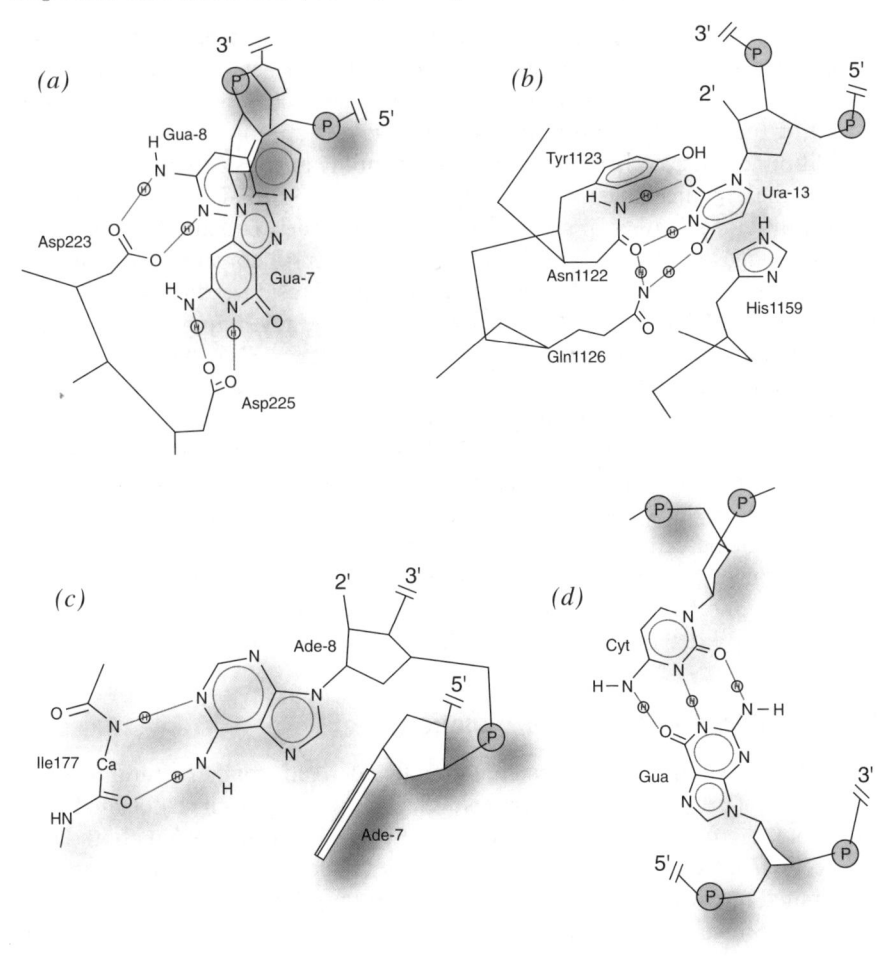

Figure 5.1 Base pair mimicry at SSNA-protein interfaces. (*a*) The sketch shows two aspartate-guanine pairs seen at the single-stranded DNA-TEBP interface (pdb id 2i0q).[28] Each aspartate-guanine pair matches hydrogen-bond donor N1 and N2 groups of the base with hydrogen-bond acceptors of the carboxylate group. (*b*) Hydrogen-bonding and stacking interactions recognize a uracil base at the RNA-pumilio Puf domain interface (pdb id 1m8y).[8] (*c*) Recognition of the branch point adenosine of intron RNA involves hydrogen bonds with main-chain NH and C=O groups of SF1 (pdb id 1k1g).[45] (*d*) For comparison, a C:G base pair is shown as found in Dickerson's dodecamer structure (pdb id 2bna).[192,193]

in the Pumilio Puf domain-RNA complex suggests that a SSNA-recognition code exists,[8] and variant Puf domain proteins with designed sequence-specificities have been successfully engineered with use of this recognition code.[8–11]

The protein code for recognizing single-stranded nucleic acid is generally more complicated than simple one-to-one amino-acid-to-base matching. Even for designed Puf-domain variants, which consistently bind best with intended cognate RNA, discrimination against non-cognate sequences is not quite on a par with the selection performance seen for the wild-type protein domain, indicating that adaptation for sequence-specific recognition involves other sites in addition to those contributing hydrogen-bonding groups.[9] In our attempts to unravel the chemical basis for recognition in the TEBP-DNA complex, we observe differing degrees of specificity even for similarly constructed aspartate-G pairs, suggesting that specificity results from several contributing elements, including but not completely determined by those residues directly hydrogen bonding to the Watson–Crick base edges (Catherine Dy, unpublished results).

Recognition strategies, especially those of the eclectic SSNA-binding proteins, also make use of main-chain carbonyl (C=O) and amide groups (N–H) of the protein for intermolecular hydrogen bonding.[40] Branch site sequences derived from intron RNA in complex with the KH/QUA2 domain of splicing factor SF1 place the chemically reactive branch site-A in a deeply buried pocket lined with hydrophobic residues.[45] Hydrogen bond acceptor (N1) and donor (N6) groups of this adenine base are matched with the amide and carbonyl groups of a residue located at the edge of a beta-sheet.[45] Purine recognition by main-chain groups at this position in other KH folds is conserved, highlighting the fact that the protein fold is an active determinant of molecular recognition.[53]

5.2.2 Folds, Evolution and Function

Molecular recognition properties are ultimately derived from structural features associated with a particular protein fold. Several folding motifs associated with SSNA-binding function are diagrammed in Figure 5.2. These include both large and deeply rooted protein families such as the OB-fold[6,54,55] as well as proteins with limited distribution such as TRAP, which is only found in *Bacillus* species of bacteria.[56] Single-stranded nucleic acid-binding surfaces (Figure 5.2) frequently involve beta-sheet surfaces and loops connecting beta-strands. The Pumilio/FBF-homology (Puf) fold provides a glaring exception to the generalization that all SSNA-binding surfaces are made with beta-sheet secondary structure. The heterogeneous ribonucleoprotein (hnRNP) K-homology (KH) domain constructs its single-stranded RNA or DNA-binding surface with both beta-sheet and alpha helical elements. The following sections introduce several (but not all[‡]) of the protein folds with SSNA-binding function.

5.2.2.1 OB-fold

The oligonucleotide/oligosaccharide/oligopeptide-binding (OB)-fold defines a large super-family of proteins, a significant subset of which have functions related to recognition of single-stranded nucleic acid (for a review see ref. 6). In the OB-fold, five beta-strands are arranged with a conserved topology (Figure 5.2a).[54] Although the beta-strands often form a closed beta-barrel capped by a connecting

---

[‡] Notable omissions include the Zn-knuckle motif.

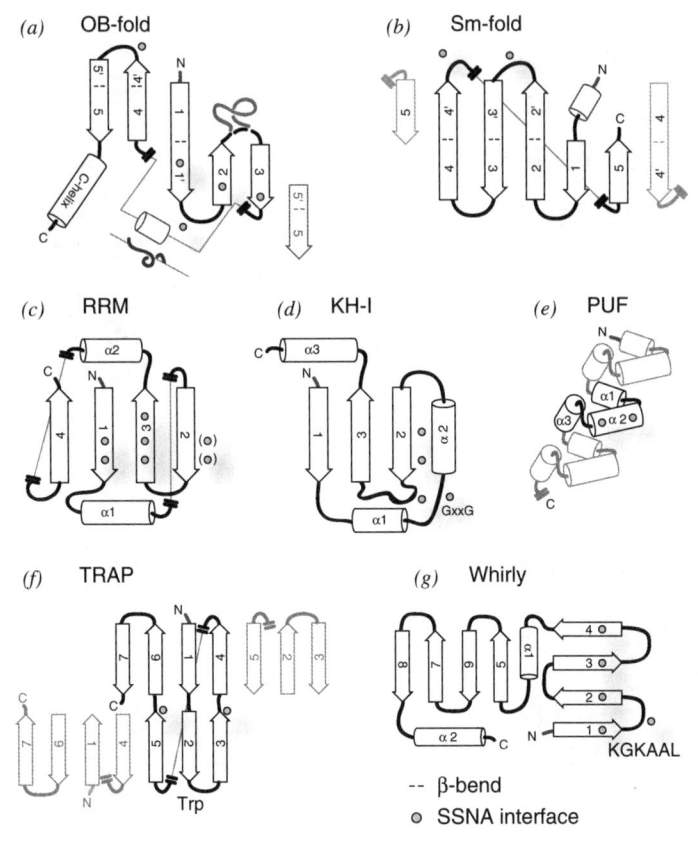

Single-stranded Nucleic Acid (SSNA)-binding Proteins

Figure 5.2 Protein folds for SSNA recognition. Filled circles highlight regions of each protein fold that bind with single-stranded nucleic acid. Thin lines denote connected elements that have been separated to simplify depiction of structure. Dashed lines indicate bends in beta-strands. (*a*) In the OB-fold, closure of the beta-barrel is indicated by duplication of strand s5. In most OB-folds, a short turn connects strands s2 and s3, but for the OB-fold of Cdc13 this region harbors a 29-amino-acid segment that adopts a structured pretzel-like conformation. (*b*) Sm-folds form continuous beta-sheet interactions with neighboring subunits, indicated by repetition of strands s5 and s4 at either edge of the s4:s3:s2:s1:s5 beta-sheet. (*c*) The RNA recognition motif (RRM) typically interacts with RNA with use of a beta-sheet face (strands s1 and s3); however, the RRM found in La contacts RNA along the edge of the beta-sheet (filled circles in parentheses). (*d*) The KH-I domain recognizes RNA with a beta-sheet edge and connecting loops. The Gly-x-x-Gly sequence motif shown here is also found in the topologically distinct KH-II fold (not shown). (*e*) Puf proteins use multiple repeats of a three-helix module to interact with RNA in a sequence-specific manner. (*f*) Edge-to-edge beta-sheet interactions seen for TRAP subunits create seven-stranded all-anti-parallel beta-sheets with RNA-binding loops and tryptophan-binding loops located at distinct sites. (*g*) The Whirly fold comprises two beta-sheets and connecting alpha helices with a Lys-Gly-Lys-Ala-Ala-Lys sequence motif noted in ssNA-binding proteins involved in transcription regulation in plants and kinetoplastid RNA editing in trypanosomes.

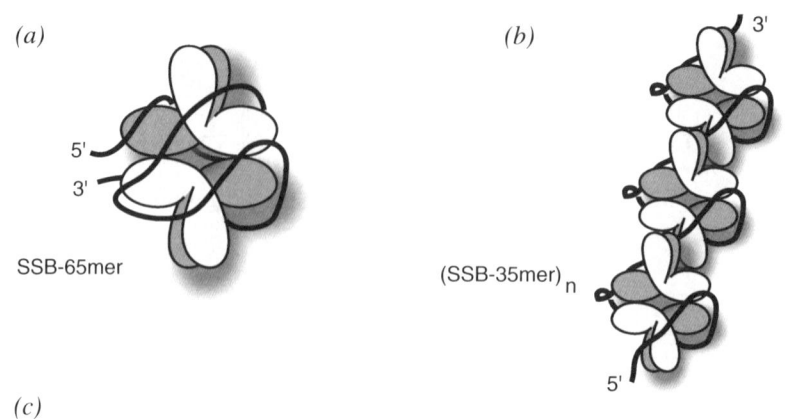

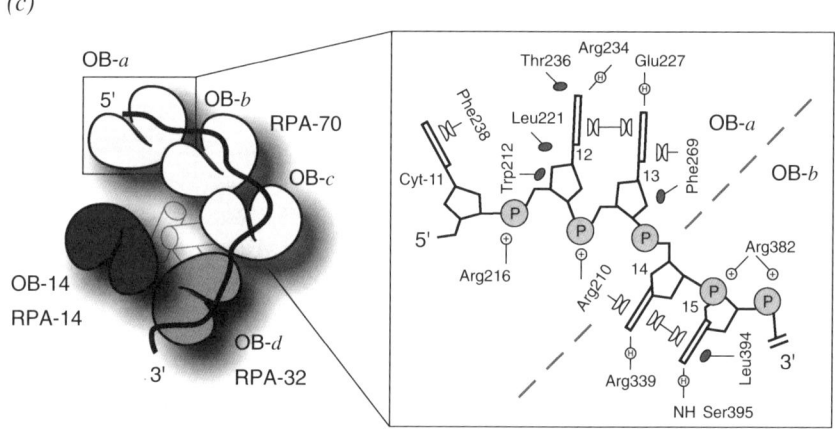

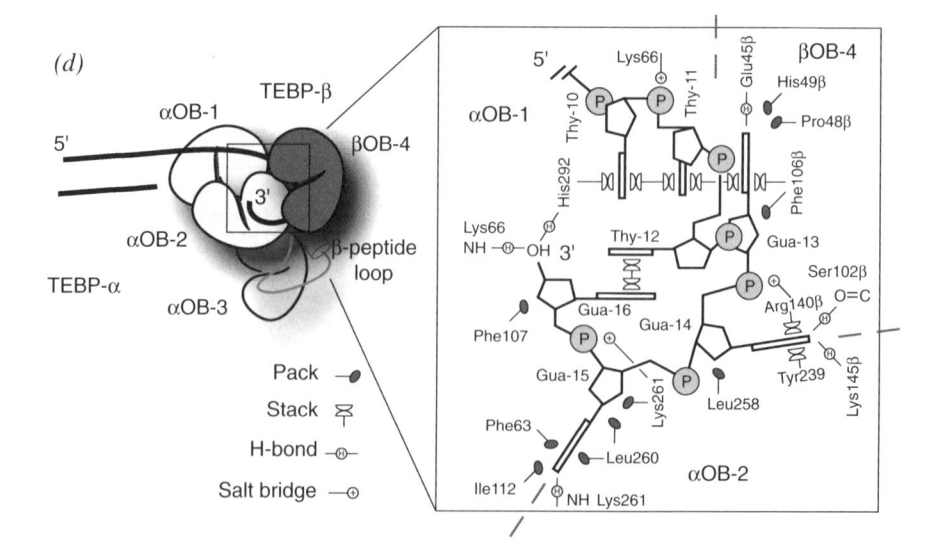

(a)

SSB-65mer

(b)

(SSB-35mer)$_n$

(c)

OB-*a*
OB-*b*
RPA-70
OB-*c*
OB-14
RPA-14
OB-*d*
RPA-32

Thr236 Arg234 Glu227
Phe238
Leu221
OB-*a*
Cyt-11 Trp212
12
Phe269 OB-*b*
5'
Arg216
13
Arg382
14
Arg210
15
Arg339
Leu394
NH Ser395
3'

(d)

TEBP-β
αOB-1
βOB-4
5'
3'
αOB-2
β-peptide loop
TEBP-α
αOB-3

Pack —●
Stack —⋈
H-bond —⊕
Salt bridge —⊖

Lys66 βOB-4
5'
Thy-10 Thy-11 Glu45β
His49β
αOB-1
Pro48β
His292
Phe106β
Lys66 Thy-12 Gua-13
NH—OH 3'
Ser102β
Phe107 Gua-16 Gua-14 Arg140β O=C
Gua-15
Tyr239
Lys145β
Phe63 Lys261 Leu258
Ile112 Leu260
NH Lys261
αOB-2

helix, variations are abundant with beta-strand extensions and differently sized loops common. These structural variations, compounded with lack of a conserved consensus sequence, make identification of OB-folds uncertain except through structure determination. The OB-fold is the principle building block for construction of single-stranded DNA-binding (SSB) proteins (Figure 5.3a), including those from bacteriophage,[57–59] eubacteria,[60] and mitochondria.[61,62] Replication protein A (RPA), the eukaryotic SSB conserved from yeast to humans, has at least six OB-folds arranged asymmetrically among three unequal subunits (Figure 5.3c).[38,63] Telomere end-binding proteins (TEBPs), which are specialized RPA-like complexes that cap linear chromosomes in *Sterkiella nova* and related protists, are constructed from four OB-folds unequally distributed between two subunits (Figure 5.3d).[27]

Evolutionary relationships among single-stranded DNA-binding proteins are reflected in the roles played by SSB, RPA and TEBP. SSB and RPA are essential components required for DNA-replication and repair reactions in eubacteria/mitochondria and eukarya, respectively (reviewed in refs. 64–66). SSB in most eubacteria and mitochondria is a stable tetramer of equivalent subunits.[60–62,67,68] Gene duplication has created an alternate, less-symmetrical OB-fold organization for SSB in *Deinococcus radiodurans* and *Thermus aquaticus*; however, four OB folds with pseudo D_2 symmetry are nonetheless preserved for these two-subunit/two-OB-folds-per-subunit proteins.[69–71] Biochemical studies of *E. coli* SSB indicate multiple DNA-binding modes with

Figure 5.3 OB-folds in eubacterial and eukaryotic single-stranded DNA-binding proteins. (*a*) Single-stranded DNA wraps around an individual *E. coli* SSB tetramer in a model for the $n = 65 \pm 5$, "limited" cooperativity binding mode. (*b*) SSB tetramers are positioned in close proximity along single-stranded DNA in a model for the $n = 35 \pm 3$, "unlimited" cooperativity binding mode. These models were each derived by application of symmetry operations to [extensive] DNA fragments seen in a co-crystal structure (pdb id 1eyg).[20] (*c*) A model of human RPA's three subunits, RPA-70, RPA-32, and RPA14, was derived from two separate crystal structures: that of the major single-stranded DNA-binding domain consisting of OB-*a* and OB-*b* in complex with d(C)$_8$ (pdb id 1jmc),[38] and that of the trimerization core consisting of OB-*c*, OB-*d* and OB-14 (pdb id 1l1o).[63] The collage sketch shown here reconstructs how ~30 nucleotides of single-stranded DNA might bind with four of the OB-folds. Inset: partial contact map of DNA-protein interactions within the major DNA-binding domain of RPA. Interaction types are indicated by differently shaped pointers with ovals for van der Waals contacts, balanced half-circles for stacking contacts, circled *H*s for hydrogen-bonds, and circled + symbols for salt bridges. (*d*) Structure of the telomere DNA-TEBP complex shows the 3′-terminal single-strand DNA bound with α and β subunits of TEBP from *Sterkiella nova* [pdb id 1otc (2.8 Å), 2i0q (1.91 Å)].[27,28] The protein-protein interface is constructed from the C-terminal αOB-3 from α and an alpha helix/peptide loop contributed by β. Double-stranded DNA is shown here to highlight the biological chromosome-capping function for TEBP-α/β but was not included with co-crystals. Inset: DNA-protein interactions for nucleotides 10–16 positioned at the so-called "α/β"-binding binding site where three OB-folds converge.

different degrees of binding cooperativity.[19] The crystal structure of *E. coli* SSB in complex with d(C)$_{35}$ showed how DNA wraps around protein tetramers and allowed construction of molecular models for each of two binding modes characterized by "limited" and "unlimited" cooperativity (Figure 5.3a and b).[20] Another OB-fold protein in *E. coli*, PriB, binds single-stranded DNA cooperatively, and participates in the multi-enzyme replication restart primosome complex to rescue replication forks following DNA-damage.[72,73] PriB forms stable homodimers,[74] not tetramers, and binds single-stranded DNA without the pronounced wrapping observed in the SSB-single-stranded DNA complex.[75]

RPA was first described as a factor required for cell-free SV40 DNA replication reactions.[76–79] Based on analogous *in vivo* function, biochemical behavior, sequence conservation and structural similarity, RPA is considered to be derived from SSB by way of gene duplication and diversification.[80] The co-crystal structure of a DNA-complex showed how single-stranded DNA interacts with two tandemly arrayed OB-folds found within the major DNA-binding domain of RPA.[38] Comparison with uncomplexed structures of the RPA single-stranded DNA-binding domain indicated that relative positions of the two OB-folds are dictated by DNA,[81] consistent with the idea that DNA-dependent structural changes in RPA facilitate downstream events in the processes of DNA replication and repair (reviewed in ref. 82).

The large subunit of RPA (RPA70) recruits additional subunits (RPA32 and RPA14) to single-stranded DNA, and the crystal structure of RPA-derived protein fragments showed how three OB-folds interface to form the trimerization core.[63] Figure 5.3(c) provides a collage view of RPA as it might assemble with single-stranded DNA. The model is derived from insights provided by both the major single-stranded DNA-binding domain co-crystal structure[38] and the RPA trimerization core structure,[63] and attempts to convey a sense of the asymmetrical disposition of subunits and DNA-binding OB-folds of RPA which contrasts with the high degree of symmetry seen for SSB.

A simpler monomeric, single-OB-fold RPA is found in the crenarchaea *Sulfolobus solfataricus*. Although this single-stranded DNA-binding protein has a domain structure similar to the eubacterial SSB protomer, a crystal structure revealed striking features shared by eukaryotic OB-folds from RPA and telomere end-binding proteins.[83] Notable among these features are intra-subunit interactions mediated by a D-x-(T/S/Y) motif that is not found in eubacterial SSB proteins but is conserved in eukaryotic RPA-associated OB-folds and in the OB-folds from TEBP, Pot1 and Cdc13.[83] The intra-subunit interactions mediated by the D-x-(T/S/Y) motif preclude homo-oligomerization, suggesting that this sequence motif was fixed during evolution in going from a symmetrical (D_2) homo-tetramer characteristic of SSB to create the asymmetrical RPA and TEBP heteromers extant in eukaryotes.

SSB and RPA bind and stabilize single-stranded DNA with only modest sequence preferences. Further specialization could have adapted an RPA-like precursor for sequence and structure-specific recognition of 3'-terminal single-stranded DNA found at telomere ends of most eukaryotes. Single-stranded DNA at telomere ends is a necessary substrate for telomerase-mediated extension of telomere DNA, yet would be highly recombinogenic because of its resemblance with broken DNA strands if not handled with special precautions. Specialized nucleoprotein complexes identify authentic and functional telomeres, stabilize the T/G-rich single-stranded DNA, and suppress DNA-damage responses.[84] The crystal structure for such a telomere-specific nucleoprotein complex from *Sterkiella nova* (formerly *Oxytricha nova*) shows how four OB-folds contained within α and β telomere end-binding protein (TEBP) subunits mediate single-stranded DNA protection and protein association (Figure 5.3d).[27,28]

As supported by analogous structures, functional behavior, and recently described evolutionary relationships, RPA-derived telomere-capping structures are widely distributed in eukarya (reviewed in ref. 85). Reconstructions of OB-fold evolution place RPA and telomere related branches deeply rooted with the SSB family.[55] Telomere protective single-stranded DNA-binding proteins called Pot1 are found in fission yeast, humans and plants.[86,87] The Pot1 proteins share sequence similarity with the N-terminal DNA-binding domain of TEBPs found in protists such as *Sterkiella nova* and *Moneuplotes crassus*. Structure determination of the DNA-binding domains of Pot1 from fission yeast and humans confirmed that telomere single-stranded DNA recognition is achieved by means of OB-fold units.[29,30] A human Pot1-interacting protein, Tpp1, functions cooperatively with Pot1 to strongly promote processive telomerase-catalyzed telomere extension reactions *in vitro*,[88] analogous to the activity of RPA in cell-free DNA replication reactions.[76–78] Sequence profiling and the recently determined structure of Tpp1 revealed a suggestive resemblance with TEBP-β, the second subunit of the TEBP-α/β heterodimer in protozoa,[88,89] indicating heteromeric, RPA-like telomere complexes are conserved in humans and protists.

A parallel story is unfolding for budding yeast (*Saccharomyces cerevisiae*) telomere end protection. In budding yeast, Cdc13 functions to stabilize telomere ends and keeps these ends from eliciting inappropriate DNA-damage responses.[90,91] The minimal DNA-binding domain of Cdc13 is made up of a single OB-fold,[92] with little or no sequence resemblance to Pot1 proteins or TEBP subunits. Nevertheless, recently described results show that Cdc13 is part of a heteromeric complex that resembles RPA[93] and by analogy TEBP–α/β and Pot1/Tpp1. Cdc13-interacting proteins Stn1 and Ten1[94,95] have weak but measurable single-stranded DNA-binding activity.[93] Furthermore, the predicted OB-fold contained within Stn1 can substitute for the OB-fold contained within the medium-sized RPA subunit of budding yeast.[93] These results, along with bioinformatics arguments, strongly suggest that a complex in *S. cerevisiae* consisting of Cdc13, Stn1 and Ten1 functions as an RPA-like assembly dedicated to telomere-related functions.[93]

5.2.2.2 Sm-fold

Sm and Sm-like (Lsm) proteins define another large family with members in eubacteria, archaea and eukarya, functioning in many aspects of RNA metabolism, including small nuclear ribonucleoprotein (snRNP) biogenesis and splicing (see refs. 4 and 96 for reviews). The Sm-fold consists of an N-terminal alpha helix followed by five beta-strands (Figure 5.2b).[97] Beta-bends in three of the strands allow portions of these to curve over the larger five-stranded sheet, forming a clam shell-like structure around a hydrophobic core. Two recognizable amino-acid sequence motifs[98,99] correspond with s1–s3 and s4–s5 portions of the Sm fold.[97]

In eukaryotes, Sm and Lsm proteins comprise a family of about 30 different gene products that oligomerize to form specific heteromeric dimers, trimers, pentamers and heptamers. Sm heptamers (Sm$_7$) assemble with Sm-sites, r(AAU UUGUGG), found in U1, U2, U4 and U5 snRNAs.[100] An Lsm hetero-heptamer (Lsm$_7$) specifically recognizes U6,[101,102] the only RNA polymerase III transcript of the U snRNA family. Loops connecting strands s2 and s3' and strands s4' and s5 within Sm subunits make contact with RNA as judged by crosslinking experiments.[103] Toroidal structures with single-stranded RNA-binding loops pointed toward the center of the ring were predicted for Sm$_7$ and Lsm$_7$ heteromers on the basis of crystal structures of stable heterodimeric sub-complexes.[97] The predicted Sm$_7$ ring was confirmed in 10 Å-resolution EM-reconstructions of the U1-specific small nuclear ribonucleoparticle.[104]

Conserved amino-acid sequence motifs characteristic of Sm and Lsm proteins in eukarya are also present in orthologs found in archaea.[98,99] The archaeal proteins form homo-heptameric[105,106] or homo-hexameric[107] ring structures, indicating that the Sm$_7$ and Lsm$_7$ quaternary structures of eukaryotes are derived from a simpler but highly homologous form through gene duplication and diversification.[108] Crystal structures of Sm1 proteins from *Archaeoglobus fulgidus*[52] and *Pyrococcus abyssi*[109] in complex with r(U)-rich single-stranded oligos revealed base-specific binding pockets lining the inner rim of Sm$_7$ rings (Figure 5.4a). These structures of the archaeal Sm proteins in complex with RNA provide a structural framework for understanding the RNA-binding properties of eukaryotic homologues, for which high-resolution RNA co-complexed structures are currently unavailable. Immunoprecipitation experiments using antibodies directed to Sm1 and Sm2 in *Archaeoglobus fulgidus* showed that these proteins associate with RNase P RNA, suggesting a role in tRNA maturation.[52] Conservation of Sm fold, heptamer ring architecture, and RNA-binding activity within archaea and eukarya leads to the conclusion that ancestral Sm proteins provided a critical RNA-related function.

The eubacterial member of the Sm family is Hfq, initially characterized as a bacterial host factor required for replication of the RNA phage Qβ (see ref. 110 for a review of Hfq). In addition to its role in Qβ replication, Hfq facilitates degradation of certain RNAs, efficient translation of other RNAs, and annealing reactions that are responsible for translation regulation by small non-coding RNAs.[111,112] The Hfq-Sm relationship was established on the basis of structure determination and sequence similarity.[51,111] Crystal structures of Hfq show a homo-hexameric ring structure (Figure 5.4b) very similar in architecture to the archaeal homomeric Sm$_7$ rings and the heteromeric rings predicted for eukaryotic Sm$_7$ and Lsm$_7$ proteins.[51,113] The RNA co-crystal complex of Hfq with r(AUUUUUG) single-stranded RNA revealed U-specific pockets lining the inner rim that are analogous with those found in the archaeal Sm

Single-stranded Nucleic Acid (SSNA)-binding Proteins 105

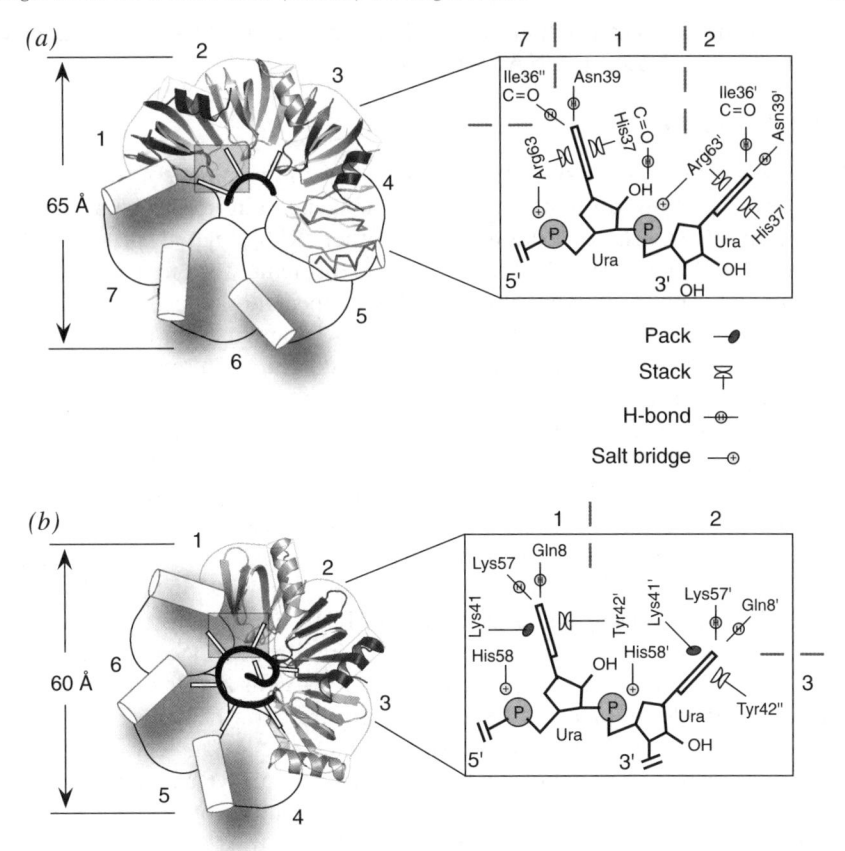

Figure 5.4 RNA complexed with archaeal Sm1 and eubacterial Hfq ring structures. (a) A homomeric seven-subunit ring architecture is shown for archaeal Sm1 protein from *Archaeoglobus fulgidus* with single-stranded RNA located at the central cavity (pdb id 1i5l).[52] A similar RNA complex was also seen for the Sm1 protein from *Pyrococcus abyssi* (pdb id 1m8v).[109] Subunits labeled 1–3 are shown as ribbon representations, subunit 4 is shown as an alpha-carbon trace, subunits 5–7 are shown as simplified shapes. Only three connected nucleotides are depicted. Additional but disconnected U nucleotides are apparent at the remaining binding pockets within subunits 4–7 (not shown). (b) A homo-hexameric Hfq structure in complex with r(AUUUUUG) RNA is shown with three subunits as ribbon representations and three subunits as simple shapes (pdb id 1kq2).[51] The 3′-terminal base points towards the central cavity, but the phosphodiester backbone remains on the surface and does not thread to the other side. Inset panels show details of the U-binding pockets with highly analogous interaction types found at the Sm1 and Hfq RNA interfaces. For these inset panels, subunit boundaries are indicated by dashed lines and prime or double-prime notations distinguish residues of neighboring subunits.

proteins – further evidence that the Hfq/Sm/Lsm ancestor originally functioned in some critical aspect of RNA metabolism.[51]

5.2.2.3 RRM

The RNA recognition motif RRM [also called the ribonucleoprotein (RNP) motif] is found in proteins involved in splicing, alternative splicing, and regulation of translation among many other RNA-related processes (for reviews of RRM see refs. 1, 2 and 5). RRM proteins were first identified on the basis of two conserved amino-acid sequence motifs called RNP1 and RNP2. A crystal structure of the RNA-binding domain from the splicesomal protein U1A showed that the RRM fold consists of an α/β sandwich with two alpha helices packing against one face of a four-stranded anti-parallel beta-sheet with s4:s1:s3:s2 topology (Figure 5.2c).[114] The central two strands, s1 and s3, correspond with RNP2 and RNP1, respectively, and these strands make up a substantial portion of the RNA-binding surface as seen in RNA co-crystal structures of U1A and U2B″/U2A′.[115,116] As is the case for many RRM proteins, these structures showed the RRM complexed with structured and partially duplexed forms of RNA. Nevertheless, examples of the RRM interacting with extended single-stranded forms of RNA merit inclusion of the RRM with SSNA-binding folds. In the case of sex-lethal protein (Sxl), which directs alternative splicing of transformer (*tra*) pre–mRNA for sex-determination and female-specific development, two RRMs each present a beta-sheet to interact with extended RNA in a sequence-specific manner.[117] La, which sequesters the 3′-ends of RNA polymerase III transcripts, uses an RRM augmented with a winged-helix motif, the La motif, for recognition of vicinal hydroxyl groups characteristic of 3′-terminal RNA. In contrast with most RRMs, the RRM in La uses the beta-sheet edge, not face, for binding with RNA (Figure 5.5a).[118]

5.2.2.4 KH

The KH domain was first identified on the basis of amino-acid sequence similarity with heterogeneous nuclear ribonucleoprotein K (hnRNP-K), a character shared by several proteins involved with RNA-related functions.[119,120] The most conserved portion of the KH domain contains a nearly invariant G-x-x-G motif also found in ribosomal protein S3.[119] On the basis of this G-x-x-G motif and overall 36% sequence identity, it had been assumed that S3 and hnRNP K shared a common structure. Comparison of KH domain structures showed, however, that the G-x-x-G sequence motif is embedded within two topologically distinct folds, now called KH type I and KH type II, which may be distantly related by way of a common ancestor.[121] As first seen for NMR structures of KH domains derived from vigilin and FRM1,[122,123] the KH-I fold found in hnRNP-K and several other proteins consists of three helices and a three-stranded anti-parallel beta-sheet with s1:s3:s2 topology (Figure 5.2d). The KH-II fold of ribosomal protein S3 (not shown) also has three helices and a three-stranded beta-sheet, but the beta-sheet is a mixture of

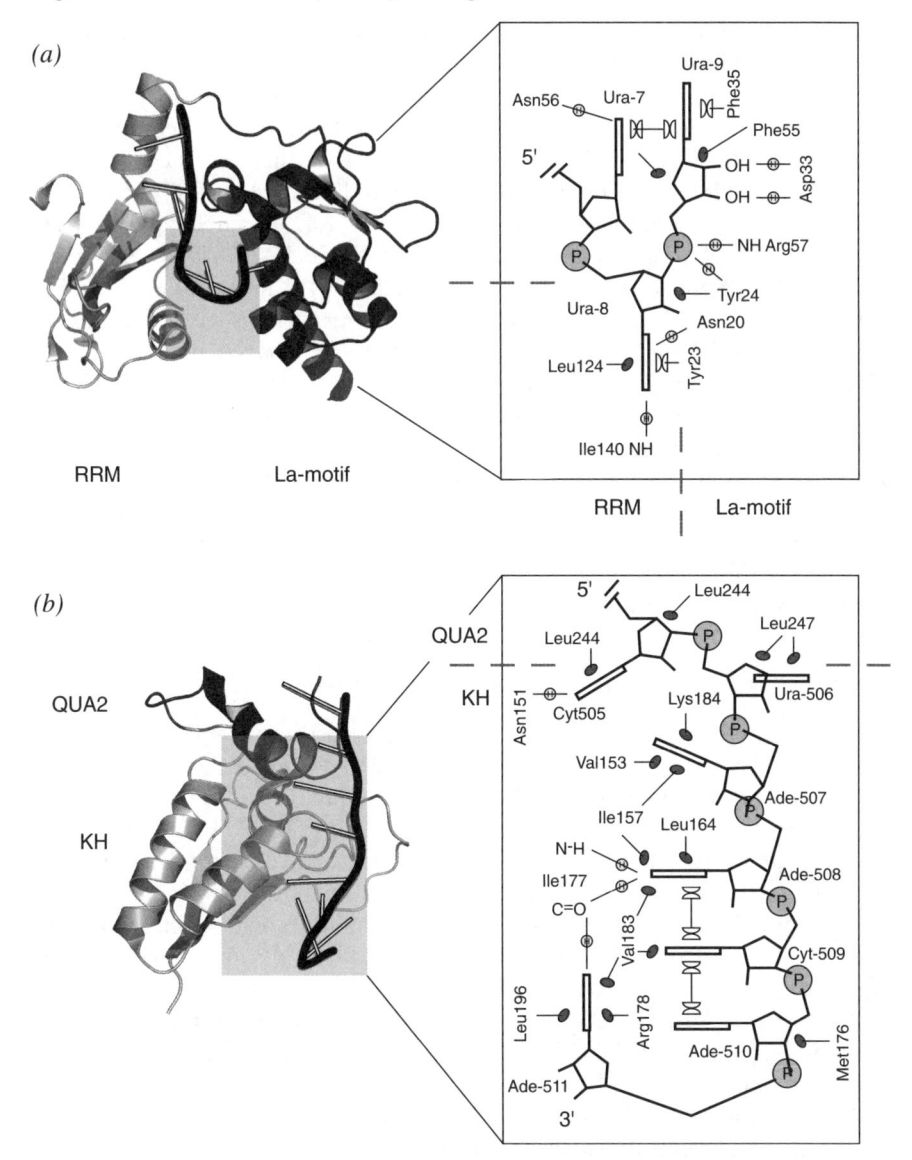

Figure 5.5 RRM and KH domains in complex with single-stranded RNA. (*a*) La protein binds with 3′-termini of polymerase III RNA transcripts with use of an RRM (light gray) and La-motif (dark gray). Inset: details of 3′-OH/2′-OH recognition involving steric complementarity and hydrogen bonding with the carboxylate group of an aspartate residue (Asp33, pdb 1zh5).[194] (*b*) SF1 binds with a single-stranded form of intron branch site RNA, r(UAUACUA-A^{508}-CAA, branch site A italicized), by means of a KH domain (light gray) augmented with a C-terminal QUA2-motif (dark gray). Inset: details of branch point A recognition (see also Figure 5.1c) that includes several packing interactions with hydrophobic residues and hydrogen bonds with main-chain amide and carbonyl groups (pdb 1k1g).[45]

parallel/anti-parallel strands with s1:s2:s3 topology and the intervening helices appear in a different order.[124]

The KH-I domain is a true single-stranded nucleic acid-binding motif (see refs. 2, 125 and 126 for reviews). Structures of KH domains in complex with single-stranded RNA or DNA substrates show that the beta-sheet edge and alpha helix α3 along with connecting loops including the G-x-x-G motif make up the nucleic-acid interface (Figure 5.2d).[45,127,128] KH domains typically work in teams of two or more, as observed for hnRNP-K, FMR1, vigilin, and FBP.[122,127–133] Splicing factor 1 (SF1, also called BBP in yeast) uses a single KH-domain augmented by a C-terminal QUA2 motif for specific recognition of intron branch site sequences (Figure 5.5b).[134–137] RNA-binding for the KH-QUA2 domain derived from SF1 is very weak (apparent $K_{D\text{-RNA}} \approx$ 30 mM).[134] Interactions with RNA are enhanced, to a degree, by non-specific zinc-knuckle/finger motifs found within the yeast BBP,[134] and through cooperative interactions with other *trans* acting splicing factors such as U2AF65.[138] Weak but nonetheless specific recognition of the intron branch site sequence may have evolved so as to allow for regulated alternative splicing and dynamic exchange of this target sequence with snRNP U2, a necessary step for splicesome assembly and catalysis.[136]

5.2.2.5 Others: Pumilio, TRAP and Whirly

To give a sense of diversity in protein folds involved with recognition of single-stranded nucleic acid, Figure 5.2 includes three additional protein folds: the alpha helix-repeat Pumilio/FBF-homology domain (Puf, Figure 5.2e), the tryptophan and RNA-binding attenuation protein (TRAP, Figure 5.2f), and Whirly (Figure 5.2g). Pumilio and TRAP structures have been reviewed in ref. 126. Pumilio from *Drosophila melanogaster* and *fem*-3 mRNA binding factor (FBF) from *Caenorhabditis elegans* defined the Puf or Pumilio-homology domain (also known as PUM-HD) protein family,[139,140] members of which bind 3′ untranslated regions (3′ UTR) of mRNA in a sequence-specific manner to regulate translation, stability and localization of target mRNA in diverse multicellular eukaryotes.[141–143] Each of eight three-helix repeats (Puf repeats) recognizes one nucleotide base with hydrogen-bonding residues supported by stacking interactions (Figure 5.6).[8] These RNA-interacting residues are contained within the central recognition helix of each Puf repeat that corresponds with a conserved consensus sequence.[139,140]

TRAP is an 11-subunit RNA-binding protein (Figure 5.7) that effects transcription attenuation to regulate *trp* operon expression in *Bacillus subtilis* and related *Bacillus* species.[144–147] Each subunit of TRAP is folded into a three-stranded sheet of s5:s2:s3 topology and a four-stranded sheet of s7:s6:s1:s4 topology (Figure 5.2f).[148] The two sheets pack with each other within each subunit and form extended seven-stranded all-anti-parallel beta-sheets with neighboring subunits that are repeated 11-fold in a closed-ring architecture (Figure 5.7). Loops connecting beta-strands and located on the outer rim of the ring form the RNA-binding surfaces specific for $r((G/U)\text{-A-G-N}_{n=2,3})$ repeats

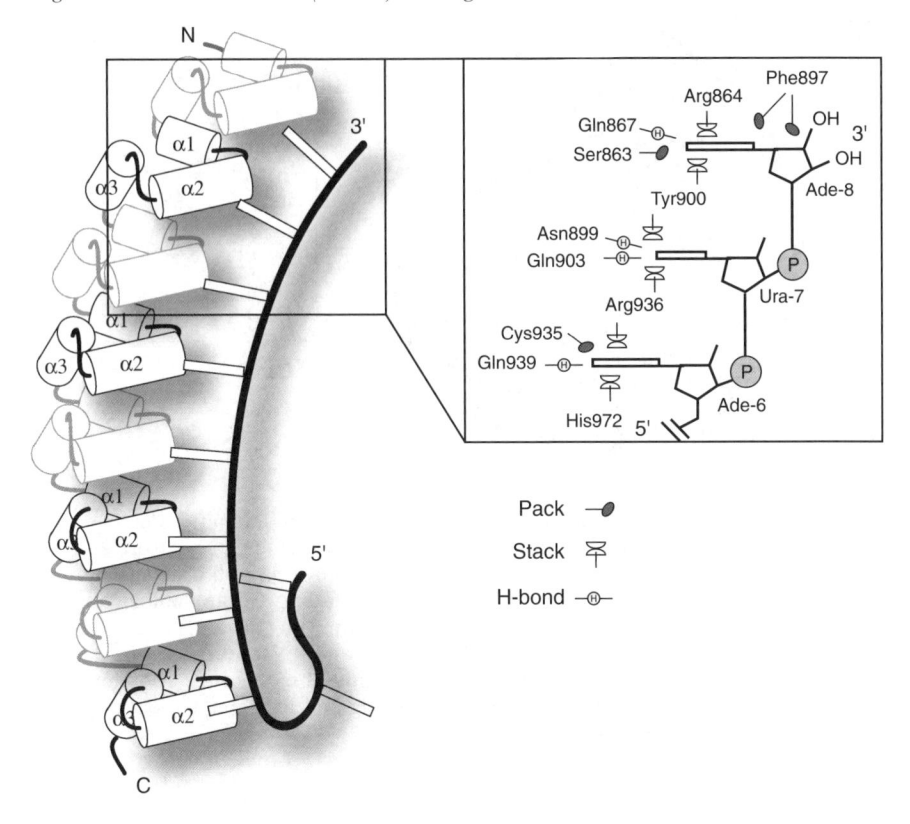

Figure 5.6 Pumilio Puf domain-single-stranded RNA interactions. Human pumilio 1 presents eight Puf repeats to recognize the 3′ untranslated region (3′UTR) of target mRNAs in a modular and sequence-specific manner. Each Puf repeat is made up of three alpha helices, which are labeled for alternate repeats. RNA bases interact with residues presented by α2. Inset: details for recognition of r(AUA) towards the 3′-end of the binding site with use of hydrogen bonds, steric packing interactions and stacking of heteroaromatic groups (pdb id 1m8y).[8] Similar interaction types are observed for each of the other modules, and there are no phosphate salt bridges in this complex.

contained within the 5′-leader sequence of *trp* operon mRNA transcripts.[36] In response to occupancy of L-tryptophan binding sites found in another portion of each subunit, TRAP binds with mRNA to disrupt an anti-terminator RNA structure otherwise formed by this section of the transcript and thereby directs termination of transcription.[147]

The Whirly protein fold (Figure 5.2g) is a newcomer to the SSNA-binding protein group; it was first identified in p24 and related plant transcription factors that bind single-stranded forms of DNA enhancer sites.[149,150] The Whirly fold is also seen in the mitochondrial RNA-binding protein MRP1/MRP2, which binds guide RNA (gRNA) as an essential intermediate

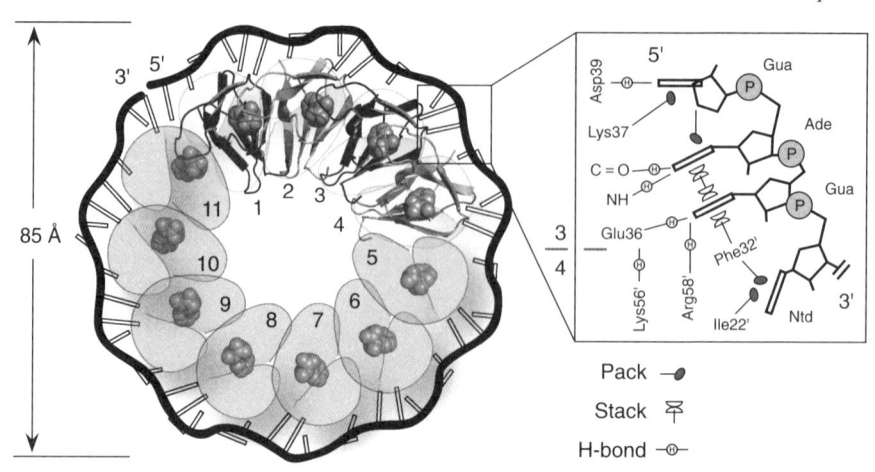

Figure 5.7 TRAP-single-stranded RNA complex. Eleven subunits assemble into a doughnut shape with single-stranded RNA-binding loops presented on the outer perimeter. Subunits 1–4 are shown as ribbon representations. Subunits 5–11 are shown as simplified shapes. L-Tryptophan, an allosteric effector molecule for this protein, is shown with van der Waals spheres for all atoms. Inset: how r((G/U)-A-G-N$_{n=2,3}$) RNA repeats in the 5′-leader region of *trp* operon mRNA are recognized by hydrogen-bonding interactions supported by aromatic stacking and hydrophobic packing interactions (pdb id 1c9s).[36]

encountered in kinetoplastid RNA editing particular to trypanosomes.[35] Whirly proteins associate as tetramers, with each subunit folded into two four-stranded all-anti-parallel beta-sheets that pack together in a crisscrossed manner (Figure 5.8). Alpha helices found in elements connecting these beta-sheets and at N- and C-terminal regions form the tetramerization core. The beta-sheets radiate from this core in such a way as to impart a whirligig appearance with C$_4$ symmetry in the case of p24's homo-tetramer or pseudo-symmetry for the (MRP1/MRP2)$_2$ hetero-tetramer. Beta-sheet surfaces from each of the four subunits form an extensive platform for single-stranded nucleic acid interaction as first deduced on the basis of deletion and amino-acid substitution analysis for p24,[149] and subsequently observed directly for the gRNA/MRP1/MRP2 co-crystal complex.[35]

5.3 Emergent Properties

Combinations of simple interaction units lead to complex molecular physiology that cannot be predicted on the basis of elemental properties. Biology is replete with similar examples where interaction networks built from simple elements such as neuronal synapses and individuals result in emergent properties like learning/memory and social colonies. Emergent properties of SSNA-binding

Single-stranded Nucleic Acid (SSNA)-binding Proteins 111

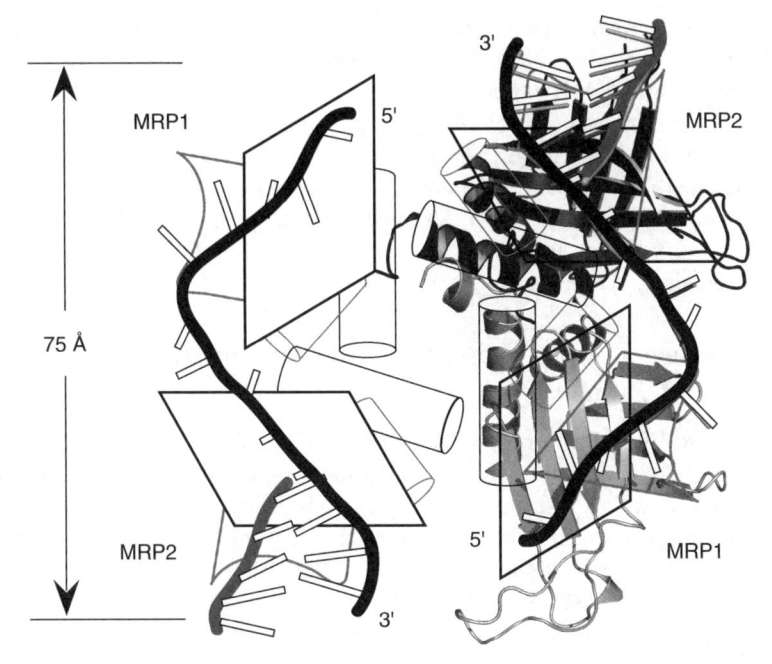

Figure 5.8 Guide RNA/MRP1/MRP2 complex. *M*itochondrial *R*NA binding *p*roteins MRP1 and MRP2 assemble as a hetero-tetramer with pseudo-C_4 symmetry (pdb id 2gje).[35] RNA binds on beta-sheet platforms provided by each MRP1/MRP2 subunit. The 3' portion of the guide RNA (gRNA) forms a stem-loop structure, the double-stranded portion of which was seen in electron density maps. The 5' portion of the gRNA was single-stranded in the protein-bound complex, indicating that association with protein was coupled with disruption of a second stem-loop structure expected in this region of the gRNA.[35]

proteins include molecular recognition specificity, cooperativity and allostery. Here I discuss aspects of these ideas with illustrations from recent progress in the field.

5.3.1 Molecular Recognition: Specificity, Adaptability and Degeneracy

Molecular specificity is intimately related to structure and function of SSNA-binding proteins. While looking at molecular models of SSNA-binding proteins and their complexes with single-stranded RNA or DNA, the eye is drawn to those networks of intermolecular contacts indicative of molecular recognition specificity, especially sequence specificity. Our views are further biased towards systems with a certain degree of sequence-specific recognition because these lend themselves to structural analysis. Only a handful of high-resolution structures have been determined for generalist proteins that bind single-stranded nucleic acid in a non-sequence-specific manner,[20,38,75,151] likely because non-sequence-specific proteins readily form heterogeneous molecular mixtures that hamper crystallization and NMR experiments. In this section, the recognition properties of four sequence-specific SSNA-binding proteins are examined. One of these belongs to the modular category and the other three employ eclectic strategies for interaction with single-stranded nucleic acid. To varying degrees, all of the so-called sequence-specific proteins demonstrate tolerance to sequence variation, meaning recognition is adaptable so that SSNA-binding proteins can bind with highly degenerate sequence motifs, in some cases.

5.3.1.1 Specific yet Adaptable Recognition by Modular Puf Proteins

Pumilio's modular Puf domain structure establishes a one-to-one base-to-recognition-helix correspondence so that each base group is carefully examined by hydrogen-bonding residues contributed by one Puf repeat.[8] Pair-wise tests of several Puf proteins matched with known target RNA binding sites assessed by a yeast three-hybrid assay showed that Puf proteins bind best with their cognate RNAs, suggesting optimized interactions at the molecular level.[10,152] Designed mutant versions of Puf-repeats in Pumilio[9] and in FBF[10] have specificities consistent with those predicted on the basis of a recognition code deduced from the wild-type Puf domain-RNA structure.[9] Pumilio-GFP fragment fusion proteins have been engineered for tracking specific RNA molecules in cells.[11] These observations lead to the impression that modular scaffold-like structures have very exacting and predictable recognition properties.

A slightly more complicated picture emerges from systematic analysis of two sets of RNA targets recognized by FBF and PUF-8, two distantly related Puf-proteins from *Caenorhabditis elegans*. Variant RNAs generated by design or selected through screening of randomized RNA libraries showed robust binding despite nucleotide substitutions at certain positions within and flanking the core RNA sequence, indicating a certain degree of recognition adaptability.[152] For FBF, an extra spacer nucleotide at position 5 of the nine-nucleotide sequence keeps flanking nucleotides in register with specificity determinants on the protein.[10] These studies showed that the molecular recognition properties of Puf proteins are actually rather intricate despite the apparent simplicity and regular repeating patterns seen in the RNA-Puf domain crystal structure.

Two emergent properties highlighted by these studies with Puf proteins are adaptability and specificity. Adaptable recognition is reflected in tolerance to nucleotide substitutions. Recognition adaptability makes sense in terms of biological requirements since each PUF protein binds with a number (40–200) of mRNAs with related but non-identical 3'UTR sequences.[142] Furthermore, adaptability likely contributes to expansion of this protein family in highly evolved multicellular organisms since new adaptive functions could arise without destroying previously established binding properties.

Even though a certain degree of degeneracy characterizes the Puf protein-RNA binding system, specificity in recognition is nonetheless evident, with

Puf proteins discriminating among different classes of RNA sequences. The emergent property of specificity is puzzling in that hydrogen-bonding groups seen in the pumilio Puf-domain-RNA structure are highly conserved among different homologs, yet they nonetheless differentiate among different classes of RNA sequences *in vivo*. Hydrogen bonding groups may be repositioned through effects exerted by non-contact residues that vary more extensively among Puf-family members. Other differential cues could also derive from RNA-binding partners[10] and through competing or collaborating stem-loop RNA structures particular to each target RNA. Structural characterization of new Puf-RNA complexes could elaborate on an adaptable code for single-stranded RNA recognition specificity.

5.3.1.2 A "Hot-spot" for Recognition of Telomere DNA by Cdc13

Proteins of the eclectic variety also exhibit adaptable recognition properties. One such example is provided by Cdc13, the major single-stranded telomere DNA-binding protein in budding yeast.[90,91] The DNA-binding domain derived from Cdc13 comprises a single OB-fold.[92] Association reactions indicated a high-affinity DNA complex with $K_{D\text{-}DNA} = 3$ pM.[153,154] The NMR determined co-complex structure delineated the single-stranded DNA-protein contact surface as encompassing 11 nucleotides of single-stranded DNA, which is much larger than other OB-fold domains that typically contact 4–6 nucleotides. The more extensive single-stranded DNA contact site for Cdc13 is constructed by means of a 30-amino-acid residue loop with an ordered pretzel-like conformation inserted between strands s2 and s3 of the core OB-fold.[39]

Contributions to binding affinity made by amino-acid residues and nucleotide base groups at the single-stranded DNA-Cdc13 interface have been probed by single-alanine replacement[33] and single-nucleotide substitution.[155] These studies indicate a "hot-spot" for recognition focused towards the core OB-fold. Although all contacts with the d($G^1TG^3T^4GGGTGTG$) single-stranded DNA (hot-spot nucleotides are numbered) are important for overall high-affinity binding, substitutions at contact points involving nucleotides G^1, G^3, and T^4 had significantly stronger effects on binding free energy, which is consistent with exacting recognition requirements at these points and more adaptable recognition properties for other sites.[33,155] Focused recognition of just three nucleotides by Cdc13 is a well-suited strategy for dealing with heterogeneous sequences characteristic of budding yeast telomeres[156,157] and affords a reasonable explanation for genetic patterns of tolerance to telomerase RNA template mutations.[155]

5.3.1.3 "Nucleotide Shuffling" and TEBP-α/β

The telomere end binding protein from *Sterkiella nova* (formerly *Oxytricha nova*) binds d($TTTT^4GGGG^8TTTT^{12}GGGG^{16}$) single-stranded DNA as an α/β heterodimer.[27,158,159] Adaptability in single-stranded DNA recognition has been probed in this system by means of structure determination for ten non-cognate

DNA-protein complexes.[160] DNA variants included single-nucleotide substitutions, nucleotide extensions, truncations and abasic residues. For some of these, solvent played a dominant role in accommodation of substitutions or missing groups with little or no disturbance to neighboring sites.[160] In several other cases, dramatic structural perturbations involving repositioning of nucleotides were observed. One of the non-cognate DNA-protein structures revealed that addition of an extra base at the 3′-end could be accommodated by a combination of register slippage and nucleotide looping so as to maximize cognate-like DNA-protein interactions at most of the binding pockets.[160] Other structures showed how variant DNA ligands with base groups removed by sequence truncation or through inclusion of abasic residues could also maintain a significant fraction of native-like interactions by nucleotide looping and register slippage, a mechanism aptly called "nucleotide shuffling".[160]

Repositioning of nucleotides with respect to the expected order seen in these non-cognate structures offers a molecular interpretation of energetic effects, also measured in the study, that is completely different from expectations that would otherwise have been deduced from inspection of the cognate DNA-TEBP structure.[160] Evidence for nucleotide shuffling can be seen in $|Fo| - |Fo|$ difference electron density maps calculated for TEBP-α/β in complex with 5-bromodeoxy-uracil-substituted DNAs and also in $2|Fo| - |Fc|$ simulated-annealing OMIT maps for complexes with unmodified DNA, suggesting that single-stranded DNA in this complex has an intrinsic inclination for nucleotide shuffling even without strongly perturbing sequence alterations (S.C. Schultz and M.P.H., unpublished results). Notably, nucleotide looping seen in this system occurs at several different positions along the telomere DNA, meaning that single-stranded nucleic acid is susceptible to nucleotide shuffling independent of sequence or other cues potentially provided by the protein. Nucleotide shuffling may explain, at least in part, how SSNA-binding proteins recognize heterogeneous and even highly degenerate single-stranded nucleic acid sequences,[160] as in the next example for intron branch site recognition.

5.3.1.4 *Degeneracy in Splicing Branch Site Identification*

Branch site sequences are highly degenerate in mammalian introns. While branch sites in yeast introns can be located on the basis of an invariant r(UACUA-*A*-C, branch A italicized), corresponding branch sites in human introns are variable, specified by r(CU̲R-*A̲*-Y, R = purine, Y = pyrimidine) with only the two underlined nucleotides absolutely required.[134,136,161] Determinants for branch site recognition in yeast are apparently exclusively determined by branch point-binding protein (BBP), which prefers the cognate sequence to the exclusion of alternatives.[137] The mammalian homologue, SF1, binds best with canonical yeast-like sequences yet intron function tolerates substitutions within the SF1 binding site to a surprising degree. A complex of the KH/QUA2 RNA-binding domain derived from SF1 in complex with r(UAUACUA-*A*-CAA) revealed close examination of the branch point *A* base through numerous van der Waals and hydrogen-bonding contacts but otherwise few base-specific interactions (Figure 5.5b).[45]

Degeneracy in branch site recognition likely arises through cooperative interactions with additional splicing factors such as U2AF65[138] as well as a diverse array of other *trans*-acting proteins that use loosely defined enhancer sequences as a platform for encouraging splicesomal assembly. Many of these factors lack unique binding specificity of their own, yet in concert with other partners can nonetheless direct splicing at specific sites. The apparent paradox of specificity arising from non-specific RNA-binding modules interacting with highly degenerate RNA sequences has recently been discussed in a cogent review by Singh and Valcarcel.[3] These authors argue convincingly that a lock-and-key analogy is inappropriate for the dynamic and shape-shifting molecular picture emerging for recognition operational during splicing.[3]

5.3.2 Cooperativity

Cooperativity is another emergent property evident for SSNA-binding proteins. Precedents set by the single-stranded DNA-binding proteins T4 gene 32 protein[162–165] and filamentous bacteriophage gene V protein[166–168] imparted the general impression that single-stranded nucleic acid-binding proteins associate with substrates in a highly cooperative manner whereby extended continuous clusters of nucleic acid-bound protein are favored over isolated protein-bound sites. Evolution has arrived at comparable examples such as *E. coli* SSB and PriB, which exhibit different degrees of homotypic positive binding cooperativity.[19,75] Many counter-examples for which homotypic cooperativity is either absent or negative are also apparent. RPA hetero-trimers bind single-stranded DNA without marked cooperativity.[65] Such is also the case for TEBP-α, which binds multiple telomere single-stranded DNA repeats in an anti-cooperative manner. The TEBP-α homolog in fission yeast, Pot1, was first described as binding single-stranded DNA cooperatively on the basis of binding properties measured for a single-OB-fold portion of the DNA-binding domain.[31] However, measurements recently obtained with the full-length Pot1 protein are consistent with substantially less cooperativity or non-cooperative binding.[169] Loss of homotypic cooperativity during evolution from a SSB-like ancestor may have been a necessary innovation to establish adaptive positive heterotypic interactions observed for TEBP-α and TEBP-β in ciliated protozoa and the analogous proteins in humans, Pot1 and Tpp1, that bind cooperatively to telomere DNA as heteromers.

5.3.2.1 SSB and Multiple Cooperativity Modes

The major single-stranded DNA-binding protein from *Escherichia coli*, SSB, exhibits several different binding modes characterized by differing occluded site sizes, binding affinities, and the degree and extent of homotypic cooperativity (reviewed in ref. 19). Under low-salt conditions SSB binds with a site size of 35 ± 3 nucleotides per SSB tetramer and a high degree of cooperativity such that proteins prefer binding immediately adjacent to already occupied sites over an isolated site by a factor of 100 000.[170] This "unlimited" cooperativity binding mode likely correlates with long, continuous and smoothly contoured complexes seen by electron microscopy.[171] Under high-salt conditions, SSB binds with an occluded site size of 65 ± 5 nucleotides per tetramer with only "limited" cooperativity. In this binding mode, SSB binds either at isolated sites or together with at most one SSB neighbor, leaving gaps of un-complexed single-stranded DNA between tetramers or octamers. Beaded structures discernable by electron microscopy likely correspond with these SSB octamers.[19,171] The crystal structure of SSB in complex with d(C)$_{35}$ suggested plausible models of both 65 ± 5 nucleotide "limited" cooperativity and 35 ± 3 nucleotide "unlimited" cooperativity modes for single-stranded DNA association (Figure 5.3a and b).[20] It is not completely clear whether both of these cooperativity modes observed *in vitro* are operational *in vivo*. A mutant allele associated with a greater tendency to adopt the 35 ± 3 nucleotide binding mode even under high-salt conditions is defective for DNA replication and UV repair,[172,173] suggesting that switching among different binding modes could be important for SSB performance inside of bacterial cells.

5.3.2.2 Anti-cooperativity and TEBP-α

The major DNA-binding domain derived from eukaryotic single-stranded DNA-binding proteins can associate with single-stranded DNA in the absence of other hetero-partners. For RPA, this major single-stranded DNA-binding domain consists of two OB-folds, OB-*a* and OB-*b* (Figure 5.3c).[38,174] Similarly, two OB-folds make up the DNA-binding domain of human Pot1[30,86] and the DNA-binding domain derived from *Sterkiella nova* TEBP-α (Figure 5.3d).[27,175,176]

In the absence of its hetero-partner, TEBP-α has binding characteristics very similar to those of the N-terminal DNA-binding domain, TEBP-αN.[16,177] A crystal structure of TEBP-α in complex with d(TTTTGGGG) revealed an extensive α–α interface suggestive of an α$_2$ homodimer interface.[178] The crystal-packing arrangement seen for TEBP-α in complex with d(TTTTGGGG) has led to the general, but probably incorrect, view that TEBP-α binds multiple telomere repeats as a symmetrical α$_2$ dimer. In solution, TEBP-α does not dimerize in either the presence or absence of single-stranded d(TTTTGGGG) DNA, as judged by analytical ultracentrifugation and several other biophysical methods (O.B. Peersen, personal communication; P. Buczek, unpublished results).

When presented with longer d(TTTTGGGG)$_{n=2}$ telomere-derived DNA fragments, TEBP-αN prefers binding to the 3′-terminal repeat but can also bind the 3′-distal repeat, albeit more weakly.[16,176] Binding transitions monitored by isothermal titration calorimetry (ITC) indicated an initial strong binding event with $K_{D\text{-DNA-A}} = 13 \pm 4$ nM and a second weaker transition with $K_{D\text{-DNA-B}} = 5600 \pm 600$ nM.[16] In the context of d(TTTTGGGGTTTT), binding to a single 3′-distal repeat (underlined) was not quite as weak as the second transition observed with d(TTTTGGGG)$_{n=2}$, with $K_{D\text{-DNA}} = 1000 \pm 50$ nM. These results indicate a certain degree of negative cooperativity governs site-to-site interactions for tandemly arrayed TEBP-αN domains.[16] Negative cooperativity is similarly observed by ITC for full-length TEBP-α binding with d(TTTTGGGG)$_{n=2 \text{ or } 3}$ (P. Buczek, unpublished results). Our studies with longer telomere-derived d(TTTTGGGG)$_{n=3, 6 \text{ or } 9}$ DNA fragments indicated that it is very difficult to completely saturate all binding sites (A. Hansen, unpublished results). Negative homotypic cooperativity may be important so as to provide access to internal telomere sites for endonucleolytic processing of long telomere-extension intermediates and could additionally guide mature telomere ends towards a heteromeric ternary complex governed by strongly positive heterotypic cooperativity described in the following section.

5.3.2.3 Positive Heterotypic Cooperativity at Telomere Ends

Together, TEBP-α and TEBP-β bind telomere derived DNA fragments with higher affinity than expected from binary α + β, α + DNA, and β + DNA reactions, meaning that positive heterotypic cooperativity governs this ternary DNA-α-β system.[159,177] Binding transitions monitored by fluorescence anisotropy[177] or by ITC,[28] two methods that do not require post-equilibrium handling steps to extract equilibrium binding constants, could be modeled with a binding linkage cycle involving −3 or −4 kcal mol^{-1} of energy ascribed to positive cooperativity. As demonstrated by the binding behaviors of protein domains and engineered fusion proteins, cooperativity in this system is entirely dependent on protein-protein interactions mediated by a C-terminal OB-fold of TEBP-α and an internal protein segment in TEBP-β comprising an alpha helix and peptide loop (Figure 5.3d).[159,177,179] Apparently, protein-protein interactions at a remote site within the complex exert an influence over DNA-binding affinity, suggesting a trigger point for complex dissassembly.[177]

What is the source of energetic coupling that links high-affinity DNA binding with TEBP α-β protein-protein association? One possibility relates to proximity effects and effective concentration,[180,181] whereby the high effective concentration of two proteins, each binding with closely adjoined sites on one DNA molecule, enhances protein-protein interactions. Conversely, DNA-binding portions of the two proteins might be expected to bind DNA more effectively when joined through subunit association. Proximity cannot be the only source for cooperative binding behavior, however, since covalently linking the DNA-binding portions of each subunit does not, on its own, lead to high-affinity DNA binding.[177] Structural reorganization involving both the DNA ligand and protein-protein interactions seems to also contribute to energetic coupling, as reflected by a strongly negative heat capacity change (ΔC_p) characterizing complete ternary complex formation.[28] As described further in the next section, our interpretation of these data is that structural reorganization and energetic coupling establish an allosteric switch with important implications for telomere biology.[28]

5.3.3 Allostery

Allosteric control of single-stranded nucleic acid binding provides a mechanism by which down-stream DNA or RNA processes are coordinated.[28,82,182,183] A biological rationale for such coordinated control becomes clear if one considers the potential risks associated with stochastic release of the single-stranded ligand. A carefully orchestrated hand-off would maintain the required single-stranded structure and confer protection against opportunistic chemical and nucleolytic attack. Dissociation at the wrong time could be potentially devastating, triggering genome destabilizing recombination events and dead-end cellular responses such as replicative senescence and apoptosis.

Allostery provides coordinated regulation of SSNA-binding and release under the influence of signals provided by two classes of effector molecules. One class of effectors consists of small molecules, ions and protons whose binding and release is energetically coupled to structural changes influencing DNA or RNA-binding properties. In the second class, I include whole proteins as effectors. These protein effectors do not bind with single-stranded nucleic acid on their own but nonetheless exert an influence on the structure and binding energetics of other proteins that have intrinsic SSNA-binding activity.

5.3.3.1 Small Molecule Effectors and SSNA-binding

Allostery developed historically as an idea to explain control over macromolecular behavior governed by small molecule binding events at sites removed from the site of substrate binding and catalysis. Familiar examples include hemoglobin oxygen-binding behavior under the influence of pH.[184,185] Binding reactions involving SSB and single-stranded DNA are coupled with uptake and release of protons and ions, making the DNA-binding behavior of SSB under the control of changes in pH and salt concentration, potentially through an allosteric mechanism.[17,21,22] Especially intriguing is the idea that transitions among "limited" and "unlimited" cooperativity modes observed for SSB *in vitro* are driven by allosteric effectors inside of cells to modulate biological function.[19]

Another example of allostery is provided by TRAP, which binds single-stranded RNA in response to L-tryptophan occupancy at effector sites. Allostery provides the basis for negative feed-back control over tryptophan biosynthesis in this system particular to *Bacillus* species of bacteria.[147] TRAP assembles as 11-subunit oligomers in both the presence and absence of L-tryptophan. Occupancy of only a few effector sites is sufficient for RNA-binding activity.[186] The mechanism of allosteric control is not fully understood, in part because a high-resolution structure of the apo-TRAP in absence of L-tryptophan is unavailable. NMR measurements are consistent with an altered conformation for the tryptophan-bound form characterized by more limited dynamic properties.[187] Flexibility in apo-TRAP likely discourages RNA binding (and crystallization) since this process is associated with a large entropic penalty. Favorable entropy and enthalpy associated with binding of L-tryptophan could compensate for some of the entropic cost associated with RNA binding and thereby tip the energetic balance in favor of RNA-association.[187] According to this model, ligand-induced protein rigidification establishes allosteric control over RNA-binding.

5.3.3.2 Proteins as Allosteric Effectors for Binding and Release of SSNA

In ciliated protists, TEBP-α binds telomere single-stranded DNA and recruits the TEBP-β hetero-partner.[188] For the telomere DNA-TEBP system studied *in vitro*, TEBP-β behaves as an allosteric effector. Although TEBP-β does not bind single-stranded DNA on its own, TEBP-β modulates and reshapes interactions between TEBP-α and DNA.[158] TEBP-α binds single-stranded DNA with $K_{\text{D-DNA}} = 220 \pm 20$ nM.[177] In the presence of excess TEBP-β, the apparent DNA-protein dissociation constant approaches a limiting value of $K_{\text{D-DNA}} = 1.5 \pm 0.5$ nM, as also measured for a TEBP-α-TEBP-β fusion protein.[177] Increased affinity for single-stranded DNA could derive from additional DNA-protein contacts that are available for the α/β heterodimer. Favorable interactions with β and new contacts with α are apparently offset by other energetic terms since a fusion protein, constructed by linking DNA-binding portions of α and β, binds DNA with the same affinity measured for TEBP-α alone.[177] High-affinity DNA binding could be restored for fusion proteins by adding back the missing protein-protein interaction domains. Remarkably, one of the protein-protein interaction domains could be added in *trans* to restore high-affinity DNA binding.[159,177] DNA-binding and protein-protein interaction sites define distinct locations. Consequently, protein *trans* complementation behavior in the TEBP-α/β system is analogous to a small molecule allosteric effector binding to an effector site. Although the picture is far from complete, we believe that our thermodynamic characterizations of TEBP-α/β-DNA-binding transitions reflect an authentic allosteric system for which structural transitions involving protein-protein interactions are energetically coupled with DNA-protein affinity.[28]

In addition to binding affinity, allostery also controls position of the single-stranded DNA which can chose between two alternate binding registers. As inferred from crystal structures of telomere-derived single-stranded DNA in complex with TEBP-α,[178] TEBP-αN,[175] and the TEBP-α/β heteromer,[27] β association is coupled to slippage of an eight-nucleotide 3'-terminal d(TTTTGGGG) repeat from a site on α to a new site at the α/β interface. Single-stranded DNA register is therefore altered through the action of the β effector protein.

Single-stranded DNA register is likely an important factor in determining productive interactions with telomerase, as demonstrated for human Pot1-DNA complexes.[189,190] Pot1 normally sequesters the 3'-terminus and blocks DNA access to telomerase. Repositioning of Pot1 in such a way as to leave an eight-nucleotide 3'-terminal portion of the single-stranded DNA exposed, converts Pot1-DNA complexes into especially efficient substrates for telomerase.[190] The mechanism for this DNA-register interconversion is not fully understood. Pot1, a TEBP-α paralog, collaborates with Tpp1, whose close evolutionary relationship with TEBP-β has recently been established.[88,89] *In vitro*, Pot1 and Tpp1 bind telomere DNA cooperatively and potentiate highly processive telomerase catalyzed extension reactions.[88] By analogy with the TEBP-α/β system, Tpp1 may serve as an allosteric effector that somehow maintains a DNA-binding register conducive for telomerase activity.

Given the evolutionarily conserved structure and function characters apparent for TEBP-α/β, Pot1/Tpp1, and RPA, it is likely that allostery in each of these systems functions to coordinate downstream processes catalyzed by polymerases.

5.4 Conclusion and Perspective

The field of SSNA-binding protein structural biology has matured considerably over recent years. Structures of a reasonably large number of single-stranded DNA and RNA-protein complexes are now available, and we can begin to identify common elements as well as unique features in these systems. The basic elements include a limited set of interaction types and protein folds. Despite the restricted repertoire, extensive interaction networks with complicated emergent properties are nonetheless evident. These emergent properties include specific yet adaptable molecular recognition, cooperative or anti-cooperative binding behaviors and allosteric switches that are likely critical for biological function of SSNA-binding proteins.

The challenge is to now build a more complete mechanistic picture for these biological functions and the role played by SSNA-binding proteins. How do molecular determinants of specificity identified under equilibrium conditions operate for systems under dynamic flux? What are the thermodynamic gears and levers that operate to coordinate downstream DNA and RNA processing events? These questions are particularly difficult to answer for SSNA-binding proteins since these proteins do not catalyze chemical reactions of their own, yet play an important supporting role to enable proper function of polymerases, helicases and recombinases. Certainly one strategy that should be pursued is to determine larger more complete structures for SSNA-binding protein-RNA/DNA complexes together with these DNA and RNA processing enzymes, and steps in this direction are being made.[191] Another feature of SSNA-binding proteins complicates this strategy, however, in that most of the interesting events involve dynamic transitions between structural states. Given these challenges I predict the richest rewards will come from integrated approaches whereby insights from structural biology can complement those arrived at from genetics, mutational analyses, solution-based biochemistry, thermodynamics, and single-molecule biophysics to synthesize a complete picture. Currently we "see" significant yet fragmented portions of what is at play; in the near future the views will be large panoramas revealing the whole true forms of complicated machineries – the splicesome, DNA replication fork, and DNA repair complexes – acting on SSNA and coordinated by SSNA-binding proteins.

Acknowledgements

I thank Catherine Dy for help with figure preparation. This work was supported in part through a grant from the NIH (R01 GM067994).

References

1. G. Varani and K. Nagai, *Annu. Rev. Biophys. Biomol. Struct.*, 1998, **27**, 407–445.
2. S.D. Auweter, F.C. Oberstrass and F.H. Allain, *Nucleic Acids Res.*, 2006, **34**, 4943–4959.
3. R. Singh and J. Valcarcel, *Nat. Struct. Mol. Biol.*, 2005, **12**, 645–653.
4. P. Khusial, R. Plaag and G.W. Zieve, *Trends Biochem. Sci.*, 2005, **30**, 522–528.
5. C. Maris, C. Dominguez and F.H. Allain, *FEBS J.*, 2005, **272**, 2118–2131.
6. D.L. Theobald, R.M. Mitton-Fry and D.S. Wuttke, *Annu. Rev. Biophys. Biomol. Struct.*, 2003, **32**, 115–133.
7. J.E. Croy and D.S. Wuttke, *Trends Biochem. Sci.*, 2006, **31**, 516–525.
8. X. Wang, J. McLachlan, P.D. Zamore and T.M. Hall, *Cell*, 2002, **110**, 501–512.
9. C.G. Cheong and T.M. Hall, *Proc. Natl. Acad. Sci. U.S.A.*, 2006, **103**, 13635–13639.
10. L. Opperman, B. Hook, M. DeFino, D.S. Bernstein and M. Wickens, *Nat. Struct. Mol. Biol.*, 2005, **12**, 945–951.
11. T. Ozawa, Y. Natori, M. Sato and Y. Umezawa, *Nat. Methods*, 2007, **4**, 413–419.
12. C. Baumann, J. Otridge and P. Gollnick, *J. Biol. Chem.*, 1996, **271**, 12269–12274.
13. C. Baumann, S. Xirasagar and P. Gollnick, *J. Biol. Chem.*, 1997, **272**, 19863–19869.
14. T. Lundback, H. Hansson, S. Knapp, R. Ladenstein and T. Hard, *J. Mol. Biol.*, 1998, **276**, 775–786.
15. A.G. Kozlov and T.M. Lohman, *Biochemistry*, 1999, **38**, 7388–7397.
16. P. Buczek and M.P. Horvath, *J. Mol. Biol.*, 2006, **359**, 1217–1234.
17. L.B. Overman, W. Bujalowski and T.M. Lohman, *Biochemistry*, 1988, **27**, 456–471.
18. W. Bujalowski and T.M. Lohman, *J. Mol. Biol.*, 1989, **207**, 269–288.
19. T.M. Lohman and M.E. Ferrari, *Annu. Rev. Biochem.*, 1994, **63**, 527–570.
20. S. Raghunathan, A.G. Kozlov, T.M. Lohman and G. Waksman, *Nat. Struct. Biol.*, 2000, **7**, 648–652.
21. L.B. Overman and T.M. Lohman, *J. Mol. Biol.*, 1994, **236**, 165–178.
22. T.M. Lohman, L.B. Overman, M.E. Ferrari and A.G. Kozlov, *Biochemistry*, 1996, **35**, 5272–5279.
23. J. Wyman and S.J. Gill, *Binding and Linkage. Functional Chemistry of Biological Macromolecules*, University Science Books, Mill Valley, CA, 1990.
24. D.E. Gottschling and V.A. Zakian, *Cell*, 1986, **47**, 195–205.
25. C.M. Price and T.R. Cech, *Biochemistry*, 1989, **28**, 769–774.
26. C.M. Price and T.R. Cech, *Genes Dev.*, 1987, **1**, 783–793.
27. M.P. Horvath, V.L. Schweiker, J.M. Bevilacqua, J.A. Ruggles and S.C. Schultz, *Cell*, 1998, **95**, 963–974.

28. P. Buczek and M.P. Horvath, *J. Biol. Chem.*, 2006, **281**, 40124–40134.
29. M. Lei, E.R. Podell, P. Baumann and T.R. Cech, *Nature*, 2003, **426**, 198–203.
30. M. Lei, E.R. Podell and T.R. Cech, *Nat. Struct. Mol. Biol.*, 2004, **11**, 1223–1229.
31. M. Lei, P. Baumann and T.R. Cech, *Biochemistry*, 2002, **41**, 14560–14568.
32. J.E. Croy, E.R. Podell and D.S. Wuttke, *J. Mol. Biol.*, 2006, **361**, 80–93.
33. E.M. Anderson, W.A. Halsey and D.S. Wuttke, *Biochemistry*, 2003, **42**, 3751–3758.
34. Y.C. Lin, Y.H. Wu Lee and J.J. Lin, *Biochem. J*, 2007, **403**, 289–295.
35. M.A. Schumacher, E. Karamooz, A. Zikova, L. Trantirek and J. Lukes, *Cell*, 2006, **126**, 701–711.
36. A.A. Antson, E.J. Dodson, G. Dodson, R.B. Greaves, X. Chen and P. Gollnick, *Nature*, 1999, **401**, 235–242.
37. M. Ruff, S. Krishnaswamy, M. Boeglin, A. Poterszman, A. Mitschler, A. Podjarny, B. Rees, J.C. Thierry and D. Moras, *Science*, 1991, **252**, 1682–1689.
38. A. Bochkarev, R.A. Pfuetzner, A.M. Edwards and L. Frappier, *Nature*, 1997, **385**, 176–181.
39. R.M. Mitton-Fry, E.M. Anderson, D.L. Theobald, L.W. Glustrom and D.S. Wuttke, *J. Mol. Biol.*, 2004, **338**, 241–255.
40. J. Allers and Y. Shamoo, *J. Mol. Biol.*, 2001, **311**, 75–86.
41. R.A. Kumpf and D.A. Dougherty, *Science*, 1993, **261**, 1708–1710.
42. D.A. Dougherty, *Science*, 1996, **271**, 163–168.
43. J.P. Gallivan and D.A. Dougherty, *Proc. Natl. Acad. Sci. U.S.A.*, 1999, **96**, 9459–9464.
44. S. Jones, D.T. Daley, N.M. Luscombe, H.M. Berman and J.M. Thornton, *Nucleic Acids Res.*, 2001, **29**, 943–954.
45. Z. Liu, I. Luyten, M.J. Bottomley, A.C. Messias, S. Houngninou-Molango, R. Sprangers, K. Zanier, A. Kramer and M. Sattler, *Science*, 2001, **294**, 1098–1102.
46. W. Kauzmann, *Adv. Protein Chem.*, 1959, **14**, 1–63.
47. P.L. Privalov and S.J. Gill, *Adv. Protein Chem.*, 1988, **39**, 191–234.
48. E.T. Kool, *Annu. Rev. Biochem.*, 2002, **71**, 191–219.
49. E.T. Kool, *Annu. Rev. Biophys. Biomol. Struct.*, 2001, **30**, 1–22.
50. X. Wang, P.D. Zamore and T.M. Hall, *Mol. Cell*, 2001, **7**, 855–865.
51. M.A. Schumacher, R.F. Pearson, T. Moller, P. Valentin-Hansen and R.G. Brennan, *EMBO J.*, 2002, **21**, 3546–3556.
52. I. Toro, S. Thore, C. Mayer, J. Basquin, B. Seraphin and D. Suck, *EMBO J.*, 2001, **20**, 2293–2303.
53. P.H. Backe, A.C. Messias, R.B. Ravelli, M. Sattler and S. Cusack, *Structure*, 2005, **13**, 1055–1067.
54. A.G. Murzin, *EMBO J.*, 1993, **12**, 861–867.
55. D.L. Theobald and D.S. Wuttke, *J. Mol. Biol.*, 2005, **354**, 722–737.
56. R.J. Hoffman and P. Gollnick, *J. Bacteriol.*, 1995, **177**, 839–842.

57. M.M. Skinner, H. Zhang, D.H. Leschnitzer, Y. Guan, H. Bellamy, R.M. Sweet, C.W. Gray, R.N. Konings, A.H. Wang and T.C. Terwilliger, *Proc. Natl. Acad. Sci. U.S.A.*, 1994, **91**, 2071–2075.
58. P.J. Folkers, M. Nilges, R.H. Folmer, R.N. Konings and C.W. Hilbers, *J. Mol. Biol.*, 1994, **236**, 229–246.
59. R.H. Folmer, M. Nilges, R.N. Konings and C.W. Hilbers, *EMBO J.*, 1995, **14**, 4132–4142.
60. S. Raghunathan, C.S. Ricard, T.M. Lohman and G. Waksman, *Proc. Natl. Acad. Sci. U.S.A.*, 1997, **94**, 6652–6657.
61. G. Webster, J. Genschel, U. Curth, C. Urbanke, C. Kang and R. Hilgenfeld, *FEBS Lett.*, 1997, **411**, 313–316.
62. C. Yang, U. Curth, C. Urbanke and C. Kang, *Nat. Struct. Biol.*, 1997, **4**, 153–157.
63. E. Bochkareva, S. Korolev, S.P. Lees-Miller and A. Bochkarev, *EMBO J.*, 2002, **21**, 1855–1863.
64. J.W. Chase and K.R. Williams, *Annu. Rev. Biochem.*, 1986, **55**, 103–136.
65. M.S. Wold, *Annu. Rev. Biochem.*, 1997, **66**, 61–92.
66. C. Iftode, Y. Daniely and J.A. Borowiec, *Crit. Rev. Biochem. Mol. Biol.*, 1999, **34**, 141–180.
67. K. Saikrishnan, J. Jeyakanthan, J. Venkatesh, N. Acharya, K. Sekar, U. Varshney and M. Vijayan, *J. Mol. Biol.*, 2003, **331**, 385–393.
68. K. Saikrishnan, G.P. Manjunath, P. Singh, J. Jeyakanthan, Z. Dauter, K. Sekar, K. Muniyappa and M. Vijayan, *Acta Crystallogr. Sect. D: Biol. Crystallogr.*, 2005, **61**, 1140–1148.
69. D.A. Bernstein, J.M. Eggington, M.P. Killoran, A.M. Misic, M.M. Cox and J.L. Keck, *Proc. Natl. Acad. Sci. U.S.A.*, 2004, **101**, 8575–8580.
70. R. Fedorov, G. Witte, C. Urbanke, D.J. Manstein and U. Curth, *Nucleic Acids Res.*, 2006, **34**, 6708–6717.
71. R. Jedrzejczak, M. Dauter, Z. Dauter, M. Olszewski, A. Dlugolecka and J. Kur, *Acta Crystallogr., Sect. D: Biol. Crystallogr.*, 2006, **62**, 1407–1412.
72. M.M. Cox, M.F. Goodman, K.N. Kreuzer, D.J. Sherratt, S.J. Sandler and K.J. Marians, *Nature*, 2000, **404**, 37–41.
73. C.J. Cadman, M. Lopper, P.B. Moon, J.L. Keck and P. McGlynn, *J. Biol. Chem.*, 2005, **280**, 39693–39700.
74. J.H. Liu, T.W. Chang, C.Y. Huang, S.U. Chen, H.N. Wu, M.C. Chang and C.D. Hsiao, *J. Biol. Chem.*, 2004, **279**, 50465–50471.
75. C.Y. Huang, C.H. Hsu, Y.J. Sun, H.N. Wu and C.D. Hsiao, *Nucleic Acids Res.*, 2006, **34**, 3878–3886.
76. M.P. Fairman and B. Stillman, *EMBO J.*, 1988, **7**, 1211–1218.
77. M.S. Wold and T. Kelly, *Proc. Natl. Acad. Sci. U.S.A.*, 1988, **85**, 2523–2527.
78. S.J. Brill and B. Stillman, *Nature*, 1989, **342**, 92–95.
79. S.J. Brill and B. Stillman, *Genes Dev.*, 1991, **5**, 1589–1600.
80. D. Philipova, J.R. Mullen, H.S. Maniar, J. Lu, C. Gu and S.J. Brill, *Genes Dev.*, 1996, **10**, 2222–2233.

81. E. Bochkareva, V. Belegu, S. Korolev and A. Bochkarev, *EMBO J.*, 2001, **20**, 612–618.
82. E. Fanning, V. Klimovich and A.R. Nager, *Nucleic Acids Res.*, 2006, **34**, 4126–4137.
83. I.D. Kerr, R.I. Wadsworth, L. Cubeddu, W. Blankenfeldt, J.H. Naismith and M.F. White, *EMBO J.*, 2003, **22**, 2561–2570.
84. A. Smogorzewska and T. de Lange, *Annu. Rev. Biochem.*, 2004, **73**, 177–208.
85. M.P. Horvath, Evolution of telomere binding proteins, in *Origin and Evolution of Telomeres*, ed. L. Tomaska and J. Nosek, Landes Bioscience, Austin, TX, 2008, in the press – available online/www.eurekah.com/chapter/3586/ ISBN: 978-1-58706-309-1.
86. P. Baumann and T.R. Cech, *Science*, 2001, **292**, 1171–1175.
87. E.V. Shakirov, Y.V. Surovtseva, N. Osbun and D.E. Shippen, *Mol. Cell Biol.*, 2005, **25**, 7725–7733.
88. F. Wang, E.R. Podell, A.J. Zaug, Y. Yang, P. Baciu, T.R. Cech and M. Lei, *Nature*, 2007, **445**, 506–510.
89. H. Xin, D. Liu, M. Wan, A. Safari, H. Kim, W. Sun, M.S. O'Connor and Z. Songyang, *Nature*, 2007, **445**, 559–562.
90. C.I. Nugent, T.R. Hughes, N.F. Lue and V. Lundblad, *Science*, 1996, **274**, 249–252.
91. J.J. Lin and V.A. Zakian, *Proc. Natl. Acad. Sci. U.S.A.*, 1996, **93**, 13760–13765.
92. R.M. Mitton-Fry, E.M. Anderson, T.R. Hughes, V. Lundblad and D.S. Wuttke, *Science*, 2002, **296**, 145–147.
93. H. Gao, R.B. Cervantes, E.K. Mandell, J.H. Otero and V. Lundblad, *Nat. Struct. Mol. Biol.*, 2007, **14**, 208–214.
94. N. Grandin, C. Damon and M. Charbonneau, *EMBO J.*, 2001, **20**, 1173–1183.
95. N. Grandin, S.I. Reed and M. Charbonneau, *Genes Dev.*, 1997, **11**, 512–527.
96. C. Kambach, S. Walke and K. Nagai, *Curr. Opin. Struct. Biol.*, 1999, **9**, 222–230.
97. C. Kambach, S. Walke, R. Young, J.M. Avis, E. de la Fortelle, V.A. Raker, R. Luhrmann, J. Li and K. Nagai, *Cell*, 1999, **96**, 375–387.
98. H. Hermann, P. Fabrizio, V.A. Raker, K. Foulaki, H. Hornig, H. Brahms and R. Luhrmann, *EMBO J.*, 1995, **14**, 2076–2088.
99. B. Seraphin, *EMBO J.*, 1995, **14**, 2089–2098.
100. V.A. Raker, K. Hartmuth, B. Kastner and R. Luhrmann, *Mol. Cell Biol.*, 1999, **19**, 6554–6565.
101. T. Achsel, H. Brahms, B. Kastner, A. Bachi, M. Wilm and R. Luhrmann, *EMBO J.*, 1999, **18**, 5789–5802.
102. V.P. Vidal, L. Verdone, A.E. Mayes and J.D. Beggs, *RNA*, 1999, **5**, 1470–1481.
103. H. Urlaub, V.A. Raker, S. Kostka and R. Luhrmann, *EMBO J.*, 2001, **20**, 187–196.

104. H. Stark, P. Dube, R. Luhrmann and B. Kastner, *Nature*, 2001, **409**, 539–542.
105. B.M. Collins, S.J. Harrop, G.D. Kornfeld, I.W. Dawes, P.M. Curmi and B.C. Mabbutt, *J. Mol. Biol.*, 2001, **309**, 915–923.
106. C. Mura, D. Cascio, M.R. Sawaya and D.S. Eisenberg, *Proc. Natl. Acad. Sci. U.S.A.*, 2001, **98**, 5532–5537.
107. I. Toro, J. Basquin, H. Teo-Dreher and D. Suck, *J. Mol. Biol.*, 2002, **320**, 129–142.
108. J. Salgado-Garrido, E. Bragado-Nilsson, S. Kandels-Lewis and B. Seraphin, *EMBO J.*, 1999, **18**, 3451–3462.
109. S. Thore, C. Mayer, C. Sauter, S. Weeks and D. Suck, *J. Biol. Chem.*, 2003, **278**, 1239–1247.
110. C.J. Wilusz and J. Wilusz, *Nat. Struct. Mol. Biol.*, 2005, **12**, 1031–1036.
111. T. Moller, T. Franch, P. Hojrup, D.R. Keene, H.P. Bachinger, R.G. Brennan and P. Valentin-Hansen, *Mol. Cell*, 2002, **9**, 23–30.
112. A. Zhang, K.M. Wassarman, C. Rosenow, B.C. Tjaden, G. Storz and S. Gottesman, *Mol. Microbiol.*, 2003, **50**, 1111–1124.
113. C. Sauter, J. Basquin and D. Suck, *Nucleic Acids Res.*, 2003, **31**, 4091–4098.
114. K. Nagai, C. Oubridge, T.H. Jessen, J. Li and P.R. Evans, *Nature*, 1990, **348**, 515–520.
115. C. Oubridge, N. Ito, P.R. Evans, C.H. Teo and K. Nagai, *Nature*, 1994, **372**, 432–438.
116. S.R. Price, P.R. Evans and K. Nagai, *Nature*, 1998, **394**, 645–650.
117. N. Handa, O. Nureki, K. Kurimoto, I. Kim, H. Sakamoto, Y. Shimura, Y. Muto and S. Yokoyama, *Nature*, 1999, **398**, 579–585.
118. C. Alfano, D. Sanfelice, J. Babon, G. Kelly, A. Jacks, S. Curry and M.R. Conte, *Nat. Struct. Mol. Biol.*, 2004, **11**, 323–329.
119. H. Siomi, M.J. Matunis, W.M. Michael and G. Dreyfuss, *Nucleic Acids Res.*, 1993, **21**, 1193–1198.
120. T.J. Gibson, J.D. Thompson and J. Heringa, *FEBS Lett.*, 1993, **324**, 361–366.
121. N.V. Grishin, *Nucleic Acids Res.*, 2001, **29**, 638–643.
122. G. Musco, G. Stier, C. Joseph, M.A. Castiglione Morelli, M. Nilges, T.J. Gibson and A. Pastore, *Cell*, 1996, **85**, 237–245.
123. G. Musco, A. Kharrat, G. Stier, F. Fraternali, T.J. Gibson, M. Nilges and A. Pastore, *Nat. Struct. Biol.*, 1997, **4**, 712–716.
124. B.T. Wimberly, D.E. Brodersen, W.M. Clemons, Jr., R.J. Morgan-Warren, A.P. Carter, C. Vonrhein, T. Hartsch and V. Ramakrishnan, *Nature*, 2000, **407**, 327–339.
125. S. Adinolfi, C. Bagni, M.A. Castiglione Morelli, F. Fraternali, G. Musco and A. Pastore, *Biopolymers*, 1999, **51**, 153–164.
126. A.C. Messias and M. Sattler, *Acc. Chem. Res.*, 2004, **37**, 279–287.
127. D.T. Braddock, J.L. Baber, D. Levens and G.M. Clore, *EMBO J.*, 2002, **21**, 3476–3485.

128. D.T. Braddock, J.M. Louis, J.L. Baber, D. Levens and G.M. Clore, *Nature*, 2002, **415**, 1051–1056.
129. H. Siomi, M.C. Siomi, R.L. Nussbaum and G. Dreyfuss, *Cell*, 1993, **74**, 291–298.
130. R.E. Dodson and D.J. Shapiro, *J. Biol. Chem.*, 1997, **272**, 12249–12252.
131. T. Tomonaga and D. Levens, *J. Biol. Chem.*, 1995, **270**, 4875–4881.
132. T. Tomonaga and D. Levens, *Proc. Natl. Acad. Sci. U.S.A.*, 1996, **93**, 5830–5835.
133. R. Duncan, L. Bazar, G. Michelotti, T. Tomonaga, H. Krutzsch, M. Avigan and D. Levens, *Genes Dev.*, 1994, **8**, 465–480.
134. J.A. Berglund, K. Chua, N. Abovich, R. Reed and M. Rosbash, *Cell*, 1997, **89**, 781–787.
135. J.C. Rain, Z. Rafi, Z. Rhani, P. Legrain and A. Kramer, *RNA*, 1998, **4**, 551–565.
136. J.A. Berglund, M.L. Fleming and M. Rosbash, *RNA*, 1998, **4**, 998–1006.
137. S.M. Garrey, R. Voelker and J.A. Berglund, *J. Biol. Chem.*, 2006, **281**, 27443–27453.
138. J.A. Berglund, N. Abovich and M. Rosbash, *Genes Dev.*, 1998, **12**, 858–867.
139. P.D. Zamore, J.R. Williamson and R. Lehmann, *RNA*, 1997, **3**, 1421–1433.
140. B. Zhang, M. Gallegos, A. Puoti, E. Durkin, S. Fields, J. Kimble and M.P. Wickens, *Nature*, 1997, **390**, 477–484.
141. M. Wickens, D.S. Bernstein, J. Kimble and R. Parker, *Trends Genet.*, 2002, **18**, 150–157.
142. A.P. Gerber, D. Herschlag and P.O. Brown, *PLoS Biol.*, 2004, **2**, E79.
143. A.P. Gerber, S. Luschnig, M.A. Krasnow, P.O. Brown and D. Herschlag, *Proc. Natl. Acad. Sci. U.S.A.*, 2006, **103**, 4487–4492.
144. H. Shimotsu, M.I. Kuroda, C. Yanofsky and D.J. Henner, *J. Bacteriol.*, 1986, **166**, 461–471.
145. A.A. Antson, A.M. Brzozowski, E.J. Dodson, Z. Dauter, K.S. Wilson, T. Kurecki, J. Otridge and P. Gollnick, *J. Mol. Biol.*, 1994, **244**, 1–5.
146. P. Babitzke, *Curr. Opin. Microbiol.*, 2004, **7**, 132–139.
147. P. Gollnick, P. Babitzke, A. Antson and C. Yanofsky, *Annu. Rev. Genet.*, 2005, **39**, 47–68.
148. A.A. Antson, J. Otridge, A.M. Brzozowski, E.J. Dodson, G.G. Dodson, K.S. Wilson, T.M. Smith, M. Yang, T. Kurecki and P. Gollnick, *Nature*, 1995, **374**, 693–700.
149. D. Desveaux, J. Allard, N. Brisson and J. Sygusch, *Nat. Struct. Biol.*, 2002, **9**, 512–517.
150. D. Desveaux, A. Marechal and N. Brisson, *Trends Plant Sci.*, 2005, **10**, 95–102.
151. Y. Shamoo, A.M. Friedman, M.R. Parsons, W.H. Konigsberg and T.A. Steitz, *Nature*, 1995, **376**, 362–366.
152. D. Bernstein, B. Hook, A. Hajarnavis, L. Opperman and M. Wickens, *RNA*, 2005, **11**, 447–458.
153. T.R. Hughes, R.G. Weilbaecher, M. Walterscheid and V. Lundblad, *Proc. Natl. Acad. Sci. U.S.A.*, 2000, **97**, 6457–6462.

154. E.M. Anderson, W.A. Halsey and D.S. Wuttke, *Nucleic Acids Res.*, 2002, **30**, 4305–4313.
155. A.M. Eldridge, W.A. Halsey and D.S. Wuttke, *Biochemistry*, 2006, **45**, 871–879.
156. K. Forstemann and J. Lingner, *Mol. Cell Biol.*, 2001, **21**, 7277–7286.
157. M.T. Teixeira, M. Arneric, P. Sperisen and J. Lingner, *Cell*, 2004, **117**, 323–335.
158. J.T. Gray, D.W. Celander, C.M. Price and T.R. Cech, *Cell*, 1991, **67**, 807–814.
159. G. Fang, J.T. Gray and T.R. Cech, *Genes Dev.*, 1993, **7**, 870–882.
160. D.L. Theobald and S.C. Schultz, *EMBO J.*, 2003, **22**, 4314–4324.
161. H. Peled-Zehavi, J.A. Berglund, M. Rosbash and A.D. Frankel, *Mol. Cell Biol.*, 2001, **21**, 5232–5241.
162. B.M. Alberts and L. Frey, *Nature*, 1970, **227**, 1313–1318.
163. S.C. Kowalczykowski, N. Lonberg, J.W. Newport, L.S. Paul and P.H. von Hippel, *Biophys. J.*, 1980, **32**, 403–418.
164. N. Lonberg, S.C. Kowalczykowski, L.S. Paul and P.H. von Hippel, *J. Mol. Biol.*, 1981, **145**, 123–138.
165. J.L. Villemain, Y. Ma, D.P. Giedroc and S.W. Morrical, *J. Biol. Chem.*, 2000, **275**, 31496–31504.
166. B. Alberts, L. Frey and H. Delius, *J. Mol. Biol.*, 1972, **68**, 139–152.
167. S.J. Cavalieri, K.E. Neet and D.A. Goldthwait, *J. Mol. Biol.*, 1976, **102**, 697–711.
168. C.W. Gray, *J. Mol. Biol.*, 1989, **208**, 57–64.
169. K.M. Trujillo, J.T. Bunch and P. Baumann, *J. Biol. Chem.*, 2005, **280**, 9119–9128.
170. M.E. Ferrari, W. Bujalowski and T.M. Lohman, *J. Mol. Biol.*, 1994, **236**, 106–123.
171. S. Chrysogelos and J. Griffith, *Proc. Natl. Acad. Sci. U.S.A.*, 1982, **79**, 5803–5807.
172. L.E. Carlini, R.D. Porter, U. Curth and C. Urbanke, *Mol. Microbiol.*, 1993, **10**, 1067–1075.
173. U. Curth, J. Greipel, C. Urbanke and G. Maass, *Biochemistry*, 1993, **32**, 2585–2591.
174. R.A. Pfuetzner, A. Bochkarev, L. Frappier and A.M. Edwards, *J. Biol. Chem.*, 1997, **272**, 430–434.
175. S. Classen, J.A. Ruggles and S.C. Schultz, *J. Mol. Biol.*, 2001, **314**, 1113–1125.
176. S. Classen, D. Lyons, T.R. Cech and S.C. Schultz, *Biochemistry*, 2003, **42**, 9269–9277.
177. P. Buczek, R.S. Orr, S.R. Pyper, M. Shum, E. Kimmel, I. Ota, S.E. Gerum and M.P. Horvath, *J. Mol. Biol.*, 2005, **350**, 938–952.
178. O.B. Peersen, J.A. Ruggles and S.C. Schultz, *Nat. Struct. Biol.*, 2002, **9**, 182–187.
179. G. Fang and T.R. Cech, *Proc. Natl. Acad. Sci. U.S.A.*, 1993, **90**, 6056–6060.

180. D.M. Crothers and H. Metzger, *Immunochemistry*, 1972, **9**, 341–357.
181. R. Schleif and C. Wolberger, *Protein Sci.*, 2004, **13**, 2829–2831.
182. G. Mer, A. Bochkarev, R. Gupta, E. Bochkareva, L. Frappier, C.J. Ingles, A.M. Edwards and W.J. Chazin, *Cell*, 2000, **103**, 449–456.
183. D. Jackson, K. Dhar, J.K. Wahl, M.S. Wold and G.E. Borgstahl, *J. Mol. Biol.*, 2002, **321**, 133–148.
184. C. Bohr, K.A. Hasselbalch and A. Krogh, *Skand. Arch. Physiol.*, 1904, **16**, 402–412.
185. K. Imai and T. Yonetani, *J. Biol. Chem.*, 1975, **250**, 2227–2231.
186. P.T. Li and P. Gollnick, *J. Biol. Chem.*, 2002, **277**, 35567–35573.
187. C. McElroy, A. Manfredo, A. Wendt, P. Gollnick and M. Foster, *J. Mol. Biol.*, 2002, **323**, 463–473.
188. K. Paeschke, T. Simonsson, J. Postberg, D. Rhodes and H.J. Lipps, *Nat. Struct. Mol. Biol.*, 2005, **12**, 847–854.
189. A.J. Zaug, E.R. Podell and T.R. Cech, *Proc. Natl. Acad. Sci. U.S.A.*, 2005, **102**, 10864–10869.
190. M. Lei, A.J. Zaug, E.R. Podell and T.R. Cech, *J. Biol. Chem.*, 2005, **280**, 20449–20456.
191. S. Sun, L. Geng and Y. Shamoo, *Proteins*, 2006, **65**, 231–238.
192. H.R. Drew, R.M. Wing, T. Takano, C. Broka, S. Tanaka, K. Itakura and R.E. Dickerson, *Proc. Natl. Acad. Sci. U.S.A.*, 1981, **78**, 2179–2183.
193. R. Wing, H. Drew, T. Takano, C. Broka, S. Tanaka, K. Itakura and R.E. Dickerson, *Nature*, 1980, **287**, 755–758.
194. M. Teplova, Y.R. Yuan, A.T. Phan, L. Malinina, S. Ilin, A. Teplov and D.J. Patel, *Mol. Cell*, 2006, **21**, 75–85.

CHAPTER 6
DNA Junctions and their Interaction with Resolving Enzymes

DAVID M.J. LILLEY

Cancer Research UK Nucleic Acid Structure Research Group, MSI/WTB Complex, The University of Dundee, Dundee, DD1 5EH, UK

6.1 The Four-way Junction in Genetic Recombination

The four-way junction in DNA (a 4H junction[1]) is a branchpoint in which four double-helical segments are connected by the covalent continuity of the four component strands. The four-way DNA junction is the central intermediate in homologous genetic recombination (the Holliday junction),[2–5] and the tyrosine recombinase family of site-specific recombination.[6–12] Branch migration occurs at the four-way junction, which is ultimately resolved enzymatically to revert to duplex species.

Genetic recombination occurs between homologous DNA molecules, resulting in re-wired DNA connectivities. This ancient and ubiquitous process is important for the repair of double-stranded breaks, the restart of stalled replication forks and for the creation of genetic diversity.

6.2 Structure and Dynamics of DNA Junctions

A four-way DNA junction adopts an extended geometry with pseudo C_4 symmetry in the absence of added metal ions, in which the arms are directed towards the corners of a square with an open central region.[13,14] The junction is

130

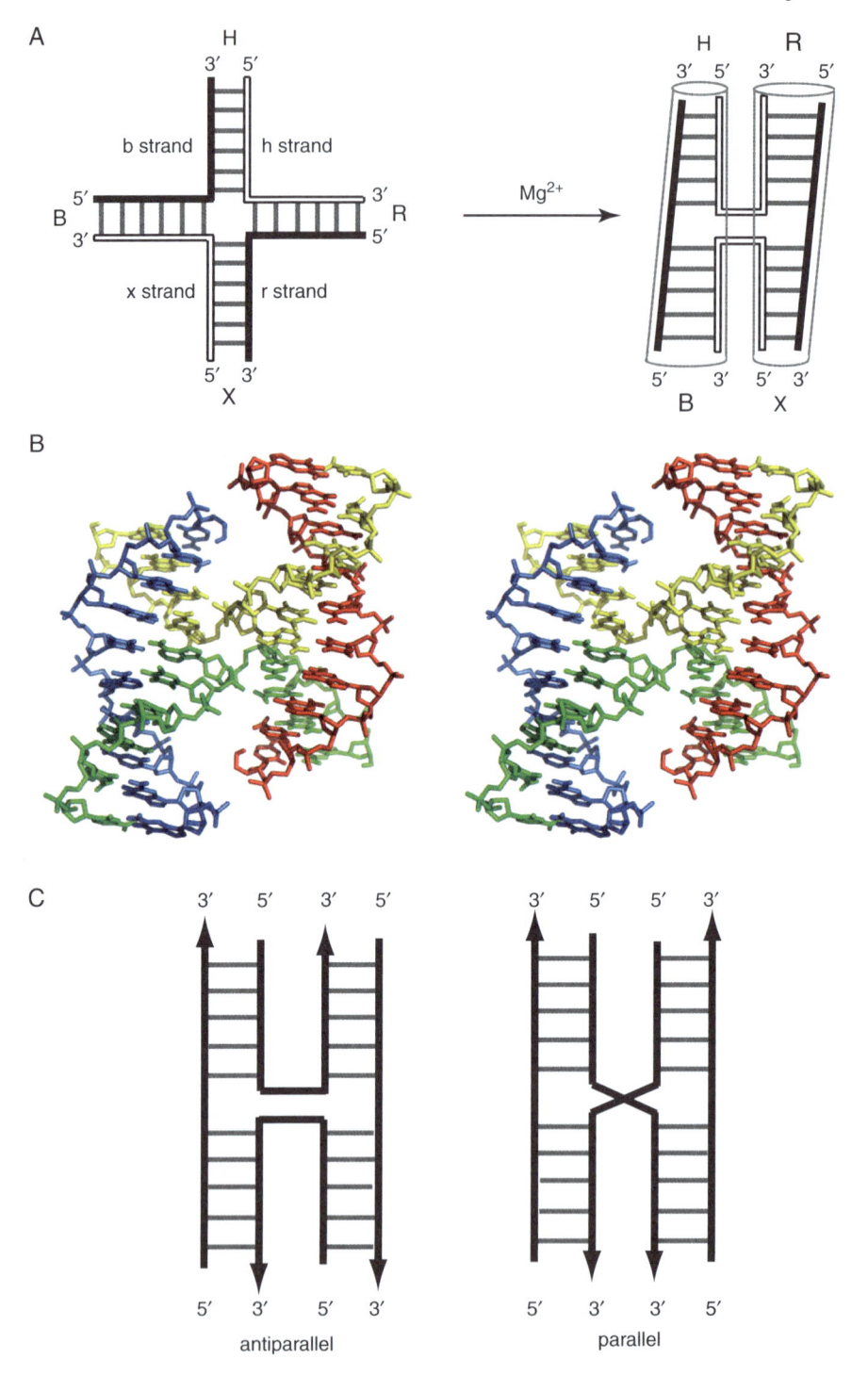

Chapter 6

induced to fold into the stacked X-structure on addition of divalent metal ions such as Mg^{2+} (Figure 6.1A).[13,15] This structure is formed by the pairwise, coaxial stacking of helices to adopt a right-handed, antiparallel cross.

The structure was first proposed on the basis of electrophoretic,[13,16,17] chemical probing,[18] hydrodynamic[19] and FRET studies,[5,20] together with molecular modelling.[21] Essentially all the conformational principles of this structure have been confirmed by crystallographic studies[22–28] (Figure 6.1B).

6.2.1 Dynamics of the Four-way Junction

Formation of the stacked X-structure involves a lowering of symmetry from four-fold in the open-square structure to two-fold. This generates two distinct conformers of the structure that are equivalent if the nucleotide sequence is ignored. This is illustrated for the junction shown schematically in Figure 6.2(A). The two-fold symmetry of the stacked X-structure divides the component strands of the junction into two classes, and the nature of every strand changes on switching between the conformers. The continuous strands run the full length of the coaxially-paired helices, while the exchanging strands cross between the stacks, forming an exchange region at the centre of the junction. A single junction can sample both stacking conformers, with continual conformational exchange between structures. This was first suggested by experiments in bulk solution,[29,30] and subsequently demonstrated directly by single-molecule FRET spectroscopy[31,32] (Figure 6.2B). The relative stability of the two conformers is a function of the base sequence around the point of strand exchange.

6.2.2 Metal Ions and the Electrostatics of the Four-way Junction

Nucleic acids are polyelectrolytes, and their folding is strongly influenced by electrostatic forces. Electrostatics are extremely important in the folding of the

---

Figure 6.1 The four-way junction in DNA. (*A*) Schematic showing the organization of a four-way junction, and its metal ion-induced folding process. In the absence of added metal ions the junction exists as an extended structure, with four helices directed towards the corners of a square. In the presence of divalent metal ions the junction fold into the stacked X-structure by the pairwise coaxial stacking of arms (e.g. helices H and B). This generates two different kinds of strand. The continuous strands are shown here as bold lines, while the exchanging strands are open lines. (*B*) Parallel-eye stereoscopic view of the crystal structure of a four-way DNA junction.[24] The four strands are differentiated by colour. (*C*) Antiparallel *vs.* parallel conformations of a four-way junction. The conformation of four-way junctions may be defined according to the relative orientation of the axes of the coaxially-stacked helices. The extreme forms are illustrated. In the antiparallel conformation the continuous strands run in opposing directions, while in the parallel conformation they run in the same direction. Although Holliday junctions are often depicted as parallel in textbooks, the natural conformation of DNA junctions is closer to the antiparallel geometry. However, RNA junctions can sample the parallel conformation.

four-way DNA junction, as demonstrated by the key role for metal ions in the folding process. Calculations indicate that there is a very high local electrostatic potential in the region of the strand exchange of the DNA junction,[1,33,34] and this is likely to be stabilized by charge screening from metal ions. Stable formation of the stacked X-structure requires either divalent metal ions,[13,15] or very high concentrations of monovalent ions,[20,35] and the rate of conformer exchange is strongly dependent on salt concentration.[31,32]

We have explored this by the substitution of phosphate groups in the junction by electrically-neutral methyl phosphonates.[36] We found that if the central phosphate groups of the exchanging strands were rendered neutral by substitution, the resulting junction was substantially folded into the stacked X-structure in the absence of added metal ions. Even the phosphate groups immediately 3' and 5' to the point of strand exchange exerted a significant influence on the folding process. These effects can dominate the relative stability of the alternative stacking conformers; we found that when the central phosphate groups of the continuous strands of a junction were replaced by methyl phosphonates, the population shifted towards the alternative conformer such that the modified phosphates would now lie at the point of strand exchange.[36] The conformational switch that interconverts the conformers results in a marked change of environment for the central phosphates as they move between the duplex-like conformation of the continuous strands (P–P distance = 34 Å) to the point of strand exchange (P–P distance = 6.2 Å). However, these effects are more subtle than simple charge neutralisation, and the stereochemical environment of the non-bridging oxygen atoms is important in the conformation of the junction.[37] Substitution of the *pro*-R oxygen atom by methyl leads to conformational transition to the stacking conformer that places this phosphate at the point of strand exchange, whereas modification of the *pro*-S oxygen destabilizes this conformation of the junction. It is likely that the stereochemical location of the methyl group affects the interaction with metal ions in the centre of the junction.

6.2.3 Branch Migration

During homologous recombination the four-way junction formed can undergo many steps of branch migration. This occurs as a random walk process, and therefore the time taken increases as the square of the length of DNA migrated. It would be expected that the exchange of basepairing would be much easier in the open conformation that exists in the absence of added metal ions, and early bulk-phase experiments showed that branch migration was slowed 1000-fold in the presence of Mg^{2+} ions.[38,39] Using single-molecule methods we were able to observe stacking conformer transitions and single steps of branch migration simultaneously from individual junctions. The behaviour of junctions was very similar to that of fixed junctions for as long as they remained at a single branch point, exhibiting sequence-dependent stacking conformer bias. Multiple conformer transitions occurred before a branch migration step was observed.

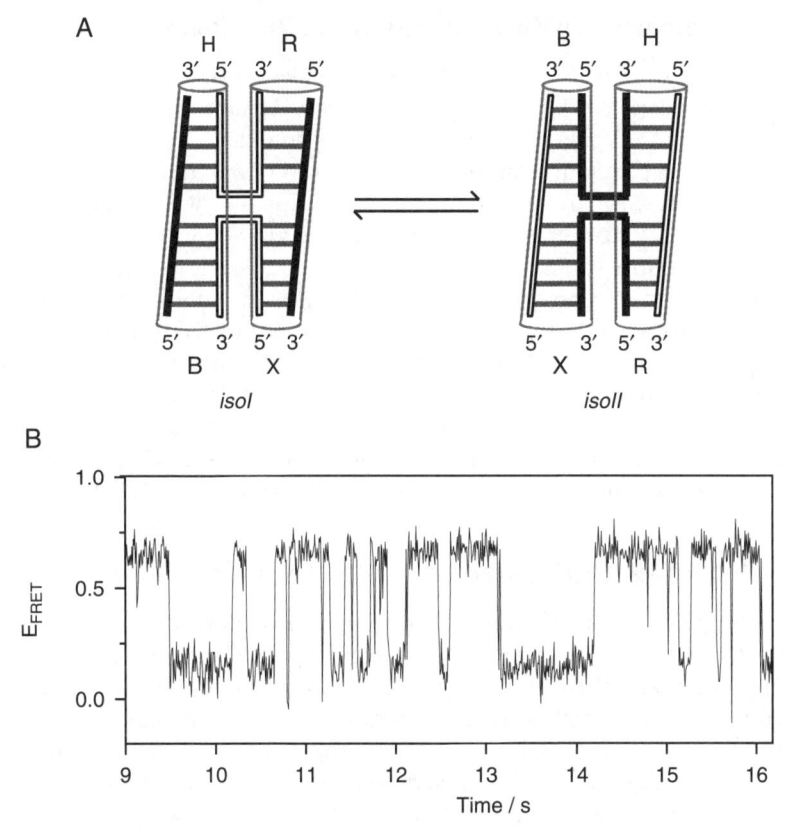

Figure 6.2 Exchange between alternative stacking conformers of a four-way DNA junction. (*A*) Schematic illustrating the process of conformer exchange. This involves the exchange of stacking partners. In the *isoI* conformer helix B is coaxially stacked with helix H, while in *isoII* helix B becomes stacked with helix X. In the process, continuous strands become exchanging strands and *vice versa*. (*B*) A single DNA junction molecule undergoing conformer exchange.[31] DNA junctions with fluorophores attached to the ends of the B and H helices were tethered to a glass slide and the efficiency of fluorescence resonance energy transfer (E_{FRET}) in single junction molecules was measured. The time trace reveals repeated interconversion between the *isoI* conformation (giving low E_{FRET}) and *isoII* (high E_{FRET}). These data were obtained in the presence of a high Mg^{2+} concentration; the exchange process is very much faster at lower ionic concentrations. The sequence of the junction determines the conformational equilibrium between the two states; in this example (junction 7) the two states are equally populated.

The data revealed a surprisingly rugged and highly sequence-dependent energy landscape of branch migration, and conformer exchange and branch migration share a common intermediate that should resemble the open state. The process of branch migration requires acceleration *in vivo*, and this is brought about by specific proteins such as the RuvAB complex of *Escherichia coli*.[40]

6.2.4 Comparison with Four-way RNA Junctions

RNA junctions are widespread in RNA molecules, and play key architectural roles in those species. For example, a perfect 4H junction is critical to the folding of the hairpin ribozyme under physiological conditions.[41-44] The properties of RNA junctions differ significantly from those of their DNA cousins, however. Firstly, unlike the DNA junction, 4H RNA junctions retain their coaxial stacking in the absence of added metal ions. Second, they exhibit a wider range of structural polymorphism. Single-molecule studies of the junction of the hairpin ribozyme have revealed that it samples not only both stacking conformers but a parallel conformation too (Figure 6.1C).[45] By contrast we know that DNA junctions essentially never adopt a parallel structure.[32] Furthermore, most natural junctions in RNA are not perfectly basepaired, but elaborated with extra formally unpaired nucleotides. This can greatly expand the conformational possibilities as well as the dynamics of these structures as exemplified by a four-way junction found in the HSV IRES.[46]

6.3 Proteins that Interact with DNA Junctions

Several proteins exhibit structure-selective binding to DNA junctions. Many of these are involved in the processing of Holliday junctions, either to promote branch migration or to resolve them into duplex species. In this chapter we concentrate on the junction-resolving enzymes. Branch migration proteins are widespread, and exemplified by the RuvAB complex of *E. coli*. As discussed above, branch migration proceeds very much faster in the absence of divalent metal ions, where the extended structure is expected to facilitate the required exchange of basepairing. RuvA is a tetramer that binds selectively to the four-way junction, distorting the structure into an open-square conformation. The process can be accelerated further by providing directionality. RuvB is a hexameric protein with helicase motifs that effectively "pumps" the DNA in one direction.

There are also many reports of other proteins with a less obvious connection to recombination that exhibit selective binding to DNA junctions.[47-49] However, the enhanced binding of this class of proteins may be to some degree artifactual, resulting from the widened minor groove of the DNA at the point of strand exchange. A good example is provided by the HMG box proteins,[47,50] where it has been shown that binding occurs to the open geometry of the DNA junction, not the stacked X-form.[51]

A very important class of proteins that bind to four-way DNA junctions consists of enzymes involved in the integrase family of site-specific recombination, exemplified by Cre (refs. 11 and 52) and Flp.[12] However, these are reviewed in Chapter 12, and so will not be discussed further here.

6.4 Junction-resolving Enzymes

The junction-resolving enzymes are essentially nucleases that are targeted to the structure of the four-way junction. They introduce paired cleavages at four-way helical branchpoints, such that the product can be dissociated and ligated into perfect duplex species. All such enzymes are relatively small, basic dimeric proteins that require divalent metal ions for function.

6.4.1 Occurrence of the Junction-resolving Enzymes

Junction-resolving enzymes have been isolated from most kinds of organism. They have been found in bacteriophage-infected *Escherichia coli*,[53,54] eubacteria,[55,56] euryarchaea,[57] crenarchaea[58] and yeast.[59–63] There is increasing evidence that the mammalian equivalent may be the complex of Rad51C and XRCC3,[64,65] but this is still some distance from being a purified junction-resolving enzyme. Pox viruses such as vaccinia catalyse active recombination in the cytoplasm of host cells. The pox viral resolving enzyme (encoded by gene A22 in vaccinia) was identified by its weak sequence similarity to the Cce1 and RuvC enzymes[66,67] and confirmed biochemically.[66]

6.4.2 Phylogeny

Some interesting phylogenetic connections have emerged between the different junction-resolving enzymes. It was first noted that RuvC, the fungal mitochondrial enzymes such as Cce1 and the pox viral enzymes shared critical catalytic and metal binding residues, and very likely a similar core fold. They were placed in a superfamily that included retroviral integrase protein (see Chapter 11).[66,67] A second grouping emerged that included the archaeal resolving enzymes and bacteriophage T7 endonuclease I; these were placed in a superfamily that included type I, II and III restriction endonucleases, λ-exonuclease, MutH and Vsr.[67] There are also similarities between the metal binding sites of the integrase and nuclease superfamilies themselves, and it is possible that both derived from an ancestral metal binding motif.

6.4.3 Junction-resolving Enzymes are Dimeric

The junction-resolving enzymes bind four-way DNA junctions in dimeric form,[61,68–70] with dissociation constants in the low nM range. They exist mainly in dimeric form in free solution,[56,68,71] but with widely varying rates of subunit exchange.

6.4.4 Structures of the Junction-resolving Enzymes

There are crystallographic structures for several the junction-resolving enzymes (Figure 6.3). The first to be determined was that of RuvC of *E. coli*[72] (Figure 6.3A). It is a dimer of 19 kDa subunits with a relatively simple dimer interface consisting of a helical bundle and no exchange of segments. The monomers consist of a mixed β-sheet sandwiched between two α-helical sections, with an overall fold similar to that of *E. coli* RNaseH (this fold is also discussed in Chapters 11 and 13). The putative active site consists of a cleft containing essential[73] acidic amino acid side

A. RuvC

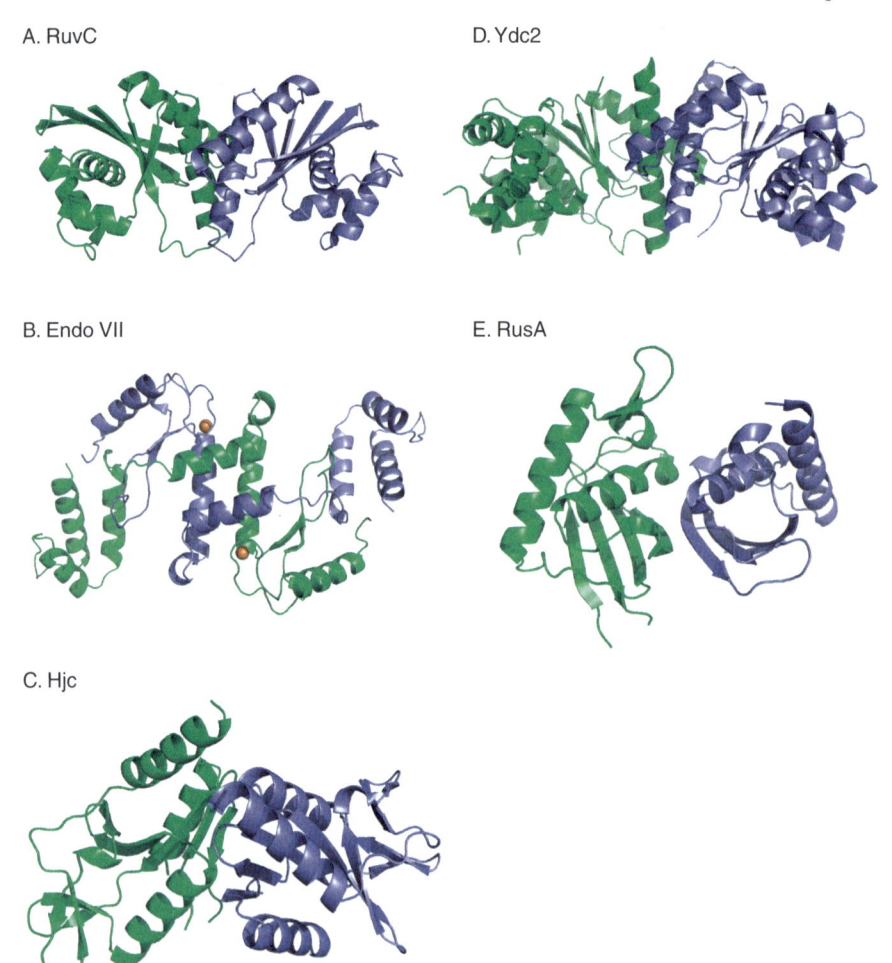

D. Ydc2

B. Endo VII

E. RusA

C. Hjc

Figure 6.3 Crystal structures of resolving enzymes. The two polypeptides of each enzyme are differentiated in green and blue. (*A*) Structure of RuvC of *E. coli*.[72] (*B*) Structure of phage T4 endonuclease VII.[74] The two zinc ions are shown as orange spheres. Note the extensive subunit interface. (*C*) Structure of Hjc of *S. solfataricus* (78). (*D*) Structure of Ydc2 of *S. pombe*.[80] Note the similarity of the overall fold to that of RuvC. (*E*) Structure of RusA of *E. coli*.[81]

chains. This structure can be regarded as the archetype for the resolvase family junction-resolving enzymes.

The second structure to be solved was that of phage T4 endonuclease VII. It has an elongated, flat dimeric structure with extensive α-helices but very little β-sheet (Figure 6.3B).[74] The dimer is formed by an extensive association between the two polypeptides, where the dimer interface is virtually the entire molecule. Despite this however, subunit exchange is rapid in solution.[68] Four cysteine side chains tetrahedrally coordinate a zinc ion, stabilizing the structure of each monomer, particularly in the region of the probable active site; mutation of individual cysteine residues led to loss of activity and destabilization of the protein.[75] The probable active sites consist of several critical residues[76,77] located in a cleft formed by a β-finger and the α-helix to its C-terminal side. The structure of a second phage enzyme, T7 endonuclease I, is discussed below; it has almost no similarity to the T7 endonuclease VII.

Hjc of *Sulfolobus solfataricus* is another α/β protein with a central mixed β-sheet flanked by three α-helices[78] (Figure 6.3C). This enzyme is an example of the nuclease family of junction-resolving enzymes. The protein has a compact structure with a dimer interface formed primarily by hydrophobic interactions between the core β-sheets in each subunit. The dimer has a highly basic flat surface that is the probable junction binding interface. Acidic residues form two pockets for the catalytic metal ions approximately 29 Å apart, identifying the probable active sites. The structure of Hjc from *Pyrococcus furiosus*[79] is very similar to that of the *Sulfolobus* enzyme.

The largest of the junction-resolving enzymes for which a structure has been determined is the mitochondrial Ydc2 enzyme of *Schizosaccharomyces pombe*.[80] Unsurprisingly in view of the phylogeny (it is a member of the integrase family), the overall fold of this protein is strikingly similar to that of RuvC, with extra helical segments apparently fused at one end (Figure 6.3D). The close similarity extends to the putative active site, where three acidic side chains corresponding to their equivalents in RuvC coordinate a metal ion, together with the phenolic hydroxyl of a tyrosine side-chain. Cce1 is the homolog of Ydc2 in budding yeast, but unfortunately there is no structure yet available of this enzyme, despite valiant attempts at crystallization.

By contrast RusA is the smallest of the junction-resolving enzymes at a mere 120 amino acids. It was isolated from *E. coli*, where it is encoded by a cryptic lambdoid prophage,[81] but it has a family of homologs in other strains. Like all the resolving enzymes it binds to DNA junctions as a dimer, and is dimeric in free solution.[70,82] The structure of the free protein was solved by X-ray crystallography at 1.9 Å resolution[83] (Figure 6.3E). It is dimeric in the crystal, consisting of compact monomers with a mixed β-sheet flanked on one side by α-helices. It dimerizes by fusion of the β-sheets and packing of the α-helices. The helical face is basic, with a cleft running down the centre formed by the dimer interface. This contains three essential[70,82] acidic side chains, forming the presumed active sites.

The structure of the RecU junction-resolving enzyme from the Gram positive bacterium *Bacillus subtilis* has been determined.[84] Like Hjc and T7 endonuclease I it is clearly related to the restriction enzymes.

6.5 Molecular Recognition and Distortion of the Structure of DNA Junctions by Resolving Enzymes

Binding to DNA junctions is fundamentally structure-selective. Complexes of a dimer of junction-resolving enzymes with a four-way DNA junction are typically not significantly displaced by a 1000-fold excess of duplex DNA of the same sequence.[61,68–70] The resolving enzymes are all characterised by high calculated p*I* values (typically ≥ 9.4), and the known structures reveal predominantly basic surfaces except for the active sites. The DNA junction probably contacts the protein at many points, such that the enzyme fits to the global shape of the junction.

6.5.1 Sequence Specificity of the Junction-resolving Enzymes

In addition to structural selectivity, several junction-resolving enzymes also exhibit a pronounced sequence specificity. Enzymes such as RuvC and Cce1 bind to junctions of any sequence with equal affinity, but will only cleave particular sequences that must be correctly positioned with respect to the point of strand exchange.[69,85] For example, Cce1 cleaves most efficiently at the optimal sequence ACT⇓A, where ⇓ represents the position of cleavage, which should be located immediately at the point of strand exchange for the fastest cleavage rate.[86] Systematic analysis of the sequence preference reveals that the central two basepairs are most discriminatory, such that any change of sequence results in a ≥ 1000-fold slower rate of cleavage. Since these changes affect the rate of cleavage but not the strength of binding, they must reflect contacts made in the transition state. Further analysis has revealed functional groups that appear to be particularly involved. The N7 of the guanine base-paired with the cytosine of the optimal sequence is particularly important for example, and replacing the nitrogen with a C–H increases the free energy of activation ($\Delta G^{\ddagger}$) by 20 kJ mol^{-1}. Similar analyses have revealed that the optimal sequence for RuvC is (A ~ T)TT⇓(C > G ~ A),[87] where the central two basepairs are again the most discriminatory positions. Cleavage must again occur at the point of strand exchange for the fastest rate. Sequence specificity for cleavage may have the effect of filtering out structures other than Holliday junctions formed by strand exchange between homologous sequences, because there is a much higher probability of finding the target sequence if the junction is scanning many sequences by branch migration.

6.5.2 Structural Distortion of DNA Junctions by the Junction-resolving Enzymes

In view of the strong selectivity for the structure of the junction it is perhaps surprising that the junction-resolving enzymes distort the DNA structure in a major way on binding, both globally and locally. This is true for all such enzymes studied, including T7 endonuclease I,[88] T4 endonuclease VII,[68] RusA,[70] RuvC,[89,90] Cce1,[91] Ydc2,[92] and Hjc.[90] In most cases, the distortion involves an opening of the structure into a geometry that is closer to that of the unbound DNA in the absence of divalent metal ions. In general this involves some degree of the loss of coaxial stacking of helical arms. For RuvC,[89,90] and Hjc,[91] the structure is an unstacked X-shape with two-fold symmetry. The yeast enzymes Cce1[91] and Ydc2[92] open the structure into one where the arms are directed towards the corners of a square, and 2-aminopurine fluorescence experiments suggest that basepairing is also disrupted around the point of strand exchange.[93]

6.5.3 Coordination of the Resolution Process

The imposed distortion of the junction may be related to another facet of the resolution process, *i.e.* the strand cleavage reaction. A productive resolution of the junction requires something approximating to coordinated cleavage of the two strands. Active subunits continue to cleave junctions unilaterally when part of a heterodimer with an inactive mutant subunit,[94–96] showing that there is no direct coupling between the active sites. Experiments using cruciform substrates stabilized by negative supercoiling showed that the two cleavages occurred within the lifetime of the enzyme-junction complexes of T4 endonuclease VII,[94] T7 endonuclease I[71] and Cce1.[96] However, examination of the products at short times revealed that a unilaterally-cleaved species existed as a transient intermediate, indicating a mechanism by which the enzyme makes the two cleavages sequentially without releasing the substrate in between. Detailed analysis of the kinetics of the process[96] showed that the rate of the cleavage of the second strand was accelerated by a factor of 5–10-fold over the rate that the same site would be cleaved in an initial attack. An equivalent analysis of RuvC cleavage indicated a 150-fold acceleration of second strand cleavage.[97] The net effect of these accelerations is that the second cleavage follows hard on the heels of the first, before the enzyme can be released from the junction, thereby ensuring a productive resolution event.

The origin of the faster rate of second strand cleavage is not known. We have suggested that this could result from a loss of conformational strain developed on binding the junction after the first cleavage reaction occurs.

6.6 T7 Endonuclease I

One of the most intensely studied of the junction-resolving enzymes is endonuclease I of the coliphage T7, and this will be discussed as a case study for these proteins. Phage T7 DNA undergoes genetic recombination during infection (Studier, 1969), and encodes its own junction-resolving enzyme called endonuclease I[54,98,99] that is encoded by gene *3*.[100–102] Mutants in gene *3* are deficient in recombination,[103,104] and accumulate branched DNA intermediates.[105]

6.6.1 Biochemistry of Endonuclease I

The T7 gene encoding endonuclease I has been cloned and expressed.[98] The protein is a highly basic ($pI_{calc} = 9.5$), 149 amino-acid nuclease with a high specificity for branched DNA. It binds to four-way junctions with a dissociation constant of ~ 1 nM, with a slow dissociation rate corresponding to a half-life of the complex of ~ 150 min.[106] Endonuclease I is dimeric in free solution, with an undetectably slow rate of subunit exchange in the absence of denaturing agents. Like all the junction-resolving enzymes, it also binds to junctions in dimeric form,[71] which is consistent with the bilateral cleavage required for a productive resolution. Simultaneous cleavage is not required, since an active monomer can cleave a DNA junction unilaterally in heterodimeric combination with an inactive partner. For the fully-active enzyme the two cleavages are introduced sequentially during the lifetime of a stable DNA-protein complex that maintains the integrity of the structure of the junction.[71]

Endonuclease I exhibits a relatively weak selectivity for base sequence, in contrast to enzymes like RuvC and Cce1. It cleaves the continuous strands of the junction,[107] at the phosphate groups one nucleotide to the 5'-side of the centre. In common with the restriction enzymes, endonuclease I catalyses the hydrolysis of the P–O3' bond[108] to give a 5'-phosphate product. The nuclease activity requires divalent metal ions, but is strongly inhibited by Ca^{2+} ions.[109]

6.6.2 Structure of Endonuclease I

The structure of dimeric endonuclease I in the absence of DNA was determined by X-ray crystallography at 2.1 Å resolution[110] (Figure 6.4). Its fold has little similarity with any of the other junction-resolving enzymes that have been solved. It forms an intimately associated symmetrical homodimer consisting of two domains. Each domain is formed by residues 17–44 from one subunit and residues 50–145 from the other, and the two domains are connected by a bridge that forms part of an extended β-sheet. Each domain consists of a central five-stranded mixed

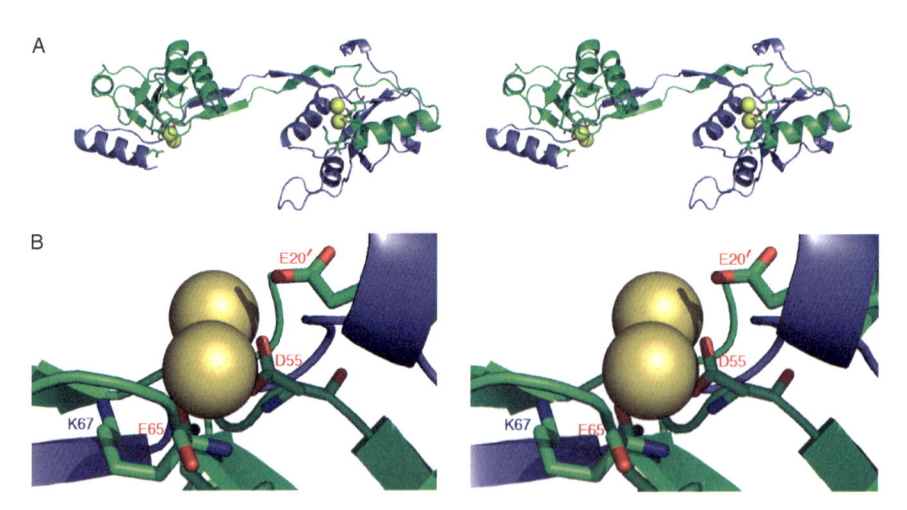

Figure 6.4 Crystal structure of phage T7 endonuclease I.[110,117] (*A*) Parallel-eye stereoscopic view of the structure of the dimer, with the two polypeptides differentiated by colour. The metal ions bound in the active sites are indicated by the yellow spheres. (*B*) Parallel-eye stereoscopic view of the structure of the active site. The side chains of the key residues D55, E65, K67 and E20' (the prime denotes that this is contributed by the other polypeptide) are highlighted in stick form.

β-sheet, flanked by five α-helices. A striking feature of the structure is that one strand and helix in each domain is contributed by the other subunit in the dimer.

6.6.3 The Active Site

The active sites were identified in the structure originally as a cluster of essential acidic amino acids.[111] The enzyme is a member of the nuclease family, and the active site sequence conforms to the restriction enzyme consensus[112] (D/E20)... P54D55... (D/E65)XK67 (the residue numbers being mapped onto the sequence of endonuclease I). The α and β carbon atoms of the four critical side chains (E20, D55, E65 and K67) plus the conserved proline (P54) can be superimposed with the structure of the corresponding residues of the restriction enzyme *Bgl*I (ref. 113) with an RMSD of 0.49 Å. Similar catalytic motifs also exist in λ-exonuclease,[114] MutH,[115] TnsA,[116] and the archaeal junction-resolving enzymes.[78,79] Two Ca^{2+} ions were observed coordinated in the active site of *Bgl*I with a bound DNA fragment in the crystal structure,[113] and two Mn^{2+} ions were found in equivalent positions in the crystal structure of wild-type sequence endonuclease I in the absence of DNA.[117] Binding of two metal ions to the protein was also indicated by isothermal titration calorimetry,[117] and Fe(II)-mediated cleavage of DNA indicated a very similar location of metal ions in the complex[109] (see Chapter 13 for further discussion of restriction endonucleases).

Interestingly, the extensive subunit domain interface generates a mixed active site, consisting of amino acids provided by both polypeptides of the dimer, *i.e.* E20 derived from one polypeptide and D55, E65 and K67 from the other. This was confirmed by the demonstration that a heterodimer of different active site mutants could exhibit *in vitro* complementation, generating a unilaterally-active enzyme.[111] A similar mixed active site exists in RusA,[83] and perhaps suggests that this may be a mechanism for avoiding promiscuous nuclease action.

6.6.4 Catalysis of Phosphodiester Bond Hydrolysis

The nature of the active site and the close similarity with type II restriction enzymes suggests that hydrolysis is achieved by nucleophilic attack of a water molecule coordinated directly to a metal ion. The anionic transition state could be stabilized by a combination of a metal ion and the active site lysine side-chain. The pH dependence of the hydrolysis reaction is log–linear, corresponding to the titration of a group with a $pK_A > 9$.[108] A gradient of 0.9 indicated the transfer of a single proton, and the data are consistent with attack by a hydroxyl ion bound to a metal ion.

Substitution of a non-bridging oxygen atom by a methyl group at the scissile phosphate was observed to give strong kinetic effects on hydrolysis at that site, with rates reduced by > 100-fold.[108] Importantly, there was a marked difference in the magnitude of the effect between substitution of the individual prochiral oxygen atoms, with the larger effect resulting from substitution of the *pro*-S oxygen atom, whereupon cleavage was virtually undetectable. This observation was entirely consistent with a model of the enzyme bound to a DNA junction

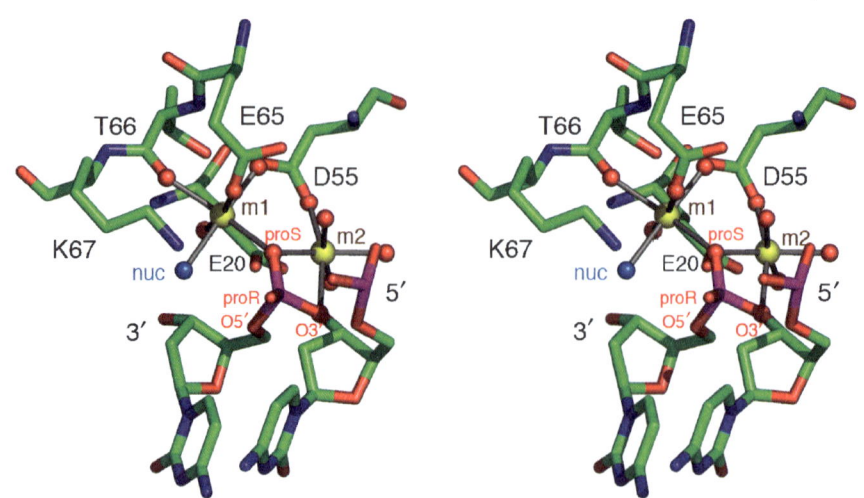

Figure 6.5 Model of DNA bound in the active site of phage T7 endonuclease I.[108] Parallel-eye stereoscopic view of the active site with two metal ions and their associated water molecules. The water molecule that acts as the putative nucleophile (nuc) in phosphodiester bond hydrolysis is coloured blue.

(see below), in which the *pro-S* oxygen atom is coordinated to both metal ions,[107] while the *pro*-R does not interact with either (Figure 6.5). This geometry is fully supported by the recent crystallographic structure of the complex (see below).

Endonuclease I and the type II restriction enzymes are clearly closely similar nucleases, in keeping with their phylogenetic relationship.[67] Both proceed by an in-line attack on the scissile phosphate of a metal bound hydroxyl group opposite the 3′-oxygen atom, and the *pro*-S oxygen of the DNA is coordinated to both metal ions in the transition state.

6.6.5 Interaction between Endonuclease I and DNA Junctions

Like the other junction-resolving enzymes, endonuclease I significantly distorts the structure of the DNA junction on binding.[107] Electrophoretic experiments have shown a major global change in conformation in which the helices remain approximately coaxial, but their axes become perpendicular. 2-Aminopurine fluorescence and permanganate probing experiments also reveal that there is a significant local change in conformation at the centre of the junction, showing that base stacking around the point of strand exchange becomes disrupted. Despite the perturbation of DNA structure, the geometry remains two-fold symmetric, and two alternative conformers are present in solution, accounting for two sets of cleavage products that are observed. However, the complex conformer bias can be markedly altered by shortening two arms, showing that the enzyme needs to contact the two arms containing the 5′ ends of the continuous strands.[106] These experiments, together with others in which sections of helix were permuted around the junction, indicated that the enzyme contacts ~6 bp in these two arms, a result that was consistent with protection against hydroxyl radical attack.

We constructed a detailed model of the structure of the complex of endonuclease bound to a four-way DNA junction.[107] It was based upon our knowledge of the global shape of the junction when bound to the enzyme, the known positions of enzymatic cleavage on the continuous strands, positions of protection against hydroxyl radicals, and the structure of the *Bgl*I-DNA complex with its closely similar active site geometry. The helices remain coaxially stacked in pairs, but with a 90° angle subtended between axes. The two protein domains are located primarily on the outer face of the junction, where the continuous strands could be accommodated into the active sites, and the β-bridges cross the major groove face of the junction. The model explained how the enzyme could be selective for the structure of a four-way junction, such that both continuous strands could be accommodated into the two active sites, leading to a productive resolution of the junction. It was subsequently found to agree very well with the stereospecific effects of methylphosphonate substitution at the scissile phosphate on cleavage rates,[108] as discussed above.

Very recently we have solved a crystal structure of a complex between a four-way DNA junction and endonuclease I,[118] at 3.1 Å resolution. Simultaneously, the structure of a complex of T4 endonuclease VII bound to a junction has also been solved, by Yang and coworkers,[119] and the two structures make an interesting contrast, as expected. The overall geometry of the endonuclease I complex is in excellent agreement with our earlier model discussed above,[107] with coaxial pairs of helices and perpendicular axes (Figure 6.6). The extent of the central distortion is greater than we anticipated. Stacking between basepairs at the point of strand exchange is broken, as if the coaxial helices have been drawn apart to leave a gap of ~9 Å, thus explaining the earlier chemical and spectroscopic data. The two active sites are separated by 35 Å, allowing them to contact the continuous strands on the outer face of the junction. In addition to the marked manipulation of the DNA structure by the protein, we also observed a significant change in the protein structure on binding to the junction. To some degree the enzymes closes around the DNA, resulting in the active sites becoming ~4 Å closer together in the bound form. The organization within the active sites is in good agreement with expectation, showing the *pro*-S non-bridging oxygen of the scissile phosphate bonded to the two metal ions. The protein makes extensive contacts with the backbone of the helices, particularly in the arms containing the 5' ends of the continuous strands. This is consistent with biochemical studies,[106] and these interactions are the basis of the recognition of the structure of the junction. The protein generates two DNA-binding channels, 30 Å long, that are approximately perpendicular (Figure 6.6B). These grip the duplexes in a specific relative orientation where the angle subtended between these two arms is 80°, thus ensuring that binding is selective for branched DNA that can achieve this geometry. Interestingly, the interaction of T4 endonuclease VII with a DNA

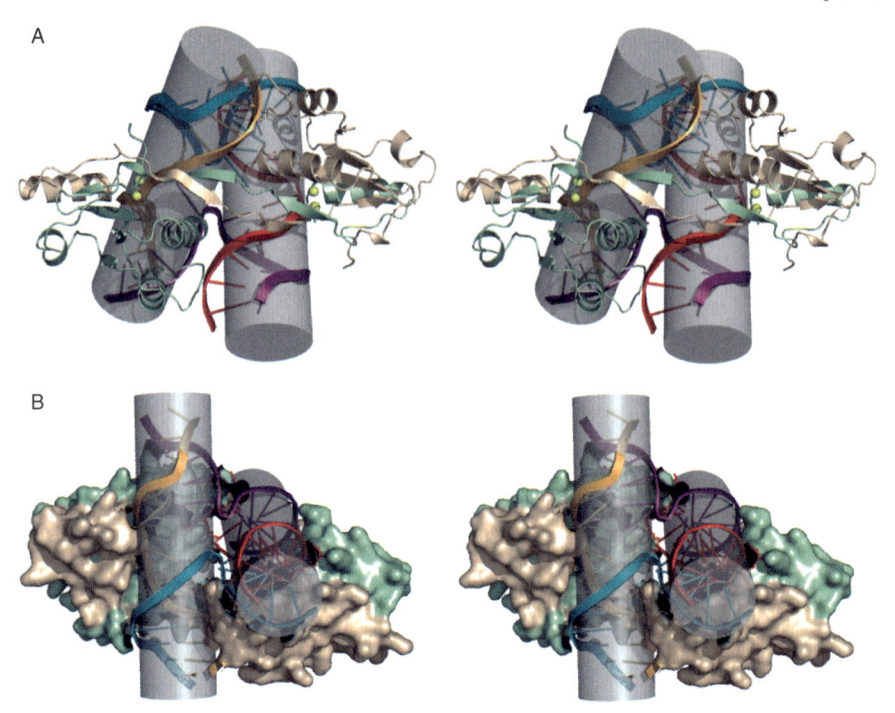

Figure 6.6 Crystal structure of complex between a DNA junction and phage T7 endonuclease I.[118] Two parallel-eye stereoscopic views of the structure, with the two polypeptides coloured green and fawn. The four strands of the junction are differentiated by colour. (*A*) View looking down onto the major groove side of the junction, with the protein shown in cartoon form. The coaxially-aligned DNA helices of the junction are shown by the cylinders. The metal ions bound in the active sites are indicated by the yellow spheres. (*B*) View of the minor groove side of the junction, showing the structure-selective binding of the enzyme. The surface of the protein is shown. The helices lie in 30 Å-long hemi-cylindrical channels, oriented approximately perpendicular to each other.

junction is quite different.[119] The protein binds an open, X-shaped junction over a relatively flat protein surface, employing a much more two-dimensional recognition process. Comparing the structure of DNA junctions bound by endonuclease VII, RuvC and Cce1, it is likely that this is a more general strategy for the structure-selective binding to junctions. However, the comparison of these two enzymes emphasizes that there is more than one way to achieve the recognition of branched DNA species.

6.7 In Conclusion

The junction-resolving enzymes are an important class of nucleases, both because of the role they have in recombination and repair and also because of what they can teach us about the recognition and manipulation of DNA structure on a rather large scale. The most striking property of these proteins is their ability to recognize the global structure of branched DNA molecules, and they can arguably be regarded as a paradigm for molecular recognition of this kind. Yet, remarkably, it turns out that all the enzymes distort the very structure that they recognize. And the distortion of the DNA seems intimately connected to the cleavage process, ensuring a productive resolution and not a mess of randomly nicked DNA. Thus, to understand fully the function of these enzymes we need to incorporate all these aspects into our description of their operation. We are hopeful that with the first crystal structures of complexes of resolving enzymes bound to DNA junctions that we will be able to dissect these processes down to the atomic level in due course.

Acknowledgements

The author gratefully acknowledges the major intellectual and experimental contributions of his coworkers and collaborators to these studies, especially Drs Anne-Cécile Déclais, Jia Liu and Alasdair Freeman (Dundee), Jon Hadden and Simon Phillips (Leeds) and Sean McKinney and Taekjip Ha (University of Illinois, Urbana-Champaign). Cancer Research UK is thanked for financial support of work in Dundee.

References

1. D.M.J. Lilley, *Q. Rev. Biophys.*, 2000, **33**, 109–159.
2. R. Holliday, *Genet. Res.*, 1964, **5**, 282–304.
3. H. Potter and D. Dressler, *Proc. Natl. Acad. Sci. U.S.A.*, 1976, **73**, 3000–3004.
4. T.L. Orr-Weaver, J.W. Szostak and R.J. Rothstein, *Proc. Natl. Acad. Sci. U.S.A.*, 1981, **78**, 6354–6358.
5. A. Schwacha and N. Kleckner, *Cell*, 1995, **83**, 783–791.
6. P.A. Kitts and H.A. Nash, *Nature*, 1987, **329**, 346–348.
7. S.E. Nunes-Düby, L. Matsomoto and A. Landy, *Cell*, 1987, **50**, 779–788.
8. R. Hoess, A. Wierzbicki and K. Abremski, *Proc. Natl. Acad. Sci. U.S.A.*, 1987, **84**, 6840–6844.
9. M. Jayaram, K.L. Crain, R.L. Parsons and R.M. Harshey, *Proc. Natl. Acad. Sci. U.S.A.*, 1988, **85**, 7902–7906.
10. R. McCulloch, L.W. Coggins, S.D. Colloms and D.J. Sherratt, *EMBO J.*, 1994, **13**, 1844–1855.
11. D.N. Gopaul, F. Guo and G.D. Van Duyne, *EMBO J.*, 1998, **17**, 4175–4187.
12. Y. Chen, U. Narendra, L.E. Iype, M.M. Cox and P.A. Rice, *Mol. Cell*, 2000, **6**, 885–897.
13. D.R. Duckett, A.I.H. Murchie, S. Diekmann, E. von Kitzing, B. Kemper and D.M.J. Lilley, *Cell*, 1988, **55**, 79–89.
14. R.M. Clegg, A.I.H. Murchie, A. Zechel and D.M.J. Lilley, *Biophys. J.*, 1994, **66**, 99–109.

15. A.I.H. Murchie, R.M. Clegg, E. Kitzing von, D.R. Duckett, S. Diekmann and D.M.J. Lilley, *Nature*, 1989, **341**, 763–766.
16. G.W. Gough and D.M.J. Lilley, *Nature*, 1985, **313**, 154–156.
17. J.P. Cooper and P.J.J. Hagerman, *Mol. Biol.*, 1987, **198**, 711–719.
18. A. Kimball, Q. Guo, M. Lu, R.P. Cunningham, N.R. Kallenbach, N.C. Seeman and T.D. Tullius, *J. Biol. Chem.*, 1990, **265**, 6544–6547.
19. J.P. Cooper and P.J. Hagerman, *Proc. Natl. Acad. Sci. U.S.A.*, 1989, **86**, 7336–7340.
20. R.M. Clegg, A.I.H. Murchie, A. Zechel, C. Carlberg, S. Diekmann and D.M.J. Lilley, *Biochemistry*, 1992, **31**, 4846–4856.
21. E. von Kitzing, D.M.J. Lilley and S. Diekmann, *Nucleic Acids Res.*, 1990, **18**, 2671–2683.
22. J. Nowakowski, P.J. Shim, G.S. Prasad, C.D. Stout and G.F. Joyce, *Nat. Struct. Biol.*, 1999, **6**, 151–156.
23. M. Ortiz-Lombardía, A. González, R. Erijta, J. Aymamí, F. Azorín and M. Coll, *Nat. Struct. Biol.*, 1999, **6**, 913–917.
24. B.F. Eichman, J.M. Vargason, B.H.M. Mooers and P.S. Ho, *Proc. Natl. Acad. Sci. U.S.A.*, 2000, **97**, 3971–3976.
25. B.F. Eichman, B.H.M. Mooers, M. Alberti, J.E. Hearst and P.S.J. Ho, *Mol. Biol.*, 2001, **308**, 15–26.
26. J.M. Vargason and P.S. Ho, *J. Biol. Chem.*, 2002, **277**, 21041–21049.
27. F.A. Hays, J.M. Vargason and P.S. Ho, *Biochemistry*, 2003, **42**, 9586–9597.
28. J.H. Thorpe, B.C. Gale, S.C. Teixeira and C.J. Cardin, *J. Mol. Biol.*, 2003, **327**, 97–109.
29. S.M. Miick, R.S. Fee, D.P. Millar and W.J. Chazin, *Proc. Natl. Acad. Sci. U.S.A.*, 1997, **94**, 9080–9084.
30. R.J. Grainger, A.I.H. Murchie and D.M.J. Lilley, *Biochemistry*, 1998, **37**, 23–32.
31. S.A. McKinney, A.-C. Déclais, D.M.J. Lilley and T. Ha, *Nat. Struct. Biol.*, 2003, **10**, 93–97.
32. C. Joo, S.A. McKinney, D.M.J. Lilley and T. Ha, *J. Mol. Biol.*, 2004, **341**, 739–751.
33. M.C. Olmsted and P.J. Hagerman, *J. Mol. Biol.*, 1994, **243**, 919–929.
34. M.O. Fenley, G.S. Manning, N.L. Marky and W.K. Olson, *Biophys. Chem.*, 1998, **74**, 135–152.
35. D.R. Duckett, A.I.H. Murchie and D.M.J. Lilley, *EMBO J.*, 1990, **9**, 583–590.
36. J. Liu, A.-C. Déclais and D.M.J. Lilley, *J. Mol. Biol.*, 2004, **343**, 851–864.
37. J. Liu, A.-C. Déclais, S.A. McKinney, T. Ha, D.G. Norman and D.M.J. Lilley, *Chem. Biol.*, 2005, **12**, 217–228.
38. I.G. Panyutin and P. Hsieh, *Proc. Natl. Acad. Sci. U.S.A.*, 1994, **91**, 2021–2025.
39. I.G. Panyutin, I. Biswas and P. Hsieh, *EMBO J.*, 1995, **14**, 1819–1826.
40. K. Hiom and S.C. West, *Cell*, 1995, **80**, 787–793.
41. A.I.H. Murchie, J.B. Thomson, F. Walter and D.M.J. Lilley, *Mol. Cell*, 1998, **1**, 873–881.

42. N.G. Walter, J.M. Burke and D.P. Millar, *Nat. Struct. Biol.*, 1999, **6**, 544–549.
43. Z.-Y. Zhao, T.J. Wilson, K. Maxwell and D.M.J. Lilley, *RNA*, 2000, **6**, 1833–1846.
44. E. Tan, T.J. Wilson, M.K. Nahas, R.M. Clegg, D.M.J. Lilley and T. Ha, *Proc. Natl. Acad. Sci. U.S.A.*, 2003, **100**, 9308–9313.
45. S. Hohng, T.J. Wilson, E. Tan, R.M. Clegg, D.M.J. Lilley and T. Ha, *J. Mol. Biol.*, 2004, **336**, 69–79.
46. S.E. Melcher, T.J. Wilson and D.M.J. Lilley, *RNA*, 2003, **9**, 809–820.
47. M.E. Bianchi, *EMBO J.*, 1988, **7**, 843–849.
48. A. Pontiggia, A. Negri, M. Beltrame and M.E. Bianchi, *Mol. Microbiol.*, 1993, **7**, 343–350.
49. P. Varga-Weisz, K.E. Van Holde and J. Zlatanova, *J. Biol. Chem.*, 1993, **268**, 20699–20700.
50. S. Ferrari, V.R. Harley, A. Pontiggia, P.N. Goodfellow, R. Lovell-Badge and M.E. Bianchi, *EMBO J.*, 1992, **11**, 4497–4506.
51. J.R.G. Pöhler, D.G. Norman, J. Bramham, M.E. Bianchi and D.M.J. Lilley, *EMBO J.*, 1998, **17**, 817–826.
52. F. Guo, D.N. Gopaul and G.D. Van Duyne, *Nature*, 1997, **389**, 40–46.
53. B. Kemper and M. Garabett, *Eur. J. Biochem.*, 1981, **115**, 123–131.
54. B. de Massey, F.W. Studier, L. Dorgai, F. Appelbaum and R.A. Weisberg, *Cold Spring Harbor Symp. Quant. Biol.*, 1984, **49**, 715–726.
55. B. Connolly, C.A. Parsons, F.E. Benson, H.J. Dunderdale, G.J. Sharples, R.G. Lloyd and S.C. West, *Proc. Natl. Acad. Sci. U.S.A.*, 1991, **88**, 6063–6067.
56. H. Iwasaki, M. Takahagi, T. Shiba, A. Nakata and H. Shinagawa, *EMBO J.*, 1991, **10**, 4381–4389.
57. K. Komori, S. Sakae, R. Fujikane, K. Morikawa, H. Shinagawa and Y. Ishino, *Nucleic Acids Res.*, 2000, **28**, 4544–4551.
58. M. Kvaratskhelia and M.F. White, *J. Mol. Biol.*, 2000, **295**, 193–202.
59. S.C. West, C.A. Parsons and S.M. Picksley, *J. Biol. Chem.*, 1987, **262**, 12752–12758.
60. L. Symington and R. Kolodner, *Proc. Natl. Acad. Sci. U.S.A.*, 1985, **82**, 7247–7251.
61. M.F. White and D.M.J. Lilley, *Mol. Cell. Biol.*, 1997, **17**, 6465–6471.
62. M.C. Whitby and J. Dixon, *J. Mol. Biol.*, 1997, **272**, 509–522.
63. M. Oram, A. Keeley and I. Tsaneva, *Nucleic Acids Res.*, 1998, **26**, 594–601.
64. Y. Liu, J.Y. Masson, R. Shah, P. O'Regan and S.C. West, *Science*, 2004, **303**, 243–246.
65. Y. Liu, M. Tarsounas, P. O'Regan and S.C. West, *J. Biol. Chem.*, 2007, **282**, 1973–1979.
66. A.D. Garcia, L. Aravind, E. Koonin and B. Moss, *Proc. Natl. Acad. Sci. U.S.A.*, 2000, **97**, 8926–8931.
67. D.M.J. Lilley and M.F. White, *Proc. Natl. Acad. Sci. U.S.A.*, 2000, **97**, 9351–9353.

68. J.R.G. Pöhler, M.-J.E. Giraud-Panis and D.M.J. Lilley, *J. Mol. Biol.*, 1996, **260**, 678–696.
69. M.F. White and D.M.J. Lilley, *J. Mol. Biol.*, 1996, **257**, 330–341.
70. M.-J.E. Giraud-Panis and D.M.J. Lilley, *J. Mol. Biol.*, 1998, **278**, 117–133.
71. M.J. Parkinson and D.M.J. Lilley, *J. Mol. Biol.*, 1997, **270**, 169–178.
72. M. Ariyoshi, D.G. Vassylyev, H. Iwasaki, H. Nakamura, H. Shinagawa and K. Morikawa, *Cell*, 1994, **78**, 1063–1072.
73. A. Saito, H. Iwasaki, M. Ariyoshi, K. Morikawa and H. Shinagawa, *Proc. Natl. Acad. Sci. U.S.A.*, 1995, **92**, 7470–7474.
74. H. Raaijmakers, O. Vix, I. Toro, S. Golz, B. Kemper and D. Suck, *EMBO J.*, 1999, **18**, 1447–1458.
75. M.-J.E. Giraud-Panis, D.R. Duckett and D.M.J. Lilley, *J. Mol. Biol.*, 1995, **252**, 596–610.
76. M.-J.E. Giraud-Panis and D.M.J. Lilley, *J. Biol. Chem.*, 1996, **271**, 33148–33155.
77. S. Golz, A. Christoph, K. Birkenkamp-Demtroder and B. Kemper, *Eur. J. Biochem.*, 1997, **245**, 573–580.
78. C.S. Bond, M. Kvaratskhelia, D. Richard, M.F. White and W.N. Hunter, *Proc. Natl. Acad. Sci. U.S.A.*, 2001, **98**, 5509–5514.
79. T. Nishino, K. Komori, D. Tsuchiya, Y. Ishino and K. Morikawa, *Structure*, 2001, **9**, 197–204.
80. S. Ceschini, A. Keeley, M.S.B. McAlister, M. Oram, J. Phelan, L.H. Pearl, I.R. Tsaneva and T.E. Barrett, *EMBO J.*, 2001, **20**, 6601–6611.
81. A.A. Mahdi, G.J. Sharples, T.N. Mandal and R.G. Lloyd, *J. Mol. Biol.*, 1996, **257**, 561–573.
82. S.N. Chan, S.D. Vincent and R.G. Lloyd, *Nucleic Acids Res.*, 1998, **26**, 1560–1566.
83. J.B. Rafferty, E.L. Bolt, T.A. Muranova, S.E. Sedelnikova, P. Leonard, A. Pasquo, P.J. Baker, D.W. Rice, G.J. Sharples and R.G. Lloyd, *Structure (Camb)*, 2003, **11**, 1557–1567.
84. N. McGregor, S. Ayora, S. Sedelnikova, B. Carrasco, J.C. Alonso, P. Thaw and J. Rafferty, *Structure*, 2005, **13**, 1341–1351.
85. R. Shah, R.J. Bennett and S.C. West, *Cell*, 1994, **79**, 853–864.
86. M.J. Schofield, D.M.J. Lilley and M.F. White, *Biochemistry*, 1998, **37**, 7733–7740.
87. J.M. Fogg, M.J. Schofield, M.F. White and D.M.J. Lilley, *Biochemistry*, 1999, **38**, 11349–11358.
88. D.R. Duckett, A.I.H. Murchie, M.-J.E. Giraud-Panis, J.R. Pöhler and D.M.J. Lilley, *Philos. Trans. R. Soc. London [Biol.]*, 1995, **347**, 27–36.
89. R.J. Bennett and S.C. West, *J. Mol. Biol.*, 1995, **252**, 213–226.
90. J.M. Fogg, M. Kvaratskhelia, M.F. White and D.M.J. Lilley, *J. Mol. Biol.*, 2001, **313**, 751–764.
91. M.F. White and D.M.J. Lilley, *J. Mol. Biol.*, 1997, **266**, 122–134.
92. M.F. White and D.M.J. Lilley, *Nucleic Acids Res.*, 1998, **26**, 5609–5616.
93. A.-C. Déclais and D.M.J. Lilley, *J. Mol. Biol.*, 2000, **296**, 421–433.

94. M.-J.E. Giraud-Panis and D.M.J. Lilley, *EMBO J.*, 1997, **16**, 2528–2534.
95. R.P. Birkenbihl and B. Kemper, *EMBO J.*, 1998, **17**, 4527–4534.
96. J.M. Fogg, M.J. Schofield, A.-C. Déclais and D.M.J. Lilley, *Biochemistry*, 2000, **39**, 4082–4089.
97. J.M. Fogg and D.M.J. Lilley, *Biochemistry*, 2000, **39**, 16125–16134.
98. B. de Massey, R.A. Weisberg and F.W. Studier, *J. Mol. Biol.*, 1987, **193**, 359–376.
99. P. Dickie, G. McFadden and A.R. Morgan, *J. Biol. Chem.*, 1987, **262**, 14826–14836.
100. M.S. Center and C.C. Richardson, *J. Biol. Chem.*, 1970, **245**, 6285–6291.
101. M.S. Center and C.C. Richardson, *J. Biol. Chem.*, 1970, **245**, 6292–6299.
102. P.D. Sadowski, *J. Biol. Chem.*, 1971, **246**, 209–216.
103. C. Kerr and P.D. Sadowski, *Virology*, 1975, **65**, 281–285.
104. A. Powling and R. Knippers, *Mol. Gen. Genet.*, 1976, **149**, 63–71.
105. Y. Tsujimoto and H. Ogawa, *J. Mol. Biol.*, 1978, **125**, 255–273.
106. A.C. Déclais, J. Liu, A.D.J. Freeman and D.M.J. Lilley, *J. Mol. Biol.*, 2006, **359**, 1261–1276.
107. A.-C. Déclais, J.M. Fogg, A. Freeman, F. Coste, J.M. Hadden, S.E.V. Phillips and D.M.J. Lilley, *EMBO J.*, 2003, **22**, 1398–1409.
108. J. Liu, A.C. Déclais and D.M.J. Lilley, *Biochemistry*, 2006, **45**, 3934–3942.
109. A.D. Freeman, A.-C. Déclais and D.M.J. Lilley, *J. Mol. Biol.*, 2003, **333**, 59–73.
110. J.M. Hadden, M.A. Convery, A.-C. Déclais, D.M.J. Lilley and S.E.V. Phillips, *Nat. Struct. Biol.*, 2001, **8**, 62–67.
111. A.-C. Déclais, J.M. Hadden, S.E.V. Phillips and D.M.J. Lilley, *J. Mol. Biol.*, 2001, **307**, 1145–1158.
112. A. Pingoud and A. Jeltsch, *Nucleic Acids Res.*, 2001, **29**, 3705–3727.
113. M. Newman, K. Lunnen, G. Wilson, J. Greci, I. Schildkraut and S.E.V. Phillips, *EMBO J.*, 1998, **17**, 5466–5476.
114. R.A. Kovall and B.W. Matthews, *Proc. Natl. Acad. Sci. U.S.A.*, 1998, **95**, 7893–7897.
115. C. Ban and W. Yang, *EMBO J.*, 1998, **17**, 1526–1534.
116. A.B. Hickman, Y. Li, S.V. Mathew, E.W. May, N.L. Craig and F. Dyda, *Mol. Cell*, 2000, **5**, 1025–1034.
117. J.M. Hadden, A.-C. Déclais, S.E.V. Phillips and D.M.J. Lilley, *EMBO J.*, 2002, **21**, 3505–3515.
118. J.M. Hadden, A.-C. Déclais, S. Carr, D.M.J. Lilley and S.E.V. Phillips, *Nature*, 2007, **449**, 621–624.
119. C Biertümpfel, W. Yang and D. Suck, *Nature*, 2007, **449**, 616–620.

CHAPTER 7
RNA-protein Interactions in Ribonucleoprotein Particles and Ribonucleases

HONG LI

Department of Chemistry and Biochemistry, and Institute of Molecular Biophysics, Florida State University, Tallahassee, FL 32306, USA

7.1 Introduction

There is an increasing appreciation that biological molecules perform extraordinary functions by forming multicomponent assemblies. The complexes, consisting of protein and RNA, are major players in the processes of gene expression and regulation. These processes are essential not only to basic survival of all cells but also to cell differentiation and development of the higher eukaryotes. The importance of RNA-protein assemblies is accentuated by the fact that over 40% of the human genome is transcribed into RNA and less than 2% is translated into proteins.[1] It is now widely appreciated that RNA molecules act both within and beyond the central dogma of molecular biology.

In comparison with structural diversities exhibited in proteins, RNA has overall simpler structures that are built upon several structural classes or motifs.[2] These include, but are not limited to, single-stranded RNA, double-stranded A-form helix, stem loops (hairpins), and internal loops. Both internal and external loops can be further divided into distinct sequence dependent motifs. RNA tertiary structures are often made of these smaller structural motifs. Correspondingly, to a first approximation, RNA-interacting proteins can be categorized according to the RNA motifs with which they interact.

RSC Biomolecular Sciences
Protein-Nucleic Acid Interactions: Structural Biology
Edited by Phoebe A. Rice and Carl C. Correll
© Royal Society of Chemistry 2008

Although divisions by protein secondary structures are not absolute, β-sheets or extended surfaces made from repeating units appear to interact preferentially with single-stranded RNA; loop regions and α-helices appear to interact with double stranded or stem loop regions of the RNA. The specificity of RNA-protein interactions can be further enhanced by the recognition of nucleotide sequences within structural motifs owing to RNA bases' ability for hydrogen bonding and base stacking.

The remarkable flexibility of the RNA chain is often apparent in RNA-protein interactions. The RNA phosphate backbone has six degrees of freedom while that of proteins has only two. This contributes to the greater flexibility of the RNA backbone than that of proteins. Binding flexible RNA molecules entails a thermodynamically unfavorable loss of entropy that is often compensated by favorable interactions in the final RNA-protein complexes. This unique property of RNA molecules is fully exploited by proteins in controlling the order of assembly for multicomponent RNA-protein systems and in engaging RNA molecules in catalysis.

Ribonucleoprotein particles (RNPs) are RNA-protein molecular assemblies responsible for various essential functions. The RNA components of RNPs often associate tightly with proteins throughout their functional cycles. RNA-protein interactions also take place in the complexes of RNA and protein enzymes that chemically alter RNA. In these complexes, enzymes are bound with RNA until after the chemical reactions take place. Both kinds of RNA-protein complexes share structural features pertinent to affinity and specificity. This chapter summarizes RNA-protein interaction features recently observed in ribonucleoprotein particles and those involving ribonucleases. Many excellent reviews on general RNA-protein interactions[3–8] and on specific types of RNA-protein interactions[9–15] can be found elsewhere. In addition, Chapters 5, 9, and 14 contain descriptions of RNA-protein interactions for single-stranded RNA, for RNA chaperones, and for RNA modifying enzymes, respectively.

7.2 Experimental Methods used to Determine RNA-protein Complex Structures

Obtaining three dimensional structures of RNA and protein complexes aids the efforts in deciphering chemical principles of RNA-protein interactions. To identify the forces that drive the specific association of proteins with RNA, this structural information must be combined with thermodynamic and kinetic analyses of these RNA-protein interactions. The techniques currently used to determine RNA-protein complex structures at atomic resolutions are primarily X-ray crystallography and nuclear magnetic resonance (NMR). Both techniques require milligram amounts of stable complexes at high concentrations (typically above $10\,\text{mg}\,\text{mL}^{-1}$) in aqueous solutions. In addition, single crystals of the RNA-protein complexes, and often the heavy atom derivatives of the crystals, are required for *de novo* determination of their structures by X-ray crystallography. These conditions may not be simultaneously satisfied for every complex owing to the intrinsic flexibility of RNA and its instability. The flexibility often leads to conformational heterogeneity, which is undesirable for crystallization. RNA is susceptible to digestion by contaminating nucleases or hydrated divalent metal ions that are potentially present throughout the sample preparation processes. NMR methods are also limited, even though they don't require crystallization, because of the slow tumbling rate for molecules beyond ~30 000 daltons in molecular weight and of the need for tedious isotopic labeling procedures.[16] These technological limitations are often circumvented by working with core fragments or with simpler homologues of RNA-protein complexes.

The steady increase in the number of RNA-protein complex structures is also facilitated by improvements in sample preparation, especially for RNA molecules, and in structural determination methods themselves. Potential problems due to heterogeneity in RNA termini resulting from the *in vitro* transcription method have been alleviated by several strategies and the cost for synthetic RNA of up to 45 nucleotides in length has reduced considerably. The use of synchrotron X-ray sources, the improved crystallization and preservation methods, and new crystallographic software[17–19] have significantly accelerated the growth in RNA-protein crystal structures. With NMR, the lack of signal in the backbone and lack of long-range distance constraints have been addressed by isotope labeling and dipolar coupling data.[16,20]

7.3 RNA-protein Interactions in Ribonucleoprotein Particles

RNPs perform some of the most essential functions in gene expression and regulation.[21,22] These include but are not limited to protein synthesis by ribosomes, chromosome end maintenance by telomerase RNP, removal of introns from mRNA by spliceosome, removal of 5′ leader sequences in tRNA by RNase P, biogenesis of ribosomal and small nuclear RNAs by small nucleolar (sno)RNPs, protein translocation by signal recognition particles (SRPs), RNA editing by the editing complex, and, finally, gene silencing by miRNPs or siRNPs, which contain microRNAs and small interfering RNAs, respectively. The non-coding RNAs found in these RNPs often play one or all of the three major functional roles: (1) catalysis; (2) substrate recognition; and (3) RNP organization. The RNAs provide the catalytic power of the ribosome and RNase P. While not yet certain, RNA is also expected to catalyze splicing in the spliceosome. In snoRNPs, miRNPs, and siRNPs, the RNA components recruit RNA substrates through Watson–Crick base-pairing. In these cases, correctly positioned substrate RNAs are subsequently modified or cleaved by a protein catalytic subunit. The telomerase RNA also acts as a template for DNA synthesis. The RNA components of RNPs possess structural or sequence characteristics recognized by proteins, which facilitate the (often) step-wise assembly of the entire particle. Although the same types of fundamental forces drive the assembly of all complexes, each complex integrates a set of interactions to achieve specificity and affinity in a different way.

Although the spliceosomal protein U1A was among the first to be crystallized with RNA,[23] there are a relatively small number of spliceosomal protein-RNA complexes for which either crystal or NMR structures are known. One reason for the paucity of structural data may be that the spliceosome is an exclusively eukaryotic enzyme, which sometimes makes it difficult to prepare proteins in sufficient quantity. The most information available on RNA-protein structures found in the spliceosome pertains to complexes involving the RNA recognition motif (RRM) similar to that of U1A. Since there have been excellent and detailed reviews on the structural principles of RRM motifs elsewhere[24,25] and in Chapter 5 of this book, readers are referred to these references for spliceosomal complexes.

7.3.1 Ribosome

The ribosome is the large RNP enzyme responsible for protein synthesis in all organisms. Recent crystal structures of the whole ribosome[26–28] and those of the large[29–31] and the small[29,30] subunits of increasingly high resolution have allowed detailed examinations of this extraordinary RNA-protein complex. Ribosomal RNAs (rRNAs) comprise nearly 60% of the total mass and are thus the core component of the ribosome. Based on secondary structures, rRNAs are arranged into domains (six for 23S and four for 16S) of a few hundred nucleotides that fold into compact structures. Ribosomal proteins mostly bind to the exterior of the ribosome and, in some cases, act as cement to seal interdomain interfaces. As high resolution ribosome structures are now available, systematic examination of molecular interactions is possible. Detailed analysis provides a structural basis for order of assembly and for protein based structural stabilization.[31]

rRNA structures contain a surprisingly high incidence of non-Watson–Crick base pairs ($\sim 40\%$). These non-helical structures introduce bends and twists that allow a high degree of structural curvature in the ribosome.[32] In addition, many rRNA structural elements pack themselves in the interior of the ribosome that is devoid of proteins. An analysis of nucleotide identities of the non-helical region showed that about 62% of the nucleotides are adenosine, suggesting the unique suitability of adenosine for unpaired regions,[32] which is consistent with the observed special tertiary interactions involving adenosine such as the A-minor interactions[33,34] and the high incidence of proteins contacting A-patches.[33]

Although at least one-third of the ribosomal proteins can be individually deleted without disrupting cell growth, the association of ribosomal proteins with rRNA improves the efficiency and accuracy of assembly and, more importantly, of translation.[31,35] The crystal structures of the ribosome and several landmark assembly studies with individual components show that ribosomal proteins can stage the order of assembly by stabilizing correctly folded rRNA.[35–37]

RNA-protein interactions play important roles in the assembled ribosome as revealed by the computed buried solvent accessible surface area. The average surface area buried between RNA and a ribosomal protein is 3400 Å^2 and the largest is 8000 Å^2. In contrast, the largest buried solvent accessible surface area between two ribosomal proteins is only 749 Å^2. A significant portion of RNA surfaces that contact proteins are helical; ribosomal proteins interact with both the minor and major grooves, depending on the accessibility of each groove. In structures of the ribosome, more proteins interact with the minor than with the major groove. In minor groove recognition, networks of hydrogen bonding are formed between residues and the edges of consecutive Watson–Crick base pairs (Figure 7.1A). The 2-amino group of the GC pair is the moiety most frequently recognized *via* the minor groove. Many proteins use their backbone groups to contact the 2-amino of guanine. Major-groove recognition is more likely near helical junctions, internal loops, and hairpin loops where this groove is wide enough to permit contacts to bases by proteins (Figure 7.1B). This mode of recognition is particularly suited for the narrow extensions of ribosomal proteins (see below).

Ribosome structures beautifully illustrate how previously identified RNA motifs[38] participate in RNA tertiary interactions[33] or RNA-protein interactions[33,39] in large RNPs. The observation of these RNA motifs in rRNA reinforces the idea that defining motifs in RNA hold the key to characterizing RNA-protein interactions. For example, common and distinctive features of the kink-turn (K-turn) RNA motif were revealed by comparing its nine occurrences in the structure of the ribosome.[39] The K-turn is formed by an asymmetric internal loop containing conserved tandem sheared GA pairs and is an important site for protein binding and formation of RNA tertiary structures. The K-turn is observed in other RNPs as well.[40] A more detailed description of the K-turn RNA and its interaction with one ribosomal protein, L7Ae, in box C/D RNP assembly is given below.

Recognition of non-helical regions of the rRNA exhibits uniquely interesting structural principles in RNA-protein interactions. Non-helical regions either form distinct motifs that can fit directly into relatively rigid pockets in proteins or interact with protein residues through base stacking. The latter kind of interaction often requires the bases to be extruded. The former type of interactions can fit the "indirect read-out" mechanism where base identity is required to form the motif that is in turn recognized by the protein. Alternatively, "direct readout" is possible when the motif presents the distinctive Watson–Crick or major groove edge of the base for recognition.

The impressive amount of covered surface area on rRNA is also due to the large basic extensions on the otherwise small ribosomal proteins.[31] The penetration of ribosomal protein extensions into the interior of the folded rRNA is an unusual protein feature not previously observed (Figure 7.2). These extensions, characterized as "idiosyncratically folded", usually lack residues that can form a hydrophobic core typical of globular proteins and are high in glycine, arginine and lysine residues. Glycine fills tight spaces both because of its size and its unrestricted torsion angles. On the other hand, the high contents of arginine and lysine are explained by their favorable electrostatic interactions with the negatively charged RNA phosphate backbone. The importance of basic residues in contacting RNA was first observed in "arginine-rich" peptide-RNA interactions.[4]

RNA-protein Interactions in Ribonucleoprotein Particles and Ribonucleases 155

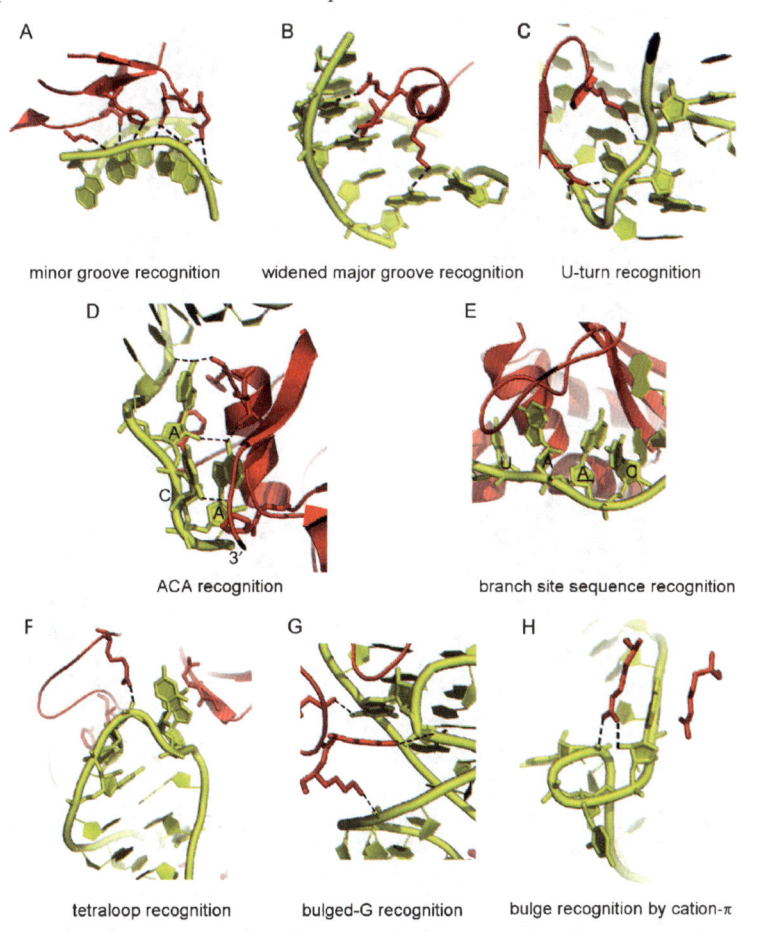

Figure 7.1 Examples of RNA motifs interacting with proteins. Dashed lines indicate close contacts between protein and RNA atoms. (*A*) minor groove recognition. The structure was taken from the ribosomal protein L21e bound in ribosome (PDB code 1S72). (*B*) Major groove recognition. The structure was taken from the ribosomal protein L15 bound in ribosome (PDB code: 1S72). (*C*) U-turn motif recognition. Structure taken from that of Alu domain of SRP (PDB code: 1E80). (*D*) ACA sequence recognition. Structure taken from that of the archaeal H/ACA RNP (PDB code: 2HVY). The ACA sequence is labeled next to nucleotides. (*E*) Branch site sequence recognition. Structure taken from that of the KH-QUA2 domain of splicing factor 1 bound with branch site RNA (PDB code: 1K1G). The UAAC sequence is labeled next to nucleotides. (*F*) Tetraloop recognition. Structure taken from that of the restrictocin-RNA complex (PDB code: 1JBS). (*G*) Bulged-G recognition. Structure taken from that of the restrictocin-RNA complex (PDB code: 1JBS). (*H*) Bulge recognition by cation-π. Structure taken from that of the splicing endonuclease bound with RNA (PDB code: 1GJW).

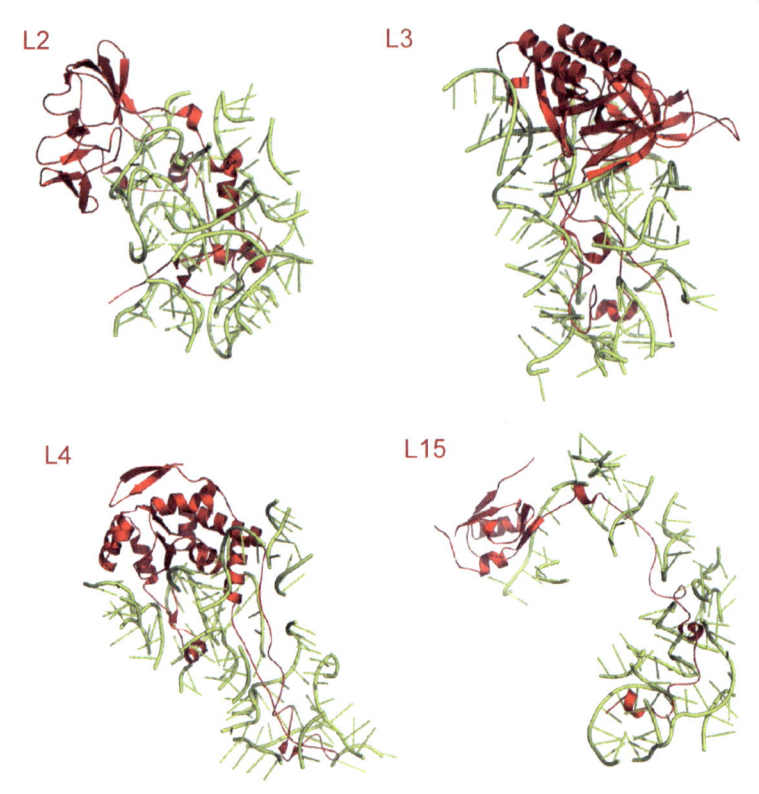

Figure 7.2 Examples of the extensive interactions formed between the unfolded basic extensions of ribosomal protein (red) and ribosomal RNA (yellow) (PDB code: 1S72).

Ribosome assembly makes very economic use of protein and RNA surfaces in creating this intricate and graceful ribonucleoprotein machine. A single protein fold may be exploited to bind RNA in different manners. For instance, the major group of large subunit proteins is of the $\alpha + \beta$ type with $\beta\alpha\beta\beta$ topology similar to the RRM typified by the splicing protein U1A. U1A uses its flat β-sheet side to interact with an RNA loop.[41,42] However, the RRM-like ribosomal proteins bind RNA differently from U1A and from one another. Both the flat β-sheet side and the α-helix side are observed to contact RNA. The β-barrel group of ribosomal protein resembles an SH3-like domain. These proteins use the loops that connect the β-strands to contact RNA.[31] Similarly, the same RNA motif interacts with ribosomal proteins differently. This is best illustrated by analyzing the interactions of the K-turn RNA motif and proteins.[39]

7.3.2 RNAi Complexes

RNA interference (RNAi) or post-transcriptional gene silencing (PTGS) is an evolutionarily conserved cellular response to the presence of small (21–23 nt) double-stranded RNAs containing 2-nucleotide 3′ overhangs.[43] Throughout the life of a small interference RNA (siRNA) or microRNA (miRNA) it interacts with a multitude of proteins, many of which share the common property of recognizing double-stranded RNA and their unique ends.

A siRNA/miRNA is the product of endonucleolytic cleavage of longer precursors by Drosha and Dicer. The primary precursors are cleaved by Drosha to release ~70 nt pre-siRNA/miRNAs that are subsequently processed by Dicer to generate mature ~22 nt siRNA/miRNAs. Following cleavage by Drosha/Dicer, the siRNA exerts its effect on target genes, either post-transcriptionally or translationally, through a ribonucleoprotein effector called the RNA-induced silencing complex (RISC). A key player of RISC is the Argonaute protein that is identified to be responsible for loading siRNA onto the RISC as well as for cleaving target mRNA.[44] Plant viruses use suppressor proteins, which recognize short duplex RNA, to sequester siRNAs against viral transcripts, thereby suppressing the host defense responses.[45] Recent structural studies on either individual domains or simpler but intact proteins have provided mechanistic insights into the action of Dicer, Argonaute and several viral suppressor proteins, in particular regarding their specificity for 21–23 bp small duplex RNAs containing 5′ phosphates and 3′ 2-nucleotide overhangs.

Owing to the prominent presence of double-stranded RNA in cells, the dsRNA binding domain is one of the most recognizable domains that bind to A-form RNA helices. In addition to its presence in Dicer, the dsRNA binding domain is found in numerous other protein enzymes involved in the fields of virology, neurobiology, and development.[46,47] This domain contains ~65 amino acids that form a two-layered structure composed of three β-strands and two α-helices in αβββα topology. Crystal and NMR structures of dsRNA binding domains with RNA explain their sequence independent recognition.[15,48–50] The first α-helix and the first β-hairpin loop contact two successive minor grooves and the loop connecting the last β-strand and the α-helix recognizes the intervening major groove (Figure 7.3A). The entire RNA-protein interface spans over 15 bp. Non-specific recognition of the RNA is facilitated by interdigitized contact between protein residues and the RNA backbone and base functional groups. Modulation of RNA binding specificity is possible through the first α-helix, which is the least conserved in all known dsRNA binding proteins. For instance, the first α-helix in the dsRBP of Rnt1p enables it to recognize a hairpin stem loop.[51] dsRNA binding proteins often contain more than one copy of the dsRBM, which increases RNA binding affinity and/or selectivity and suggests that the placement of this domain may also influence the specificity of the protein.[49]

The PAZ domain is found in both the Dicer and the Argonaute proteins. Structural studies of the PAZ domain alone and in complex with RNA suggest an attractive model of how the two proteins specify the length of 21 nt in RNAi processing. This domain, unexpectedly, binds specifically to the 3′ overhang-containing RNA. The 3′-end of the nucleic acid is bound without sequence specificity in a hydrophobic cleft of the PAZ domain that contains the most conserved residues (Figure 7.3B).[52,53] RNA recognition is facilitated by base-aromatic amino acid stacking. A tyrosine and a histidine residue interact with

A   dsRNA binding domain   B   PAZ domain

3'OH

3'OH

C   PIWI domain   D   P19:viral suppressor

5'P

Trp bracket   Trp bracket

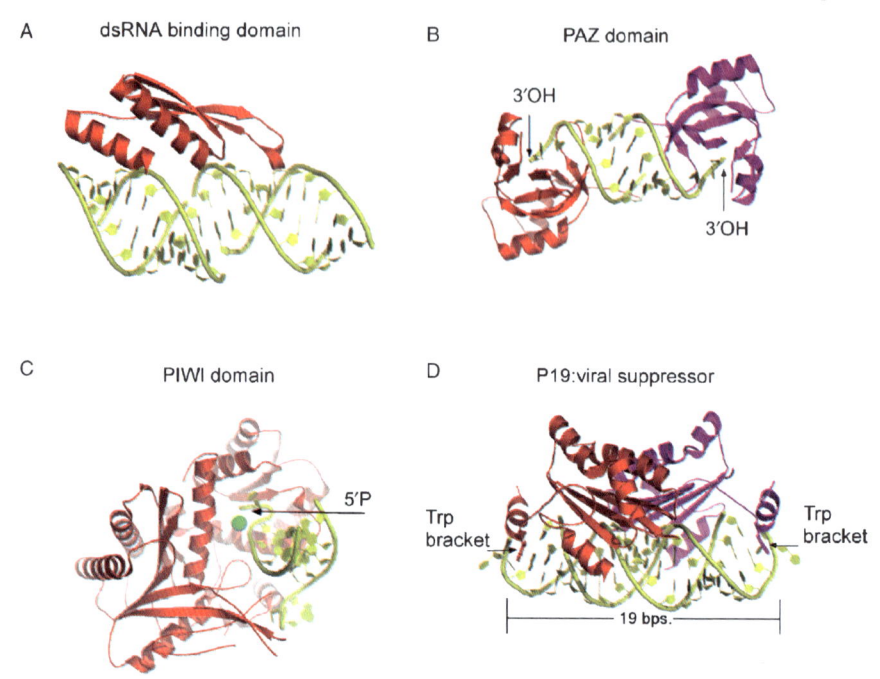

19 bps.

Figure 7.3 Structural domains and their interactions with RNA found in RNAi pathway. (*A*) Typical dsRNA binding domain (red) bound with RNA duplex (yellow) (PDB code: 1DI2). (*B*) PAZ domain (red and magenta) and RNA (yellow) complex (human Eif2c1 bound to RNA is shown, PDB code: 1SI3). 3'-OH of the siRNA, which is specifically recognized by PAZ domain, is indicated by arrows. (*C*) PIWI domain (red) and RNA (yellow) complex (PDB code: 2BGG or 1YTU). The 5' phosphate group of the siRNA is indicated by an arrow and the bound magnise is indicated by the green sphere. (*D*) Viral suppressor protein p19 (red and magenta) and siRNA duplex (yellow) (PDB code: 1RPU) "Trp bracket" refers to the stacking interaction between the tryptophane residue and the terminal base pair that specifies the length of siRNA. The distance between two terminal base pairs is indicated.

the phosphate groups of the 3'-end. Thus, recognizing the 3' end is primarily by steric exclusion because a paired RNA base would not fit into the binding pocket. Double stranded RNA is apparently not required for the recognition of the 3'-end because a single-stranded RNA can bind in the same pocket.[53] Similar PAZ domain-RNA interactions are used by both Dicer and Argonaute to define siRNA by their 3' overhang nucleotides.

The PIWI domain of the Argonaute proteins plays important roles in recognition of the 5'-ends of siRNA and miRNAs and cleavage of the target RNA. Cleavage by Dicer leaves a 5'-phosphate on the siRNA, which is subsequently recognized by RISC and other ribonucleoprotein complexes.[9] The 5' ends of siRNAs and miRNAs are important for mRNA target recognition and definition of the cleavage sites. The PIWI domain has the characteristic fold of RNase H and contains two of the conserved aspartate residues in the potential active site. The structure of PIWI bound to RNA provides insight into the recognition of the 5' end of the guide siRNA and its target RNA. The structure also identifies the location of a potential catalytic center with respect to the target RNA.[54,55] The 5'-end phosphate of the siRNA is recognized by a metal-containing pocket formed by two subdomains (Figure 7.3C). The first base of the siRNA unpairs when PIWI is bound and stacks with an aromatic residue, which explains its relative unimportance for specifying mRNA targets (Figure 7.3C). The 5'-P-binding site is 10–11 base pairs from the active site for target RNA cleavage. Together with the knowledge of how PAZ domains bind to the 3' overhanging nucleotides of siRNAs[52,53] and a full-length Argonaute structure,[56] a plausible model for mRNA cleavage by Argonaute was proposed. In this model, either the 3', the 5', or both ends of the siRNA strand are anchored by the PAZ and PIWI domains, respectively, while a bound mRNA strand passes through the active site of the protein.[56]

Although there is currently no crystal structure of Dicer bound to RNA, a reasonable model of how it generates 21–23 mer siRNA with 2-nt 3'-overhangs was derived from the crystal structure of *Giardia intestinalis* Dicer,[57] that of bacterial RNase III bound to an RNA duplex, and that of a PAZ domain bound to RNA.[52,53] In this model, Dicer contains a single processing center of two excision sites that are staggered by about two base pairs. This controls the length of the product by measuring the distance from the other end, which resulted from a previous processing event, and is bound to the PAZ domain.

p19 Provides an example of how viral suppressor proteins sequester siRNAs and prevent their incorporation into RISC.[58] p19 Forms a homodimer that binds to RNA like a vice (Figure 7.3D). In each monomer, a critical tyrosine residue from the N-terminal helix stacks with the terminal base of the siRNA against, acting as a molecular bracket. The central β-sheet of p19 interacts with the phosphate backbone of the duplex RNA. Recognition of the 2-nt overhang is apparently not essential. The apparent requirement for protein dimerization in specific recognition of the 21–23 nt duplex RNA suggests that the binding affinity and specificity are directly influenced by protein concentration.

7.3.3 Signal Recognition Particle

The signal recognition particle (SRP) is a large GTP-dependent protein-RNA assembly directing integral membrane and secretory proteins to the correct intracellular locations during translation.[59,60] SRP facilitates this process by binding to hydrophobic signal sequences at the N-termini of polypeptides as they emerge from ribosomes. During each cycle of SRP function, it recognizes the nascent peptide chain emerging from the ribosome, targets the cargo protein to the membrane *via* a membrane-bound SRP receptor, releases the ribosome-associated polypeptide into a translocon channel in the membrane, and dissociates from the receptor for another targeting cycle. Surprisingly, a conserved RNA molecule (7S RNA in eukarya and 4.5S RNA in bacteria) as well as up to another six proteins (SRP72, SRP68, SRP54 [or Ffh in bacteria], SRP19, SRP14 and SRP9) are found in the SRP. The RNA functions within the peptide-recognition domain of the RNP.[14]

An impressive amount of structural information is available on SRP, shedding light on the proteins' roles in stabilizing RNA tertiary structures.[14,61–64] An intact SRP consists of three distinct domains: the Alu domain, responsible for translation arrest, the S domain, for signal sequence binding, and a low-mass region connecting the two. Core structures of both the Alu and S domains have been determined in several organisms. The S domain comprises SRP54/Ffh bound to helix 8 of 7S RNA and to helix 6 (or domain IV in the 4.5S RNA). In eukarya, binding of SRP19 to RNA is required for SRP54 to bind. Therefore, SRP19's function is primarily architectural. The SRP54/Ffh protein includes a GTPase domain and an M-domain for binding RNA. The Alu domain consists of SRP14 and SRP9 bound cooperatively to the end opposite helix 8. SRP72 and SRP68 form a heterodimer that binds to the middle of 7S RNA.

The structure of the Alu domain of the mammalian SRP is known.[65] SRP9 and SRP14 share a similar $\alpha\beta\beta\beta\alpha$ fold and form a pseudosymmetric heterodimer. The β-strands of the two proteins form a single concave β-sheet that is the primary structural element for contacting RNA (Figure 7.4A). The SRP9/14 dimer binds specifically to the core region of the three-way junction, with the interface spanning one of the helical branches and the U-turn internal loop. The interaction between the SRP9/14 hybrid β-sheet and the helical region involves solely phosphate backbone atoms. Specificity primarily arises from base-specific contacts made from a conserved arginine residue (Arg54) and a lysine residue (Lys66) of SRP14 to the U-turn bases (Figure 7.1C). Interestingly, another SRP9/14 heterodimer is found in the asymmetric unit that is bound to Alu RNA at the ends where the 5' and 3' helices join. However, this interaction is non-specific and is thus non-functional.[65]

The structures of a SRP54/Ffh domain (M domain) bound to domain IV of the SRP RNA with or without SRP19 are also known.[66,67] The M domain interacts with an unusual internal loop in helix 8 that includes a sheared GG and a reverse Hoogsteen AC pair. The conserved residues in the M domain interact extensively with an extruded adenosine (Figure 7.4B). A network of metal ions and water-molecules at the RNA-protein interface was observed in the co-crystal structure and its importance was confirmed experimentally.[66,68] SRP19, which binds to the conserved GNRA (N stands for A, G, C or U; R for A or G) tetraloop of domains III and IV of 4.5S RNA, appears to function solely as a stabilizer for helix 6 that in turn stabilizes the bound form of the asymmetric loop *via* A-minor base triples.[67,69,70] In bacteria where SRP19 is absent, the RNA is alternatively stabilized by ions.[64] SRP assembly is driven by the formation of new RNA-RNA contacts that are stabilized by proteins.

7.3.4 s(no)RNPs

snoRNPs are small RNPs found in the nucleolus of the eukaryotic cells. The archaeal homologues of snoRNPs are called small ribonucleoprotein particles (sRNPs). These RNA-protein assemblies catalyze three types of RNA modification and RNA processing reactions on precursor rRNAs: (1) pseudouridylation; (2) 2'-O-methylation, and (3) RNA cleavage. Based on the three functions and on secondary structural features, major s(no)RNPs are divided into three types. A large number of H/ACA RNPs are responsible for site-specific isomerization of uridine of rRNAs and snRNAs. The box C/D RNPs are primarily for site specific 2'-O-methylation of rRNAs. The RNaseP-like MRP RNP (note: MRP RNP is different from the mitochondrial RNA binding protein complex discussed in Section 7.3.5) is required for processing of precursor rRNA at the internal transcribed sequence 1 (ITS1) and is thought to exploit catalytic RNA.[71] Interestingly, several specialized H/ACA and C/D RNPs are also responsible for guiding processing of rRNA.[72-75] In these cases, protein(s) other than the common core proteins may be responsible for the actual cleavage.

Limited structural information is known about the archaeal C/D and H/ACA RNP homologues. Archaeal box C/D RNPs contain three core proteins: Nop56/58, fibrillarin, and L7Ae, and box H/ACA RNPs contain four: Nop10, Gar1, Cbf5, and L7Ae. Fibrillarin and Cbf5 are the catalytic subunits of the two particles, respectively. For both RNPs, the structures of the L7Ae proteins bound to their respective RNA binding site are available.[76-78] Although a structure of the Nop56/58-fibrillarin complex is known,[79] no structural data on its complex with RNA are available. In contrast, a full H/ACA RNP structure, although without a bound target RNA, is available.[86] These structural studies have revealed two general features of the assembly: (1) the small RNAs act as scaffolds for protein assembly and for substrate binding; and (2) the assembly process follows a certain order that is facilitated by conformational changes in both RNA and proteins.

Both archaeal C/D and H/ACA RNAs contain sequences that form K-turn RNA motifs.[39] In general, a K-turn is a helix-internal loop-helix structure. This structure requires, minimally, sheared tandem GA pairs and three unpaired nucleotides and occurs either as a bulge flanked by two helices or as part of a terminal loop. The K-turn motif in box C/D and H/ACA RNA interacts specifically with the L7Ae protein, acting as a common link between the ribosome and sRNPs. In eukaryotes, L7Ae is replaced by specialized homologues. For instance, L7Ae is replaced by Nhp2p in yeast H/ACA RNPs and by Snu13p in yeast (15.5 K protein in human) C/D RNPs.

The conserved box C (UGAUGA) and D (CUGA) sequences fold together to form the K-turn motif (Figure 7.4F). The tandem GA pairs critical to the K-turn motif are formed by the first GA in box C and the GA in box D. The bulged nucleotide adjacent to the GA pairs is extruded and its phosphate group is the site of the most severe kink. The L7Ae protein and its eukaryotic homologues specifically recognize the base-specific features of the two guanidine nucleotides presented on the major groove side of the noncanonical stem (NC-stem) comprising the tandem GA pairs (Figure 7.4F). The most conserved protein motif NExxK in L7Ae forms specific hydrogen bonds with the two guanosine nucleotides. This mode of interaction is supported by results from mutagenesis and nucleotide analog interference modification (NAIM) studies.[80-84] The two adenosine residues of the GA pairs do not form base-specific interactions with

L7Ae residues at all, yet their mutation disrupted L7Ae-box C/D RNA interactions, suggesting the importance of the GA pair in maintaining the structural integrity of the K-turn for L7Ae binding. The other flanking stem, the canonical stem (C-stem), and the two adjacent bulged nucleotides interact with a non-conserved protein loop. Since the C-stem is frequently replaced by a short terminal loop in the internal box C'/D' motif in many box C/D RNAs, the sequence variation of this loop accounts for the different ability of the eukaryotic and archaeal proteins in binding the internal box C'/D' motif.[77,85]

Comparison of the L7Ae-RNA complex structures with that of their eukaryotic homologue, a 15.5 K protein bound to a stem-loop of U4 snRNA,[40,76–78] suggests a conformational "adaptability" of K-turn RNAs in binding to the 15.5 K/L7Ae family of proteins. Despite the differences in primary structures among the various K-turn RNAs, they form a similar K-turn structure. However, crystal structures

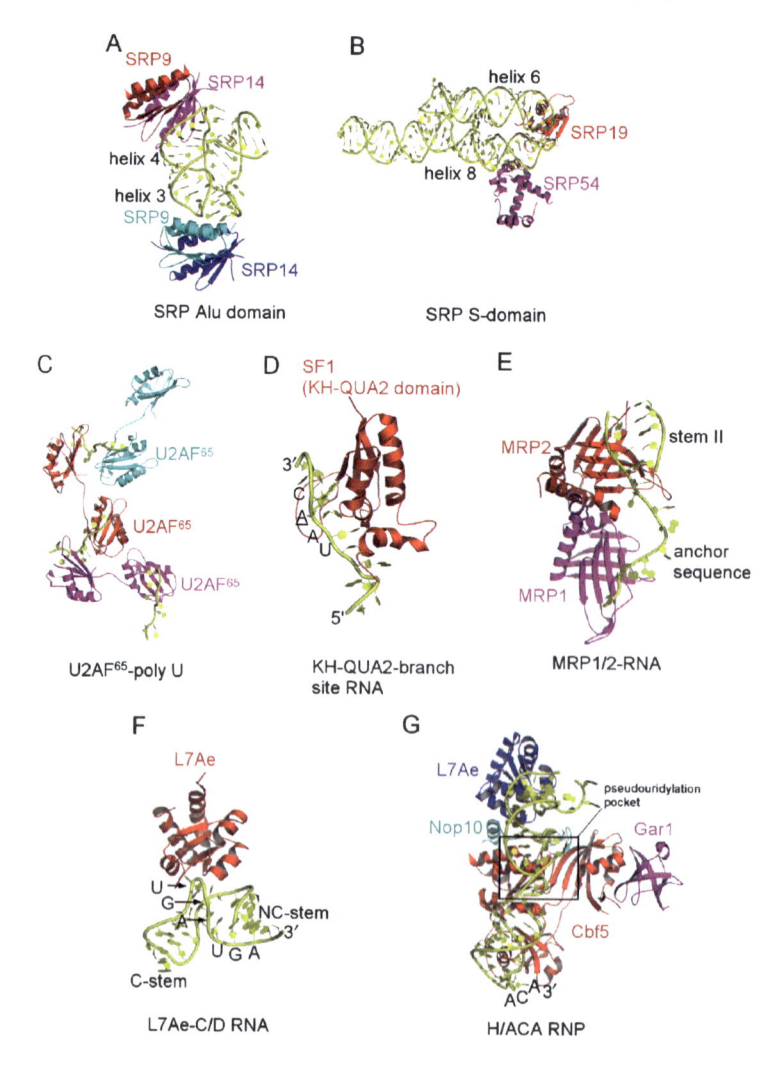

RNA-protein Interactions in Ribonucleoprotein Particles and Ribonucleases 163

can not reveal dynamic information accompanying the binding process. A recent study by electrophoretic mobility and by time-resolved fluorescence resonance energy transfer shows that a K-turn RNA in solution is polymorphic, with a less kinked and a tightly kinked structure in equilibrium even at relatively high magnesium concentrations.[87] Association of a protein with its cognate K-turn could drive the RNA conformation towards that of a tightly kinked structure, such as those observed in the crystals structure.[88] Such dynamic processes could potentially confer binding specificity to the 15.5 K/L7Ae protein for different K-turns.

A recent crystal structure of an archaeal H/ACA RNP in the absence of substrate RNA shows the roles of each protein in stabilizing the RNA and in recognizing features unique to H/ACA RNA.[86] The H/ACA RNA containing a helix-internal loop-helix-tail unit lies on a platform formed by Cbf5, Nop10 and L7Ae proteins. The structure corroborates previously known biochemical data on this RNP. The entire RNP resembles the "+" symbol where the catalytic site is approximately at the junction of the two bars and the RNA is bound along the vertical bar (Figure 7.4G). Cbf5, Nop10 and L7Ae contact the RNA throughout its helical region. Nop10 and L7Ae are at the top end of the vertical bar and cooperatively anchor the upper stem of the H/ACA RNA to the surface of Cbf5. Binding of L7Ae to the K-turn near the apical loop bends the upper stem. This causes the upper stem to interact with a loop of Nop10 that is simultaneously bound to Cbf5. The composite protein surface formed by L7Ae, Nop10, and Cbf5 for binding H/ACA RNA presumably ensures the proper sequence of events in the placement of the substrate RNA at the catalytic site.

Figure 7.4 Structural complexes found in RNPs. (*A*) Structure of the Alu domain of SRP containing a specifically and nonspecifically bound SRP9/14 heterodimer (PDB code: 1E80). The SRP9 protein is colored in red (specific) and cyan (nonspecific) and the SRP14 protein is in magenta (specific) and blue (nonspecific). (*B*) Structure of the S-domain of SRP containing SRP19 (red) and the M-domain of SRP54 (magenta) (PDB code: 1MFQ). (*C*) U2AF^{65} bound with a 7-nt polyuridine oligomer (yellow) (PDB code: 2G4B). Symmetry equivalent U2AF^{65} molecules are colored in red, magenta, and cyan. Each 7mer is bound by two U2AF^{65} molecules. (*D*) Structure of the KH-QUA2 domain (red) of splicing factor 1 (SF1) bound with an RNA containing branch site sequence (yellow) (PDB code: 1K1G). The specificity sequence UAAC is labeled next to the RNA nucleotides with the bulged A underlined. (*E*) Structure of the mitochondrial binding proteins MRP1 and MRP2 (red and magenta, respectively) bound with a model guide RNA for editing (yellow) (PDB code: 2GJE). (*F*) L7Ae-C/D RNA complex (PDB code: 1RLG). Note that L7Ae is a ribosomal protein as well as a sRNP protein found in archaeal C/D and H/ACA RNPs. The box C sequence is labeled next to the nucleotides. NC-stem denotes noncanonical stem while C-stem denotes canonical stem. (*G*) Structure of an archaeal H/ACA RNP (PDB code: 2HVY). Each H/ACA protein is colored (Cbf5: red, Nop10: cyan, Gar1: magenta, and L7Ae: blue). The pseudouridylation pocket is where the target RNA binds. The specificity sequence ACA is labeled next to nucleotides.

Gar1 is placed farthest from the H/ACA RNA at the right-hand end of the horizontal bar of the "+" symbol. Although Gar1 does not contact the guide RNA in the RNP structure, it is believed to be important for stabilization and release of the target RNA. Verification of this hypothesis awaits the structure of the holoenzyme.

The strictly conserved trinucleotide ACA comprises the tail of the H/ACA RNA and is specifically recognized by the pseudouridine synthase and archaeosine tRNA-ribosyltransferase (PUA) domain of Cbf5. The PUA domain of Cbf5 is structurally homologous to the same domain of the archaeosine tRNA-ribosyltransferase (arcTGT) that modifies G15 of tRNA. Similar to the Cbf5 PUA domain, the arcTGT PUA domain binds to the CCA trinucleotide at the 3' end of tRNA.[89] Surprisingly, however, despite the close structural homology and the common "CA" nucleotides in their RNA targets, the PUA domains in the two enzymes bind RNA in different manners. In the Cbf5 PUA domain, both adenosine residues adopt the *syn* conformation with respect to their N-glycosidic bonds and are sandwiched by both RNA and protein residues (Figure 7.1D). These results further illustrate that similar protein domains may bind RNA differently.

7.3.5 RNA Editing Complexes

The mitochondrial RNA binding protein (MRP) complex functions with a large protein enzyme complex in kinetoplastid mRNA editing. The MRP complex, consisting of MRP1 and MRP2 subunits, facilitates hybridization of a guide RNA to a target mRNA by stabilizing the guide RNA conformation suitable for its binding to its target. The guide RNA bound by the MRP1/MRP2 heterocomplex consists of two hairpin loops (stem/loop I and II) and a 3' oligo(U)-tail. The stem/loop I region of the guide RNA contains the antisense sequence complementary to the mRNA target that is designated as the "anchor sequence".

The co-crystal structure of MRP1/MRP2 in complex with a model guide RNA shows that the MRP1/MRP2 heterocomplex is a homodimer of two heterodimers.[90] Despite sharing a low sequence homology (18% sequence identity), MRP1 and MRP2 display a nearly identical fold consisting of two layers of four-stranded β-sheets intercepted by three α-helices, leading to a pseudo-four fold symmetry (Figure 7.4E). The protein fold, as well as the overall assembly of the MRP1/MRP2 complex, is analogous to the "whirly" transcription factor proteins.[91]

Each MRP1/MRP2 heterodimer creates a continuous surface made exclusively of β-sheets that interacts with guide RNA in a sequence independent manner. The negatively charged RNA phosphodiester backbone forms favorable electrostatic contacts with the electropositive RNA binding surface. This is in contrast to RNA interactions with other β-sheets such as the RRM that bind RNA through base-aromatic ring stacking. The β-sheet of MRP1 binds to the phosphate backbone of the single-stranded region of the stem/loop I. Bases in this region are extruded and face the solvent in preparation for hybridization with target RNA. The β-sheet of MRP2 binds to the phosphate backbone of the 5' strand of the stem/loop II. However, this region exhibits the most structural disorder; the 3' strand of this hairpin is loosely hybridized to the 5' strand, suggesting flexibility in this region of the RNA.[90] The flat β-sheet surfaces of the MRP heterocomplex are called "presentation platforms" for their role in presenting the anchor sequences.

The crystal structure of the MRP1/MRP2 complex with guide RNA suggests a mechanism for its functional role in promoting target annealing by (1) stabilizing an annealing-active conformation and (2) by neutralizing the negative charge on the phosphate backbone of the guide RNA, thereby decreasing the electrostatic repulsion between the RNAs. Although the MRP1 surface was observed in the crystal to bind a single RNA strand, biochemical studies indicate that it can bind double-stranded RNA,[90] suggesting the possibility that either MRP unit can mediate hybridization of the guide with the target RNA.

7.4 RNA-protein Interactions in Ribonucleases

The biochemical principles that underlie RNA cleaving enzymes include both RNA recognition and catalysis. In addition to the general methods of RNA-protein recognition required by ribonucleases to ensure substrate specificity, additional strategies are in place for ribonucleases to stabilize the transition state and to release the products after the chemical step. By exploiting reaction equilibria, enzyme mutants and substrate analogues, X-ray crystallography can capture atomic structures that mimic each reaction intermediate so that the entire RNA cleavage process may be visualized as a molecular movie. Previously, the best studied ribonucleases were from the RNase A and T1-families of enzymes, for which high-resolution structures of the enzymes and enzyme bound RNA oligomers are known. However, these ribonucleases are degradative in nature and show little discrimination of RNA structures. RNase A, for instance, cleaves after pyrimidine nucleotides. Recently, data for several ribonucleases specific for certain RNA structural motifs bound to their respective substrates or substrate analogues have been obtained. These new structural data shed light on unique aspects of RNA recognition by ribonucleases. Based on their cleavage mechanisms, ribonucleases discussed below are of two types. RNase E, RNase II, RNase III, tRNAse Z are Mg^{2+}-dependent ribonucleases that cleave phosphodiester bonds to produce 5'-phosphate and 3'-hydroxyl ends. Restrictocin and RNA splicing endonucleases do not depend on metal and produce 5'-hydroxyl and 2',3'-cyclic phosphate ends. To trap the reaction intermediates for structure determination, modifications of either the enzyme or the substrate have to be made. Therefore, the structures are often close mimics of the true reaction intermediates.

7.4.1 RNase E

As an integral component of the bacterial degradosome, RNase E is involved in mRNA decay. RNase E is also responsible for cleavage of rRNA and RNase P transcripts to produce mature RNAs.[92] RNase E is specific for 5' termini, in particular 5′-monophosphates, in binding RNA but excises at an internal site. The recently obtained crystal structure of the *Escherichia coli* RNase E-RNA complex illustrates the structural basis for 5′-monophosphate recognition and a mechanism for RNA substrate selection based on subunit organization.[93] The crystal structure reveals that RNase E is a dimer of dimers. Each RNase E monomer is divided into a small and a large domain. The large domain is responsible for the formation of the homodimer and the small domain forms the tetramerization interface. The large domain consists of three subdomains: an RNase H-like domain, a S1 domain and a DNase I domain (Chapter 13). The DNase I domain contains the actual cleavage site.[93] In each subunit, the 5′-monophosphate is bound by the RNase H-like domain while the first base after the cleavage site is bound by the S1 domain. Interestingly, RNA cleavage takes place in the DNAse I-like catalytic domain of another subunit. The physical separation between the 5′-monophosphate sensing domain and the catalytic site in the DNase I domain allows the control of RNA specificity through domain arrangement (Figure 7.5A). Furthermore, a different mode of RNA binding is observed when a longer oligomer is used, which suggests that RNase E can selectively bind different substrates. Although these structural results can not answer why RNase E does not indiscriminately cleave all RNAs containing 5′ monophosphates, they provide an intriguing hypothesis that the arrangement of protein subunits and domains structurally guides target site selection. It is believed that substrate interactions with the 5′-monophosphate sensing pocket somehow cause the S1 domain to clamp down on the RNA downstream, resulting in the activation of cleavage activity. Thus, catalysis by RNase E is expected to proceed *via* a classical induced-fit mechanism.

7.4.2 RNase II

Many RNases are required to both process and degrade functional RNAs. These RNases often contain single-stranded RNA binding domains such as the cold-shock domain and the S1 domain. The structure of RNase II bound to its RNA substrate illustrates how these domains bind single-stranded RNA nonspecifically. The concept of separating RNA binding from the catalytic sites is again apparent from this example.

RNase II is a quality-control RNA exonuclease that processively hydrolyzes RNA in the 3′ to 5′ direction, releasing 5′ monophosphates. Structures of *Escherichia coli* RNase II with and without a bound 13-mer single-stranded RNA are now available.[94] RNase II contains four distinct domains: two cold-shock domains, a S1 domain, and a catalytic domain. The deep cleft formed by the two cold-shock domains and the S1 domain is referred to as the anchor region that binds to the 5′ end of the single-stranded RNA. Interestingly, the canonical oligonucleotide-binding motifs (OB, Chapter 5) of the two cold-shock domains use their noncanonical RNA binding surfaces to interact with the downstream RNA. The hydroxyl groups of the RNA nucleotides in this region form a network of hydrogen bonds with the protein. This explains why the cold-shock domain can also bind but not cleave DNA. The 3′ end of the

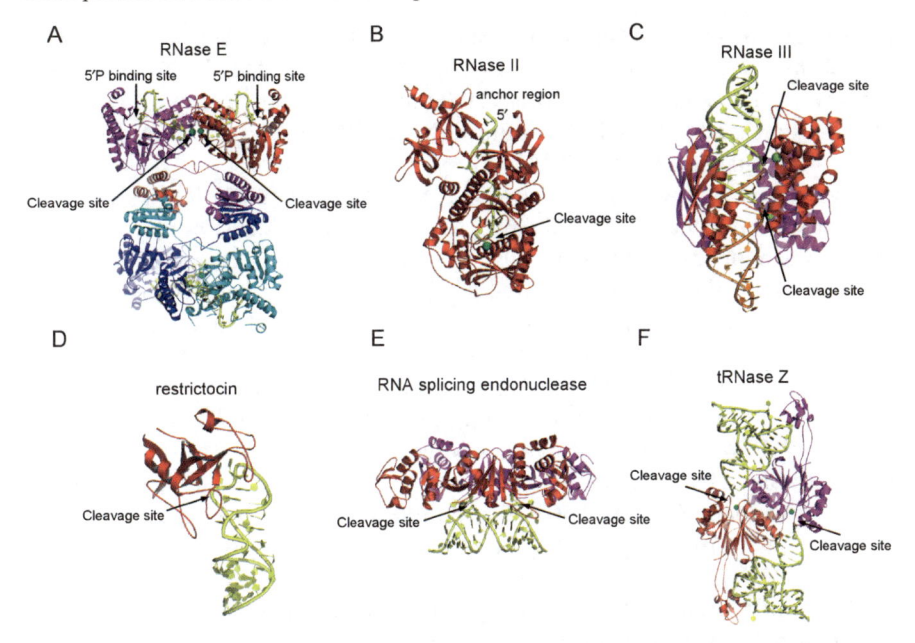

Figure 7.5 Structures of ribonucleases bound with RNA. (*A*) RNase E bound with a 15mer RNA (yellow) (PDB code: 2C0B). Each subunit of the tetrameric enzyme is colored (red, magenta, cyan and blue). Two of the four cleavage sites are indicated by arrows and the bound magnesium (green spheres). (*B*) RNase II (red) bound with a single-stranded RNA (yellow) (PDB code: 2IX1). The cleavage site is indicated by an arrow and the bound magnesium (green sphere). (*C*) RNase III (red and magenta) bound with a cleaved duplex RNA (yellow and orange) (PDB code: 2EZ6). The cleavage sites are indicated by arrows and the bound magnesium (green spheres). (*D*) Structure of restrictocin (red) bound with a misdocked SRL RNA (yellow) (PDB code: 1JBS). (*E*) RNA splicing endonuclease (red and magenta) bound with a bulge-helix-bulge RNA (yellow) (PDB code: 2GJW). (*F*) tRNase Z (red and magenta) bound with tRNA^{Thr} (yellow) (PDB code: 2FK6). The cleavage sites are indicated by arrows and bound magnesium (green spheres).

RNA bound at the catalytic site is nearly eight nucleotides away from the 5' end (Figure 7.5B). Based on the single-stranded binding mode, a translocation mechanism is proposed that involves threading single-stranded RNA through the center of the enzyme.[94] However, the mechanism of how RNA binding to the anchor domain regulates RNA cleavage by the catalytic domain remains to be determined.

7.4.3 RNase III

RNase III is a universally conserved ribonuclease that is specific for double-stranded RNA.[92] Members of the RNase III family of ribonucleases have at least one endonuclease domain and a double-stranded RNA binding domain.

RNase III is involved in many aspects of RNA metabolism in various organisms. Bacterial RNase III cleaves double-stranded RNA in rRNA biogenesis. The eukaryotic RNase III family of proteins includes Dicer and Drosha that are involved in maturation of siRNA and miRNA for gene expression and regulation.[43,57,95]

Crystal structures of *Aquifex aeolicus* RNase III with duplex RNA substrates have now been determined.[96] The bacterial RNase III forms a homodimer that binds to two identical 13 base pair RNA hairpins (Figure 7.5C). The specificity for double-stranded RNA is ensured by the two symmetrically placed dsRNA binding domains that interact with the phosphate backbone of the RNA (Figure 7.3A). Even though a catalytically deficient mutant RNase III was used in crystallization, each RNA is cleaved in the crystal at the site 11 base pairs away from the apical loop, leading to a 2-nt overhang.[96] Because the two RNA hairpins form an end-to-end continuous helix, the two overhang nucleotides are able to base pair in the crystal, leaving the two cleavage sites staggered by two base pairs (Figure 7.5C). The double-stranded RNA binding domain can bind to RNA of any length but a hairpin is required to lock the RNA in place for cleavage. Once the RNA is placed at the catalytic cleft, displacement of the double-stranded RNA binding domain from the endonuclease domain determines the two sites of cleavage.

RNase III exhibits high cooperativity in binding and cleaving double-stranded RNA. The co-crystal structure shows that, for catalysis to take place, the two catalytic domains have to form two composite active sites. At each active site, the RNA binding region 3 from one subunit holds the RNA in place for it to be cleaved by the catalytic domain of the other subunit.[96] This mechanism of cooperativity is also observed in multimeric RNase E, tRNase Z and the splicing endonuclease (see below).

7.4.4 Restrictocin

Restrictocin is a family member of ribonuclease toxins that function by cleaving one phosphodiester bond in rRNA, resulting in cell death by apoptosis.[97] Restrictocin is specific for the universal sarcin/ricin loop (SRL) of 23-28S rRNA that is characterized by a GNRA tetraloop proceeded by the S-turn of a bulged-G motif. Structures of both restrictocin bound to an SRL RNA and of the SRL RNA alone illustrate how restrictocin confers binding specificity and the structural changes in RNA required for catalysis.[38,98–100] In the restrictocin-RNA complex (Figure 7.5D), restrictocin simultaneously contacts the bulged-G motif and tetraloop. The tetraloop is completely remodeled (Figure 7.1F). The primary site of recognition, the bulged-G, is contacted by conserved lysine residues (Figure 7.1G) and is 12 Å away from the scissile phosphate bond, suggesting that separate elements in the enzyme are used for "anchoring" and "cleaving". Whereas the RNA misdocks by one nucleotide in the active site due to modifications necessary for crystallization, near inline geometry involving the attacking 2'-oxygen, the scissile phosphate and the leaving 5'-oxygen is adopted at the misdocked nucleotide. This "near-attack" conformation is a direct result of base-flipping mediated by the enzyme. It is suggested that all structure-specific ribonucleases catalyzing intramolecular phosphotransfer reactions utilize this mechanism of substrate remodeling to align the reactive groups.[98]

7.4.5 RNA Splicing Endonucleases

A fraction of tRNAs (5–25%) contain introns that are recognized and excised by a protein enzyme called the splicing endonuclease. This mechanism of intron removal is distinct from that of mRNA intron removal catalyzed by the spliceosome. The splicing endonuclease, first identified and isolated from yeast,[101,102] is responsible for the recognition and excision of introns from precursor nuclear tRNAs and archaeal RNAs. A key feature of splicing endonuclease function is organism-specific substrate recognition.[103] A splicing endonuclease recognizes a folded RNA motif (rather than base sequences) that contains the two intron–exon junctions. In Archaea, the bulge-helix-bulge (BHB) motif consists of two three-nucleotide bulges separated by four base pairs. Variations from the canonical BHB motif are observed in archaeal substrates and can be clearly correlated with archaeal phyla.[104–106] In Eukarya, introns are invariably located one base 3′ to the anticodon but lack a defined local motif. In this case, the recognition motif extends the entire precursor tRNA and the enzyme exploits a "ruler mechanism" and locates the intron–exon junctions by measurement.[107]

The splicing endonucleases known so far are classified into α_4, β_2, $(\alpha\beta)_2$, and $\alpha\beta\gamma\delta$ families. The first three families are found in Archaea, and the $\alpha\beta\gamma\delta$ family appears only in Eukarya. Regardless of the number of subunits, the enzymes form four homologous units – two catalytic and two structural. The two catalytic units are symmetrically placed with the active sites ~28 Å diagonally apart through a cyclic interaction among the units (structural → catalytic → structural → catalytic). This enzyme architecture is absolutely essential for the recognition of the two cleavage sites and, more importantly, for stabilization of the transition state for catalysis. A recently determined crystal structure of a dimeric (β_2) splicing endonuclease bound to a BHB RNA (Figure 7.5E) revealed that two catalytic units interact with one cleavage site in a surprisingly coordinated fashion.[108] One catalytic unit holds the bulge that is cleaved by the other (Figure 7.5E). The interdependence of the two catalytic units for binding a pseudosymmetric substrate prevents the enzyme from binding and cleaving RNA promiscuously and thus explains the critical importance of enzyme assembly for the control of catalytic activity.

The inherent structural flexibility in RNA allows the recognition of the three-nucleotide bulge and the excision of the phosphodiester bond. Remarkably, all three bulged nucleotides are extruded, with only the first bulged nucleotide interacting significantly with the endonuclease (Figure 7.1H). The nucleobase of this nucleotide is sandwiched by a pair of arginine residues from the opposing catalytic subunit (Figure 7.1H). The critical importance and nature of the conserved cation–π interaction explains well why the nucleobases of the bulged nucleotides are not strictly conserved.[108] The phosphate backbone of the bulge is significantly bent to complement the relatively rigid active site of the enzyme (Figure 7.1H). This recognition mechanism also serves catalysis because it enables the three atoms involved in the phosphotransfer reaction to be aligned in a near attack conformation, as previously proposed.[98,108,109]

7.4.6 tRNase Z

tRNase Z is a universally conserved enzyme that endonucleolytically trims the 3′ end of tRNA. Recent structural studies on tRNase Z highlight a new principle of tRNA upper stem recognition by composite binding sites.[110,111] Similar to the splicing endonucleases, tRNase Z exhibits sequence-independent but organism-specific substrate specificity for precursor tRNA. For instance, *Thermotoga maritima* tRNase Z cleaves after CCA, some archaeal enzymes cleave after CC, and the *Bacillus subtilis* enzyme cleaves strictly after the discriminator nucleotide immediately upstream of the CCA.[112] The crystal structure of *B. subtilis* tRNase Z (with a catalytic His to Ala mutation) bound to a pre-tRNA^{Thr} that contains 3′ extensions[110] shows that both monomers of the dimeric enzyme recognize the precursor tRNA in a coordinated fashion (Figure 7.5F). A domain attached to the catalytic domain from one subunit rests on the platform formed by the T-stem, but it is the adjacent catalytic subunit that interacts with the first base pair (1 and 72) and the discriminator base 73. U73 of tRNA^{Thr} unstacks from the acceptor stem and is co-stabilized by both subunits of the enzyme in the active site. A tyrosine residue from one subunit and a phenylalanine residue from the other sandwich the U73 nucleobase. Unfortunately, no structural information is available for the 3′ extension nucleotides, which prevents detailed analysis of the catalytic mechanism. Based on the composite RNA binding site and the apparent flexibility of the 3′ end of tRNA, however, it has been proposed that subtle differences in the dimer arrangement and accessibility of the active site could account for the different effects of CCA on the enzymes.[112] These differences cannot explain why some enzymes can cleave both CCA-less and CCA-plus substrates, however.

7.5 Concluding Remarks

Recent structural studies of RNA-protein complexes have revealed how proteins interact with single and double-stranded RNA, with stem and internal loops, and with specific end groups or sequences of RNA. This knowledge extends that gained from previously studied tRNA-synthetase complexes while reinforcing the general features of RNA-protein interaction shared by most complexes. Regardless of the RNA-protein complexes, positively charged amino acids, protein backbone atoms, and RNA phosphate backbone atoms dominate the RNA-protein interfaces.[113–115] Similar functional groups were observed at RNA-protein interfaces and at crystal packing surfaces involving RNA and proteins.[116]

The examples of RNA-protein structures discussed in this chapter show that the specificity of RNA-protein interactions can be achieved by multiple strategies. Sequence specific recognition is often aided by hydrogen bonds formed between RNA bases and protein residues. Shape specific recognition can be achieved through the molecular ruler strategy, in which proteins form unique platforms that bind RNA of defined sizes. The concept of molecular ruler is also exploited by several processing enzymes. These enzymes, typically of multi-subunit or of multi-domain architecture, anchor on one RNA site by its distinct chemical features and process another site away from the anchoring site. The implications of how these various interaction types effect binding kinetics and catalysis await further quantitative studies.

Acknowledgements

The author is grateful for scholarly discussions with Carl Correll and Phoebe Rice. The author thanks the National Institutes of Health (R01 GM66958-01) and National Science Foundation (MCB-0517300) for financial support.

References

1. J. Cheng, P. Kapranov, J. Drenkow, S. Dike, S. Brubaker, S. Patel, J. Long, D. Stern, H. Tammana, G. Helt, V. Sementchenko, A. Piccolboni, S. Bekiranov, D.K. Bailey, M. Ganesh, S. Ghosh, I. Bell, D.S. Gerhard and T.R. Gingeras, *Science*, 2005, **308**, 1149–1154.
2. P.B. Moore, *Annu. Rev. Biochem.*, 1999, **68**, 287–300.
3. R.N. De Guzman, R.B. Turner and M.F. Summers, *Biopolymers*, 1998, **48**, 181–195.
4. D.E. Draper, *J. Mol. Biol.*, 1999, **293**, 255–270.
5. J.M. Perez-Canadillas and G. Varani, *Curr. Opin. Struct. Biol.*, 2001, **11**, 53–58.
6. Y. Chen and G. Varani, *FEBS J.*, 2005, **272**, 2088–2097.
7. S. Cusack, *Curr. Opin. Struct. Biol.*, 1999, **9**, 66–73.
8. D. Moras and A. Poterszman, *Curr. Biol.*, 1995, **5**, 249–251.
9. T.M. Hall, *Structure*, 2005, **13**, 1403–1408.
10. T.M. Hall, *Curr. Opin. Struct. Biol.*, 2005, **15**, 367–373.
11. A.C. Messias and M. Sattler, *Acc. Chem. Res.*, 2004, **37**, 279–287.
12. R.E. Collins and X. Cheng, *FEBS Lett.*, 2005, **579**, 5841–5849.
13. A. Lingel and M. Sattler, *Curr. Opin. Struct. Biol.*, 2005, **15**, 107–115.
14. J.A. Doudna and R.T. Batey, *Annu. Rev. Biochem.*, 2004, **73**, 539–557.
15. R. Stefl, L. Skrisovska and F.H. Allain, *EMBO Rep.*, 2005, **6**, 33–38.
16. C.D. Mackereth, B. Simon and M. Sattler, *Chembiochem*, 2005, **6**, 1578–1584.
17. T.C. Terwilliger and J. Berendzen, *Acta Crystallogr. Sect. D: Biol. Crystallogr.*, 1999, **55**, 849–861.
18. Collaborative Computational Project, Number 4. "The CCP4 Suite: Programs for Protein Crystallography", *Acta Crystallogr., Sect. D: Biol. Crystallogr.*, 1994, **50**, 760–763.

19. A. Perrakis, M. Harkiolaki, K.S. Wilson and V.S. Lamzin, *Acta Crystallogr. Sect. D: Biol. Crystallogr.*, 2001, **57**, 1445–1450.
20. H. Wu, L.D. Finger and J. Feigon, *Methods Enzymol.*, 2005, **394**, 525–545.
21. G. Storz, *Science*, 2002, **296**, 1260–1263.
22. A.G. Matera, R.M. Terns and M.P. Terns, *Nat. Rev.*, 2007, **8**, 209–220.
23. C. Oubridge, N. Ito, P.R. Evans, C.H. Teo and K. Nagai, *Nature*, 1994, **372**, 432–438.
24. C. Maris, C. Dominguez and F.H. Allain, *FEBS J.*, 2005, **272**, 2118–2131.
25. C.L. Kielkopf, S. Lucke and M.R. Green, *Genes Dev.*, 2004, **18**, 1513–1526.
26. A. Korostelev, S. Trakhanov, M. Laurberg and H.F. Noller, *Cell*, 2006, **126**, 1065–1077.
27. M. Selmer, C.M. Dunham, F.V.T. Murphy, A. Weixlbaumer, S. Petry, A.C. Kelley, J.R. Weir and V. Ramakrishnan, *Science*, 2006, **313**, 1935–1942.
28. B.S. Schuwirth, M.A. Borovinskaya, C.W. Hau, W. Zhang, A. Vila-Sanjurjo, J.M. Holton and J.H. Cate, *Science*, 2005, **310**, 827–834.
29. J.M. Ogle, F.V. Murphy, M.J. Tarry and V. Ramakrishnan, *Cell*, 2002, **111**, 721–732.
30. F. Schluenzen, A. Tocilj, R. Zarivach, J. Harms, M. Gluehmann, D. Janell, A. Bashan, H. Bartels, I. Agmon, F. Franceschi and A. Yonath, *Cell*, 2000, **102**, 615–623.
31. D.J. Klein, P.B. Moore and T.A. Steitz, *J. Mol. Biol.*, 2004, **340**, 141–177.
32. H.F. Noller, *Science*, 2005, **309**, 1508–1514.
33. P. Nissen, J.A. Ippolito, N. Ban, P.B. Moore and T.A. Steitz, *Proc. Natl. Acad. Sci. U.S.A.*, 2001, **98**, 4899–4903.
34. D.J. Battle and J.A. Doudna, *Proc. Natl. Acad. Sci. U.S.A.*, 2002, **99**, 11676–11681.
35. G.M. Culver, *Biopolymers*, 2003, **68**, 234–249.
36. M.W. Talkington, G. Siuzdak and J.R. Williamson, *Nature*, 2005, **438**, 628–632.
37. W.A. Held, B. Ballou, S. Mizushima and M. Nomura, *J. Biol. Chem.*, 1974, **249**, 3103–3111.
38. N.B. Leontis and E. Westhof, *Curr. Opin. Struct. Biol.*, 2003, **13**, 300–308.
39. D.J. Klein, T.M. Schmeing, P.B. Moore and T.A. Steitz, *EMBO J.*, 2001, **20**, 4214–4221.
40. I. Vidovic, S. Nottrott, K. Hartmuth, R. Luhrmann and R. Ficner, *Mol. Cell*, 2000, **6**, 1331–1342.
41. F.H. Allain, C.C. Gubser, P.W. Howe, K. Nagai, D. Neuhaus and G. Varani, *Nature*, 1996, **380**, 646–650.
42. F.H. Allain, P.W. Howe, D. Neuhaus and G. Varani, *EMBO J.*, 1997, **16**, 5764–5772.
43. Y. Tomari and P.D. Zamore, *Genes Dev.*, 2005, **19**, 517–529.
44. M.A. Carmell, Z. Xuan, M.Q. Zhang and G.J. Hannon, *Genes Dev.*, 2002, **16**, 2733–2742.
45. M.B. Wang and M. Metzlaff, *Curr. Opin. Plant Biol.*, 2005, **8**, 216–222.
46. M. Doyle and M.F. Jantsch, *J. Struct. Biol.*, 2002, **140**, 147–153.

47. P.A. Beal, *Chembiochem*, 2005, **6**, 257–266.
48. B. Tian, P.C. Bevilacqua, A. Diegelman-Parente and M.B. Mathews, *Nat. Rev. Mol. Cell Biol.*, 2004, **5**, 1013–1023.
49. K.Y. Chang and A. Ramos, *FEBS J.*, 2005, **272**, 2109–2117.
50. J.M. Ryter and S.C. Schultz, *EMBO J.*, 1998, **17**, 7505–7513.
51. H. Wu, A. Henras, G. Chanfreau and J. Feigon, *Proc. Natl. Acad. Sci. U.S.A.*, 2004, **101**, 8307–8312.
52. J.B. Ma, K. Ye and D.J. Patel, *Nature*, 2004, **429**, 318–322.
53. A. Lingel, B. Simon, E. Izaurralde and M. Sattler, *Nat. Struct. Mol. Biol.*, 2004, **11**, 576–577.
54. J.S. Parker, S.M. Roe and D. Barford, *Nature*, 2005, **434**, 663–666.
55. J.B. Ma, Y.R. Yuan, G. Meister, Y. Pei, T. Tuschl and D.J. Patel, *Nature*, 2005, **434**, 666–670.
56. J.J. Song, S.K. Smith, G.J. Hannon and L. Joshua-Tor, *Science*, 2004, **305**, 1434–1437.
57. I.J. Macrae, K. Zhou, F. Li, A. Repic, A.N. Brooks, W.Z. Cande, P.D. Adams and J.A. Doudna, *Science*, 2006, **311**, 195–198.
58. J.M. Vargason, G. Szittya, J. Burgyan and T.M. Tanaka Hall, *Cell*, 2003, **115**, 799–811.
59. H.G. Koch, M. Moser and M. Muller, *Rev. Physiol. Biochem. Pharma*, 2003, **146**, 55–94.
60. M.R. Pool, *Mol. Membr. Biol.*, 2005, **22**, 3–15.
61. J. Luirink and I. Sinning, *Biochim. Biophys. Acta*, 2004, **1694**, 17–35.
62. M. Halic and R. Beckmann, *Curr. Opin. Struct. Biol.*, 2005, **15**, 116–125.
63. P.B. Rupert and R. Ferre-D'amare, *Structure*, 2000, **8**, R99–104.
64. P.F. Egea, R.M. Stroud and P. Walter, *Curr. Opin. Struct. Biol.*, 2005, **15**, 213–220.
65. O. Weichenrieder, K. Wild, K. Strub and S. Cusack, *Nature*, 2000, **408**, 167–173.
66. R.T. Batey, R.P. Rambo, L. Lucast, B. Rha and J.A. Doudna, *Science*, 2000, **287**, 1232–1239.
67. A. Kuglstatter, C. Oubridge and K. Nagai, *Nat. Struct. Biol.*, 2002, **9**, 740–744.
68. R.T. Batey and J.A. Doudna, *Biochemistry*, 2002, **41**, 11703–11710.
69. T. Hainzl, S. Huang and A.E. Sauer-Eriksson, *Nature*, 2002, **417**, 767–771.
70. K. Wild, I. Sinning and S. Cusack, *Science*, 2001, **294**, 598–601.
71. A. Fatica and D. Tollervey, *Curr. Opin. Cell Biol.*, 2002, **14**, 313–318.
72. S. Kass, K. Tyc, J.A. Steitz and B. Sollner-Webb, The U3 small nucleolar ribonucleoprotein functions in the first step of preribosomal RNA processing, *Cell*, 1990, **60**, 897–908.
73. J. Venema and D. Tollervey, Processing of pre-ribosomal RNA in Saccharomyces cerevisiae, *Yeast*, 1995, **11**, 1629–1650.
74. G.L. Eliceiri, The vertebrate E1/U17 small nucleolar ribonucleoprotein particle, *J. Cell Biochem.*, 2006, **98**, 486–495.

75. W.Q. Liang and M.J. Fournier, U14 base-pairs with 18S rRNA: a novel snoRNA interaction required for rRNA processing, *Genes Dev.*, 1995, **9**, 2433–2443.
76. T. Moore, Y. Zhang, M.O. Fenley and H. Li, *Structure*, 2004, **12**, 807–818.
77. T. Hamma and A.R. Ferre-D'Amare, *Structure*, 2004, **12**, 893–903.
78. J. Suryadi, E.J. Tran, E.S. Maxwell and B.A. Brown II, *Biochemistry*, 2005, **44**, 9657–9672.
79. M. Aittaleb, R. Rashid, Q. Chen, J.R. Palmer, C.J. Daniels and H. Li, Structure and function of archaeal box C/D sRNP core proteins, *Nat. Struct. Biol.*, 2003, **10**, 256–263.
80. N.J. Watkins, V. Segault, B. Charpentier, S. Nottrott, P. Fabrizio, A. Bachi, M. Wilm, M. Rosbash, C. Branlant and R. Luhrmann, *Cell*, 2000, **103**, 457–466.
81. N.J. Watkins, R.D. Leverette, L. Xia, M.T. Andrews and E.S. Maxwell, *RNA*, 1996, **2**, 118–133.
82. J.F. Kuhn, E.J. Tran and E.S. Maxwell, Archaeal ribosomal protein L7 is a functional homolog of the eukaryotic 15.5 kD/Snu13p snoRNP core protein, *Nucleic Acids Res.*, 2002, **30**, 931–941.
83. R. Rashid, M. Aittaleb, Q. Chen, K. Spiegel, B. Demeler and H. Li, Functional requirement for symmetric assembly of archaeal box C/D small ribonucleoprotein particles, *J. Mol. Biol.*, 2003, **333**, 295–306.
84. L.B. Szewczak, S.J. DeGregorio, S.A. Strobel and J.A. Steitz, *Chem. Biol.*, 2002, **9**, 1095–1107.
85. S. Oruganti, Y. Zhang and H. Li, *Biochem. Biophys. Res. Commun.*, 2005, **333**, 550–554.
86. L. Li and K. Ye, *Nature*, 2006, **443**, 302–307.
87. T.A. Goody, S.E. Melcher, D.G. Norman and D.M. Lilley, *RNA*, 2004, **10**, 254–264.
88. B. Turner, S.E. Melcher, T.J. Wilson, D.G. Norman and D.M. Lilley, *RNA*, 2005, **11**, 1192–1200.
89. R. Ishitani, O. Nureki, N. Nameki, N. Okada, S. Nishimura and S. Yokoyama, *Cell*, 2003, **113**, 383–394.
90. M.A. Schumacher, E. Karamooz, A. Zikova, L. Trantirek and J. Lukes, *Cell*, 2006, **126**, 701–711.
91. D. Desveaux, J. Allard, N. Brisson and J. Sygusch, *Nat. Struct. Biol.*, 2002, **9**, 512–517.
92. A.W. Nicholson, *FEMS Microbiol. Rev.*, 1999, **23**, 371–390.
93. A.J. Callaghan, M.J. Marcaida, J.A. Stead, K.J. McDowall, W.G. Scott and B.F. Luisi, *Nature*, 2005, **437**, 1187–1191.
94. C. Frazao, C.E. McVey, M. Amblar, A. Barbas, C. Vonrhein, C.M. Arraiano and M.A. Carrondo, *Nature*, 2006, **443**, 110–114.
95. I.J. Macrae and J.A. Doudna, *Curr. Opin. Struct. Biol.*, 2007, **17**, 138–145.
96. J. Gan, J.E. Tropea, B.P. Austin, D.L. Court, D.S. Waugh and X. Ji, *Cell*, 2006, **124**, 355–366.

97. J. Lacadena, E. Alvarez-Garcia, N. Carreras-Sangra, E. Herrero-Galan, J. Alegre-Cebollada, L. Garcia-Ortega, M. Onaderra, J.G. Gavilanes and A. Martinez del Pozo, *FEMS Microbiol. Rev.*, 2007, **31**, 212–237.
98. X. Yang, T. Gerczei, L.T. Glover and C.C. Correll, *Nat. Struct. Biol.*, 2001, **8**, 968–973.
99. C.C. Correll and K. Swinger, *RNA*, 2003, **9**, 355–363.
100. C.C. Correll, I.G. Wool and A. Munishkin, *J. Mol. Biol.*, 1999, **292**, 275–287.
101. C.L. Peebles, P. Gegenheimer and J. Abelson, *Cell*, 1983, **32**, 525–536.
102. C.R. Trotta, F. Miao, E.A. Arn, S.W. Stevens, C.K. Ho, R. Rauhut and J.N. Abelson, *Cell*, 1997, **89**, 849–858.
103. G. Knapp, J.S. Beckmann, P.F. Johnson, S.A. Fuhrman and J. Abelson, *Cell*, 1978, **14**, 221–236.
104. K. Calvin, M.D. Hall, F. Xu, S. Xue and H. Li, *J. Mol. Biol.*, 2005, **353**, 952–960.
105. G.D. Tocchini-Valentini, P. Fruscoloni and G.P. Tocchini-Valentini, *Proc. Natl. Acad. Sci. U.S.A.*, 2005, **102**, 8933–8938.
106. C. Marck and H. Grosjean, *RNA*, 2003, **9**, 1516–1531.
107. V.M. Reyes and J. Abelson, *Cell*, 1988, **55**, 719–730.
108. S. Xue, K. Calvin and H. Li, *Science*, 2006, **312**, 906–910.
109. G.A. Soukup and R.R. Breaker, *RNA*, 1999, **5**, 1308–1325.
110. I. Li de la Sierra-Gallay, N. Mathy, O. Pellegrini and C. Condon, *Nat. Struct. Mol. Biol.*, 2006, **13**, 376–377.
111. I.L. de la Sierra-Gallay, O. Pellegrini and C. Condon, *Nature*, 2005, **433**, 657–661.
112. A. Vogel, O. Schilling, B. Spath and A. Marchfelder, *Biol Chem.*, 2005, **386**, 1253–1264.
113. S. Jones, D.T. Daley, N.M. Luscombe, H.M. Berman and J.M. Thornton, *Nucleic Acids Res.*, 2001, **29**, 943–954.
114. J. Allers and Y. Shamoo, *J. Mol. Biol.*, 2001, **311**, 75–86.
115. M. Treger and E. Westhof, *J. Mol. Recognit.*, 2001, **14**, 199–214.
116. K.R. Phipps and H. Li, *Proteins: Stuct. Funct. Gene*, 2007, **67**, 121–127.

CHAPTER 8
Bending and Compaction of DNA by Proteins

REID C. JOHNSON, STEFANO STELLA AND JOHN K. HEISS

Department of Biological Chemistry, David Geffen School of Medicine at UCLA, 615 Charles Young Drive South, Los Angeles, CA 90095-1737, USA

8.1 Introduction

Essentially all active and passive reactions involving DNA require the duplex to undergo considerable and often dramatic gymnastics. Active processes include transcription, replication and recombination reactions where nucleoprotein complexes that usually require DNA bending both within and between DNA binding proteins are assembled. Passive processes include the folding and wrapping of DNA that is essential to compact chromosomes within the eukaryotic cell nucleus or prokaryotic cell nucleoid. How is this accomplished, given the stiffness of the DNA duplex over short distances? This chapter discusses protein-induced bending of DNA primarily with respect to the molecular features of the DNA duplex that influence the nature and position of structural distortions induced by proteins. Selected examples of protein-DNA complexes that induce significant DNA bending will then be presented to illustrate these points, and, in some cases, their long-range effects on compaction of DNA molecules are described.

It is well recognized that DNA has remarkable, but not unrestricted, flexibility that enables it to adopt many different conformations when it associates with a protein surface. Whereas earlier views of protein-induced DNA bending often emphasized highly localized severe deformations

RSC Biomolecular Sciences
Protein-Nucleic Acid Interactions: Structural Biology
Edited by Phoebe A. Rice and Carl C. Correll
© Royal Society of Chemistry 2008

introduced upon complex formation,[1-4] there are increasing numbers of examples reported where DNA smoothly adapts to a rigid protein surface. In the latter cases, the cumulative effects of many modest distortions add up to generate a substantial change in helix trajectory. This is perhaps best illustrated by the structure of the nucleosome, which will be referred to throughout the chapter. Different DNA sequences can adopt a common local structure when associated with the same protein surface, and identical DNA sequences can adopt different structures in the same local environment to give the same overall curvature to the segment. These properties all point to a highly malleable DNA molecule.

Structural flexibility of DNA is often a critical aspect of protein recognition and binding. For most sequence-specific binding proteins, targeted binding requires both direct and indirect readout mechanisms. Direct readout is largely mediated through complementary hydrogen bonding groups between amino acids and bases, and can include intermediary water molecules. Indirect readout mechanisms involve conformational changes in the DNA, which involve sequence-dependent features, such as an intrinsic bend or distortion that approaches or mimics the structure in the bound complex, or simply the ease in which a certain sequence can adopt a particular structure to conform to the protein surface. As discussed below, there are examples where a preformed structure enhances binding affinity, but sequence-dependent flexibility of DNA appears to be the dominant factor driving indirect recognition.

The importance of indirect recognition of DNA sequences by conformation becomes evident when one considers the small number of direct base-specific contacts found in most complexes, combined with the degenerate "code" specified by the hydrogen bonding groups on amino acids and bases.[5] Moreover, there are many examples where the identity of bases is critical for target site recognition even though they are not directly contacted by the protein. In a recent survey, Gromiha *et al.* attempted to quantify the contributions of direct versus indirect recognition mechanisms in 51 structurally different protein-DNA complexes.[6] They estimated that indirect readout by DNA conformational changes provided an equal or greater component to the binding energy than direct intermolecular contacts in about 60% of the complexes. Likewise, about 90% of the non-redundant collection of protein-DNA complexes analyzed by Dickerson exhibited significant bending of the DNA segment.[1,2,7]

In addition to binding site recognition, DNA bending has been shown to be a critical determinant of DNA catalysis. For example, the restriction enzyme EcoRV introduces a sharp bend at the position of the scissile phosphates in the center of its binding site.[8-10] However, this bend was absent in a catalytically inactive complex of the enzyme bound to a noncognate site containing a 2 bp substitution at the center of the recognition sequence. In the inactive structure the scissile phosphates were not in position to form the Mg^{2+}-coordinated contacts with residues in the active site that are required to initiate hydrolysis.

Essentially all chromatin- and nucleoid-associated proteins that bind DNA in a largely sequence-neutral manner introduce strong bends into DNA. In many cases, their sole function is to bend DNA to facilitate interactions between nearby DNA-bound proteins in the assembly of higher-order nucleoprotein complexes or to promote chromosome condensation. Prominent examples are the monomeric eukaryotic HMGB proteins that can induce bends of up to 115° within a 10 bp region, histone dimers which as units introduce 130–140° of bending within a 30 bp region, and the dimeric prokaryotic HU/IHF proteins that induce bends of up to 180° over a 40 bp region. Each of these proteins appears to recognize structure-specific features of the DNA helix to target binding, as highlighted by their enhanced affinities for predistorted or more easily distorted DNA segments. For example, 1000-fold increases in affinity have been measured for binding of HMGB proteins to DNA segments containing a cisplatin-induced kink and for HU protein to DNA segments containing strategically placed nicks. In the case of DNA binding by histones to generate nucleosomes, 1000-fold differences between DNA segments of constant length have been measured even though there are essentially no sequence-specific interactions between the protein and DNA. Thus, a close coupling exists between DNA sequence, DNA structure, and protein binding, particularly with respect chromatin/nucleoid-associated proteins.

8.2 Forces Controlling DNA Rigidity

8.2.1 DNA Elasticity and the Influence of DNA Sequence

The stiffness of DNA is often discussed in terms of its persistence length, which can be generally defined as the chain length in which an initial direction of the chain axis remains unaltered on average. That is, the persistence length is the length where a DNA segment can be physically modeled as a straight rod. For DNA of mixed sequence, the persistence length has been traditionally accepted to be 45–50 nm or about 150 bp.[11] At any given time, however, the DNA ends will be closer to each other due to thermally-induced fluctuations, and, indeed, mixed sequence DNA of around 150 bp will begin to support ligation into a monomeric circle, provided the correct helical phasing is present. DNA lengths increasingly shorter than the persistence length are correspondingly more resistant to significant bending, and DNA lengths longer than this value increasing behave as flexible coils, reaching maximum cyclization efficiency at about 150 nm or 450–500 bp. Early studies showed that the composition of the DNA chain affects its apparent persistence length,[11] and more recent experiments have further demonstrated that alternating AT polymers are 20–30% more flexible than mixed sequence DNA.[12,13]

Recent experiments by Cloutier and Widom have provided evidence that DNA may be far more flexible than traditionally assumed.[14] They have shown that DNA lengths of 94 bp can be efficiently ligated into circles without the aid of a DNA bending protein. The overall curvature within these microcircles is in the same range as for DNA wrapped in a nucleosome. They further demonstrated that formation of these thermally-excited circular structures is strongly influenced by their DNA sequence and correlates well to the 25-fold range of binding affinities obtained for H3/H4 histone tetramer binding to the different sequences.

Cloutier and Widom calculate that the spontaneous formation of these DNA circles is up to 10^5 times greater than allowed by current theories. The molecular basis for the formation these tight DNA microcircles is controversial, with some proposing that local regions of flipped-out bases or denatured duplexes are responsible.[15,16] Alternatively, these findings indicate that bending an intact DNA duplex by conventional mechanisms (see below) is not nearly as energetically costly as most theories assume, leading investigators to attempt to model the elastic properties of DNA as a kinkable wormlike chain.[17,18]

As discussed in more detail below, the high-resolution crystal structures of Richmond, Lugar, and colleagues have shown that DNA within nucleosomes is not evenly bent but rather unevenly distorted to adapt to the donut-like surface of the histone octamer core. It has been known since the initial observations by Travers and co-workers in the mid-1980s that the wrapping of DNA in nucleosomes is strongly influenced by particular sequence motifs that are phased with the helical repeat of DNA.[19,20] Numerous subsequent studies analyzing nucleosome binding or positioning sequences from yeast to mammals, and from *in vitro* selected random sequences, have confirmed these results (reviewed in refs 18 and 21). In general, they show that A-A (or T-T) and T-A dinucleotides (reading $5'$ to $3'$ on the top strand) are overrepresented at positions where the minor groove faces the histone octamer, and that the dinucleotide G-C is overrepresented when the minor groove is facing away from the octamer. The DNA sequence effects on nucleosome assembly strongly imply that the structural properties of a DNA sequence, but not the precise identity of the base sequence itself, can strongly influence binding by reducing the energy cost for adopting a particular conformation. Additional examples of this are discussed in Section 8.4.

8.2.2 Base Stacking Primarily Controls Helix Rigidity

The forces responsible for the rigidity of the DNA double helical structure are primarily base stacking interactions and, to a lesser degree, the Watson–Crick base pairing on the inside and phosphate–phosphate repulsions on the outside of the helix. A striking illustration of the dominant effect of base stacking on the rigidity of the helix comes from the experiment of Mills and Hagerman.[22] They initially constructed a 100 bp gapped molecule with the central 24 residues of oligo dT unpaired. The gap was filled by addition of the adenine base (actually N^6-methyladenine to eliminate the possibility of alternative base paired structures) in solution to form a duplex lacking the sugar-phosphate backbone over the stretch of 24 adenines. Helix rigidity was monitored by transient electric birefringence, which provides a measure of persistence length. Whereas the gapped molecule was highly flexible, the molecule with the noncovalently associated adenine displayed a rigidity that was nearly indistinguishable from an intact duplex.

Studies with molecules containing single-strand nicks have also shown that the integrity of the phosphodiester backbone is not a primary factor in helix bending. X-Ray crystal and NMR structures of duplexes containing a strand discontinuity have revealed little effect on the helix structure.[23,24] In addition, solution experiments employing gel electrophoresis and cyclization assays have found that DNA duplexes containing single-strand nicks do not exhibit significantly increased bending flexibility, although increases in torsional (rotational) flexibility have been noted in some experiments.[12,25] Recently, Frank-Kamenetskii and co-workers have compared the effects of all possible combinations of dinucleotide steps across a nick by their mobilities on polyacrylamide gels.[25,26] A well-stacked nicked DNA duplex with or without a 5' phosphate group migrates almost immediately to the intact duplex, whereas a duplex with a flexible joint originating from a gap of two or more nucleotides migrates more slowly. Different dinucleotide sequences spanning a nick exhibit different migratory properties that can be related to their stacking energies. These experiments not only provide additional evidence that base stacking is the overwhelming stabilizing component of helix stability and that Watson–Crick base pairing contributes minimally, but show that duplex stability is dependent upon nearest neighbors. For example, the dinucleotide step TG is more prone to flexure at a nick than AT, and CG is more prone to flexure than GT.[25] The rank order of dinucleotide stacking energies generally fits with their flexibilities based on atomic structures (see below). Most traditional thermodynamic analyses that have measured denaturation properties of different duplexes have generally concluded that duplex stability largely varies with respect to the GC content. Many of these studies have also concluded that stacking interactions between nearest neighbors are an important determinant of helix stability, but the rank order of stability of nearest neighbor pairs from these studies do not always match the properties of dinucleotide steps inferred from X-ray data.[27–29]

8.2.3 Electrostatic Forces Modulate DNA Bending

Electrostatic effects mediated by mutual repulsion of phosphates on the DNA surface have been proposed to be a significant factor contributing to helix rigidity.[30] Bending of DNA will alter the distance between the anionic phosphates, leading to electrostatic repulsion upon close approach. This would be particularly important across the minor groove where separation between the van der Waals surfaces of proximal phosphates normally averages around 6 Å but can compress to ≤ 3 Å in both free DNA and protein-DNA structures. Manning calculated that a large fraction, about 75%, of the phosphates is neutralized by a concentrated (1 M) cloud of hydrated counterions that extends ∼7 Å from the DNA surface[31] (see also ref. 32). It is generally agreed that divalent cations or polyamines are more effective in screening the charge on the phosphates than monovalent ions.[33–36]

Theoretical considerations by Manning, Rich, and colleagues[37,38] also predicted that asymmetrically localized neutralization of backbone phosphates would encourage bending of the helix in the direction of the neutralized

Bending and Compaction of DNA by Proteins 181

phosphates. Compelling experimental evidence for this idea was provided by Maher and coworkers, who introduced racemic mixtures of methylphosphonates at selected positions within a 21 bp duplex.[39] Bending was measured by evaluating the electrophoretic mobility of linear multimers generated by ligation, where bends present in the modified DNA could be phased relative to a static bend directed by an A-tract sequence on the same oligonucleotide segment.[40,41] Based on a value of 18° for an A_{5-6}-tract bend, partial neutralization of three pairs of consecutive phosphates on either side of the minor groove by the methylphosphonate substitutions resulted in a calculated curvature of ~20° into the minor groove (Figure 8.1A), irrespective of whether the modified sequence was over GC or AT-rich regions. In the presence of 6 mM Mg^{2+} or 1 mM spermidine, this value was reduced to ~7° or 10°, respectively, which is consistent with the change due solely to electrostatic effects.[39] When the three pairs of neutralizing methylphosphonate substitutions alternated with

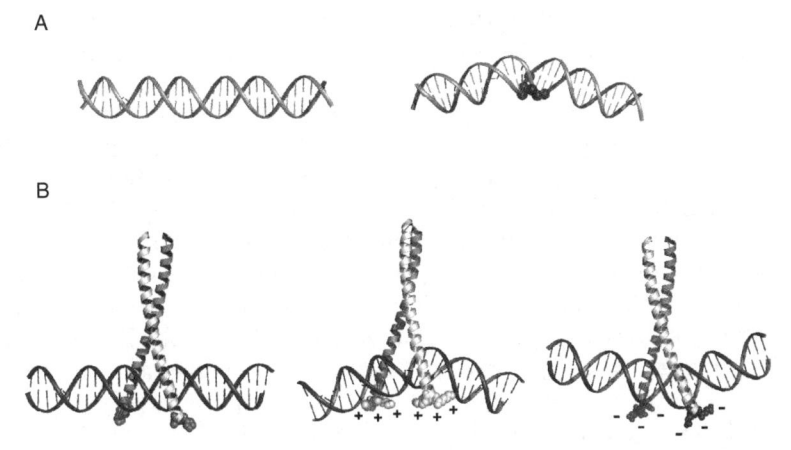

Figure 8.1 Electrostatic contributions to DNA bending. (A) Neutralization of phosphate charges across the minor groove promotes modest DNA bending. The left-hand panel represents a standard straight B-DNA molecule. Strauss and Maher introduced three consecutive methylphosphonates (spheres) across the minor groove from each other in both phosphodiester backbones.[39] The methylphosphonate nucleotides contain one of the non-esterified oxygens replaced with a methyl group, thereby partially neutralizing the negative charge. This results in a bend into the minor groove as schematically illustrated in the right-hand panel. (B) Charged amino acids within the N-termini of bZIP DNA binding domains alter the trajectory of DNA emanating from the protein complex. The left-hand panel shows the GCN4-DNA complex (pdb code 1YSA) with the uncharged N-terminal residues (Pro-Pro-Ala) rendered in a space-filling view. When lysine residues are substituted at the N-termini, the DNA becomes dynamically bent away from the leucine zipper dimerization motif. When glutamic acid residues are substituted at the N-termini, the DNA becomes dynamically bent towards the leucine zipper dimerization motif.[52,53] Similar effects of neutral, positively or negatively charged residues in the N-termini of Fos/Jun and CRE-BP1 have been observed.[49–51]

unmodified nucleotides, the magnitude of the bend was reduced to ~13°, indicating that neutralization of a contiguous string of phosphates is important.[42] In a related strategy, six ammonium cations were tethered to the 5 position (in the major groove) of cytosine or uracil, using three- or six-carbon linkers.[43,44] The NH_3^+ groups were positioned to form ion pairs with phosphates across the minor groove in a manner that would qualitatively mimic a patch of six lysines from a bound protein. These modifications introduced 4–8° of bending in the direction of the minor groove. Control molecules where the amine was neutralized by acetylation exhibited no introduced bend, again supporting the conclusion that the effects were electrostatic and not due to a structural consequence of the tether. Although the effects are modest, especially in the contexts of much larger bends introduced by some proteins, the experiments of Maher and coworkers on oligonucleotides demonstrate that localized asymmetric neutralization of phosphates can promote directional bending of the DNA helix.

Related experiments employing basic-leucine zipper (bZIP) proteins have provided strong evidence that electrostatic effects can alter the trajectory of DNA when bound by proteins. bZIP dimeric proteins like Fos-Jun or GCN4 bind DNA by means of highly basic N-terminal ends of α-helices from each subunit running through the major groove on either side of the helix axis, which then associate with each other over a coiled-coil leucine zipper to form the protein dimer (Figure 8.1B). Even though Jun-Jun homodimers and Fos-Jun heterodimers interact with a common binding site (AP-1) in a similar manner, Jun-Jun dimers bend DNA in the opposite direction from Fos-Jun dimers.[45–47] A survey of DNA complexes containing Fos and Jun family proteins found that the trajectories of DNA segments extended in various directions with apparent bend angles ranging from 0° to over 40°.[48] These bends are produced by charged residues located immediately N-terminal to the basic helix as it enters the major groove. Elimination or swapping of the series of N-terminal basic or acidic residues present in Fos and Jun, respectively, has been shown to have a corresponding effect on the direction and magnitude of DNA bend trajectories[49,50] (see also ref. 51). Likewise, introducing 1–3 basic or acidic residues onto the N-terminal ends of the bZIP homodimer GCN4 results in the normally unbent DNA being bent towards or away from the N-termini, respectively (Figure 8.1B).[52,53] Most of these experiments relied on gel electrophoretic mobility assays, where the protein binding site is positioned relative to a statically bent A-tract, to evaluate bending, but recent confirmation of an effect by charged residues on DNA bending has been obtained from FRET-based assays.[54,55] Although this solution method cannot distinguish bend direction, FRET-measurements showed that both acidic and basic substitutions led to a shortening of the distance between DNA ends. This independent evidence for bending is important since the DNA in X-ray crystal structures of bZIP proteins like Fos-Jun and GCN4 complexes is unbent.[56,57] As discussed further below, crystal packing and other considerations result in X-ray structures that do not always reflect the extent of bending within a protein complex, particularly where electrostatic interactions that dynamically modulate the structures of DNA segments flanking the primary interface are operating.

8.3 Bending of DNA at High Resolution

Most of the precise structural information on DNA helix structure comes from X-ray crystallography, although recent advances in NMR, such as the use of residual dipolar couplings, are providing increasingly higher resolution views of nucleic acids in solution.[58] Many of the X-ray structures of duplex oligonucleotides have been determined at resolution limits of <2 Å and some even around the 1 Å range, generating a wealth of atomic level data. Ever since the first single-crystal structure of B-DNA over 25 years ago,[59] researchers have tried to relate DNA sequence to structure in the hopes of establishing predictable relationships. These efforts have met with mixed success, particularly when applied to protein-DNA complexes. While there are some trends and relationships that will be summarized below, few hard "rules" have emerged. Perhaps this is not surprising, given the pliable nature of DNA structure and since accumulation of energies gained from multiple protein-DNA contacts can easily overcome modest energetic barriers to deforming individual DNA base pair steps. Moreover, protein binding clearly amplifies the range of structural distortions; distortions that are rare or uncommon in the oligonucleotide database are well represented among protein-DNA complexes. Nevertheless, many of the molecular features revealed from analysis of oligonucleotide structures have been maintained in protein-DNA complexes,[7] providing encouragement that information revealed from naked DNA is applicable to protein-DNA complexes.

It is worth noting some of the limitations of the huge database of information on oligonucleotide structures. For nearly a decade after the first B-DNA X-ray structure, almost all structures were of dodecamers that not only packed in the crystals in an identical fashion but were highly related in DNA sequence, particularly at their GC-rich ends.[60,61] In these structures the DNA molecules are aligned end-to-end with the terminal 2–3 bp on each end inserted into the minor grooves of their neighbors. Most, but not all, of the decamer structures that followed have their DNA segments aligned as nearly straight pseudocontinuous helices through the crystals. The effects of crystal packing on DNA structure have been discussed in several papers and reviews, with most crystallographers arguing that oligonucleotide structures are informative because local base stacking forces dominate.[61,62] Nevertheless, crystal environmental forces no doubt dampen the magnitude of deformations that could form in solution. The structure of DNA molecules bound by proteins are primarily or exclusively influenced by the protein over the binding interface, but DNA segments extending from the protein again very often align with neighboring molecules to form pseudocontinuous helices in the crystal. In many cases, the coaxial stacking of DNA segments also dampens the full extent of bending, particularly where dynamic bending over positively or negatively charged surfaces is involved.

Williams and Maher also argue that the under-representation of counterions in most of the current crystal structures leads to a non-physiological imbalance of charge and an under appreciation of the importance of electrostatic forces.[30,63] Monovalent cations are usually too delocalized to reveal their presence by X-ray crystallography, but several high-resolution X-ray structures and NMR studies have highlighted bound divalent cations and polyamines that have been shown to influence DNA structure.[64–67] For example, the 1 Å structures of Chiu and Dickerson[65] revealed up to 22 Mg^{2+} or Ca^{2+} ions bound to a decamer sequence. Some 50–75% of these were bound to one or both of the non-esterified oxygens of backbone phosphates through either direct ionic bonds or more commonly through intermediary water molecules, thus partially neutralizing the negative charge. Hydrated Mg^{2+} or Ca^{2+} ions were also found bound in a sequence-specific manner to hydrogen bonding groups on bases within the major groove; in some cases, Ca^{2+} was found to be directly bonded to bases, such as to adjacent N7 and O6 atoms of stacked guanines. In these cases, the metal ion bonding results in a compression of the major groove with an associated smooth bend of 10–20° spread over 2–3 bp. Hydrated divalent ions can also stably associate within the minor groove by hydrogen bonding to base atoms on the floor of the groove and to the O4' position of the sugar. However, in these cases, they are not believed to have a significant effect on bending.[65,67]

8.3.1 Helix Parameters Controlling DNA Structure

8.3.1.1 Roll and Tilt

DNA helix structure is defined at the atomic level by several parameters;[61,68] those most relevant to this discussion are shown in Figure 8.2 and described below. Changes in the direction of the helix axis occur primarily by deviations

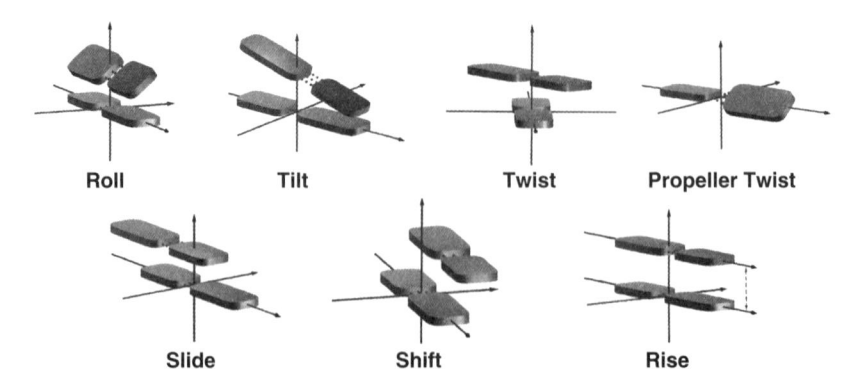

Figure 8.2 DNA helix parameters. Helix parameters[68] discussed in this chapter are depicted schematically. Note that the rise, or separation between base pairs, seldom significantly deviates from a mean of around 3.34 Å, except for dinucleotides subject to intercalation by amino acid side-chain(s).

Bending and Compaction of DNA by Proteins

in roll or tilt, or a combination of the two. Roll describes the rotation about the long axis of the base pair with the convention being that (+) roll designates a bend into and thus compression of the major groove, and (−) roll angle refers to a bend into and thus compression of the minor groove. Tilt is rotation about the short axis of a base pair. An appreciable change of tilt angle that increases the distance from a neighboring base will almost immediately cause a clash with the base on the opposite side. By contrast, modest changes in roll will have much less of a consequence on nearest neighbor stacking and hence it is the overwhelmingly preferred manner in which the helix bends. Roll angle deviations of up to 10° in either direction within naked DNA crystal structures are common, implying that roll angles extending to 10° and probably up to 15° are tolerated with little energetic cost, especially considering that crystal packing forces are limiting natural flexibility of DNA. Numerous surveys of both naked DNA structures and protein-DNA complexes have all emphasized that changes in the helix axis observed in oligonucleotide or protein-bound DNA structures are dominated by roll deviations over tilt.[2–4,7,69–71] For large bends or kinks, resulting in >20° localized deflection of the helix axis, roll changes are almost exclusively observed. For smaller bends, roll still predominates about 2:1 over changes in tilt, but the latter becomes increasingly more significant.[3] In the nucleosome, where DNA wrapping around the histone core is accomplished by many small bends plus a few moderate kinks, roll changes predominate over tilt changes 3:1.[72] Complete unstacking of base pairs by protein binding to create roll angles of >40° is rare and is almost always accompanied by insertion or intercalation of hydrophobic amino acid residues. To achieve an overall planar bend, roll bends of the same sign must be phased according to the ~10.5 bp/turn helical repeat of DNA (or roll bends of opposite sign will need to be introduced every half turn). Alternatively, moderate roll bends at out-of-phase positions can be balanced by tilt angle changes. Roll bends of the same sign introduced into consecutive base pairs will lead to writhing or non-planar curvature.

8.3.1.2 Twist

The twist angle (Figure 8.2) between base pairs establishes the DNA helical repeat, and averages 34.3° or 10.5 bp/turn for linear B-DNA in solution. However, twist values between base pair steps can range widely, with values of 20–55° being observed in oligonucleotides structures and values well below 20° being observed in protein complexes where amino acid side-chain intercalation causes base pair unstacking. Recent microcircle ligation experiments by Cloutier and Widom[17] also demonstrate the remarkable torsional flexibility of DNA in solution. Changes in twist are influenced by base stacking and other forces intrinsic to the double helical structure and can be significantly impacted by protein binding as twist changes are a flexible means of orienting groups on the bases for protein contact. Among all combinations of DNA structural parameters, the greatest correlation is observed between twist and roll: low twist values or helix unwinding are usually associated with (+) roll changes whereas high twist values are usually associated with (−) roll changes.[70,73,74] Over the 147 bp of nucleosomal DNA, twist values range from 24° to 50°.[72] Twist values are significantly greater than the 34.3° canonical DNA value when the minor grooves are oriented towards the histone octamer and exceed 40° at each position where the DNA exhibits negative roll values of over 15°. On the other hand, the average twist value drops to 28.5° when the roll is greater than +15°. Changes in twist can alter the width of the DNA grooves with helix overwinding leading to narrower grooves and underwinding expanding the grooves.

8.3.1.3 Propeller Twist, Slide, and Shift

DNA base pairs virtually always display some propeller twist (Figure 8.2), which maximizes their stacking interactions with their nearest neighbors but increases the length and therefore weakens Watson–Crick hydrogen bonds.[75,76] Thus, the propensity for propeller twisting is consistent with other data showing that base stacking provides more stabilization energy to the DNA duplex than hydrogen bonding. Like roll, moderate changes in propeller twist are not energetically costly and often occur as the DNA adapts to a protein surface and optimizes intermolecular hydrogen bonding. The translation parameters slide and shift refer to the relative positions of stacked base pairs along their long or short axes, respectively (Figure 8.2). Modest changes in slide and shift also occur to maximize stacking interactions but, like propeller twist, do not directly cause a change in the helix axis. Large values of propeller twist will restrict translational movements of bases, especially slide, because the bases at adjacent steps are stereochemically interlocked. Where propeller twist approaches 20° or more, diagonal or bifurcated Watson–Crick hydrogen bonding can occur with the 3' base on the opposite strand (Figure 8.3B).[74,77–79] Extended DNA segments of high propeller twist with a continuous network of bifurcated hydrogen bonds, such as found in runs of As (A-tracts), tend to be unusually stiff.[78,80,81]

Dinucleotide steps exhibiting low stacking interactions, such as found with pyrimidine-purine (Y-R) base pair steps, tend to be more prone to slide and shift changes, along with lower propeller twisting, and are consequently more prone to bending.[82] Negative values of slide are sometimes correlated with high (+) roll and positive values of slide with (−) roll bending. These correlations are particularly evident for certain Y-R steps like T-A and C-A (T-G) that tend to exhibit the highest ranges of slide.[7,70,71,73,82] For example, in the nucleosome structure all of the T-G (C-A) steps that show large negative rolls display slide values over 2.0 Å.[72] Whereas shift values tend to deviate less and are generally not perceived to significantly impact DNA bending or protein binding, Richmond and Davey have noted alternating patterns of shift where the DNA in the nucleosome is smoothly bent into the minor groove.[72]

8.3.1.4 Changes in DNA Groove Width

An important aspect of DNA structure that varies greatly and is often correlated with protein binding and DNA bending is the width of the DNA

Bending and Compaction of DNA by Proteins

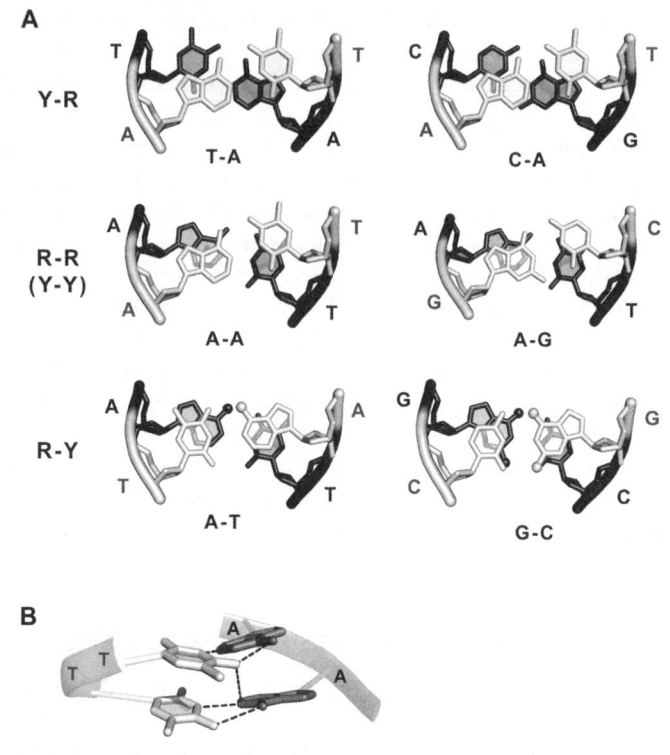

Figure 8.3 (*A*) Base pair stacking in dinucleotides looking down the helix axis. The structures are based on the canonic B-DNA fiber structure of Arnott except that the twist is 34.3°, the value for mixed sequence B-DNA in solution. In Y-R steps, only the exocyclic groups of the pyrimidine ring are positioned over the purine ring. In R-R steps, the purine rings share extensive surface contacts and the pyrimidine exocyclic groups are stacked over each other. In R-Y steps, both sets of purine and pyrimidine rings share extensive surface contacts. The N6 groups on adenine and O6 groups on guanine that are oriented towards each other and exhibit repulsive forces within the major groove are highlighted as spheres. Likewise, the N2 groups on guanine in the minor groove of the G-C dinucleotide are portrayed as spheres. Each dinucleotide is oriented such that the major groove is on top and minor groove is on the bottom. (*B*) An A-A dinucleotide step with high propeller twist and bifurcated H-bonding from the A-tract structure of Nelson *et al.* (pdb code: 1D98). In addition to the Watson–Crick hydrogen bonds, the N6 of adenine and O4 of the 3′ thymine are positioned to form a strong hydrogen bond.

grooves. In averaged B-DNA fibers the distance across the minor groove is 6.0 Å and across the major groove 11.6 Å, as measured from the closest phosphates with their van der Waals radii subtracted.[83] Groove widths in oligonucleotide crystals vary widely, especially for the minor groove, where the distances between closest phosphates can vary from <3 to 8.9 Å.[60] The widths of both grooves tend to decrease appreciably within protein interfaces,[4] though proteins that insert secondary structure elements in the minor groove almost always severely widen the groove. Narrowing of a groove can be due to roll angle changes, in which case it is directly correlated with a bend in the helix axis, and compression of one groove generally leads to an expansion of the other. However, the minor groove can be highly compressed without an accompanying deflection in the helix axis or a corresponding change in major groove width. A classic example of this is the A-tract structure where the minor groove narrows to about 3 Å, but the helix is straight and the major groove width remains fairly constant.[78,80,81] In general, minor groove widths over AT-rich regions tend to be more variable and usually narrower than over GC-rich regions.[74,84,85] This difference is influenced by the presence of the 2-amino group on the minor groove side of guanine, the ordered spine of hydration or cations that line narrow AT-rich minor grooves, and the tendency for some AT-rich sequences to exhibit high propeller twists.[7] As illustrated in Section 8.4, the ability of DNA sequences to accommodate changes in groove width, particularly with respect to the minor groove is one of the most important indirect determinants in selective protein binding and DNA bending.

8.3.2 Influence of Exocyclic Groups on Base Stacking

As discussed above, base step conformations are strongly influenced by stacking interactions or the degree of surface overlap between the bases. These interactions are modulated by positive or negative interactions between exocyclic groups on nearest neighbor bases on opposite strands and dipole effects, particularly with respect to guanine. Calladine was the first to point out that identical purines separated by one step on opposite strands (*i.e.*, G/C-C/G and A/T-TA) will exhibit a clash due to juxtaposition of their exocyclic groups in the major groove (O6 on G and N6 on A as depicted on the G-C and A-T dinucleotides in Figure 8.3A).[76] This repulsion is enhanced at G-C steps due to the relatively strong negative dipole over the O6 and N7 groups of guanine.[69,71,73,86] The charge on the other bases is more dispersed. The 2-amino group on the minor groove side of guanine (see G-C dinucleotide in Figure 8.3A) also effects local base step structure,[76] as evidenced in part by the decrease in persistence length of DNA when inosine (missing the N2) is substituted for guanine, and the increased persistence length when 2-aminopurine is substituted for adenine.[87] These electrostatic and steric repulsions contribute to twist and translational adjustments, *e.g.*, increased slide at G-C steps to separate the guanines. Because twist and slide deviations are often coupled with changes in roll bending, these nearest neighbor interactions can enhance the bendability of a base pair step.[69] On the other hand, attractive forces between diagonally opposed bases can promote bifurcated hydrogen bonding with accompanying large propeller twists and thereby limit flexibility (Figure 8.3B).[82] In addition, the close approach between the 5-methyl exocyclic group on thymine and the sugar on the 5' nucleotide has been proposed to limit compression of the major groove and hence limit positive roll.[86]

Bending and Compaction of DNA by Proteins

8.3.3 Flexibility of Dinucleotide Steps

There have been numerous surveys of base pair step conformations found in crystal structures of oligonucleotides and protein-DNA complexes. Most of these studies have categorized these at the dinucleotide step level, of which there are ten unique combinations.[3,69–71,82,88,89] Analyses of 3 bp steps (32 different combinations)[90] and 4 bp steps (136 different combinations)[74,91] have also been attempted to better reflect the contributions of nearest neighbors. However, since most oligonucleotide crystal structures in the database are of a non-random set of short segments with confounding issues of end crystal packing effects, meaningful analysis of the longer base pair sets has been limited. Some of the major features are summarized below in the context of dinucleotide family groupings. Those features that appear to have predictive value are highlighted though, as noted above, there are always examples that contradict general trends, particularly when applied to protein-DNA complexes. The recent high-resolution structure of the nucleosome has provided a large increase in data from protein bound structures, but of course the DNA is wrapped in a superhelix, which imparts a structural bias. This may contribute to Richmond and Davey's finding that their analysis of nucleosomal DNA fits only moderately well to energetics of dinucleotides inferred from oligonucleotide crystal structures.[72]

8.3.3.1 *Pyrimidine-purine (Y-R) Steps*

Probably the most important sequence-dependent conformational feature of DNA is the inherent flexibility of Y-R steps, a property that has been emphasized in nearly all modern reviews of DNA structure.[2–4,7,62,70,71] This property makes structural sense since Y-R steps stack with the least overlap between them; in fact, in most cases only polar exocyclic groups of the 5' pyrimidine ring are in van der Waals contact with the 3' purine ring (Figure 8.3A). Thus, Y-R steps exhibit the greatest dispersion of roll, twist, and slide values and usually low propeller twists. Likewise, DNA bends within protein structures containing roll angle kinks of $\geq 15°$ in either direction occur most often at Y-R steps. In the nucleosome, Y-R steps account for half of all bends exhibiting $\geq 10°$ roll angles, with these about equally distributed between positive and negative directions (local base pair step parameters calculated using 3DNA[92]). Of course, most of the Y-R steps in the nucleosome or other protein-DNA structures exhibit only small roll deviations. A feature of many Y-R steps that are sharply kinked into the major groove is that the sugar conformation switches from B-like (C2'-endo) to A-like (C3'-endo) at the pyrimidine base on each strand.[93] Examples of this are most often found in complexes where the interactions are through the minor groove, as with HMGB proteins, TBP, and a region of PurR.

8.3.3.2 *Purine-purine (R-R) or Pyrimidine-pyrimidine (Y-Y) Steps*

R-R (Y-Y) base pair steps are relatively well stacked (Figure 8.3A), particularly over the purine rings, and the oligonucleotide crystallographic data show them to be intermediate in flexibility, with mean values and dispersions of roll, twist, and slide being relatively tightly clustered around those of canonical DNA. However, in protein structures there are many examples where R-R (Y-Y) steps are at the locus of major unstacking by insertions of amino acids from the minor groove side. For example, IHF inserts a proline into A-A (see below), yeast TBP inserts phenylalanines into A-A or A-G,[94–96] pUT3 inserts an isoleucine and valine into G-G,[97] and the HMGB proteins HMGD and Nhp6A insert hydrophobic groups into G-A and A-A, respectively (see below). In each of these complexes rolls from 20° to 60° are present. Approximately one-third of the roll deviations that are $\geq 10°$ in nucleosomal DNA are at R-R (Y-Y) steps with most of these (7/8) being towards the major groove and at positions devoid of direct protein interactions.

Runs of >4 As on the same strand represent a special case where crystal structures[78,80,81] and a high-resolution NMR structure[98] show the DNA segment to be unusually inflexible. The crystal structures reveal this rigidity is at least in part due to high propeller twisting with accompanying bifurcated hydrogen bonding that stereochemically interlocks the bases over each other (Figure 8.3B). The tightly stacked structures over the segments containing 5–6 consecutive A/T base pairs exhibit high twist, low slide, and an extremely narrow minor groove. Solution biochemistry experiments have shown that A-tracts generate ~18° of static bending in the direction of the minor groove, which is additive when A-tracts are phased with the helical repeat.[40,99,100] The rigid and unbent structures of A-tracts make it likely that the bending originates at the junctions of the A-tract and intervening DNA. This is observed in the NMR structure, where a set of small roll and tilt bends primarily around the junctions combine to generate an overall bend of 16–19° directed towards the minor groove at the center of the A-tract.[98,101] Extended A-tracts have been correlated with nucleosome-free regions in chromatin, presumably because their inflexibility inhibits DNA wrapping around the octamer.[102,103]

8.3.3.3 Purine-pyrimidine (R-Y) Steps

R-Y dinucleotide steps share the greatest surface area of the three classes (Figure 8.3A) and consequently tend to be unbent in DNA crystal structures and not typically associated with large bends in protein complexes. R-Y steps are by far the least represented (14%), among dinucleotide steps exhibiting roll deviations of $\geq 10°$ in the nucleosome. However, exceptions have been observed; for example, the large (−) roll kink induced by EcoR1 and the large (+) roll kinks in the Hbb-DNA complex are at A-T steps.[104,105]

Another special case of a static bend occurs at the sequence (G)GGCC(C), where several independent experimental approaches employing gel electrophoresis, minicircle formation, and DNase I cutting have shown that this motif generates a pronounced bend into the major groove.[106–108] The magnitude of the bend varies among different reports from only a few degrees to about the same as an A-tract.[44,107] The direction of bending at (G)GGCC(C) is opposite to that generated by an A-tract and, unlike an A-tract, bending is dependent upon the presence of divalent cations. Oligonucleotide crystal structures have revealed up to $+24°$ of roll centered at the G-C step of this motif.[109,110] Interestingly, the G-C dinucleotide is the most underrepresented of the ten within high-resolution protein-DNA structures;[7] thus the significance of an intrinsic structure of the GGCC motif on protein-DNA interactions remains to be established.

8.4 Examples of DNA Bending Proteins

In the section below, a few selected examples of bending and global DNA compaction by proteins are discussed with the focus on DNA structural changes rather than on mechanisms of sequence recognition. These examples serve to illustrate several general themes. First, complexes that result in a major change in the DNA helix trajectory are often best characterized by relatively smooth local changes to the helix axis rather than one or more sharp kinks, especially when the primary protein-DNA interactions are within the major groove (see also ref. 111). In these cases, the overall curvature is a function of many small deformations that alter the helix axis in the same direction and thus reinforce each other. Second, different DNA sequences or even the same DNA sequence can adapt to a protein surface by different mechanisms, thus highlighting the malleable nature of the double helix structure. Third, a prominent feature of most of these examples is their radically changing minor groove widths that are nearly always accompanied by changes in twist. Thus, the DNA undergoes an induced fit by altering groove widths and capturing small bends or localized flexure to conform to the protein surface and to generate the constellation of contacts present in the final complex. These deformations may or may not be at the positions of intimate protein contacts. On the other hand, large localized changes in DNA conformation are always induced by direct interactions with one or more amino acid side-chains at the position of the kink. Most of the time these deformations involve radical unstacking of adjacent base pairs by insertion(s) of aliphatic amino acid side-chain(s) from within the minor groove.

8.4.1 Histone Binding to DNA

The nucleosome has received enormous attention over the past several decades because of its importance in chromosome packaging and its effects on all DNA transactions in eukaryotic cells. Atomic structures of the nucleosome have now been refined down to 1.9 Å by Richmond, Davey, Lugar, and co-workers and provide a high-resolution view of a long segment of bent DNA.[72,112,113] The ~ 210 kDa structure contains 147 bp of DNA wrapped 1.67 turns around the histone octamer to form a left-handed superhelix. About 80 bp effectively forms a circle with a diameter of 84 Å, which corresponds to a $\sim 4.5°$ bend per dinucleotide step if distributed evenly. The compact disk shaped histone octamer consists of two H3-H4 heterodimers that dock to form a stable tetramer and two H2A-H2B heterodimers that interact with the H3-H4 tetramer. The H3-H4 tetramer binds over the center of the DNA, with each dimer contacting 28 bp on each side of the dyad axis (Figure 8.4). The DNA extending from each end of the H3-H4 tetramer associates with an H2A-H2B dimer in a manner similar to the H3-H4 dimer and then exits the nucleosome. Each histone dimer thus interacts with about 2.5 turns of DNA and another half turn of spacer region bridges each dimer to make a total interface over the histone octamer of about 130 bp. Multiple contacts between histone residues and the DNA backbone occur at each position where the minor groove faces the protein, and these result in 130–140° of bending over each dimer (Figure 8.4). There are roughly an equivalent number of direct (116) and water-bridged (121) hydrogen bonds from the histones to the DNA over the nucleosome.[113] Hydrogen bonds from peptide backbone amides within the protein octamer to phosphate oxygens on the DNA backbone are believed to be especially important for locking the DNA in place. Eighteen of these

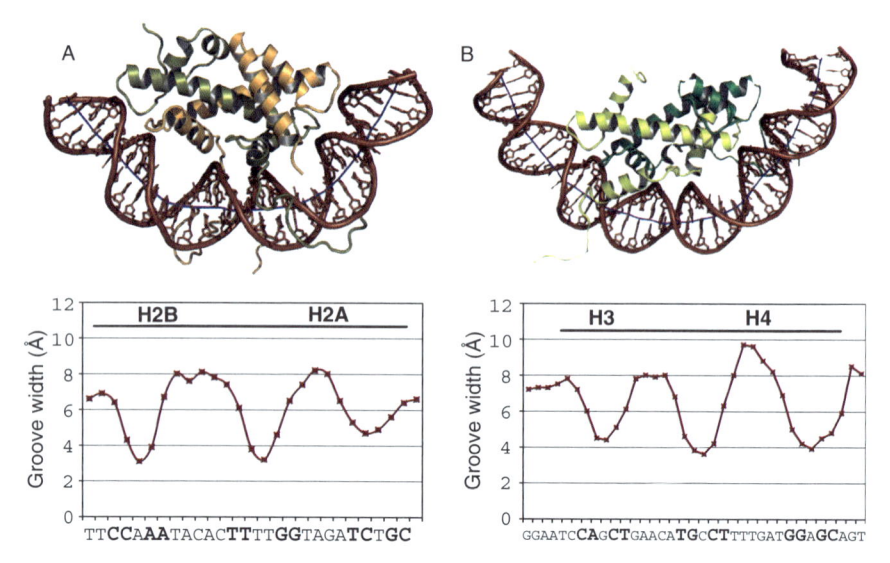

Figure 8.4 Histone dimers bound to DNA. (*A*) An H2A (olive green)-H2B (gold) dimer and (*B*) an H3 (yellow)-H4 (green) dimer bound to 30–40 bp DNA segments are shown. In each case the contacts with DNA occur where the minor groove faces the protein, but the DNA is relatively smoothly curved throughout. The widely varying minor groove widths, as measured from the closest phosphates minus their van der Waals surfaces,[92] are plotted below the structures. The bold nucleotides denote positions of phosphate contacts with histones on either DNA strand across the minor groove.[72] The minor groove width in canonical B-DNA is ~6 Å. The DNA complexes of the histone dimers are from a structure of the nucleosome (pdb code: 1KX5). In this and subsequent figures of protein-DNA complexes, the DNA axis, as calculated by CURVES,[207] is shown as a blue line. Some of the N-terminal tail segments that extend away from the DNA are not shown.

hydrogen bonds are from N-terminal residues of α-helices and are thus enhanced by helix dipole forces. There are no direct base-specific hydrogen bonds, but one or more arginines point into the minor groove at each position where it faces the octamer. While most of these arginines are oriented towards DNA backbone oxygens, some are H-bonded to N3 or O2 acceptors through intermediary waters. The extensive water-mediated H-bonding network may contribute to the ability of different DNA sequences to adapt to the octamer surface by providing spatial flexibility.

The DNA is relatively smoothly bent as it wraps over each histone dimer in the octamer (Figure 8.4).[72] As would be expected, negative roll values overwhelmingly predominate over the minor groove interfaces, but over the 12 minor groove facing blocks, only three contain roll angle changes exceeding −15°. Likewise, varying degrees of positive roll are present over the major groove facing blocks that are not in intimate contact with the octamer, but these only exceed 15° at three locations. Over half of the steps displaying moderate kinks in either direction are at Y-R dinucleotides. The short DNA segments connecting the dimers in the nucleosome generally maintain the overall curvature of the DNA, with a moderate kink (17°) present at a CT dinucleotide in the spacer connecting one of the H2A-H2B and H3-H4 dimers. The H2A/H2B interfaces exhibit a somewhat greater number of moderate roll angle changes than the DNA associated with the H3/H4 dimers, but Richmond and Davey[72] believe that this probably reflects differences in the nucleotide sequence rather than any significant distinction in the way the two histone dimers interact with DNA. As illustrated in Figure 8.4, the width of the minor groove oscillates in a strongly periodic fashion. Where the minor groove faces the octamer core, the groove narrows to 3.0 ± 0.55 Å, or less than half the canonical width for B-DNA; where the minor groove is oriented away from the core, it expands to 50% greater than the canonical width. The width of the major groove does not vary appreciably. The close approach of the phosphates across the backbone strands within the compressed minor groove blocks is aided by the ring of positively charged residues that line the DNA path and presumably help neutralize the charge.

Notably, although the human α-satellite sequence repeat DNA used in the crystal structures is a high affinity nucleosome binding sequence, the nucleosome positioning sequence motifs that have been identified are not well represented. In fact, only a minority of the minor groove blocks are enriched for A/T base pairs. Recently, a 3.2 Å structure of a nucleosome assembled on a DNA containing a 16 bp poly(dA•dT) sequence within the α-satellite sequence has been reported.[114] Even though A-tract sequences are normally unbent, the A_{16}-tract in the nucleosome was curved and conformed quite closely to the DNA structure of the standard α-satellite sequence. Introduction of the A_{16}-tract plus a 9 bp transcription factor binding site did detectably destabilize the structure and affected nucleosome sliding properties, which is consistent with A-tracts being overrepresented in nucleosome-free regions in yeast chromatin.[103,115] Nevertheless, the close correspondence between the structures of the A_{16}-tract and α-satellite sequence illustrates that forces generated from numerous protein-DNA interactions can dominate over intrinsic DNA energetics. More structures of histone octamers bound to different DNA sequences will be needed to gain a better understanding of how the base sequence is positioning nucleosomes.

8.4.2 Phage λ Xis Protein

Another example of smooth bending over an extended DNA segment that results in substantial overall curvature is the DNA complex formed by the phage lambda Xis protein. Xis functions as a DNA architectural element in the assembly of the intasome complex that catalyzes excision of the phage genome from the bacterial host chromosome.[116] DNA bending by Xis, together with Fis and IHF (see below), facilitate simultaneous binding by bivalent Int recombinase protomers to DNA sites separated by ∼100 bp. In a genetically separable function, Xis also directly promotes Int binding to one of its binding sites.

Xis is a 72 amino acid residue protein whose N-terminal 52 amino acid residue DNA binding domain folds into a minimal winged-helix motif consisting of 2 α-helices and 5 short β-strands.[117,118] Helix 2 inserts in the major groove while the wing formed by β-strands 3 and 4 inserts in the adjacent minor groove. Three monomers cooperatively assemble along one side of the DNA helix in a head-to-tail orientation to form a short filament-like structure. Crystal structures of a single Xis protomer (residues 1–55) bound to a 12 bp target site (Figure 8.5A) and of the 33 bp microfilament with the three protomers connected by related protein interactions (Figure 8.5B) have recently been determined.[119–121] A single Xis introduces a slight 25° bend in the DNA with no dinucleotide step exhibiting a significant roll deflection. When the three Xis monomers assemble into the microfilament, the small deformations induced by each protomer are phased with each other to generate an overall curvature of about 75° in the direction of the bound protein. The Xis protomer in the center is bound to a sequence that shares only 5/11 bp in common with the flanking Xis protomers, yet the DNA conforms to the surface of the central Xis protomer in manner similar to the flanking interfaces. Moreover, the 5 bp block within the major groove interface of the central Xis protomer can be substituted with a different sequence with little effect on formation of the microfilament. Xis binding thus highlights the conformational flexibility of DNA sequences and illustrates how a series of small deformations accumulated over several interacting protomers sum up to generate substantial curvature, which in this case is critical to the function of the regulatory element.

8.4.3 Papillomavirus E2 Protein

DNA binding by the papillomavirus E2 protein is a well studied example illustrating the importance of DNA bending on protein recognition. Work by Hegde, Shakked, and co-workers has provided crystal structure views of the

Bending and Compaction of DNA by Proteins

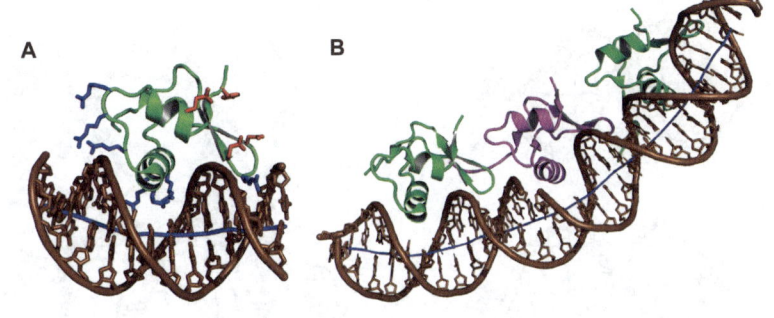

Figure 8.5 Phage λ Xis binding to DNA. (*A*) An Xis monomer, a winged-helix DNA binding protein, bound to a 14 bp segment that contains the X2 binding site on the λ genome (pdb code: 2OG0). Only a very small bend (~25°) is evident, although additional dynamic bending over the basic surface on the left side of the protein as oriented in the figure is likely when more DNA is present. Some mostly basic side-chains that are involved in DNA binding or Xis-Xis interactions are shown in blue; red side-chains, which are mostly acidic, participate in Xis-Xis interactions. (*B*) Three Xis monomers assemble to form a microfilament on a 33 bp DNA segment representing the X1/X1.5/X2 binding sites (pdb code: 2IEF). The DNA forms a smooth curve of about 75°. The central Xis protein is connected to the upstream and downstream monomers by a set of related protein contacts involving residues in the wing of one protein and adjoining basic surface of the other.

human papillomavirus (HPV) E2 protein bound to high and low affinity sites and of the respective unbound DNA segments (Figure 8.6), which have been correlated with their relative binding affinities.[122–126] The E2 protein binds to a 12 bp DNA target whose outer 3–4 bp are conserved among binding sites but whose central or spacer 4 bp segments differ.[127] Binding studies have shown that the identity of the base pairs in the spacer can modulate the affinity over 100-fold,[128–130] and the ordered binding to four sites with different spacer sequences within the HPV genome is believed to be important for viral development. However, the spacer segment is not contacted either directly or indirectly by residues of the E2 protein. Nevertheless, the most prominent deformations to the DNA helix upon E2 binding occur within this region.

The antiparallel four-stranded β-sheet from each 80 amino acid subunit form an eight-stranded β-barrel in the dimer, with many large hydrophobic side-chains positioned in the center to further stabilize the dimer interface.[127] The DNA recognition α-helices from each subunit, which are tightly packed onto the β-barrel core, insert into the major groove and make sequence-specific contacts with the conserved bases. There is no change in the protein structure, but a ~45° bend is introduced into the DNA upon binding.[124] The relatively smooth bend arises from small positive roll changes distributed within the major groove interface and negative roll changes peaking at the center of the spacer region. The minor groove of the spacer region also undergoes a dramatic compression (Figure 8.6E), which is in part caused by overtwisting of the base

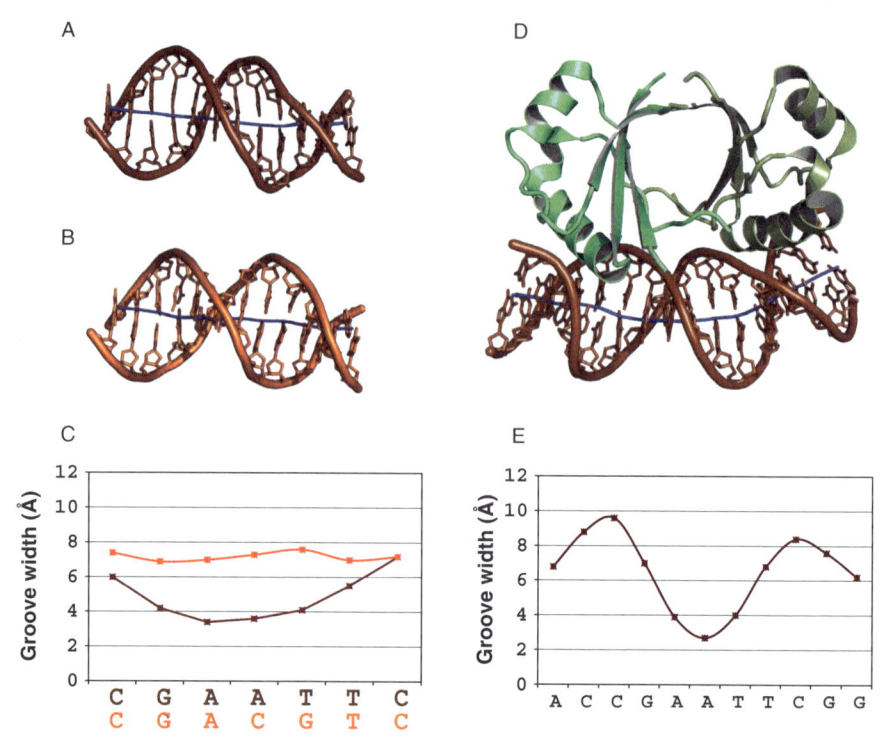

Figure 8.6 The human papillomavirus virus E2 protein binding to DNA. (*A*) A dodecamer oligonucleotide structure representing the central region of the high affinity E2 DNA binding site containing the sequence AATT in the center (pdb code: 1ILC). The DNA exhibits a slight bend into along with narrowing of the minor groove over the AATT sequence. (*B*) A dodecamer oligonucleotide structure representing the central region of the low affinity E2 DNA binding site containing the sequence ACGT in the center (pdb code: 423D). This DNA exhibits a slight bend into the major groove over the ACGT sequence. (*C*) Plot of the minor groove widths of the central regions of the AATT (brown line) and ACGT (orange line) DNA structures. (*D*) The HPV E2 homodimer bound to a 16 bp DNA containing the high affinity AATT sequence (pdb code: 1JJ4). Contacts between the protein and DNA only occur within the major groove. Modest (−) roll angle changes over the central minor groove and (+) roll angle changes over the flanking major groove protein interfaces combine to generate ∼45° curvature in the DNA. (*E*) Minor groove width changes across the E2 binding site in the structure in (*D*).

pairs together with the roll bending. This compression, in which phosphates across the minor groove are separated by just 2.7 Å, is critical for binding because the recognition helices that insert into the adjacent major grooves are separated by only about 25 Å. Most dimeric proteins that bind DNA by inserting α-helices into the major groove on one side of the helix have a spacing of 32–34 Å between the centers of the recognition helices, which is consistent with the pitch of average B-DNA.

Bending and Compaction of DNA by Proteins 197

Several crystal structures have been solved of the unbound target DNA dodecamers containing an AATT spacer sequence (Figure 8.6A) that specifies a high affinity binding site and a ACGT spacer sequence (Figure 8.6B) that causes a 100-fold reduction in affinity.[124–126] The AATT DNA exhibits a small intrinsic bend of 8–11° centered into the minor groove of the spacer and the width of the minor groove is compressed to within an angstrom of the spacer in the complex (Figure 8.6C and E). Thus, the structure of the unbound DNA of the high affinity target already partially conforms to the surface of the E2 binding site structure of the DNA in the complex. By contrast, the DNA binding site containing the ACGT spacer exhibits a slight bend in the opposite direction over the spacer region and the minor groove width is ~ 1 Å wider than found in canonical B-DNA. Nevertheless, the conformation of the ACGT DNA target in the complex bound by HPV E2, including the bending and narrow groove width of spacer region, was nearly identical to the AATT target DNA. Thus, the 100-fold difference in binding affinities between the different targets can be explained by the substantially larger structural change in the DNA with the ACGT spacer that is required to support complex assembly.

Zhang *et al.* correlated binding affinities to 16 different targets containing various 4 bp spacer sequences with their bending and twisting properties as measured in solution by a FRET-based assay.[130] The AT-rich sequences all bound considerably better than GC-rich sequences with affinities ranging over a nearly 2000-fold range from the best (TAAT) to the worst (CGCG). The relative binding affinities correlated well with magnitudes of intrinsic bending into the minor groove for the respective sequence. The GC-rich spacers, including ACGT, all exhibited mild bends into the major groove and consequently bound poorly. The AT-rich spacers could be divided into two classes. The high affinity class, which included AATT and ATAT, exhibited pronounced bends into the minor groove, whereas others, which included TTAA and TATA, displayed only slight bending into the minor groove and were at least ten-fold reduced in binding affinity. This study confirms the conclusions from the high-resolution crystal structures. However, the widely varying properties of the related sequences suggests caution in predicting binding site affinities based on the structures of dinucleotide units from the various compilations or the frequencies of dinucleotides within binding sites where DNA structure is a primary determinant. For example, the dinucleotides TA, AA, and TT are correlated with high-affinity binding sites of nucleosomes at the positions of the minor groove blocks. Even though the minor groove blocks of nucleosomes and the E2 complex share similar structural features, such as groove narrowing and negative rolls, the mere presence of these dinucleotides only partially correlates with binding affinities of E2 targets.

8.4.4 *Escherichia coli* Fis Protein

Fis is among the most abundant nucleoid-associated proteins in enteric bacteria under nutrient-rich rapid growth conditions.[131–133] Fis regulates transcription, recombination, and replication reactions and has been proposed to function in chromosome condensation because of its DNA bending and loop-stabilization activities.[133–136] Fis forms unstable complexes on random sequence DNA with nanomolar affinities but can generate stable DNA complexes with specific sites that share a weak consensus sequence. The most conserved feature of specific Fis binding sites is a G/C and C/G separated by 14 bp, of which the center 5 bp is A/T-rich.

The two N-terminal α-helices from each subunit fold into a four-helix bundle to form the core of the Fis dimer.[137–139] N-Terminal β-hairpin arms that function to regulate site-specific DNA inversion reactions extend from one side of the dimer core, and the C-terminal helix-turn-helix DNA binding motifs from each subunit protrude from the other side. X-Ray structures of Fis bound to 27 bp DNA fragments have recently been determined.[140] The DNA is again best described as being relatively smoothly bent to generate an overall curvature in the crystals of about 65° (Figure 8.7A). Although Fis is a helix-turn-helix protein, its mode of binding bears similarities to the HPV E2 protein. Like E2, its recognition helices are spaced about 25 Å apart, but they are oriented more perpendicularly to the DNA helix axis when inserted into the major grooves. Also like E2, this spacing is accommodated by severe compression and high base pair twist values within the intervening minor groove, which in the Fis complex narrows to about 3.3 Å between phosphates on the two DNA strands (Figure 8.7B). However, the DNA in the Fis spacer region is relatively straight, unlike the E2 spacer. Fis can form a complex with a binding site containing all G/C base pairs in the spacer region, but the affinity is reduced about 1000-fold. The minor groove over the spacer in a crystal structure of the G/C spacer binding site is also compressed, but not over as large a region and to the same extent as observed with A/T-rich spacers. Thus, like HPV E2, the ability of the DNA to support the required groove compression is an important indirect determinant for Fis binding.

Most of the bending in the Fis complex arises from small bends present within the major groove interface. The two locations exhibiting the greatest deflections of the helix axis are where amino acid side-chains contact bases, including at the conserved G and C. However, the relative magnitudes and distributions of these small bends vary with different DNA binding sites, including ones with changes in the spacer region. Additional variable amounts of bending within the DNA segment flanking the core binding site is predicted from gel electrophoresis and other studies.[141] This additional bending of the flanking DNA was shown to be sequence dependent and strongly influenced by arginine residues on the side of Fis (*e.g.*, Arg71) that are not contacting DNA in the current crystal forms.

8.4.4.1 Long-range DNA Condensation by Fis

DNA compaction forces have been measured for a few DNA bending proteins using magnetic tweezers on single-DNA molecules. In typical experiments, a

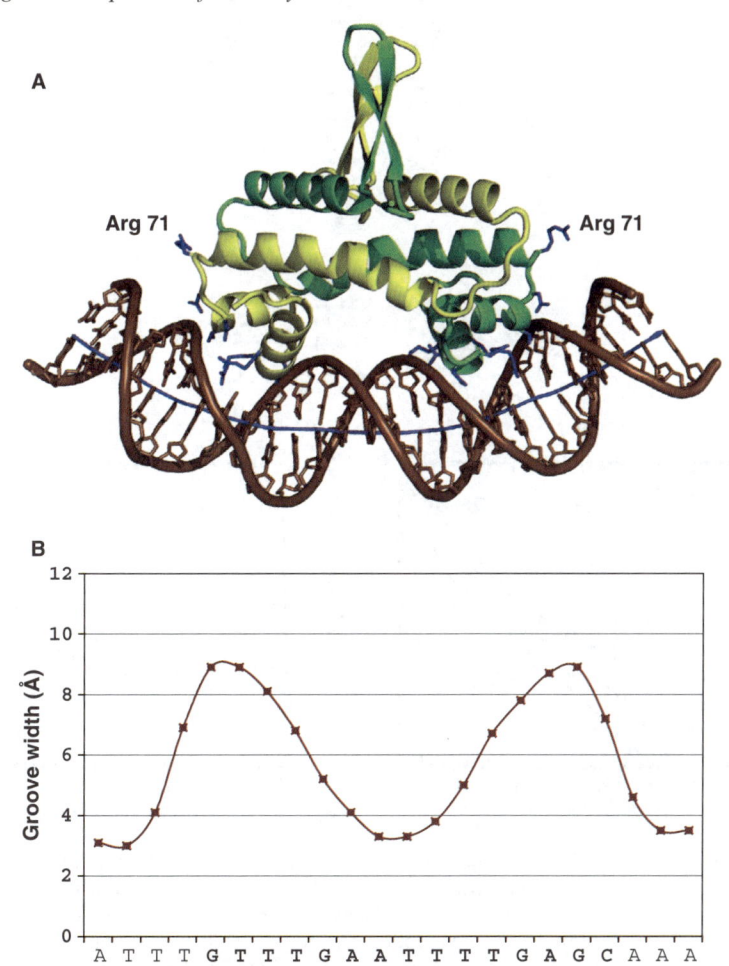

Figure 8.7 The *E. coli* Fis protein bound to DNA. (*A*) The Fis homodimer is bound to a 27 bp segment to generate an overall DNA curvature of about 65°. Most of the bending is distributed within the major groove interfaces. Side chains involved in DNA binding are shown in blue. Arginine 71 is not contacting DNA in the crystal structures but is implicated to mediate additional dynamic wrapping of flanking DNA from solution experiments. (*B*) Minor groove width changes across the Fis binding site. The 15 bp core binding sequence is highlighted in bold.

10–50 kb DNA containing a paramagnetic bead on one end is tethered to a glass slide on the other end (Figure 8.8A).[142,143] An adjustable magnet pulls on the DNA with a defined force and the length of DNA under different forces is measured to generate a force–extension curve (see the naked DNA curve in Figure 8.8B and C). Binding of a DNA bending protein leads to DNA compaction as signaled by a reduction in the DNA molecule length at a given applied force.[144] Nonspecific Fis binding to DNA results in compaction that is

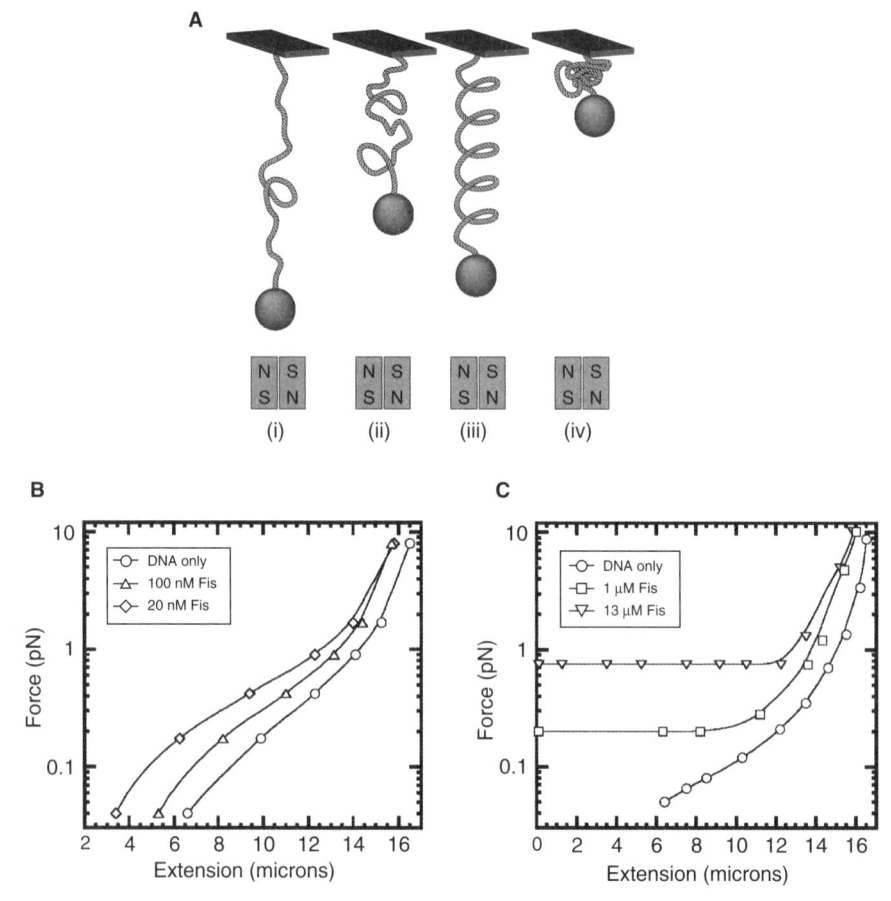

Figure 8.8 Single-DNA molecule analysis of compaction and looping by Fis. (*A*) Schematic representation of a DNA molecule pulled using "magnetic tweezers." (i) The DNA molecule is tethered to a glass slide on one end and contains a paramagnetic bead on the other end. A magnet generates a force on the bead that extends the molecule. The position of the bead relative to the glass surface is followed microscopically. The magnitude of the force exerted on the bead when the magnet is at different distances can be calibrated from the Brownian motion of the bead.[142,143] (ii) Dispersed binding by a DNA bending protein results in random bends that compact the DNA against the applied force. (iii) Formation of a protein filament can alter DNA compaction. When Fis coats the DNA, the molecule becomes slightly extended in comparison to the dispersed binding mode. (iv) Stabilization of increasing numbers of thermally-generated loops in DNA by protein binding results in collapse of the DNA against the glass slide. (*B*) Force–extension curves showing compaction of DNA by Fis binding. The applied force (in pN) is plotted against the length of the tethered phage λ DNA (in microns). Open circles represent naked DNA data points. Open diamonds denote data obtained in the presence of 20 nM Fis, where Fis is bound in a dispersed mode and compacts the DNA. Open triangles denote data measured at 100 nM Fis, where Fis coats the DNA. (*C*) Force–extension curves showing looping of DNA by Fis binding. The naked DNA is shown as circles. Binding of 1 μM (open squares) or 13 μM (open inverted triangles) Fis results in the DNA collapsing against the glass slide at a constant force of 0.2 and 0.7 pN, respectively, due to progressive trapping of DNA loops. (Data from ref. 146.)

maximal under low concentrations where Fis is forming dispersed complexes (Figure 8.8B).[145,146] For example, at a mild stretching force of about 0.2 pN, the presence of 20 nM Fis reduces the length of the 50 kb phage λ DNA ~4 µm. The extent of compaction is partially sensitive to force, but significant compaction is retained even at DNA stretching forces that exceed 10 pN. The compaction present at the high forces implies that Fis binding forms rigid bends in DNA, which fits well with the crystal structure, showing that a certain amount of bending, probably at least 50°, is required for complex formation.

At higher concentrations, Fis forms ordered arrays or filaments where a Fis dimer is bound every 21 bp along DNA.[146] Fis-DNA filaments generate less compaction at low forces, probably because the dynamic bending of DNA around the sides of the protein (*e.g.*, to contact arginine 71, Figure 8.7A) is sterically occluded. Consistent with this view, the Fis mutant Arg71Ala does not exhibit the enhanced compaction at low forces observed with the wild-type protein. At still higher concentrations of Fis, filaments assemble with increasingly greater numbers of Fis dimers that are associated by protein-protein interactions. The higher-order filaments formed at Fis concentrations exceeding 1 µM have the ability to stabilize loops of DNA formed by Brownian motion, resulting in rapid shortening of the DNA molecule against a constant force as increasing numbers of loops are trapped (Figure 8.8C). This looping–condensation activity only occurs below a force where thermally-excited loops within the Fis-coated DNA can still form. The looping activity by Fis may be one of the mechanisms responsible for the organization of bacterial DNA into topologically-discrete domains when packaged in the nucleoid.

8.4.5 *Escherichia coli* CAP Protein

The CAP (catabolite activator protein, also known as the cyclic AMP receptor protein CRP) is an important transcriptional regulator of many promoters in *E. coli*.[147,148] In the presence of the allosteric effector cAMP, CAP binds to a well-defined 22 bp palindromic consensus sequence aaaTGT-GAtct•agaTCACAttt, where the 5 bp capitalized sequence is of primary importance and the bold TG and CA indicates the primary bending sites. The 70 amino acid residue C-terminal DNA binding domain of each subunit of the homodimer folds into a variant of the helix-turn-helix motif where β-strands from an extended region between helix 1 and helix 2 and from the chain emanating from the C-terminus of the recognition helix 3 organize into a β-sheet.[149–153] The recognition helix is inserted into the major groove and the two C-terminal antiparallel β-strands, which form a wing-like structure, protrude towards the minor groove in the adjacent flanking DNA. Four different crystal forms show that binding of CAP-cAMP to DNA introduces 80–90° of overall curvature (Figure 8.9A), which is consistent with independent measurements obtained in solution by gel electrophoresis and cyclization experiments.[154–157] Most of the bending arises from distortions within the major groove where the protein specifically contacts the bases. A secondary

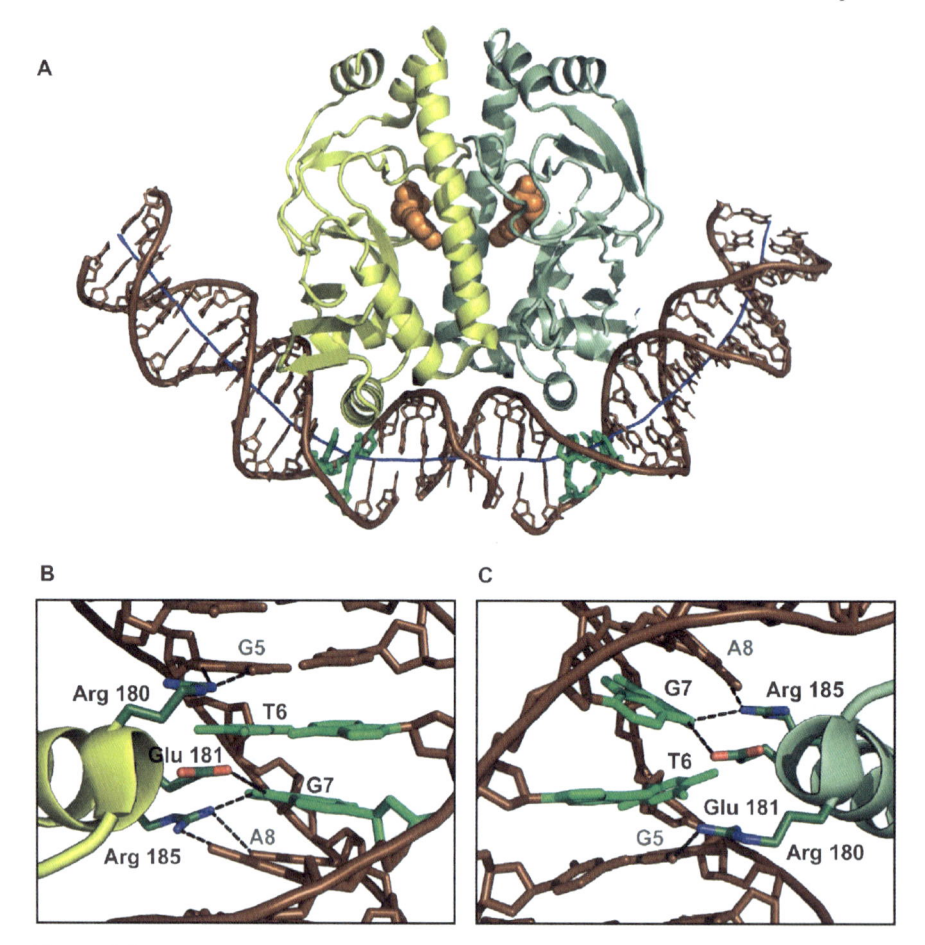

Figure 8.9 The *E. coli* CAP protein bound to DNA. (*A*) The CAP winged-helix-turn-helix homodimer with bound cAMP (orange surface) is complexed with a 38 bp DNA segment (pdb code: 1ZRC). Even though the sequences of the half-sites are identical, the major groove associated with the green subunit has a severe kink (+40° roll) localized at the T6-G7 step (colored green), whereas the major groove bound to the yellow subunit is smoothly bent. The overall DNA curvatures induced by CAP at the two half-sites are similar with ~40° on the smoothly bent side and ~50° on the kinked side.[152] The discontinuities of the DNA backbones are because the DNA duplex in the crystal was assembled from half-sites. (*B*) and (*C*) Close-up views of the major groove within the smoothly bent and kinked half-sites.

bend occurs where the flanking A/T-rich minor groove is wrapped against the sides of the protein and stabilized by a set of hydrogen bonds and salt links between the protein and DNA backbone. These contacts result in negative roll angles distributed over several base pairs together with compression of the minor groove.

Bending in the major groove occurs around the region of sequence-specific contacts, but the local structural distortions vary considerably between the different crystal structures and even between half-sites of identical sequence within the same structure.[149–150,152,153] In most cases, the DNA suffers a sharp kink at the T_6-G_7/C_{16}-A_{17} dinucleotide steps (Figure 8.9C). The unstable Y-R steps are nearly completely unstacked in these structures, leading to roll deviation of around 40° and twist values under 20°. However, a smooth bend that results in a similar overall curvature within the region has also been observed in two different crystal structures; in these cases a roll angle change of only 11.5° and only slight unwinding occurs at the T-G (C-A) steps (Figure 8.9B). Thus, the same DNA sequence can flexibly conform to the same protein surface in a nearly isoenergetic manner.

Sequence-specific recognition by CAP using direct and indirect mechanisms has been investigated in detail and is the focus of Chapter 4. With respect to recognition of the T_6-G_7 step and DNA bending it is important to emphasize that the T/A at bp 6 is not hydrogen bonded to the protein even though it is at the position of the kink, when present. Nevertheless, the identity of base pair 6 controls both binding affinity and the local structure of the kink. X-Ray structures of complexes containing T_6-G_7 substituted with the related Y-R dinucleotide C-G largely retain the kinked DNA parameters observed with the wild-type sequence.[151,152] However, structures containing the R-R substitutions A-G and G-G both reduced the magnitude of the kink by about half on one half-site ($\sim 20°$ roll and $\sim 22°$ twist) and converted the other half-site into a smooth bend with DNA parameters similar to those observed in the wild-type smooth bending conformer. The effects of these substitutions generally conform to the dinucleotide step features surmised from the oligonucleotide studies and strongly imply that changes in the energetics of bending modulate binding affinity.

8.4.6 Prokaryotic HU/IHF Protein Family

DNA binding by the HU/IHF family of prokaryotic DNA binding proteins represents a particularly dramatic example of radical bending by highly localized deformations. These proteins are major constituents of the bacterial nucleoid and function as DNA architectural agents that promote assembly of a diverse range of replication, transcription and recombination complexes.[134–135,153] There is considerable evidence that members of the HU/IHF family of proteins also function in DNA compaction *in vivo*; where tested, mutants lacking all HU/IHF proteins are difficult to construct, grow poorly, and contain decondensed nucleoids. All prokaryotic genomes thus far examined encode at least one HU/IHF homolog.[134,158] In *E. coli*, HU is present under all growth conditions and binds DNA with moderate affinity (~ 100 nM) without an obvious sequence preference. However, it can bind with affinities down to the picomolar range to DNA segments containing various structural distortions or discontinuities, such as nicks, bulges, or branches.[159–162] The

E. coli IHF protein binds to specific DNA segments at low nanomolar affinities. These sequences are related by the bipartite consensus WATCARNNNNTTR (W represents A or T; N represents any base), which corresponds to one half-site of the full 30–35 bp binding site.[163] Both HU and IHF in E. coli are heterodimers made up of α and β subunits that are about 30% related to each other, though homodimeric forms of HU are active and present at varying levels that depend on growth conditions. Some bacteria only encode one form of this protein. For example, *Borrelia burgdorferi* encodes a single known HU/IHF homolog called Hbb, which is a homodimer that binds to a specific sequence (like IHF) but also binds with moderate affinities to random sequence DNA (like HU).[164] Crystal structures of DNA complexes containing the E. coli IHF,[165] the homodimeric HU from *Anabaena*,[166] and the B. burgdorferi Hbb[105] proteins have been determined by Rice and co-workers. The Hbb-DNA crystal structure, unlike those of IHF or HU, was obtained with fully intact DNA duplexes (except for a T-T mismatch in the center) and is shown in Figure 8.10(A). These structures are all distinguished by the unusually large protein-DNA interface (especially for IHF and Hbb), the presence of a pair of severe DNA kinks due to intercalation of proline rings within the minor groove, and the absence of any major groove contacts. The extreme bending of DNA, which generates up to 180° of total curvature, is mediated by the proline-induced kinks together with additional variable amounts of wrapping around the basic surface of the protein body.

HU/IHF family members fold into a structure that is unrelated to other DNA binding proteins. The body of the dimer is composed of six α-helices covered on one side by a 6–8 stranded β-sheet that forms a concave surface where the central 20 bp reside in the DNA complex. Two of the antiparallel β-strands extend away from the body to form long β-ribbon arms that are mobile in solution. Upon DNA binding, these arms position themselves within the minor groove and wrap around the groove to the opposite side of the DNA from the protein body at their ends. Proline residues located just before the tips of these arms insert their aliphatic ring into the base pair stack to induce ∼60° roll angle deflections with extreme untwisting of the two base pairs (Figure 8.10B). These two abrupt kinks, which are separated by 9 bp, account for most of the bending in the complex. The 9 bp of DNA between kinks are relatively straight (IHF) or moderately curved (Hbb) around the surface of the β-sheet. In IHF and Hbb, the DNA strands continue to travel down the basic track on each side of the protein body in a mildly curved manner to generate the near complete "U-turn" of the 35 bp DNA segment. In fact, if the DNA path is continued into the next asymmetric units of the crystals of both IHF and Hbb, 180° of overall curvature is achieved.[105,165] Solution measurements of IHF-mediated bending by FRET, DNA cyclization, and gel electrophoresis are all consistent with a bend of this magnitude.[157,167,168] A whopping 4650 (IHF)–5235 (Hbb) Å^2 of solvent accessible surface area is buried upon binding DNA by the 18–20 kDa dimers.[105,165,168] Recent time-resolved spectroscopic analyses have provided evidence that complex formation by IHF occurs in two sequential steps, where an initial association that presumably involves

Bending and Compaction of DNA by Proteins 205

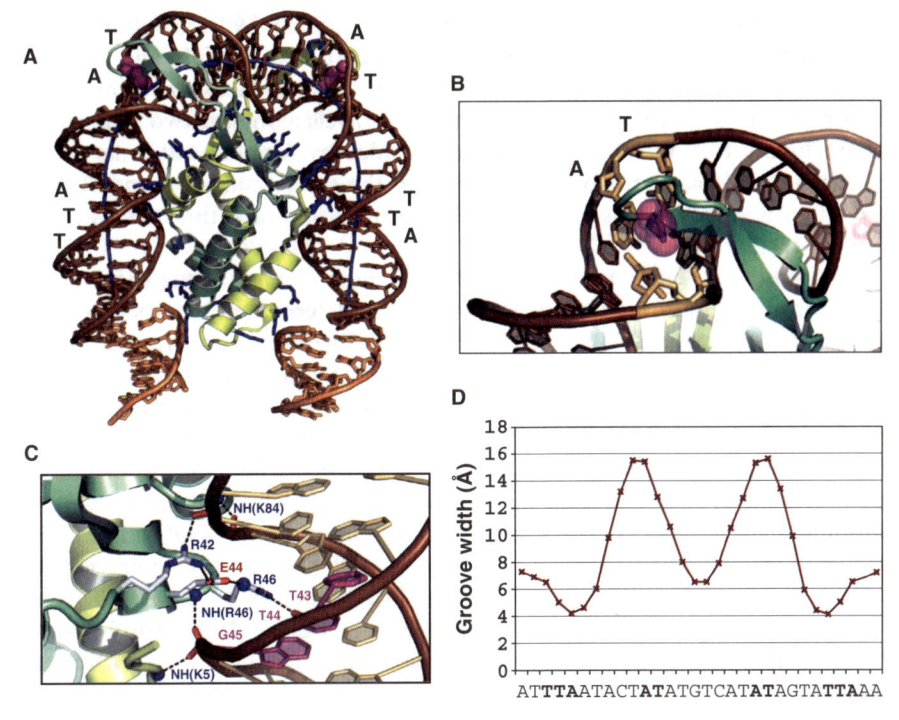

Figure 8.10 DNA binding by the HU/IHF family. (*A*) The *Borrelia burgdorferi* Hbb protein bound to a 35 bp DNA (pdb code: 2NP2). The proline atoms at the ends of the β-ribbon arms are rendered as magenta spheres. Lysine and arginine residues along the DNA path are shown in blue. The orange DNA segments at the lower end represent parts of the adjacent molecules in the crystal that engage in contacts with the protein. (*B*) Close-up of the proline-induced kink at the end of one of the β-ribbon arms of Hbb. (*C*) Close-up of the interactions between IHF and the minor groove at the conserved TTR motif (pdb code: 1IHF). The conserved DNA bases T43, T44, and G45 are colored magenta and hydrogen bonds between the protein and DNA are denoted. The highly compressed minor groove enables these hydrogen bonds to be formed. (*D*) Minor groove width changes over the Hbb binding site. The bases in bold letters correspond to the labeled bases in panel (*A*).

interactions by the β-ribbon arms is followed by wrapping of the flanking DNA along the protein body to generate the fully bent structure.[169,170]

Nearly all of the protein-DNA contacts present in the IHF and Hbb complexes are to the phosphate backbone; in addition to the prolines, there are a few sequence-independent contacts by arginines to bases in the minor groove, but none in the major groove where base identity can be distinguished. Thus, IHF and Hbb must choose their binding sites solely by the conformational properties of the DNA or through indirect recognition mechanisms. The flanking DNA segments appear to be relatively tightly connected to the sides of the IHF and Hbb body in the crystals by numerous contacts with the peptide

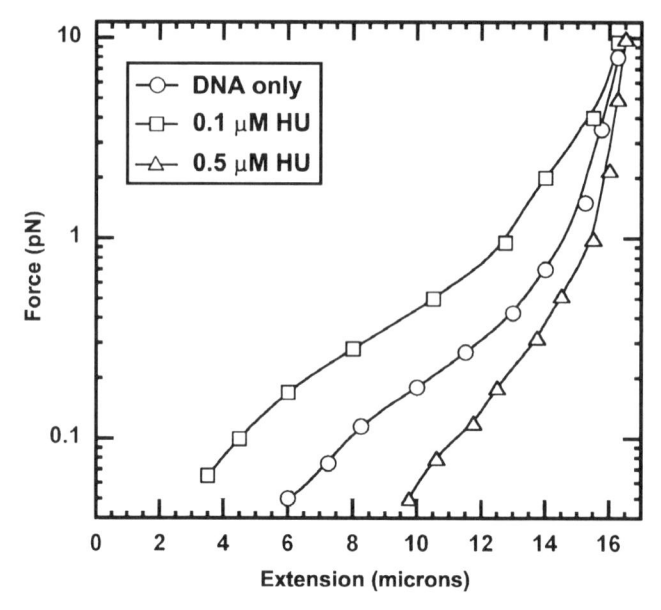

Figure 8.11 Force–extension curves generated with different amounts of HU protein. Dispersed binding by moderate levels of HU (0.1 μM, open squares) causes DNA compaction of the tethered λ DNA. Higher concentrations of HU (0.5 μM, open triangles) causes the DNA to be stiffer than naked DNA (circles) due to formation of HU-DNA filaments. (Data from ref. 175.)

consistent with earlier bulk phase solution studies showing that HU saturates DNA at 1 HU dimer per 9 bp, corresponding to the number of base pairs between the proline insertion sites.[179] Binding cooperativity could be mediated indirectly by the widened minor groove induced by insertion of the arms, and not by direct protein-protein contacts. Although HU is among the most abundant DNA binding proteins in prokaryotic cells, there are insufficient amounts to coat extensive regions in this manner. It remains to be seen whether localized regions of the chromosome could have HU bound in this stiffening regime.

Nonspecific binding by IHF has been shown to promote modest DNA compaction, but a stiffening mode has not been observed.[180] Single IHF proteins binding to a specific site under stretching forces appeared to undergo rapidly reversible binding/bending reactions that could reflect transitions between wrapped and unwrapped binding states, whereby the latter would reflect an association *via* only the β-ribbon arms.[181]

8.4.7 HMGB Protein Family

The HMGB family of DNA binding proteins can generally be divided into two groups: those that preferentially bind to specific target sequences, often in concert with other proteins, and those that bind in a sequence-neutral manner.[182,183] The former class includes HMGB domains associated with transcription factors like SRY, LEF1, and SOX, which bind to a related DNA sequence. The latter class includes the ubiquitously expressed HMGB1 and tissue-specific HMGB2 chromatin proteins in mammals, HMGD in *Drosophila*, and Nhp6A and Nhp6B in *S. cerevisiae*. The nonspecifically binding HMGB proteins are present in the nucleus at levels second only to histones. While their nuclear functions are still being elucidated, they have been found to collaborate with other factors in the assembly and function of transcription, replication, repair, and recombination complexes by various mechanisms.[184,185] They are also implicated, along with linker histones, in modulating chromatin structure and chromosome packaging.[186] In support of the latter activity, expression of HMGB proteins can suppress the chromosome condensation phenotype of *E. coli* HU mutants.[187,188]

The minimal HMGB domain consists of ∼75 amino acid residues that fold into an L- or boot-shaped structure (Figure 8.12).[189,190] The long arm of the L is formed from the N-terminal extended polypeptide strand that is connected to helix 3 and can be variable in length, particularly with respect to helix 3. Helix 1 and 2 form the short arm of the L. The extended strand and helix 1 + 2 bind within the minor groove of DNA over a 7–9 bp region, with residues at the C-terminal end of helix 3 also sometimes engaged with the DNA backbone. The DNA conforms to the concave surface of the protein by undergoing major conformational distortions that generate a non-planar bend of ∼90°. Although the details vary with different structures, including the bend angles, very large (+) roll and very low twist angles are typically present over a 3 bp region where residues from helices 1 and 2 are contacting. In the Sox2 and HMGD structures, the minor groove is expanded to a distance of 12.5 Å between closest phosphates, which nearly flattens the groove (Figure 8.12B and D). On the major groove side, the phosphates approach to <3 Å from each other. In most HMGB proteins, an extended polypeptide arm that is rich in basic residues extends from either the N- or C-terminal ends of the protein. The arm of the Sox2 HMGB domain extends from helix 3 along the minor groove (Figure 8.12A) in the tripartite crystal structure of DNA bound by Sox2 and the Oct1 POU domain and interacts with the Oct 1 binding partner.[191] The basic arm of LEF-1 also extends from helix 3, but in the NMR structure is inserted within the compressed major groove.[192] In the case of yeast Nhp6A, the basic arm is located on the N-terminal end and inserts into the major groove in a similar manner as LEF-1.[193] Where tested, the basic residues on one or both ends of the HMGB core domain have been shown to have a large effect on binding affinity. For example, an Nhp6A mutant without the 16 amino acid residue arm binds DNA with 100-fold poorer affinity *in vitro* and the protein is largely nonfunctional *in vivo*.[194]

The sequence-specific and sequence-neutral classes of HMGB proteins differ somewhat in their modes of DNA interactions within the minor groove. Proteins from both classes usually have a hydrophobic wedge near the center of the DNA interface that most often consists of a methionine (or isoleucine)

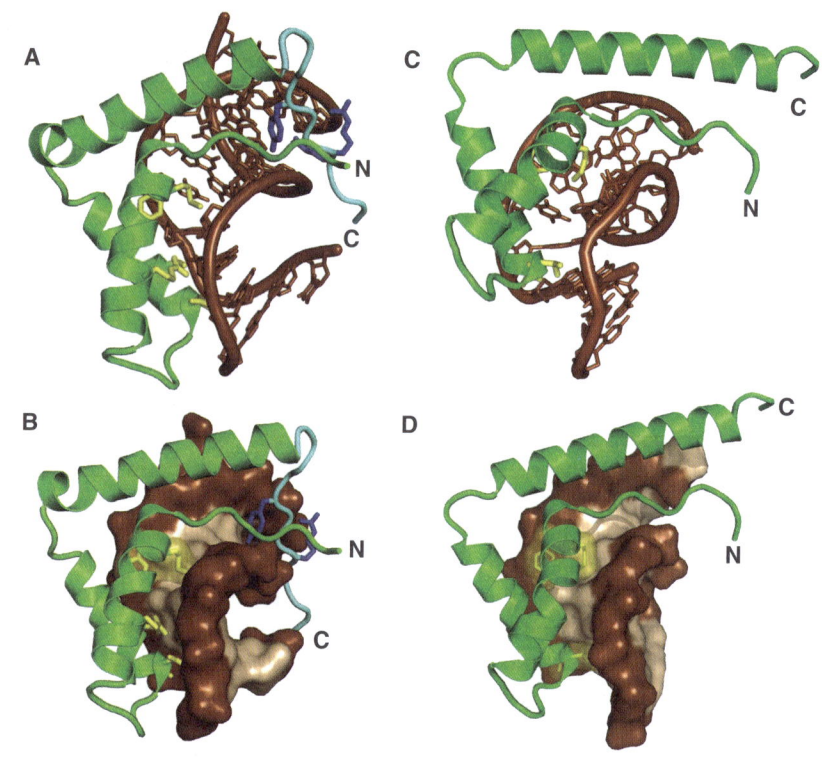

Figure 8.12 Binding of HMGB proteins to DNA. (*A*) and (*B*) Sox2 bound to DNA from the crystal structure of the Sox2/Oct1/DNA complex (pdb code: 1GT0). Sox2 is an example of a sequence-specific HMGB protein. The yellow side-chains at the beginning of helix 1 are the phenylalanine and methionine that comprise the methionine wedge. A serine plus arginine and a serine plus asparagine also protrude into the minor groove and form hydrogen bonds with bases at successive dinucleotide steps. The light blue polypeptide strand at the C-terminal end extends along the minor groove towards the Oct1 protein (not shown); two arginines and a tyrosine that interact with DNA are displayed in blue. (*C*) and (*D*) HMGD, a non-specifically binding HMGB protein, bound to a decamer (pdb code: 1QRV). The tyrosine plus methionine that form the methionine wedge and the valine plus threonine that form the exit wedge at the bottom are depicted in yellow. The basic C-terminal segment of HMGD was not present in the polypeptide used for the crystal structure. The DNA surfaces are rendered in (*B*) and (*D*) to highlight the wide and shallow minor grooves induced by binding of the HMGB domains. The methionine and exit wedges are rendered as transparent surfaces.

and a tyrosine (or phenylalanine). The "methionine wedge" forms a kink in the DNA, which in the case of Sox2 and HMGD, results in a 45–50° roll angle bend. In the two nonsequence-specific HMGB-DNA complexes, the methionine protrudes into an unstable Y-R step (T-A in HMGD and T-G in Nhp6A) and thus may be responsible for the preferential binding to a particular sequence within a short DNA segment.[193,195] An NMR structure of the unbound Nhp6A binding site did not reveal anything unusual about the DNA structure except for the poorly stacked T-G (C-A) dinucleotide. In the Sox2, SRY, and LEF-1 complexes, the aliphatic methionine side-chain is intercalated into an A-A (T-T) step.[191,192] The sequence-specific proteins have a series of polar residues that hydrogen bond to groups on bases within the floor of the minor groove over the next 2 bp, which contribute to their sequence specificity. In the nonsequence-specific class, the hydrophilic residues at the end of the interface are replaced by largely hydrophobic residues to form a second "DNA exit wedge". In Nhp6A and domain A of mammalian HMGB1, the DNA exit wedge consists of a prominently exposed phenylalanine.[193,196] This phenylalanine is responsible for the selective binding to prebent DNA formed by a cisplatin adduct by intercalating between the kinked bases. In the HMGD (Figure 8.12C and D), NHP6A, and HMGB domain A structures, the exit wedge pushes the DNA away from the protein as it leaves the complex.[193,195,196] DNA cyclization and single-DNA molecule studies have shown that the phenylalanine wedge has a more prominent effect on HMGB-induced DNA bending for Nhp6A and HMGB1 domain A than the methionine wedge, which is absent in HMGB1 domain A.[193,197–199] The exit wedge is also critical for cooperative interactions with other transcription factors or recombinase proteins.[197,198,200,201] On the other hand, the methionine wedge in HMGD has a more significant effect on DNA structure and DNA cyclization than the exit wedge, which consists of a valine and threonine.[195,202]

8.4.7.1 Single DNA Molecule Analyses of HMGB Protein Binding

The abundance and extreme DNA bending properties suggest that the nonspecific binding class of HMGB proteins could have a significant effect on DNA compaction. Single-DNA molecule experiments have confirmed that HMGB proteins strongly compact DNA.[175,203] For example, at a stretching force of 0.2 pN, saturating amounts of HMGB1 or Nhp6A compact lambda DNA to about one-third of its naked DNA length, an amount greater than observed for HU or Fis. Forces approaching 10 pN reversed the compaction induced by Nhp6A but did not remove the protein from the DNA. At high concentrations HMGB1 or an isolated HMGB2 domain A polypeptide also stably bind DNA such that unbinding into protein-free buffer under high DNA tension does not occur. Taken together with the HMGB-DNA structures, these observations support the idea that the HMGB proteins can associate with DNA in a flexible hinge-like manner and also imply that HMGB proteins can coat the DNA to form stable filament-like complexes. At high forces, achievable using optical tweezers, binding of HMGB domain A was found to increase the contour length of DNA up to 16%.[203] This increased extension of DNA is consistent with the large twist deficit and increases in rise due to amino acid side-chain intercalation observed in the atomic structures.

8.4.7.2 DNA Binding by HMGB Shares Features with TBP

The remarkable distortions induced by binding of HMGB proteins are reminiscent of the DNA structure within TBP complexes that function in eukaryotic transcription initiation.[94–96] TBP also forms a concave and largely hydrophobic DNA binding surface, albeit it is assembled from a β-sheet. Moreover, TBP has two phenylalanine wedges that partially intercalate into T-A or A-A steps from the minor groove side, generating large roll angle bends at each end of the protein-DNA interface. As observed with the HMGB-DNA complexes, the minor groove becomes tremendously expanded and flattened with accompanying compression of the major groove.

8.5 Concluding Remarks

The examples described above were chosen both to illustrate the diversity of binding mechanisms that can generate a highly curved DNA molecule and to emphasize high copy chromatin-associated proteins that function in chromosome compaction. Thus, many of these protein-DNA complexes are largely stabilized by DNA backbone and non- or weakly-specific base contacts. In the case of histone binding, DNA interactions almost exclusively consist of non-specific contacts from the minor groove side, but the bends are distributed throughout the DNA helix. Many DNA bending proteins that interact with DNA predominantly through the major groove introduce a series of relatively modest bends that together generate considerable curvature over the molecule. Fis and HPV E2 are examples of this, with some of the bends generated by HPV E2 again being remote from the sites of protein-DNA contacts. Likewise, whereas binding of a single phage λ Xis protein induces only a modest change to the helix trajectory, the cooperative binding of several phage λ Xis proteins results in considerable DNA curvature. The *E. coli* CAP protein is an unusual case where a predominantly major groove binding protein generates a large localized kink. However, recent structures show that CAP-induced bending does not always occur by sharp kinking; similar overall curvature can be distributed among several base pairs such that none are significantly unstacked.

In contrast, proteins that insert secondary structure elements into the minor groove often induce large localized deformations. In many cases these deformations involve intercalation of aliphatic side-chains between bases, thereby disrupting base stacking and generating a severe kink. Among chromatin-associated proteins, the HU/IHF and HMGB families are particularly dramatic examples of this binding-bending mode. Similar features are also found among DNA repair proteins, which have to scan through DNA in a sequence-independent fashion.[204] Introduction of kinks through minor groove interactions is not limited to nonspecifically binding proteins. The highly sequence-specific Lac and Pur repressor proteins induce kinks into the DNA by insertion of aliphatic side-chains (leucines) from α-helices within the minor groove.[205,206]

References

1. R.E. Dickerson, *Nucleic Acids Res.*, 1998, **26**, 1906.
2. R.E. Dickerson and T.K. Chiu, *Biopolymers*, 1997, **44**, 361.
3. M.A. Young, G. Ravishanker, D.L. Beveridge and H.M. Berman, *Biophys. J.*, 1995, **68**, 2454.
4. S. Jones, P. van Heyningen, H.M. Berman and J.M. Thornton, *J. Mol. Biol.*, 1999, **287**, 877.
5. N.C. Seeman, J.M. Rosenberg and A. Rich, *Proc. Natl. Acad. Sci. U.S.A.*, 1976, **73**, 804.
6. M.M. Gromiha, J.G. Siebers, S. Selvaraj, H. Kono and A. Sarai, *J. Mol. Biol.*, 2004, **337**, 285.
7. R.E. Dickerson, *Helix Structure and Molecular Recognition by B-DNA*, ed. S. Neidle, Oxford University Press, Oxford, 1999, pp. 145–191.
8. F.K. Winkler, D.W. Banner, C. Oefner, D. Tsernoglou, R.S. Brown, S.P. Heathman, R.K. Bryan, P.D. Martin, K. Petratos and K.S. Wilson, *EMBO J.*, 1993, **12**, 1781.
9. D. Kostrewa and F.K. Winkler, *Biochemistry*, 1995, **34**, 683.
10. J.J. Perona and A.M. Martin, *J. Mol. Biol.*, 1997, **273**, 207.
11. P.J. Hagerman, *Annu. Rev. Biophys. Biophys. Chem.*, 1988, **17**, 265.
12. Y. Zhang and D.M. Crothers, *Proc. Natl. Acad. Sci. U.S.A.*, 2003, **100**, 3161.
13. T.M. Okonogi, S.C. Alley, A.W. Reese, P.B. Hopkins and B.H. Robinson, *Biophys. J.*, 2000, **78**, 2560.
14. T.E. Cloutier and J. Widom, *Mol. Cell*, 2004, **14**, 355.
15. A. Travers, *Curr. Biol.*, 2005, **15**, R377.
16. J. Yan and J.F. Marko, *Phys. Rev. Lett.*, 2004, **93**, 108108.
17. T.E. Cloutier and J. Widom, *Proc. Natl. Acad. Sci. U.S.A.*, 2005, **102**, 3645.
18. H.G. Garcia, P. Grayson, L. Han, M. Inamdar, J. Kondev, P.C. Nelson, R. Phillips, J. Widom and P.A. Wiggins, *Biopolymers*, 2007, **85**, 115.
19. S.C. Satchwell, H.R. Drew and A.A. Travers, *J. Mol. Biol.*, 1986, **191**, 659.
20. H.R. Drew and A.A. Travers, *J. Mol. Biol.*, 1985, **186**, 773.
21. J. Widom, *Q. Rev. Biophys.*, 2001, **34**, 269.
22. J.B. Mills and P.J. Hagerman, *Nucleic Acids Res.*, 2004, **32**, 4055.
23. J. Aymami, M. Coll, G.A. van der Marel, J.H. van Boom, A.H. Wang and A. Rich, *Proc. Natl. Acad. Sci. U.S.A.*, 1990, **87**, 2526.
24. L. Kozerski, A.P. Mazurek, R. Kawecki, W. Bocian, P. Krajewski, E. Bednarek, J. Sitkowski, M.P. Williamson, A.J. Moir and P.E. Hansen, *Nucleic Acids Res.*, 2001, **29**, 1132.
25. E. Protozanova, P. Yakovchuk and M.D. Frank-Kamenetskii, *J. Mol. Biol.*, 2004, **342**, 775.
26. P. Yakovchuk, E. Protozanova and M.D. Frank-Kamenetskii, *Nucleic Acids Res.*, 2006, **34**, 564.
27. J. SantaLucia, Jr., *Proc. Natl. Acad. Sci. U.S.A.*, 1998, **95**, 1460.

28. T. Johnson, J. Zhu and R.M. Wartell, *Biochemistry*, 1998, **37**, 12343.
29. R. Owczarzy, P.M. Vallone, R.F. Goldstein and A.S. Benight, *Biopolymers*, 1999, **52**, 29.
30. L.D. Williams and L.J. Maher, 3rd, *Annu. Rev. Biophys. Biomol. Struct.*, 2000, **29**, 497.
31. G.S. Manning, *Q. Rev. Biophys.*, 1978, **11**, 179.
32. K. Range, E. Mayaan, L.J. Maher, 3rd and D.M. York, *Nucleic Acids Res.*, 2005, **33**, 1257.
33. I. Rouzina and V.A. Bloomfield, *Biophys. J.*, 1998, **74**, 3152.
34. C.G. Baumann, S.B. Smith, V.A. Bloomfield and C. Bustamante, *Proc. Natl. Acad. Sci. U.S.A.*, 1997, **94**, 6185.
35. C.G. Baumann, V.A. Bloomfield, S.B. Smith, C. Bustamante, M.D. Wang and S.M. Block, *Biophys. J.*, 2000, **78**, 1965.
36. M.D. Wang, H. Yin, R. Landick, J. Gelles and S.M. Block, *Biophys. J.*, 1997, **72**, 1335.
37. A.D. Mirzabekov and A. Rich, *Proc. Natl. Acad. Sci. U.S.A.*, 1979, **76**, 1118.
38. G.S. Manning, K.K. Ebralidse, A.D. Mirzabekov and A. Rich, *J. Biomol. Struct. Dyn.*, 1989, **6**, 877.
39. J.K. Strauss and L.J. Maher, 3rd, *Science*, 1994, **266**, 1829.
40. D.M. Crothers and J. Drak, *Methods Enzymol.*, 1992, **212**, 46.
41. E.D. Ross, R.B. Den, P.R. Hardwidge and L.J. Maher, 3rd, *Nucleic Acids Res.*, 1999, **27**, 4135.
42. J.K. Strauss-Soukup, M.M. Vaghefi, R.I. Hogrefe and L.J. Maher, 3rd, *Biochemistry*, 1997, **36**, 8692.
43. J.K. Strauss, T.P. Prakash, C. Roberts, C. Switzer and L.J. Maher, *Chem. Biol.*, 1996, **3**, 671.
44. J.K. Strauss, C. Roberts, M.G. Nelson, C. Switzer and L.J. Maher, 3rd, *Proc. Natl. Acad. Sci. U.S.A.*, 1996, **93**, 9515.
45. T.K. Kerppola and T. Curran, *Science*, 1991, **254**, 1210.
46. T.K. Kerppola and T. Curran, *Cell*, 1991, **66**, 317.
47. T.K. Kerppola, *Proc. Natl. Acad. Sci. U.S.A.*, 1996, **93**, 10117.
48. T.K. Kerppola and T. Curran, *Mol. Cell. Biol.*, 1993, **13**, 5479.
49. D.A. Leonard, N. Rajaram and T.K. Kerppola, *Proc. Natl. Acad. Sci. U.S.A.*, 1997, **94**, 4913.
50. V.R. Ramirez-Carrozzi and T.K. Kerppola, *J. Mol. Biol.*, 2001, **305**, 411.
51. D.N. Paolella, Y. Liu, M.A. Fabian and A. Schepartz, *Biochemistry*, 1997, **36**, 10033.
52. R.J. McDonald, J.D. Kahn and L.J. Maher, 3rd, *Nucleic Acids Res.*, 2006, **34**, 4846.
53. J.K. Strauss-Soukup and L.J. Maher, 3rd, *Biochemistry*, 1998, **37**, 1060.
54. R.J. McDonald, A.I. Dragan, W.R. Kirk, K. L. Neff, P.L. Privalov and L.J. Maher, 3rd, *Biochemistry*, 2007, **46**, 2306.
55. P.R. Hardwidge, J. Wu, S.L. Williams, K.M. Parkhurst, L.J. Parkhurst and L.J. Maher, 3rd, *Biochemistry*, 2002, **41**, 7732.
56. J.N. Glover and S.C. Harrison, *Nature*, 1995, **373**, 257.

57. T.E. Ellenberger, C.J. Brandl, K. Struhl and S.C. Harrison, *Cell*, 1992, **71**, 1223.
58. N. Tjandra, S. Tate, A. Ono, M. Kainosho and A. Bax, *J. Am. Chem. Soc.*, 2000, **122**, 6190.
59. R.E. Dickerson and H.R. Drew, *J. Mol. Biol.*, 1981, **149**, 761.
60. U. Heinemann, C. Alings and M. Hahn, *Biophys. Chem.*, 1994, **50**, 157.
61. H.M. Berman, *Biopolymers*, 1997, **44**, 23.
62. R.E. Dickerson, D.S. Goodsell and S. Neidle, *Proc. Natl. Acad. Sci. U.S.A.*, 1994, **91**, 3579.
63. L. McFail-Isom, C.C. Sines and L.D. Williams, *Curr. Opin. Struct. Biol.*, 1999, **9**, 298.
64. T.K. Chiu, M. Kaczor-Grzeskowiak and R.E. Dickerson, *J. Mol. Biol.*, 1999, **292**, 589.
65. T.K. Chiu and R.E. Dickerson, *J. Mol. Biol.*, 2000, **301**, 915.
66. C.L. Kielkopf, S. Ding, P. Kuhn and D.C. Rees, *J. Mol. Biol.*, 2000, **296**, 787.
67. N.V. Hud and J. Feigon, *Biochemistry*, 2002, **41**, 9900.
68. R.E. Dickerson, *et al.*, *Nucleic Acids Res.*, 1989, **17**, 1797.
69. M.A. El Hassan and C.R. Calladine, *Phil. Trans. R. Soc. London, Ser. A*, 1997, **355**, 43.
70. W.K. Olson, A.A. Gorin, X.J. Lu, L.M. Hock and V.B. Zhurkin, *Proc. Natl. Acad. Sci. U.S.A.*, 1998, **95**, 11163.
71. M. Suzuki, N. Amano, J. Kakinuma and M. Tateno, *J. Mol. Biol.*, 1997, **274**, 421.
72. T.J. Richmond and C.A. Davey, *Nature*, 2003, **423**, 145.
73. A.A. Gorin, V.B. Zhurkin and W.K. Olson, *J. Mol. Biol.*, 1995, **247**, 34.
74. K. Yanagi, G.G. Prive and R.E. Dickerson, *J. Mol. Biol.*, 1991, **217**, 201.
75. R.E. Dickerson, *J. Mol. Biol.*, 1983, **166**, 419.
76. C.R. Calladine, *J. Mol. Biol.*, 1982, **161**, 343.
77. U. Heinemann and C. Alings, *J. Mol. Biol.*, 1989, **210**, 369.
78. H.C. Nelson, J.T. Finch, B.F. Luisi and A. Klug, *Nature*, 1987, **330**, 221.
79. M. Coll, C.A. Frederick, A.H. Wang and A. Rich, *Proc. Natl. Acad. Sci. U.S.A.*, 1987, **84**, 8385.
80. A.D. DiGabriele, M.R. Sanderson and T.A. Steitz, *Proc. Natl. Acad. Sci. U.S.A.*, 1989, **86**, 1816.
81. A.D. DiGabriele and T.A. Steitz, *J. Mol. Biol.*, 1993, **231**, 1024.
82. M.A. El Hassan and C.R. Calladine, *J. Mol. Biol.*, 1996, **259**, 95.
83. R. Chandrasekaran and S. Arnott, *J. Biomol. Struct. Dyn.*, 1996, **13**, 1015.
84. D.G. Alexeev, A.A. Lipanov and I. Skuratovskii, *Nature*, 1987, **325**, 821.
85. C. Yoon, G.G. Prive, D.S. Goodsell and R.E. Dickerson, *Proc. Natl. Acad. Sci. U.S.A.*, 1988, **85**, 6332.
86. C.A. Hunter, *J. Mol. Biol.*, 1993, **230**, 1025.
87. J. Virstedt, T. Berge, R.M. Henderson, M.J. Waring and A.A. Travers, *J. Struct. Biol.*, 2004, **148**, 66.
88. M.J. Packer, M.P. Dauncey and C.A. Hunter, *J. Mol. Biol.*, 2000, **295**, 71.
89. M. Suzuki and N. Yagi, *Nucleic Acids Res.*, 1995, **23**, 2083.

90. I. Brukner, R. Sanchez, D. Suck and S. Pongor, *J. Biomol. Struct. Dyn.*, 1995, **13**, 309.
91. M.J. Packer, M.P. Dauncey and C.A. Hunter, *J. Mol. Biol.*, 2000, **295**, 85.
92. X.J. Lu and W.K. Olson, *Nucleic Acids Res.*, 2003, **31**, 5108.
93. M.Y. Tolstorukov, R.L. Jernigan and V.B. Zhurkin, *J. Mol. Biol.*, 2004, **337**, 65.
94. Y. Kim, J.H. Geiger, S. Hahn and P.B. Sigler, *Nature*, 1993, **365**, 512.
95. J.L. Kim, D.B. Nikolov and S.K. Burley, *Nature*, 1993, **365**, 520.
96. D.B. Nikolov, H. Chen, E.D. Halay, A. Hoffman, R.G. Roeder and S.K. Burley, *Proc. Natl. Acad. Sci. U.S.A.*, 1996, **93**, 4862.
97. K. Swaminathan, P. Flynn, R.J. Reece and R. Marmorstein, *Nat. Struct. Biol.*, 1997, **4**, 751.
98. D. MacDonald, K. Herbert, X. Zhang, T. Pologruto and P. Lu, *J. Mol. Biol.*, 2001, **306**, 1081.
99. P.J. Hagerman, *Annu. Rev. Biochem.*, 1990, **59**, 755.
100. H.S. Koo, J. Drak, J.A. Rice and D.M. Crothers, *Biochemistry*, 1990, **29**, 4227.
101. A. Barbic, D.P. Zimmer and D.M. Crothers, *Proc. Natl. Acad. Sci. U.S.A.*, 2003, **100**, 2369.
102. D. Rhodes, *Nucleic Acids Res.*, 1979, **6**, 1805.
103. K. Struhl, *Proc. Natl. Acad. Sci. U.S.A.*, 1985, **82**, 8419.
104. Y.C. Kim, J.C. Grable, R. Love, P.J. Greene and J.M. Rosenberg, *Science*, 1990, **249**, 1307.
105. K.W. Mouw and P.A. Rice, *Mol. Microbiol.*, 2007, **63**, 1319.
106. M. Dlakic and R.E. Harrington, *J. Biol. Chem.*, 1995, **270**, 29945.
107. I. Brukner, S. Susic, M. Dlakic, A. Savic and S. Pongor, *J. Mol. Biol.*, 1994, **236**, 26.
108. I. Brukner, M. Dlakic, A. Savic, S. Susic, S. Pongor and D. Suck, *Nucleic Acids Res.*, 1993, **21**, 1025.
109. M. McCall, T. Brown and O. Kennard, *J. Mol. Biol.*, 1985, **183**, 385.
110. D.S. Goodsell, M.L. Kopka, D. Cascio and R.E. Dickerson, *Proc. Natl. Acad. Sci. U.S.A.*, 1993, **90**, 2930.
111. M.A. El Hassan and C.R. Calladine, *J. Mol. Biol.*, 1998, **282**, 331.
112. C.A. Davey and T.J. Richmond, *Proc. Natl. Acad. Sci. U.S.A.*, 2002, **99**, 11169.
113. C.A. Davey, D.F. Sargent, K. Luger, A.W. Maeder and T.J. Richmond, *J. Mol. Biol.*, 2002, **319**, 1097.
114. Y. Bao, C.L. White and K. Luger, *J. Mol. Biol.*, 2006, **361**, 617.
115. G.C. Yuan, Y.J. Liu, M.F. Dion, M.D. Slack, L.F. Wu, S.J. Altschuler and O.J. Rando, *Science*, 2005, **309**, 626.
116. M. A. Azaro and A. Landy, in λ *Integrase and the* λ *Int Family*, ed. N. Craig, R. Craigie, A. Lambowitz and M. Gellert, ASM Press, Washington D.C., 2002, pp. 118–148.
117. M.D. Sam, C.V. Papagiannis, K.M. Connolly, L. Corselli, J. Iwahara, J. Lee, M. Phillips, J.M. Wojciak, R.C. Johnson and R.T. Clubb, *J. Mol. Biol.*, 2002, **324**, 791.

118. V.V. Rogov, C. Lucke, L. Muresanu, H. Wienk, I. Kleinhaus, K. Werner, F. Lohr, P. Pristovsek and H. Ruterjans, *Eur. J. Biochem.*, 2003, **270**, 4846.
119. M.D. Sam, D. Cascio, R.C. Johnson and R.T. Clubb, *J. Mol. Biol.*, 2004, **338**, 229.
120. C.V. Papagiannis, M.D. Sam, M.A. Abbani, D. Yoo, D. Cascio, R.T. Clubb and R.C. Johnson, *J. Mol. Biol.*, 2007, **367**, 328.
121. M.A. Abbani, C. V. Papagiannis, M.D. Sam, D. Cascio, R.C. Johnson and R.T. Clubb, *Proc. Natl. Acad. Sci. U.S.A.*, 2007, **104**, 2109.
122. R.S. Hegde, *J. Nucl. Med.*, 1995, **36**, 25S.
123. R.S. Hegde and E.J. Androphy, *J. Mol. Biol.*, 1998, **284**, 1479.
124. S.S. Kim, J.K. Tam, A.F. Wang and R.S. Hegde, *J. Biol. Chem.*, 2000, **275**, 31245.
125. H. Rozenberg, D. Rabinovich, F. Frolow, R.S. Hegde and Z. Shakked, *Proc. Natl. Acad. Sci. U.S.A.*, 1998, **95**, 15194.
126. J. Hizver, H. Rozenberg, F. Frolow, D. Rabinovich and Z. Shakked, *Proc. Natl. Acad. Sci. U.S.A.*, 2001, **98**, 8490.
127. R.S. Hegde, *Annu. Rev. Biophys. Biomol. Struct.*, 2002, **31**, 343.
128. C.S. Hines, C. Meghoo, S. Shetty, M. Biburger, M. Brenowitz and R.S. Hegde, *J. Mol. Biol.*, 1998, **276**, 809.
129. C.L. Bedrosian and D. Bastia, *Virology*, 1990, **174**, 557.
130. Y. Zhang, Z. Xi, R.S. Hegde, Z. Shakked and D.M. Crothers, *Proc. Natl. Acad. Sci. U.S.A.*, 2004, **101**, 8337.
131. C.A. Ball, R. Osuna, K.C. Ferguson and R.C. Johnson, *J. Bacteriol.*, 1992, **174**, 8043.
132. A.A. Talukder, A. Iwata, A. Nishimura, S. Ueda and A. Ishihama, *J. Bacteriol.*, 1999, **181**, 6361.
133. A. Travers and G. Muskhelishvili, *Curr. Opin. Genet. Dev.*, 2005, **15**, 507.
134. R.C. Johnson, L.M. Johnson, J.W. Schmidt and J.F. Gardner, in *Major Nucleoid Proteins in the Structure and Function of the Escherichia coli Chromosome*, ed. N.P. Higgins, ASM Press, Washington, D.C., 2005, pp. 65–132.
135. R.T. Dame, *Mol. Microbiol.*, 2005, **56**, 858.
136. M.S. Luijsterburg, M.C. Noom, G.J. Wuite and R.T. Dame, *J. Struct. Biol.*, 2006, **156**, 262.
137. D. Kostrewa, J. Granzin, C. Koch, H.W. Choe, S. Raghunathan, W. Wolf, J. Labahn, R. Kahmann and W. Saenger, *Nature*, 1991, **349**, 178.
138. H.S. Yuan, S.E. Finkel, J.A. Feng, M. Kaczor-Grzeskowiak, R.C. Johnson and R.E. Dickerson, *Proc. Natl. Acad. Sci. U.S.A.*, 1991, **88**, 9558.
139. M.K. Safo, W.Z. Yang, L. Corselli, S.E. Cramton, H.S. Yuan and R.C. Johnson, *EMBO J.*, 1997, **16**, 6860.
140. S. Stella, D. Cascio and R.C. Johnson, unpublished.
141. C.Q. Pan, S.E. Finkel, S.E. Cramton, J.A. Feng, D.S. Sigman and R.C. Johnson, *J. Mol. Biol.*, 1996, **264**, 675.
142. T.R. Strick, J.F. Allemand, D. Bensimon, A. Bensimon and V. Croquette, *Science*, 1996, **271**, 1835.

143. C. Bustamante, S.B. Smith, J. Liphardt and D. Smith, *Curr. Opin. Struct. Biol.*, 2000, **10**, 279.
144. J. Yan and J.F. Marko, *Phys. Rev. E: Stat. Nonlin. Soft Matter Phys.*, 2003, **68**, 011905.
145. D. Skoko, J. Yan, R.C. Johnson and J.F. Marko, *Phys. Rev. Lett.*, 2005, **95**, 208101.
146. D. Skoko, D. Yoo, H. Bai, B. Schnurr, J. Yan, S.M. McLeod, J.F. Marko and R.C. Johnson, *J. Mol. Biol.*, 2006, **364**, 777.
147. G. Gosset, Z. Zhang, S. Nayyar, W.A. Cuevas and M.H. Saier, Jr., *J. Bacteriol.*, 2004, **186**, 3516.
148. A. Kolb, S. Busby, H. Buc, S. Garges and S. Adhya, *Annu. Rev. Biochem.*, 1993, **62**, 749.
149. J.M. Passner and T.A. Steitz, *Proc. Natl. Acad. Sci. U.S.A.*, 1997, **94**, 2843.
150. S.C. Schultz, G.C. Shields and T.A. Steitz, *Science*, 1991, **253**, 1001.
151. S. Chen, J. Vojtechovsky, G.N. Parkinson, R.H. Ebright and H.M. Berman, *J. Mol. Biol.*, 2001, **314**, 63.
152. A.A. Napoli, C.L. Lawson, R.H. Ebright and H.M. Berman, *J. Mol. Biol.*, 2006, **357**, 173.
153. G. Parkinson, C. Wilson, A. Gunasekera, Y.W. Ebright, R.E. Ebright and H.M. Berman, *J. Mol. Biol.*, 1996, **260**, 395.
154. S.S. Zinkel and D.M. Crothers, *Biopolymers*, 1990, **29**, 29.
155. H.N. Liu-Johnson, M.R. Gartenberg and D.M. Crothers, *Cell*, 1986, **47**, 995.
156. J.D. Kahn and D.M. Crothers, *J. Mol. Biol.*, 1998, **276**, 287.
157. J.F. Thompson and A. Landy, *Nucleic Acids Res.*, 1988, **16**, 9687.
158. K.K. Swinger and P.A. Rice, *Curr. Opin. Struct. Biol.*, 2004, **14**, 28.
159. K.K. Swinger and P.A. Rice, *J. Mol. Biol.*, 2007, **365**, 1005.
160. A. Pontiggia, A. Negri, M. Beltrame and M.E. Bianchi, *Mol. Microbiol.*, 1993, **7**, 343.
161. D. Kamashev, A. Balandina and J. Rouviere-Yaniv, *EMBO J.*, 1999, **18**, 5434.
162. B. Castaing, C. Zelwer, J. Laval and S. Boiteux, *J. Biol. Chem.*, 1995, **270**, 10291.
163. J.A. Goodrich, M.L. Schwartz and W.R. McClure, *Nucleic Acids Res.*, 1990, **18**, 4993.
164. K. Kobryn, D.Z. Naigamwalla and G. Chaconas, *Mol. Microbiol.*, 2000, **37**, 145.
165. P.A. Rice, S. Yang, K. Mizuuchi and H.A. Nash, *Cell*, 1996, **87**, 1295.
166. K.K. Swinger, K.M. Lemberg, Y. Zhang and P.A. Rice, *EMBO J.*, 2003, **22**, 3749.
167. M. Lorenz, A. Hillisch, S.D. Goodman and S. Diekmann, *Nucleic Acids Res.*, 1999, **27**, 4619.
168. D. Sun, L.H. Hurley and R.M. Harshey, *Biochemistry*, 1996, **35**, 10815.
169. S.V. Kuznetsov, S. Sugimura, P. Vivas, D.M. Crothers and A. Ansari, *Proc. Natl. Acad. Sci. U.S.A.*, 2006, **103**, 18515.

170. S. Sugimura and D.M. Crothers, *Proc. Natl. Acad. Sci. U.S.A.*, 2006, **103**, 18510.
171. K.A. Aeling, M.L. Opel, N.R. Steffen, V. Tretyachenko-Ladokhina, G.W. Hatfield, R.H. Lathrop and D.F. Senear, *J. Biol. Chem.*, 2006, **281**, 39236.
172. T.W. Lynch, E.K. Read, A.N. Mattis, J.F. Gardner and P.A. Rice, *J. Mol. Biol.*, 2003, **330**, 493.
173. E.C. Lee, M.P. MacWilliams, R.I. Gumport and J.F. Gardner, *J. Bacteriol.*, 1991, **173**, 609.
174. D. Ussery, T.S. Larsen, K.T. Wilkes, C. Friis, P. Worning, A. Krogh and S. Brunak, *Biochimie*, 2001, **83**, 201.
175. D. Skoko, B. Wong, R.C. Johnson and J.F. Marko, *Biochemistry*, 2004, **43**, 13867.
176. J. van Noort, S. Verbrugge, N. Goosen, C. Dekker and R.T. Dame, *Proc. Natl. Acad. Sci. U.S.A.*, 2004, **101**, 6969.
177. B. Schnurr, C. Vorgias and J. Stavans, *Biophys. Rev. Lett.*, 2006, **1**, 29.
178. D. Sagi, N. Friedman, C. Vorgias, A.B. Oppenheim and J. Stavans, *J. Mol. Biol.*, 2004, **341**, 419.
179. S.S. Broyles and D.E. Pettijohn, *J. Mol. Biol.*, 1986, **187**, 47.
180. B.M. Ali, R. Amit, I. Braslavsky, A.B. Oppenheim, O. Gileadi and J. Stavans, *Proc. Natl. Acad. Sci. U.S.A.*, 2001, **98**, 10658.
181. S. Dixit, M. Singh-Zocchi, J. Hanne and G. Zocchi, *Phys. Rev. Lett.*, 2005, **94**, 118101.
182. M. Bustin, *Mol. Cell Biol.*, 1999, **19**, 5237.
183. J.O. Thomas and A.A. Travers, *Trends Biochem. Sci.*, 2001, **26**, 167.
184. A. Agresti and M.E. Bianchi, *Curr. Opin. Genet. Dev.*, 2003, **13**, 170.
185. J.O. Thomas, *Biochem. Soc. Trans.*, 2001, **29**, 395.
186. J. Zlatanova and K. van Holde, *Bioessays*, 1998, **20**, 584.
187. T.L. Megraw and C.-B. Chae, *J. Biol. Chem.*, 1993, **268**, 12758.
188. T.T. Paull and R.C. Johnson, *J. Biol. Chem.*, 1995, **270**, 8744.
189. F.V.T. Murphy and M.E. Churchill, *Structure Fold Des.*, 2000, **8**, R83.
190. A. Travers, *Curr. Opin. Struct. Biol.*, 2000, **10**, 102.
191. A. Remenyi, K. Lins, L.J. Nissen, R. Reinbold, H.R. Scholer and M. Wilmanns, *Genes Dev.*, 2003, **17**, 2048.
192. J.J. Love, X. Li, D.A. Case, K. Giese, R. Grosschedl and P.E. Wright, *Nature*, 1995, **376**, 791.
193. J.E. Masse, B. Wong, Y.M. Yen, F.H. Allain, R.C. Johnson and J. Feigon, *J. Mol. Biol.*, 2002, **323**, 263.
194. Y.M. Yen, B. Wong and R.C. Johnson, *J. Biol. Chem.*, 1998, **273**, 4424.
195. F.V. Murphy, R.M. Sweet and M.E. Churchill, *EMBO J.*, 1999, **18**, 6610.
196. U.M. Ohndorf, M.A. Rould, Q. He, C.O. Pabo and S.J. Lippard, *Nature*, 1999, **399**, 708.
197. Y. Dai, B. Wong, Y.M. Yen, M.A. Oettinger, J. Kwon and R.C. Johnson, *Mol. Cell Biol.*, 2005, **25**, 4413.
198. Q. He, U.M. Ohndorf and S.J. Lippard, *Biochemistry*, 2000, **39**, 14426.
199. D. Skoko, B. Wong, R.C. Johnson and J.F. Marko, unpublished data.

200. K. Mitsouras, B. Wong, C. Arayata, R.C. Johnson and M. Carey, *Mol. Cell Biol.*, 2002, **22**, 4390.
201. B. Wong, J.E. Masse, Y.M. Yen, P. Giannikopoulos, J. Feigon and R.C. Johnson, *Biochemistry*, 2002, **41**, 5404.
202. J. Klass, F.V. Murphy, S. Fouts, M. Serenil, A. Changela, J. Siple and M.E. Churchill, *Nucleic Acids Res.*, 2003, **31**, 2852.
203. M. McCauley, P.R. Hardwidge and L.J. Maher, 3rd and M.C. Williams, *Biophys. J.*, 2005, **89**, 353.
204. W. Yang, *DNA Repair*, 2006, **5**, 654.
205. M.A. Schumacher, K.Y. Choi, H. Zalkin and R.G. Brennan, *Science*, 1994, **266**, 763.
206. C.E. Bell and M. Lewis, *Nat. Struct. Biol.*, 2000, **7**, 209.
207. R. Lavery and H. Sklenar, *J. Biomol. Struct. Dyn.*, 1988, **6**, 63.

CHAPTER 9

Mode of Action of Proteins with RNA Chaperone Activity

SABINE STAMPFL, LUKAS RAJKOWITSCH, KATHARINA SEMRAD AND RENÉE SCHROEDER

Max F. Perutz Laboratories, University of Vienna, Dr. Bohrgasse 9/5, A-1030 Vienna, Austria

9.1 Introduction

9.1.1 RNA Folding

Most if not all functional RNAs rely on their native three-dimensional structure to achieve activity. In RNA molecules, intramolecular base–base interactions are very dominant, since non-canonical base-pairs form in addition to Watson–Crick pairings. Hence, RNA often folds into several different secondary structures. RNA folding pathways are hierarchical, with elements of the secondary structure forming first and fast, while formation of the tertiary structure can take minutes to hours.[1] Therefore, RNA molecules obtain their correct fold slowly and are easily trapped in long-lived intermediate states. Additionally, tertiary interactions, which are responsible for the stabilization of three-dimensional structures, are often not very strong, resulting in similarly stable alternative structures.[2,3] When devoid of interacting proteins, large RNA molecules, therefore, tend to misfold. Misfolding usually affords long-lived stable structures, which trap the RNA molecule in non-native and unproductive conformations. This phenomenon is known as the RNA folding problem.[4,5] In contrast to protein misfolding, which becomes prevalent at elevated temperatures, RNA misfolding manifests itself mainly at lower temperatures, where incorrect conformations cannot easily open up to permit refolding. The RNA folding problem seems to be mainly a test tube problem, although misfolding does occur *in vivo*.[6–9] Only recently, this complex problem has been addressed, and RNA quality control mechanisms are being uncovered.[10,11] Moreover, co-transcriptional folding also appears to contribute significantly to the functional RNA structure.[12,13]

9.1.2 Proteins with RNA Chaperone Activity (RCA)

Within cells, RNA molecules probably never exist isolated from proteins, and we assume that RNA misfolding *in vivo* is a rare event, suggesting that proteins provide the guidance for correct folding of RNAs. This is in accord with recent data about proteins that have the ability to assist folding of RNA molecules. This ability was named RNA chaperone activity (RCA), which is used for a wide and not well defined set of activities.[4,14] These include the capability of a protein to accelerate annealing of two complementary RNA strands, to melt a nucleic acid duplex without requiring ATP or any analogous NTP energy source, and complex activities like the acceleration of ribozyme reactions, *e.g.* group I intron splicing or hammerhead ribozyme catalysis. The term RNA chaperone activity has also been used for RNA helicases,[15] for the ribosome[6] and often also for proteins that bind specifically to RNAs with high affinity.[4] Owing to this multitude of diverse activities, different assays are required to monitor RCA. Proteins with RCA can either encompass only one or multiple of these activities. For example, the *E. coli* histone-like protein StpA accelerates RNA annealing and also melts RNA duplexes, while ribosomal protein S1 only promotes strand displacement but not annealing, and the human Ro60 protein is only capable of annealing but cannot melt secondary structures (L.R. and R.S, unpublished). Other proteins like the nucleocapsid protein of HIV (NCp) do not distinguish between DNA or RNA duplexes and destabilize both of them whereas the nucleocapsid protein of Hantavirus (NP) only melts RNA duplexes but not DNA or DNA–RNA double strands.[16] In addition, a group of RNA helicases is included in this heterogeneous collection of proteins with RCA because some helicases encompass ATP-independent RNA annealing activities.[17–19] It thus becomes evident that the two types of reaction – RNA annealing and strand displacement – follow different mechanisms and require specific activities.

The number and diversity of proteins that have RNA chaperone activity is astonishing, ranging from ribosomal proteins, histone-like proteins, viral capsids, hnRNP proteins to translation factors and cold shock proteins. For this, we have also established a website, where the activities of these proteins are presented in a comparable way (http://www.projects.mfpl.ac.at/rnachaperones) and have written a review on the different protein families in detail.[20] Consequently, this chapter focuses on the mechanism of action of proteins with RCA and not on a description of them. The strategies proteins employ to chaperone RNA folding are discussed, using some well-described examples for "annealers" and "strand-displacers". In this context we also refer to Chapter 5 of this book on single-stranded nucleic acid binding, because proteins with RCA bind predominantly to ssRNA. Proteins that interact specifically with RNA are also included because some of them display RNA chaperone-like activities during their encounter with RNA.

9.1.3 Measuring RCA

How proteins accelerate RNA folding by acting as RNA chaperones can be monitored *via* several assays.[21] They describe discrete activities and proteins perform differently in these assays. The simplest set-up monitors the RNA chaperone-mediated acceleration of the annealing of two complementary RNA oligonucleotides (Figure 9.1A). Another basic activity is the promotion of RNA strand displacement, and the corresponding assay detects the displacement of one RNA strand by a competitor (Figure 9.1B). Recently, a coupled assay, which monitors both RNA annealing and RNA strand displacement activities, has been reported.[22]

To assess nucleic acid melting activity the degree of strand dissociation can be quantified by separating double and single stranded molecules on native polyacrylamide gels.[23] In a more advanced set-up force-induced melting assays have been performed to monitor the stretching of single DNA molecules in an optical tweezers instrument.[23] These assays allow quantitative assessment of helix destabilizing activities, which is especially useful when comparing mutant proteins. Similarly, the effect of helicases on the unwinding of RNA secondary structures has been monitored by atomic force microscopy.[24]

Assays that monitor advanced RNA (re)structuring measure the acceleration of the reaction rate of group I intron *cis*- or *trans*-splicing, or the stimulation of hammerhead ribozyme-mediated RNA cleavage (Figure 9.1C–E). These *in vitro* assays generally employ artificial RNA substrates or model ribozyme systems because RNA chaperones usually show non-specific RNA binding properties and because the natural target RNA(s), if any, might not be known.

In addition to the *in vitro* assays, two *in vivo* assays have been established: a folding trap assay to examine chaperone-dependent group I intron splicing *in vivo*[25] (Figure 9.1F), and a transcriptional assay to observe chaperone assisted "melting" of a terminator stem[26] (Figure 9.1G). Several assays, especially those that measure annealing and helix destabilizing activities, have often been performed using DNA as substrate and the observed action, hence, refers to nucleic acid chaperone activity. There have been no systematic studies that compare the effect of these proteins on RNA and DNA substrates.

9.2 Mode of Action of Proteins with RCA

9.2.1 RNA Annealing Activity

The annealing reaction of two complementary RNAs to form a double-stranded RNA is slow and inefficient, most probably due to intramolecular base-pairings. Naturally occurring antisense RNAs, for example, which have to bind and complement their targets, often fold into hairpin loops, and the annealing of the antisense RNAs with their target is induced *via* loop–loop contacts termed "kissing loops".[27] Depending on the concentration of the

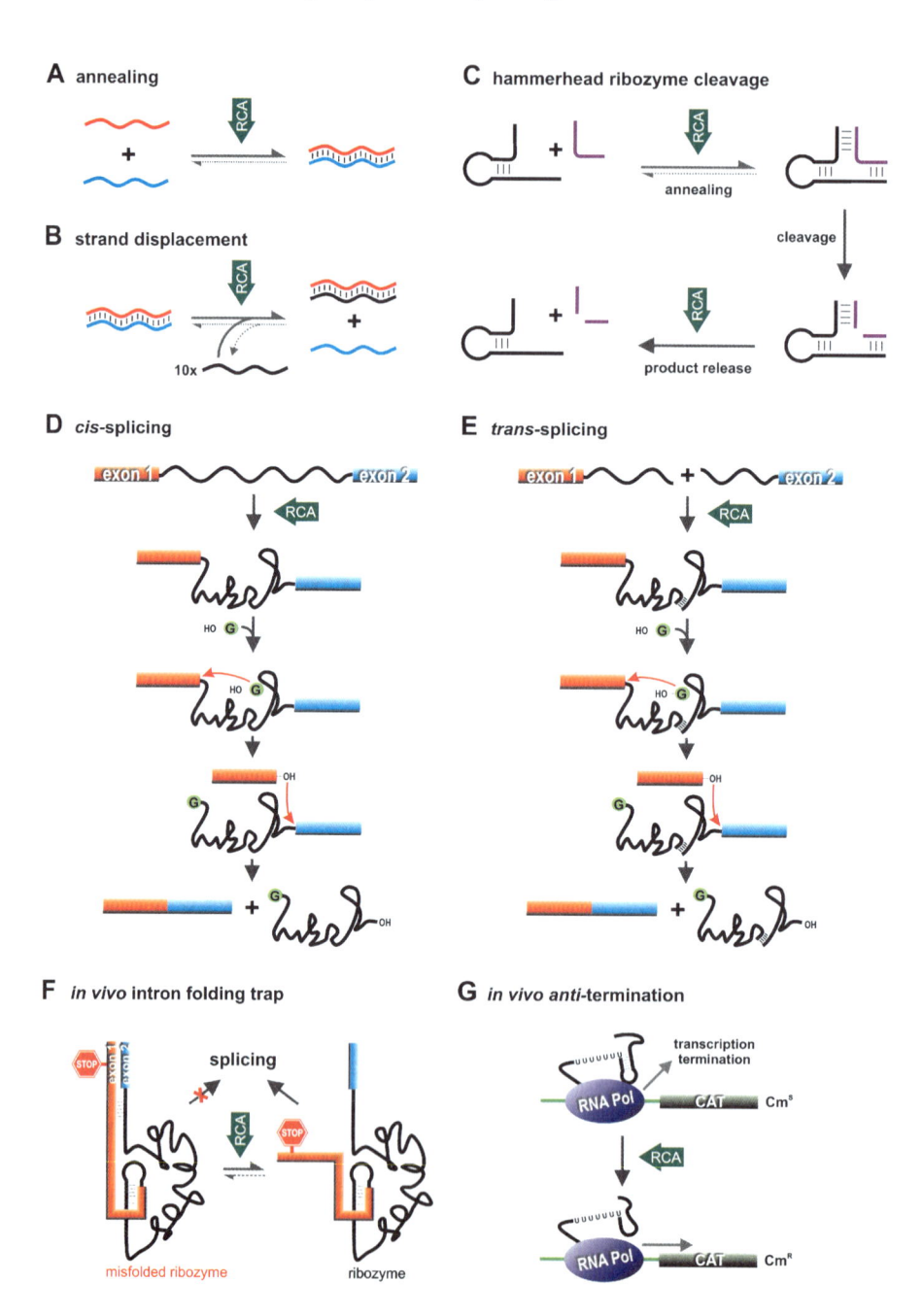

RNAs and/or on their secondary structure, RNA "annealers" are required to accelerate efficiently duplex formation. Two unstructured 21-mer oligoribonucleotides, which do not undergo intramolecular base pairing, anneal approximately with a reaction rate of $10^6 M^{-1} s^{-1}$ at 37 °C.[28] RNA chaperones can accelerate this activity reaction up to the diffusion limit of $10^8 M^{-1} s^{-1}$.[17,29]

The mechanism of RNA annealing has been studied in detail for the *E. coli* histone-like protein StpA and the global regulator protein Hfq. StpA is a small

Figure 9.1 *In vitro* (*A–E*) and *in vivo* (*F–G*) assays for RNA chaperone activity. (*A*) RNA annealing is accelerated by proteins with RNA chaperone activity (RCA, green arrow). To visualize hybridization, at least one annealing partner is usually radioactively or fluorophore-labeled. The reaction can be monitored in real-time by fluorescence methods, or the reaction is quenched and the products are analyzed after separation by gel electrophoresis. (*B*) Strand displacement of an RNA duplex by a competitor RNA can be facilitated by proteins with RCA. To distinguish the duplexes before and after the exchange reaction, strands can be of different lengths or uniquely marked with radioactive or fluorescence labels. While an excess of competitor strand positively affects the reaction kinetics, mismatches in the initial double-stranded RNA drive the reaction towards the thermodynamically favored duplex with a fully complementary competitor. (*C*) The hammerhead ribozyme assay monitors the cleavage of a substrate RNA that first has to anneal to the ribozyme. Moreover, by changing the ribozyme to substrate ratio, multiple turnover conditions can be investigated that require the release of cleavage product, which is comparable to strand dissociation. Thereby, two different activities of an RNA chaperone – annealing and strand dissociation – can be assessed. (*D*) To catalyze self-splicing, pre-RNAs containing the *thymidylate synthase* group I intron have to fold correctly. For *cis*-splicing, the purified transcript is folded by heat-renaturing. Subsequent incubation with proteins with RCA significantly increases the population of correctly folded and catalytically active molecules. The splicing reaction is then initiated by addition of a guanosine cofactor. Ligation of this cofactor to the intron 5′-end in the first transesterification step is followed by exon ligation and intron release. (*E*) In the *trans*-splicing assay, the pre-RNA is transcribed in a bipartite form and, therefore, the requirement for RNA–RNA interactions and correct folding is increased. The two parts have to anneal and obtain a catalytically active structure, which is facilitated by proteins with RCA, especially at low temperatures. (*F*) To assay RCA *in vivo*, splicing of a mutant of the *thymidylate synthase* gene is monitored. A premature stop codon upstream from the 5′ splice site in exon 1 prevents the ribosome from reaching and resolving an aberrant base pairing between 3′-terminal intron and exon 1 sequences. This folding trap can be alleviated by the co-expression of proteins with RCA. As a result the correctly folded intron splices. (*G*) In a nascent transcript, formation of a stem-loop that is followed by a poly(U) stretch can lead to pausing of the RNA polymerase. Transcription termination prevents the expression of a downstream reporter gene (chloramphenicol acetyl transferase, CAT), making the cells chloramphenicol sensitive. Co-expression of RNA chaperones leads to "melting" of the terminator stem, CAT is transcribed and the cells become chloramphenicol resistant (Cm^R).

basic protein, which consists of two functional domains: an N-terminal dimerization domain and a C-terminal nucleic acid binding domain (reviewed in ref. 30). Using fluorescently labeled non-complementary RNA molecules, it could be shown that StpA is able to bind two RNA molecules simultaneously, suggesting that the mechanism of annealing acceleration is based on crowding of the RNA molecules.[28] Additionally, it was demonstrates that a dimerization-deficient form of StpA with a mutation in the N-terminal domain lost its capability to enhance RNA annealing, and to bind two RNAs. These findings suggest that protein dimerization might be one way to achieve simultaneous binding of several RNAs, thereby increasing their concentration locally and consequently accelerating their interaction (L.R and R.S., unpublished).

The *E. coli* host factor Hfq was also shown to promote RNA annealing. Hfq is a small basic protein that probably does not exist as a monomer in the cell but has always been found to form hexameric ring structures.[31] Like StpA, Hfq can simultaneously bind at least two RNA molecules. In both cases, RNA annealing coincides with dual RNA binding, suggesting an RNA crowding mechanism. Simultaneous binding of complementary RNA strands increases the local RNA concentration, prolongs the encounter duration and thus the annealing likelihood in a matchmaker-like fashion. As both StpA and Hfq have higher affinity for single-stranded than for double-stranded RNA molecules, the newly formed duplexes probably dissociate from the protein, enabling turnover. However, multiple turnover kinetic analysis for RNA annealing acceleration reactions has not yet been performed.

The Ro60 protein, which also accelerates RNA annealing but not strand displacement, has been structurally characterized *via* X-ray crystallography in complex with different RNA molecules.[32] The monomeric Ro60 has a doughnut-shaped structure with a very large basic surface, which has been implicated in specific binding of misfolded RNAs.[33] Furthermore, it has been proposed that, *via* this surface, Ro60 scavenges different RNAs by decreasing the electrostatic repulsion.[32] RNA annealing by Ro60 could be another example of RNA crowding based on simultaneous binding of several RNAs.

9.2.1.1 Annealing of Protein-bound Guide RNAs with Target RNAs

There are several special cases of annealing of short RNAs with their target sequences, whereby the short RNAs are tightly bound to specific proteins, and as a consequence the specialized RNA/protein complexes search and efficiently bind their target RNAs. RISC is such a complex, where the siRNA is tightly bound to an Argonaute protein. The Argonaute protein accelerates target recognition by non-specific interaction with single-stranded RNA.[34] Another prominent example of protein assisted RNA–RNA annealing is the MRP1/MRP2 guide RNA complex of *trypanosoma brucei*. It was suggested that the MRP proteins act as molecular matchmakers by promoting the annealing of guide RNAs (gRNAs) with cognate pre-mRNAs.[35] The X-ray crystal structure of the MRP1/MRP2-gRNA complex reveals some mechanistic details of the annealing process.[36] The gRNAs are bound to the highly basic β sheet surface of the MRP1/MRP2 proteins *via* electrostatic interactions, thus guaranteeing non-sequence specific interactions. One stem of the gRNA is buried in the surface and the part of the gRNA that needs to interact with the target RNA is unfolded and its bases exposed in a conformation suitable for hybridization. In addition, the basic surface surrounding the gRNA in the complex reduces electrostatic repulsion of the target RNA.[36] A third example of a protein-assisted RNA annealing process occurs during ribosomal RNA processing. Cleavage of the pre-rRNA requires a small nucleolar RNA, the U3 snoRNA that has to bind to its target pre-rRNA. This processing step takes place in the context of a large RNP complex, the small subunit processome (SSUP).[37] Two yeast proteins, Imp3p and Imp4p, promote annealing of the U3 snoRNA to the pre-RNA by stabilizing the interaction between the U3 snoRNA hinge region and the external transcribed spacer of the pre-rRNA. In addition, Imp4p seems to restructure the U3 snoRNA to promote annealing with the 5′ end of the 18S rRNA.[37] These three examples seem to use slightly different strategies, but they do have in common that one RNA is bound more specifically and the bases are displayed in a way that facilitates interaction.

9.2.2 Nucleic Acid Melting Activity

The final result of nucleic acid melting or strand displacement activity is comparable to the well-described unwinding activity of helicases, but with the important difference that RNA chaperone activity-based unwinding does not require hydrolysis of ATP or any other NTP source. Hence, similar activities might rely on different mechanisms. Several proteins with RCA show only strand displacement activity but do not accelerate annealing, suggesting that these two reactions are based on two independent activities. Examples for proteins exclusively displaying strand displacement activity are the ribosomal protein S1 from *E. coli* and the main cold shock factor CspA (L.R. and R.S., unpublished). CspA is a very short protein consisting of only one cold shock domain (CSD). This CSD is a common motif, which was first detected in several proteins binding oligonucleotides and oligosaccharides and hence was termed the OB-fold (for a review see ref. 38). The *E. coli* ribosomal protein S1 consists of six copies of the CSD motif, and is thus additionally termed S1 motif. This OB/CSD/S1 motif is also found in the *E. coli* translation initiation factor 1 (IF1), which has recently been reported to have RNA chaperone activity.[29] The structure of CspA, which is identical to the cold shock domain, contains five antiparallel β strands forming a closed five-stranded β-barrel. Two RNA-binding motifs (RNP1 and RNP2) are located on the β2 and β3 strands, respectively. The protein has an overall negative surface with a positively charged nucleic acid binding region consisting of seven aromatic residues.[39] Binding of the RNA to the protein involves stacking of the aromatic side chains with the RNA bases. In CspE – a well-studied member of the CspA family – mutation of three centrally located residues does not affect RNA binding but does abolish the ability of the mutant to melt nucleic acids, to antiterminate transcription and eventually to acclimate to cold (Figure 9.1G).[40] This interaction surface might play a key role in the mode of action of these proteins in destabilizing RNA helices.

The best-studied protein with RCA is the nucleocapsid from HIV-1, NCp7, which has RNA annealing and nucleic acid destabilizing activities. It is a small, highly basic protein of 55 amino acids and two zinc finger motifs. NCps are essential for various steps in viral replication, like reverse transcription, genomic RNA packaging, virus assembly and tRNA annealing.[14,41,42] Mechanistic studies using single molecule techniques are helping to unravel the mode of action of NCp7. Using its natural substrate, the TAR RNA/DNA, it was shown that the acceleration of extended duplex formation requires enhanced annealing of the two kissing loops and also subsequent promotion of stem strand exchange. NCp accelerates annealing of these two nucleic acids 5000-fold by stabilizing an extended kissing loop intermediate 100- to 200-fold and by accelerating the subsequent strand exchange reaction 10- to 20-fold.[43] In the case of the interaction of the priming tRNA with the primer binding site, electrostatic attraction by NCp was found to be sufficient for the complete annealing reaction, including melting of the tRNA.[44] Using time-resolved single molecule fluorescence resonance energy transfer (FRET), it was shown that NCp induces secondary structure fluctuations in specific DNA and RNA hairpins.[45]

The *E. coli* protein StpA, like HIV-1 NCp, shows both RNA annealing and strand displacement activities. The RNA unfolding activity of StpA was studied *in vivo* using the self-splicing group I intron of the *thymidylate synthase* (*td*) gene of phage T4 as substrate RNA. The unspliced pre-mRNA of this gene becomes trapped in an inactive conformation in the absence of ongoing translation, because sequences from the upstream exon base pair with the very 3′ end of the intron, hindering the folding of the 3′ splice site into a splicing competent conformation.[6] StpA and several other proteins with RCA are able to resolve this kinetic folding trap, resulting in partial but efficient splicing (Figure 9.1F).[25,46] This system was used to analyze the effect of StpA on the tertiary structure of the *td* group I intron. *In vivo* dimethyl sulfate (DMS) modifications in the presence of StpA demonstrated that bases that are involved in tertiary structure interactions become more accessible, suggesting that StpA loosens the tertiary structure of the intron RNA. In contrast, a specific binding protein for group I introns, Cyt-18 that binds and stabilizes the group I intron structure, renders the tertiary structure more compact and less accessible to DMS, demonstrating that Cyt-18 does not loosen the group I intron structure but stabilizes it.[47,48]

9.3 RNA Binding and Restructuring

9.3.1 Proteins with RCA Interact with RNA only Weakly

Comparison of a specific RNA-binding protein and a protein with RCA points to the essential difference between these two modes of interaction. Specific binding of a protein to RNA leads to structural stabilization, whereas the interaction between RNAs and proteins with RCA induces structural destabilization. Conceivably, proteins with RCA do not bind tightly to the RNA, especially not to natively folded RNAs. In agreement with these ideas, an inverse correlation between RNA-binding affinity and RCA was observed. Binding of StpA to RNA is weak and StpA has a clear preference for unstructured RNAs. A mutation in the nucleic acid binding domain of StpA leads to a decrease in RNA binding compared to wild-type StpA but to a higher RNA chaperone activity. Strengthening of protein interactions with nucleic acids is expected to reduce its ability to promote refolding by structure destabilization. The low binding efficiency and affinity of StpA to RNA, as well as the fact that low amounts of mono- and divalent metal ions can easily impede protein binding, suggest that the interaction between StpA and RNA is weak and most probably highly transient.[28]

The bacterial FinO protein, which represses F-plasmid conjugative transfer, facilitates sense-antisense RNA interactions by accelerating RNA strand exchange. Like StpA, FinO binds poorly to RNA (K_d of 0.2–1 μM) and mutant variants of FinO that have partially lost their RNA strand exchange acceleration ability bind with higher affinity to RNA than the wild type. Thus, as for StpA, an inverse correlation between RNA chaperone and RNA-binding activity has been observed for FinO.[49]

It is part of the definition of an RNA chaperone that once the RNA is correctly folded the chaperone is no longer required for RNA function and is then displaced from the RNA.[4,50] For ribosomal protein S12,[50] StpA,[51] and Hfq[52] it was shown that they can be removed after folding of the RNA without the RNA losing its structure. Thus, part of the RCA might be the preference of the proteins to bind to unfolded or only loosely structures RNAs, but once the RNA finds its native fold the affinity to the RNA decreases, leading to dissociation. From all these observations, we can propose that proteins with nucleic acid destabilizing activity bind RNA weakly and non-specifically but with preference for unfolded or unstructured RNAs. The encounter of the proteins with the RNAs destabilizes misfolded structures, providing an additional chance for refolding. Once the RNAs have found their native structure, they will no longer interact with the proteins and their structure is too stable to be resolved by proteins with RCA. StpA is indeed able to sense the structural stability of the RNA.[46]

9.3.2 Proteins with Specific RNA-binding Affinity

Proteins with specific RNA-binding activity have often been defined as antagonists to the proteins with RCA because, in contrast to the RNA chaperones, they bind specifically, stabilizing the structure of their cognate RNAs. Prior to specific binding or when binding to non-cognate RNAs, these proteins might, however, act similarly to proteins with RCA. Recently, single molecule FRET has been used to monitor conformational dynamics during folding of the bI5 group I intron in the presence of its specific splicing factor CBP2.[53] Two different types of interaction could clearly be detected. Before CBP2 binds with high affinity to the RNA, a non-specific mode of interaction is observed, which causes structural fluctuations in the intron RNA corresponding to a searching process between various conformations. This non-specific mode of interaction is a mechanism that also could be exploited by proteins with RNA chaperone activity. Conceivably, many proteins that interact specifically with an RNA molecule first undergo this type of non-specific interaction, promoting conformational searching before they successfully achieve specific and high affinity binding (see also ref. 4). This might explain why so many specific RNA-binding proteins show additional RNA chaperone activity when interacting with non-cognate targets.

Similarly, the tRNA synthetase Cyt-18, which binds specifically to group I introns, also showed RNA displacement activity *in vitro* but no RNA annealing activity.[22]

9.3.3 Protein Structure and RNA Chaperone Activity

The mode of action of proteins with RCA may require a reassessment of our classical view on intermolecular interactions and their relation to protein structure. A review by Dyson and Wright suggests that the traditional correlation between protein structure and function needs reconsideration.[54] Intrinsically unstructured proteins or unstructured regions in proteins do, in contrast to the classical view, exert function. Supporting this novel concept is a model for the mode of action of proteins with RCA proposed by Tompa and Csermely. They observed that proteins with RCA belong to the class of proteins with the highest degree of structurally disordered regions.[55] Binding of a disordered protein to RNA might be accompanied by local folding of the protein and unfolding of the RNA driven by entropy exchange. The RNA can then explore additional conformations and dissociate from the chaperone. Multiple cycles of folding and unfolding transitions involving reciprocal entropy transfer between the protein and the RNA might then lead to natively folded RNA. This model still awaits confirmation but recent results of structural analysis of proteins with RCA suggest that protein–RNA interactions are very dynamic and structural rearrangements in both interacting partners might occur.[56] In agreement with this hypothesis, it was reported that the N-terminal region of the FinO protein, which is essential for strand exchange activity, was found to be unstructured, at least in the free protein.[49,57] In a recently developed algorithm, which analyses local secondary structure preformation and exposure of residues prone for interactions, StpA was found to display significantly smaller calculated compactness values than the average of proteins.[28] Most importantly, the location of these exposed preformed regions coincides with experimentally validated DNA recognition sites found for its homologue H-NS.[58] Compactness values and secondary structure parameters are being prepared for all proteins reported to have RCA and will be made available on our RCA website (http://www.projects.mfpl.ac.at/rna-chaperones). This is an emerging field that still needs a lot of input and tests, but eventually will make the prediction of RCA possible.

Acknowledgements

Work on proteins with RNA chaperone activity in our laboratory is funded by the Austrian Science Fund (FWF) grant SFB F1703.

References

1. P. Brion and E. Westhof, *Annu. Rev. Biophys. Biomol. Struct.*, 1997, **26**, 113–137.
2. D.K. Treiber, M.S. Rook, P.P. Zarrinkar and J.R. Williamson, *Science*, 1998, **279**, 1943–1946.
3. J. Pan, D. Thirumalai and S.A. Woodson, *J. Mol. Biol.*, 1997, **273**, 7–13.
4. D. Herschlag, *J. Biol. Chem.*, 1995, **270**, 20871–20874.
5. R. Russell and D. Herschlag, *J. Mol. Biol.*, 2001, **308**, 839–851.
6. K. Semrad and R. Schroeder, *Genes Dev.*, 1998, **12**, 1327–1337.
7. S.A. Jackson, S. Koduvayur and S.A. Woodson, *RNA*, 2006, **12**, 2149–2159.
8. R. Schroeder, R. Grossberger, A. Pichler and C. Waldsich, *Curr. Opin. Struct. Biol.*, 2002, **12**, 296–300.
9. R. Schroeder, A. Barta and K. Semrad, *Nat. Rev. Mol. Cell Biol.*, 2004, **5**, 908–919.
10. Z. Li, S. Reimers, S. Pandit and M. Deutscher, *EMBO J.*, 2002, **21**, 1132–1138.
11. S. Kadaba, X. Wang and J.T. Anderson, *RNA*, 2006, **12**, 508–521.
12. E.M. Mahen, J.W. Harger, E.M. Calderon and M.J. Fedor, *Mol. Cell*, 2005, **19**, 27–37.
13. A. Xayaphoummine, V. Viasnoff, S. Harlepp and H. Isambert, *Nucleic Acids Res.*, 2007, **35**, 614–622.
14. G. Cristofari and J.L. Darlix, *Prog. Nucleic Acid Res. Mol. Biol.*, 2002, **72**, 223–268.
15. S. Mohr, M. Matsuura, P.S. Perlman and A.M. Lambowitz, *Proc. Natl. Acad. Sci. U.S.A.*, 2006, **103**, 3569–3574.
16. M. Ma and A.T. Panqaniban, *J. Virol.*, 2005, **79**, 1824–1835.
17. Q. Yang and E. Jankowsky, *Biochemistry*, 2005, **44**, 13591–13601.
18. H. Uhlmann-Schiffler, C. Jalal and H. Stahl, *Nucleic Acids Res.*, 2006, **34**, 10–22.
19. C. Halls, S. Mohr, M. del Campo, Q. Yang, E. Jankowsky and A.M. Lambowitz, *J. Mol. Biol.*, 2007, **365**, 835–855.
20. L. Rajkowitsch, D. Chen, S. Stampfl, K. Semrad, C. Waldsich, O. Mayer, M.F. Jantsch, R. Konrat, U. Blaesi and R. Schroeder, *RNA Biol.*, 2007, in press.

21. L. Rajkowitsch, K. Semrad, O. Mayer and R. Schroeder, *Biochem. Soc. Trans.*, 2005, **33**, 450–455.
22. L. Rajkowitsch and R. Schroeder. *BioTechniques*, 2007, **43**, 304–310.
23. S.L. Martin, M. Cruceanu, D. Branciforte, P. Wai-Iun Li, S.C. Kwok, R.S. Hodges and M.C. Williams, *J. Mol. Biol.*, 2005, **348**, 549–561.
24. S. Marsden, M. Nardelli, P. Linder and J.E. McCarthy, *J. Mol. Biol.*, 2006, **361**, 327–335.
25. E. Clodi, K. Semrad and R. Schroeder, *EMBO J.*, 1999, **18**, 3776–3782.
26. S. Phadtare, M. Inouye and K. Severinov, *J. Biol. Chem.*, 2002, **277**, 7239–7245.
27. Y. Eguchi and J. Tomizawa, *Cell*, 1990, **60**, 199–209.
28. O. Mayer, L. Rajkowitsch, C. Lorenz, R. Konrat and R. Schroeder, *Nucleic Acids Res.*, 2007, **35**, 1257–1269.
29. V. Croitoru, K. Semrad, S. Prenninger, L. Rajkowitsch, M. Vejen, B.S. Laursen, H.U. Sperling-Petersen and L.A. Isaksson, *Biochimie*, 2006, **88**, 1875–1882.
30. C.J. Dorman, J.C. Hinton and A. Free, *Trends Microbiol.*, 1999, **7**, 124–128.
31. M.A. Schumacher, R.F. Pearson, T. Moller, P. Valentin-Hansen and R.G. Brennan, *EMBO J.*, 2002, **21**, 3546–3556.
32. G. Fuchs, A.J. Stein, C. Fu, K.M. Reinisch and S.L. Wolin, *Nat. Struct. Mol Biol.*, 2006, **13**, 1002–1009.
33. X. Chen, J.D. Smith, H. Shi, D.D. Yang, R.A. Flavell and S.L. Wolin, *Curr. Biol.*, 2003, **13**, 2206–2211.
34. S.L. Ameres, J. Martinez and R. Schroeder, *Cell*, 2007, **130**, 101–112.
35. U.F. Mueller, L. Lambert and H.U. Goeringer, *EMBO J.*, 2001, **20**, 1394–1404.
36. M.A. Schumacher, E. Karamooz, L. Zikova and J. Lukes, *Cell*, 2006, **126**, 701–711.
37. T. Gerczei and C. Correll, *Proc. Natl. Acad. Sci. U.S.A.*, 2004, **101**, 15301–15306.
38. D.L. Theobald, R.M. Mitton-Fry and D.S. Wuttke, *Annu. Rev. Biophys. Biomol. Struct.*, 2003, **32**, 115–133.
39. H. Schindelin, W. Jiang, M. Inouye and U. Heinemann, *Proc. Natl. Acad. Sci. U.S.A.*, 1994, **91**, 5119–5123.
40. S. Phadtare, S. Tyagi, M. Inouye and K. Severinov, *J. Biol. Chem.*, 2002, **277**, 46706–46711.
41. A. Rein, L.E. Henderson and J.G. Levin, *Trends Biochem. Sci.*, 1998, **23**, 297–301.
42. Z. Tsuchihashi and P.O. Brown, *J. Virol.*, 1994, **68**, 5863–5870.
43. M.N. Vo, G. Barany, I. Rouzina and K. Musier-Forsyth, *J. Mol. Biol.*, 2006, **363**, 244–261.
44. M.C. Williams, I. Rouzina, J.R. Wenner, R.J. Gorelik, K. Musier-Forsyth and V.A. Bloomfield, *Proc. Natl. Acad. Sci. U.S.A.*, 2001, **98**, 6121–6126.
45. G. Cosa, Y. Zeng, H.-W. Liu, C.F. Landes, D.E. Makarov, K. Musier-Forsyth and P.F. Barbara, *J. Phys. Chem.*, 2006, **110**, 2419–2426.

46. R. Grossberger, O. Mayer, C. Waldsich, K. Semrad and R. Schroeder, *Nucleic Acids Res.*, 2005, **33**, 2280–2289.
47. C. Waldsich, R. Grossberger and R. Schroeder, *Genes Dev.*, 2002, **16**, 2300–2312.
48. C. Waldsich, B. Masquida, E. Westhof and R. Schroeder, *EMBO J.*, 2002, **19**, 5281–5291.
49. D.C. Arthur, A.F. Ghetu, M.J. Gubbins, R.A. Edwards, L.S. Frost and J.N. Glover, *EMBO J.*, 2003, **22**, 6346–6355.
50. T. Coetzee, D. Herschlag and M. Belfort, *Genes Dev.*, 1994, **8**, 1575–1588.
51. A. Zhang, V. Derbyshire, J.L. Salvo and M. Belfort, *RNA*, 1995, **1**, 783–793.
52. I. Moll, D. Leitsch, T. Steinhauser and U. Blasi, *EMBO Rep.*, 2003, **4**, 284–289.
53. G. Bokinsky, L.G. Nivon, S.L. Liu, G. Chai, M. Hong, K.W. Weeks and X. Zhuang, *J. Mol. Biol.*, 2006, **361**, 771–784.
54. H.J. Dyson and P.E. Wright, *Nat. Rev. Mol. Cell Biol.*, 2005, **6**, 197–208.
55. P. Tompa and P. Csermely, *FASEB J.*, 2004, **18**, 1169–1175.
56. R. Ivanyi-Nagy, L. Davidovic, E.W. Khandjian and J.L. Darlix, *Cell Mol. Life Sci.*, 2005, **62**, 1409–1417.
57. A.F. Ghetu, D.C. Arthur, T.K. Kerppola and J.N. Glover, *RNA*, 2002, **8**, 816–823.
58. H. Shindo, A. Ohnuki, H. Ginba, E. Katoh, C. Ueguchi, C. Mizuno and T. Yamazaki, *FEBS Lett.*, 1999, **455**, 63–69.

CHAPTER 10
Structure and Function of DNA Topoisomerases

KEN C. DONG AND JAMES M. BERGER

Department of Chemistry, College of Chemistry, Department of Molecular and Cell Biology, University of California, Berkeley, 374D Stanley Hall, Berkeley, CA 94720-3220, USA

10.1 Introduction

The complementary, intertwined structure of deoxyribonucleic acid (DNA), while a convenient and stable storage system for genetic information, presents several challenges to the cell. For example, the chemical information coded within the nucleotide base sequence is stored in the interior of DNA, and can only be accessed during synthetic events such as replication and transcription by duplex unwinding. DNA's double-helical architecture likewise is linked to certain topological obstacles, becoming underwound, overwound, or tangled as a result of various essential cellular processes (Figure 10.1). Many of these problems were immediately apparent upon realizing the structure of DNA,[1] a molecule whose polymer properties presaged the discovery of powerful nucleic-acid reshaping enzymes known as topoisomerases.

The genome is subject to a myriad of topological problems engendered by gene expression and chromosome duplication. Accordingly, cells have evolved several classes of topoisomerases to control and resolve these potentially deleterious intermediates. Through a complex series of DNA strand breakage and rejoining reactions, topoisomerases assist numerous DNA transactions, from the maintenance of supercoiling homeostasis to the decatenation of chromosomes. Topoisomerases have been divided into two broad categories

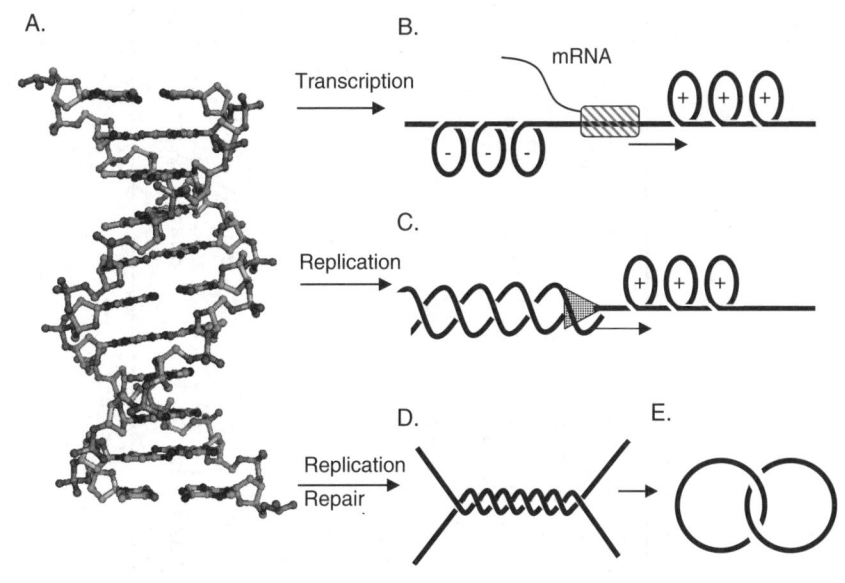

Figure 10.1 Topology problems in the cell caused by DNA transactions. (*A*) The double helical structure of DNA. (*B*) Formation of positive and negative supercoils during translocation of a DNA unwinding motor, *e.g.*, RNA polymerase. (*C*) Formation of precatenanes and positive supercoils during DNA replication. (*D*) Leftover DNA intertwinings give rise to catenanes (*E*) as replication forks meet. Recombination events likewise can generate complex DNA tangles.

according to the number of DNA strands cleaved during their reaction. Enzymes that cut only one DNA strand are termed type I topoisomerases, while those that generate double strand DNA breaks are called type II. Despite this distinction, both topoisomerase types utilize a catalytic tyrosine to attack the phosphodiester backbone of DNA, generating a transient covalent enzyme•DNA intermediate that stabilizes broken DNA ends during enzyme turnover, preventing the formation of DNA nicks and breaks.

Since the initial discovery of topoisomerases in the early 1970s,[2-4] it has become evident that there are multiple topoisomerase subtypes as well. There are currently three distinct type I topoisomerase subclasses (IA, IB, and IC), and two subclasses of type II enzymes (IIA and IIB). Type I topoisomerases differ markedly in both structure and mechanism. For example, type IB topos create a single stranded break in DNA through a 3′ phosphotyrosine intermediate, whereas type IA proteins form a 5′ phosphotyrosyl-linkage upon strand scission. For type II topoisomerases, enzymes of both subclasses use ATP, create a double-stranded break, and transport a second DNA duplex through the break; although the IIA and IIB enzymes share several catalytic domains, however, their overall tertiary and quaternary structure differs significantly. A comparison of topoisomerase types is summarized in Table 10.1 and will be discussed in detail as the focus of this chapter.

Table 10.1 Comparison of the biochemical characteristics of different topoisomerase classes.

Type	IA	IB	IIA	IIB
Subunit composition	Monomer	Monomer	Dimer (euk.) A_2 Heterodimer (prok.) A_2B_2	Heterodimer A_2B_2
Domains (examples)	Bacteria, Eukaryotes, Archea (some)	Eukaryotes, Poxviruses, Bacteria (some)	Bacteria, Eukaryotes, Archaea (some), Viruses	Archaea, Plants
Representatives	Topo I, Topo III, reverse gyrase	Topo I	Topo II, Topo IV, gyrase	Topo VI
Divalent metal requirement	Mg^{++} & Zn^{++}	None	Mg^{++}	Mg^{++}
ATP requirement	No (reverse gyrase-yes)	No	Yes	Yes
Number of strands cleaved	1	1	2	2
$|\Delta Lk|$	1	n	2	2
Phosphotyrosine intermediate	5'-DNA end	3'-DNA end	5'-DNA end	5'-DNA end
Mechanism	Strand passage	Rotary	Strand passage	Strand passage
Substrate Specificity	ssDNA	dsDNA	dsDNA	dsDNA

Structure and Function of DNA Topoisomerases 239

gap in a single strand of DNA and spanning the broken DNA ends to facilitate the passage of a second DNA duplex through the cleaved segment.[22] Structural studies of type IA topoisomerases,[23,24] together with domain crosslinking and single-molecule experiments,[25,26] have provided strong support for this model.

10.2.2 Structures and Mechanism

There are currently several structures of type IA topoisomerase proteins and domains, both in isolation and complexed with DNA.[23,24,27] The structure of the catalytic core fragment of *E. coli* topo I first highlighted the remarkable overall architecture that typifies the family.[24] The type IA core consists of four domains (I–IV), which together form a toroidal shape reminiscent of a padlock (Figure 10.3). Domain I adopts a doubly-wound α/β TOPRIM

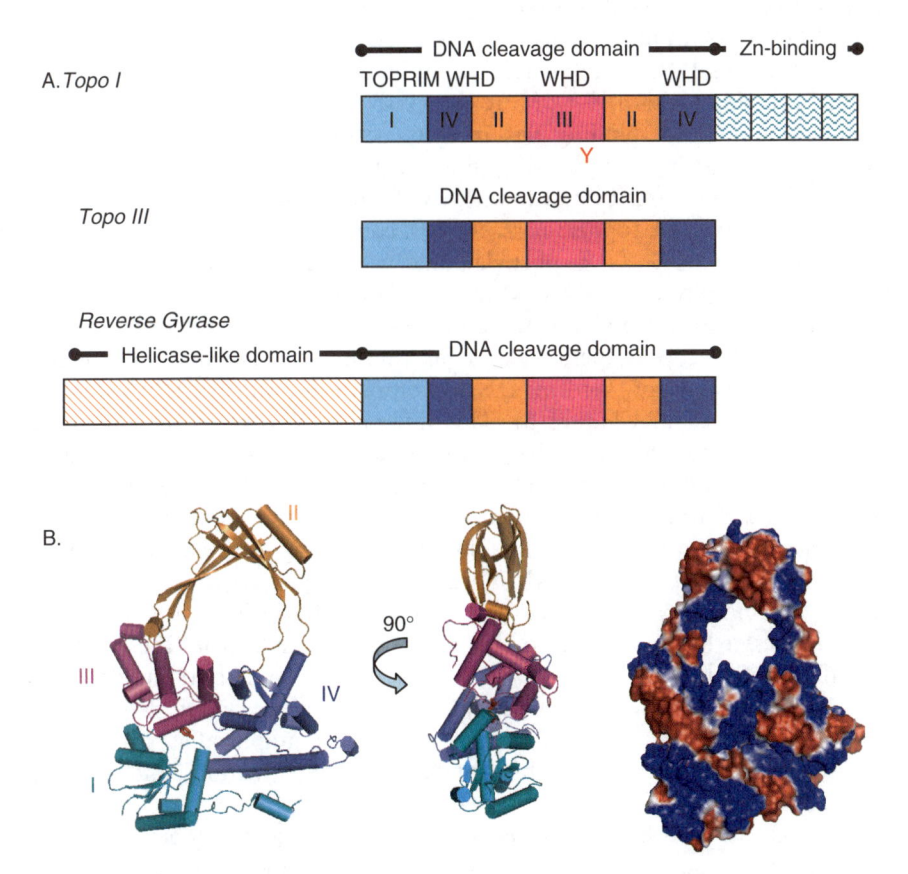

Figure 10.3 Domain organization and structure of type IA topoisomerases. (*A*) Arrangement of topo I, topo III, and reverse gyrase. (*B*) Catalytic core ("topofold") structure of *E. coli* topo I (PDB code 1ECL). A cartoon representation, left, and electrostatic surface representation, right, of the catalytic core.

(TOPoisomerase/PRIMase) fold, and serves as the site of Mg^{2+} ion binding.[16,18,28] Domain II is a linker element that sits at the top of the topo I arch, while domains III and IV comprise Winged Helix Domains (WHDs),[29,30] elements at times referred to as "CAP" domains since they resemble the WHD of the Catabolite Activator Protein.[31] Domain III has a truncated wing, but still retains a traditional Helix-Turn-Helix (HTH) fold. Domain III also is noteworthy in that it contains the active site tyrosine, and abuts both domain IV and the TOPRIM fold of domain I to form a bipartite DNA cleavage center. Interestingly, the catalytic WHD/TOPRIM fold constellation also is found in type II topoisomerases (see below), suggesting that these enzyme classes may have derived from a common ancestor.[32]

While the *E. coli* topo I structure was critical for beginning to understand the strand passage mechanism for type IA topos, the model did not immediately address how topo I engages DNA. Insights into this process initially derived from biochemical studies, which showed that while topo I can act on duplex DNA, the minimum piece of DNA necessary for binding and cleavage is a seven-base ssDNA oligonucleotide, and that thymine is strongly preferred at the scissile position.[33–35] More detailed information on the physical mechanism of DNA binding has recently become available as a result of structures comprising both full-length and catalytic core fragments of several type IA topoisomerases bound to ssDNA (Figure 10.4A). The first co-crystal structure of a type IA topoisomerase•DNA complex required the mutation of the catalytic tyrosine to a phenylalanine;[23] however, more recent crystal structures have revealed ssDNA•topo I complexes captured with this functional residue intact.[36] Each of these models shows that ssDNA binds to the active site of the topoisomerase in a conformation that is remarkably similar to a single strand of B-form DNA (compare to other ssDNA binding proteins discussed in Chapter 5). This finding helps explain the preference of type IA topos for negatively-supercoiled DNA, as the underwound substrate is likely to be melted more readily than relaxed or positively-supercoiled substrates. Concomitant with DNA binding, domain III is seen to reorient with respect to the apo structure, altering the relative juxtapositions of the domain I TOPRIM fold and the catalytic tyrosine in the active site.[23] This movement positions the tyrosine directly in line with the scissile phosphodiester linkage, and also co-localizes amino acids from the TOPRIM fold that are important for cleavage (Figure 10.4B). To date, the positions of the magnesium ions that are essential for full catalytic activity in type IA topoisomerases have not been observed crystallographically, leaving the precise chemical function of this cofactor in DNA binding and cleavage unresolved.

Together, the wealth of biochemical and structural data available for type IA topoisomerases has led to a general mechanistic model for DNA strand passage[24] (Figure 10.4C). The crystal structure of the enzyme in the absence of DNA reveals a closed-toroid state where the active site is buried and inaccessible to DNA. To first bind DNA, domain III must separate from domains I and IV, a movement that may be driven by the flexion of a hinge

Structure and Function of DNA Topoisomerases 241

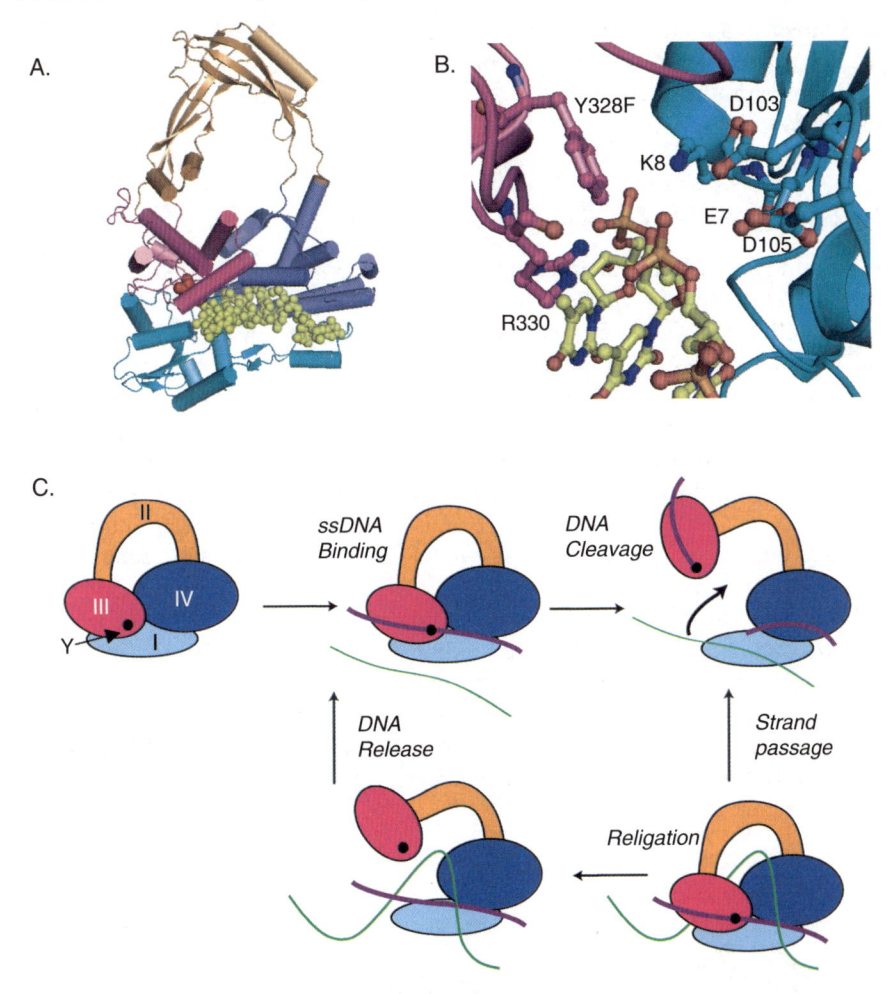

Figure 10.4 Type IA topoisomerase interactions with DNA and molecular mechanism. (*A*) Structure of *E. coli* topoisomerase III bound to ssDNA (PDB code 1I7D). (*B*) Close-up of the topo III active site complexed with DNA. The active site tyrosine was mutated to phenylalanine in this structure. (*C*) Molecular mechanism for DNA strand passage by a type IA enzyme.

element between domains II and III.[37] Once ssDNA is bound, domain III reengages the domain I/IV active site, subsequently cleaving DNA, and becoming covalently attached to 5′ end of the nucleic acid strand through its active site tyrosine. Domain III then separates from domains I and IV, opening a gap in the DNA, and allowing a second DNA segment to pass through the break and into the interior of the enzyme. Once passed, domain III reassociates with the bulk of the topoisomerase to religate the cleaved DNA; release of the DNA segment from the interior of the protein is facilitated by yet another rigid body movement of domain III away from domains I and IV. To processively catalyze the relaxation of DNA supercoils or removal of single-stranded tangles, type IA topos simply reiterate this sequence of events without dissociating from the cleavage DNA.

10.2.3 Type IA Topoisomerase Paralogs

Within the type IA topoisomerases, there are several homologous subclasses of enzymes with slightly different features (Figure 10.3A). For example, among topo I orthologs, the C-terminus can be appended with differing numbers of a Zn^{2+} binding module (domain V). The NMR structure of this domain first revealed that the zinc-binding domain forms a β-sheet motif that interacts with single-stranded DNA;[38,39] subsequent analyses indicate that this element belongs to the zinc-ribbon superfamily.[40] A recent crystal structure of the zinc ribbon domain in the context of the full length topoisomerase has been solved from *T. maritima* (Figure 10.5A),[41] and showed that some variants of the domain do not always coordinate zinc to fold properly. This finding was noteworthy because the *E. coli* topo I zinc-binding domain requires Zn^{2+} ions for full function.[42] Other biochemical studies likewise have indicated that the ssDNA binding ability of the zinc-binding domain is essential for full topo I activity, but is not important for the binding and cleavage of DNA by the catalytic core fragment.[42–44] Although the precise role of this module remains to be determined, these studies have suggested that the zinc-binding domain is a variable element that assists with DNA strand passage and modulates type IA topoisomerase function.

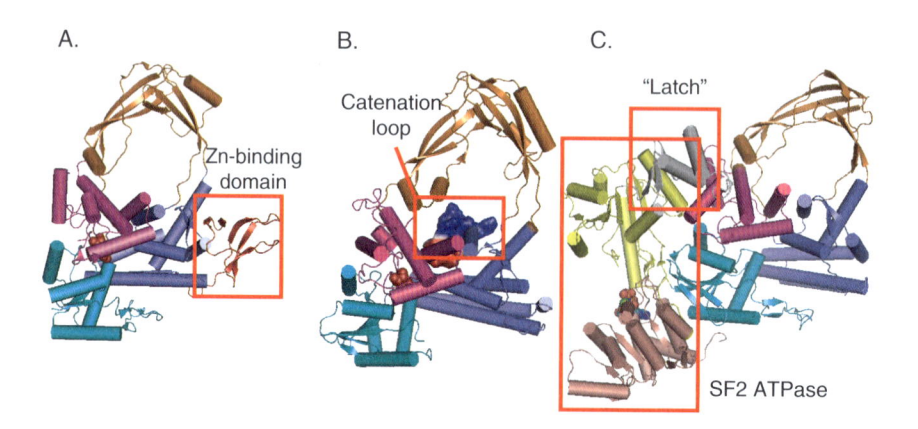

Figure 10.5 Structures of type IA topoisomerase paralogs. Domains colored as in Figure 10.3. (*A*) Structure of *T. maritima* topoisomerase I (PDB code 1GAI) with the zinc-binding domain colored red. (*B*) Structure of *E. coli* topoisomerase III, with residues of the catenation loop colored by electrostatic charge (PDB code 1D6M). (*C*) Structure of reverse gyrase (PDB code = 1GL9). The "latch" is shown in grey.

10.2.3.1 Topoisomerase III

It was not appreciated until more than a decade after the discovery of topo I that many species of bacteria harbor a second, distinct type IA paralog known as topo III.[45] Phylogenetic analyses have since established that topo III, which shares a common overall architecture with bacterial topo I termed a "topofold," is present not only in eubacteria but also in eukaryotes.[46–48] Structural studies have shown that topo III possesses a short 17 amino acid insertion in its catalytic core region that may be specifically utilized to assist with DNA catenation and decatanation reactions (Figure 10.5B).[27] Although the precise mechanism by which the decatenation loop participates in DNA strand passage reactions remains to be determined, the structure of the topo III catalytic core shows that the loop formed by this insertion is positively charged, which is consistent with a role in binding DNA.

The activities of topo I and topo III vary greatly. For example, topo I robustly removes negative DNA supercoils, whereas topoisomerase III is a relatively poor relaxase.[49] Topo III, by contrast, is a much better decatenase than topo I.[48] As a consequence, the physiological roles of these enzymes appear to be distinct, with topo I participating in supercoiling homeostasis,[50,51] and topo III acting on specialized DNA structures arising from replication and repair events.[52–55] Moreover, many of the DNA transactions that topo III takes part in appear to be carried out in collaboration with other proteins, most notably RecQ-family helicases.[56] The joint action of these two enzymes stimulates the decatenation activity of topo III and permits the topoisomerase to efficiently resolve Holliday junctions.[57–59] Still other studies have indicated that topo III is involved with the untangling of sister chromatids as replication forks converge,[54,60] resolving recombination intermediates arising from DNA damage,[61,62] and in telomere recombination.[63] Deletion of one of the two topo III paralogs in mice, topo IIIβ, leads to shortened mean lifespan and multiple types of organ lesions,[64] which is consistent with links reported between mutations in RecQ homologs and premature aging phenotypes.[65–67] A fundamental area of future research will be to define and understand how these exogenous factors precisely regulate type IA topoisomerase function to elicit such a broad range of activities on DNA.

10.2.3.2 Reverse Gyrase

Thermophilic organisms retain a third type IA topoisomerase variant, reverse gyrase, that is distinguished by an ability to generate positively-supercoiled DNA.[68] The restriction of reverse gyrase to this narrow grouping of organisms probably arises from the need to stabilize DNA against thermal denaturation.[69,70] The ability of reverse gyrase to actively supercoil DNA requires ATP, which is consumed by a superfamily-2 (SF2) helicase domain fused to the N-terminus of the topoisomerase catalytic core domain.[71] In isolation, the helicase domain fails to unwind or translocate along DNA, but is necessary for the introduction of (+) supercoils, presumably by coupling ATP hydrolysis with active DNA unwinding or directional strand transport.[72] The isolated topoisomerase domain acts much like an ordinary type IA topoisomerase, albeit less robustly. Structural data for this unique type IA topoisomerase has thus far been limited to crystallographic models determined in the presence and absence of an ATP analogue, ADPNP (adenylylimidodiphosphate) (Figure 10.5C).[73] These structures led to the discovery of a "latch" element between the ATPase domain and the topofold that turns out to be key for coordinating ATP turnover with the positive supercoiling of DNA.[74]

Recent studies have indicated that reverse gyrase may have functions beyond its positive supercoiling activity. In particular, the action of reverse gyrase has been associated with two factors, the type of nucleotide bound to the enzyme and the DNA:enzyme ratio.[75] For example, with the ATP analog adenylyl imidodiphosphate (AMPPNP), reverse gyrase negatively supercoils DNA, whereas in the presence of adenosine diphosphate (ADP) reverse gyrase almost completely relaxes DNA. The final topological state of DNA produced by these reactions seems to be highly dependent upon the relative molar stoichiometries between the protein and DNA. Moreover, reverse gyrase displays an expanded set of activities depending on the type of DNA substrate assayed;[76] for instance, annealing single-stranded DNA or positively supercoiling circular DNAs containing a single-stranded bubble. These findings have suggested that reverse gyrase may be particularly advantageous to thermophilic organisms by helping to regenerate duplex DNA regions that spontaneously melt in a harsh thermal environment.

10.3 Type IB Topoisomerases

10.3.1 Overview

Shortly after the discovery of *E. coli* topo I, Champoux and Dulbecco purified a topoisomerase activity from mouse-embryo cells.[4] This enzyme, also termed topo I, at first appeared to be similar to *E. coli* topo I in that it relaxed negatively supercoiled DNA, functioned as a monomer, and used a tyrosine to cut only one stand of DNA.[77] Additional studies, however, showed that eukaryotic topo I differed markedly from bacterial topo I in many important aspects. For example, eukaryotic topo I forms a 3′ phosphotyrosine intermediate with DNA during cleavage,[78–80] does not require Mg^{2+} ions for activity,[81] and can relax positively supercoiled DNA[4] (Table 10.1), properties later found to be exhibited by a type I topoisomerase from the *vaccinia* virus.[82] Phylogenetic studies indicated that there was no sequence homology between bacterial and either eukaryotic or poxviral topo I,[83] a distinction borne out by structural efforts revealing that the two enzymes adopt completely different folds.[24,84–86] Structures also demonstrated that the catalytic core of these enzymes is actually a variant of tyrosine recombinase/integrase superfamily[84,85] (see also Chapter 12). These distinctions led to the branching of the type I topoisomerase tree into two subtypes: IA for the bacterial enzymes and IB for the eukaryotic proteins.[87] Further inspection of genomic data has expanded this breakdown, showing that certain bacteria also contain type IB enzymes.[88] As a consequence, the distribution of type I topoisomerases is no longer

10.3.2 General Architecture

Eukaryotic type IB topoisomerases generally consist of three domains, an N-terminal region of variable length that serves as a nuclear localization and regulatory element, a central domain, and a C-terminal domain typified by an AraC/tyrosine-recombinase/-integrase fold that contains the catalytic tyrosine (Figure 10.6A).[89,90] The central domain can be further divided into two subdomains (I and II), while the catalytic region is split into two subdomains by a helical linker. The viral and bacterial type IB variants share this general

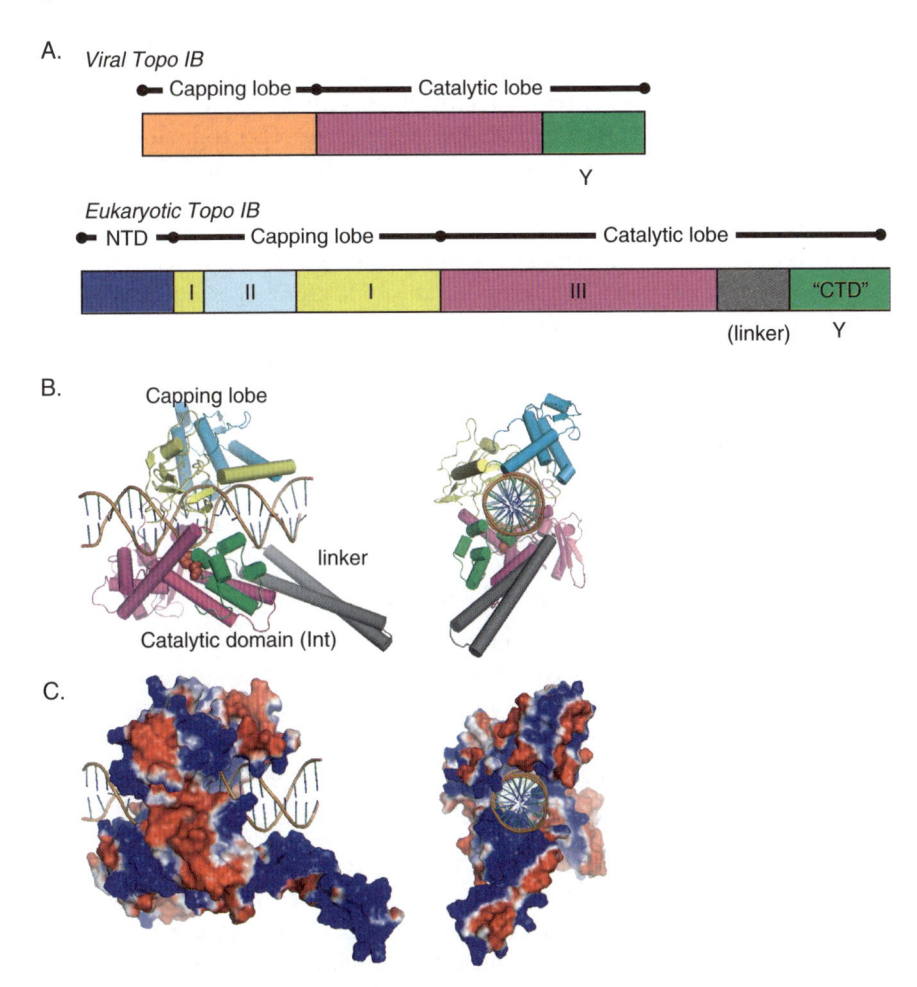

Figure 10.6 Domain arrangement and structures of type IB topoisomerases. (*A*) Domain organization. (*B*) Two orthologonal views of human topo I bound to DNA (PDB code 1A36). (*C*) Electrostatic view of the structure shown in (*B*).

organization, but lack both the N-terminal and linker domains. Interestingly, viral type IB topoisomerases are further distinguished by a high degree of sequence specificity for DNA cleavage sites compared to eukaryotic type IB topoisomerases.[91]

Several structures of type IB topoisomerases have been solved in the presence and absence of DNA. In human topo I, the two subdomains of the central core region form a capping lobe, which together with the catalytic domain enclose a large (~ 20 Å diameter), electrostatically-positive hole (Figure 10.6B, C).[85,86] Structures of bacterial and viral type IB topos reveal that these enzymes have a much smaller version of the capping element.[92] A recent structure of *variola* topo I also has highlighted the molecular determinants that provide sequence specificity to the poxvirus enzymes,[93] demonstrating that the viral type IB topos contain an additional α-helix (not found in eukaryotic homologs), which makes major groove interactions with the DNA (Figure 10.7A). These interactions show how a modest structural change has been exploited to narrow the target site selection of what is otherwise a relatively promiscuous enzyme family.

Co-crystal structures solved to date indicate that, in order to bind DNA, the catalytic and capping lobes reversibly open and close at one end of the enzyme to first admit DNA into the active site, and subsequently clamp around the duplex.[85,86] Biochemical studies, in which cysteine substitutions have been used to covalently lock the lobes together, have provided support for this conformational event, demonstrating that disulfide crosslinks between the two lobes can be used to topologically trap a closed DNA ring.[94,95] More recently, the structure of a full-length bacterial ortholog was determined in the absence of DNA,[96] demonstrating that the catalytic and capping lobes do indeed separate. This structure also revealed that active site of type IB topos remains poised for catalytic activity even in the absence of its DNA substrate, suggesting that the active site is essentially preassembled.

10.3.3 DNA Recognition and Cleavage

Insights into type IB topoisomerase mechanism, particularly DNA cleavage, have benefited greatly from a highly complementary synthesis of structure, chemistry, and detailed enzymatic analysis. For human topo I, both a noncovalent DNA–protein complex and a covalent structure using a suicide DNA substrate revealed how active site residues in the catalytic lobe converge to facilitate strand scission by the active site tyrosine.[85,86] The structure of the noncovalent complex further highlighted the architecture of the linker domain, showing that this region forms an extended helical element that juts out from the catalytic core to contact DNA outside of the active site.[86] Both models showed that when DNA is bound in the interior of the protein it retains a largely B-form conformation (Figures 10.6B, C). Several positively-charged residues manifest extensive interactions with the phosphate backbone of DNA (~ 10 bp worth) to create a large, non-sequence specific binding surface. Structural differences between the covalently bound and noncovalently bound structures are few, and largely confined to the active site,

Structure and Function of DNA Topoisomerases

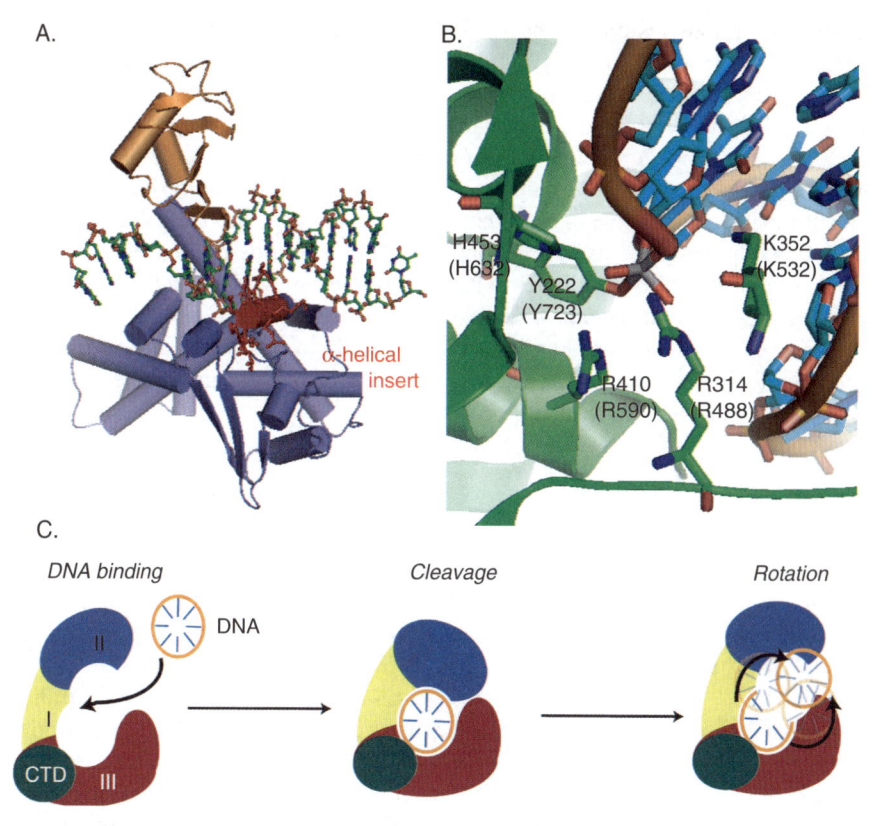

Figure 10.7 Type IB topoisomerase/DNA interactions. (*A*) Structure of the *variola* topo I•DNA complex (PDB code 2H7G). The α-helix responsible for DNA sequence recognition is highlighted in red. (*B*) Close-up of the active site of *Leishmania* topo I captured in a vanadate-trapped complex with nicked DNA (PDB code 2B9S). Key catalytic residues involved in cleavage are labeled, with the human numbering in parentheses. (*C*) Type IB topoisomerase rotary mechanism.

where the catalytic tyrosine and accompanying arginines show slight positional shifts. This finding indicates that no major conformation change is likely to occur during strand scission.

More recently, *L. donovani* topo I was captured in covalent complex with vanadate and DNA.[97] Vanadate is highly useful for studies of nucleic acid cleavage, because it forms a trigonal bipyramidal bonding arrangement that mimics the transition state of the phosphodiester backbone as it is attacked by the catalytic tyrosine. The vanadate structure, along with human topo I•DNA complexes and an extensive number of mutational studies on *vaccinia* virus topo I, has provided detailed evidence for the precise mechanism of DNA cleavage by type IB enzymes. Altogether, there is a rich network of acidic and basic residues that surround the DNA cleavage site and assist with formation of the phosphotyrosine intermediate[85,86,98] (Figure 10.7B). An invariant histidine in the catalytic center

(His632 in humans) is one of the residues essential for stabilizing the pentavalent transition state, while a nearby highly-conserved lysine (Lys532 in humans) may act as a general acid involved in protonating the 5′-oxygen leaving group.[99] Studies using methylphosphonate diastereomers have revealed that phosphates remote to the active site bind in a stereospecific manner, helping to properly assemble the active site for catalysis.[100] Analyses of the kinetic effects of non-bridging thiol substitutions further have shown that an invariant arginine (Arg590 in human topo I) is important for its interactions with both nonbridging and bridging oxygens of the scissile phosphodiester linkage.[101] Interestingly, the addition of methyl-substituted groups on the non-bridging R_p and S_p positions of the phosphodiester backbone at the site of strand scission allows methylphosphonate-topo I linkages to be rapidly hydrolyzed, indicating that the DNA phosphate anion repels water to suppress a latent nuclease activity of the enzyme.[102]

10.3.4 Mechanism

Several avenues have suggested that, unlike the strand passage mechanism utilized by type IA topoisomerases, type IB topos relax DNA supercoils through a very different means. This distinction was initially supported by the observation that type IB enzymes relax DNA in multiple integral steps ($\Delta Lk = n$), centered around a Gaussian distribution, rather than discrete steps of one.[81,103,104] Further clues to this process derive from co-crystal studies, which have shown that human and variola topo I do not bind single-stranded DNA, as type IA topoisomerases do, but rather associate with and cleave a duplex substrate. Together, these findings have indicated that type IB topos are not "enzyme bridges," but rather "swivelases," which remove superhelical strain by nicking DNA and allowing one duplex to rotate with respect to another about a single phosphodiester linkage (Figure 10.7C). This reaction does not require ATP, and instead relies on the free energy stored in overwound or underwound DNA to drive relaxation. This mechanism also accounts for the ability of type IB enzymes to change Lk in multiple steps, as many DNA rotations may occur before religating the nick to restore duplex integrity. Inspection of an ensemble of DNA•type IB topoisomerase complexes indicates that, following cleavage, the duplex end that is not covalently tethered to the catalytic tyrosine can "wiggle" in the active site,[105] indicating that the enzyme can differentially regulate the stability of its contacts with substrate to facilitate rotation. The fact that DNA is not fully relaxed immediately following cleavage suggests that the stability of these loose contacts is nonetheless restrained by steric considerations, such that rotation around the phosphodiester backbone is not completely unimpeded.

Recently, single molecule experiments have been used to further explore the biophysical mechanism of type IB activity.[106] The assay, in which a segment of DNA can be supercoiled and mechanically pulled by a rotatable magnetic bead, permits direct force/torque measurements on DNA relaxation. Researchers were able to observe and quantify changes in linking number as human topo I acted upon its substrate, as well as measure the rates of DNA unwinding.

It was observed that DNA does not freely rotate, but instead makes discrete steps where it releases several supercoils and then pauses. By taking into account multiple measurements, the data suggested that after each rotation there is a certain probability that topo I will religate the duplex, and that this likelihood is regulated by the interplay between superhelical density and the frictional drag of the enzyme on the DNA. These experiments provide further support for the idea that type IB topoisomerases operate as swivelases, and highlight the degree to which friction and torque control the activity of these enzymes.

10.4 Topoisomerase V – The Defining Member of the Type IC Topoisomerases?

Topo V is a type I topoisomerase originally identified from the hypothermophile *Methanopyrus kandleri*.[107] Although sequence analyses failed to reveal a significant homology to either type I topoisomerases, several properties of topo V, including the ability to form a 3′ phosphotyrosine intermediate with DNA and carry out relaxation in the absence of Mg^{2+} ions, initially appeared strikingly similar to type IB enzymes. Topo V also possesses non-canonical topoisomerase activities, such as the repair of apurinic/apyrimidinic sites in damaged DNA[108] and the decatanation of circular DNA at temperatures well above boiling.[109] Topo V repairs abasic sites by cleaving the phosphodiester bond and removing the sugar, creating a single-nucleotide gap that is eventually repaired by a polymerase. Obvious structural differences between topo V and type IB topoisomerases became apparent with the recent determination of its structure by X-ray crystallography.[110] In particular, topo V is significantly larger in size than bacterial type IB topoisomerases and possesses a completely different overall fold (Figure 10.8), as well as 24 Helix-hairpin-Helix (HhH) DNA binding motifs appended to its C-terminus that assist with processivity. Interestingly, the active site of topo V is buried in the interior of the protein, suggesting that large conformational changes are necessary for the enzyme to associate with DNA. Recent single molecule experiments suggest that topo V, like topo IB, also uses a rotary relaxation mechanism,[237] but the basis by which topo V binds and cleaves DNA is unknown. On the basis of these differences, and from phylogenetic analyses indicating that topo V has no obvious homologs in the non-redundant database,[111] it has been proposed that topo V represents a new type IC topoisomerase subclass. Clearly, future studies will be necessary to elucidate the reaction mechanism of this mysterious enzyme.

10.5 Type IIA Topoisomerases

10.5.1 Overview

Shortly after the discovery of both type I topoisomerase activities, Gellert *et al.* reported the characterization yet another topoisomerase, termed DNA gyrase.[3]

A.

Topoisomerase domain — HhH motifs

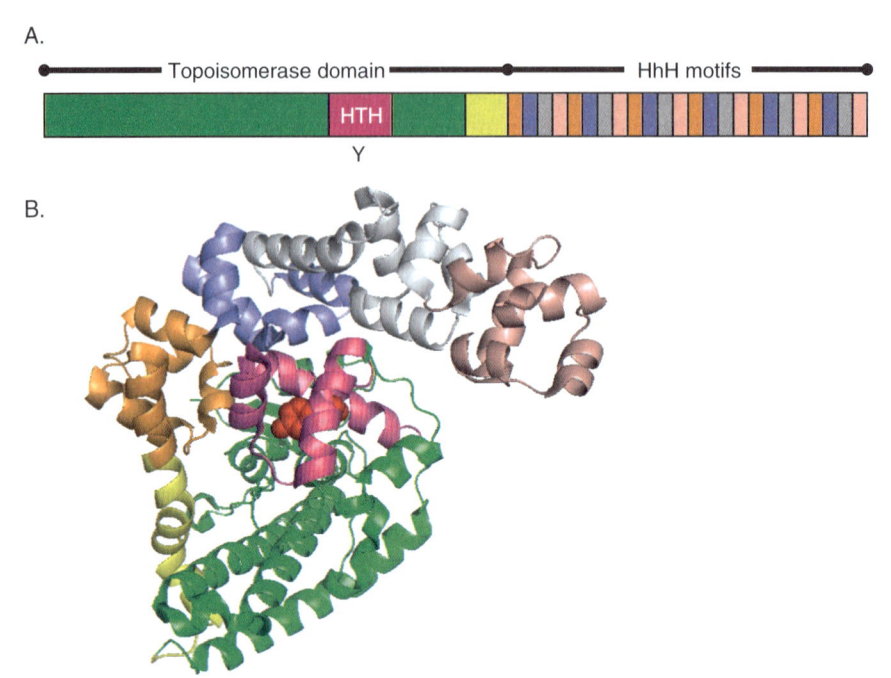

HTH
Y

B.

Figure 10.8 Type IC topoisomerase domain organization and structure. (*A*) Domain organization of topo V. (*B*) Structure of the topo V catalytic region. The active site tyrosine is shown as red spheres.

Identified initially as an activity that both required ATP and could negatively supercoil DNA, it was subsequently observed that gyrase could also decatenate and unknot intact, duplex DNA circles.[112] This latter activity, along with single-topoisomer supercoiling studies,[113] demonstrated that, unlike either type IA or IB topos, gyrase altered the linking number of DNA in discrete steps of two ($|\Delta Lk| = 2$).

Gyrase turned out be the founding member of a broad group of enzymes, the type IIA topoisomerases, which have since been found in all three domains of life. Although related at the amino acid sequence level,[114] the quaternary organization of type IIA topoisomerases can differ significantly. The eukaryotic enzyme (called topo II) is a single chain homodimer,[115,116] whereas bacterial and archaeal homologs (gyrase and topo IV) are split into two chains to form A_2B_2 heterotetramers[117,118] (Figure 10.9A); bacteriophage variants (such as phage T4) also exist as heterohexamers,[119] while gyrase orthologs have been identified in plants.[120] Despite these differences, the key structural components of type IIA topoisomerases are very similar, consisting of a GHKL (Gyrase/Hsp90/histidine-Kinase/MutL)-family ATPase domain,[121] a divalent metal-ion binding TOPRIM fold,[28] and a core DNA-binding region, which contains the active site tyrosine and is typically appended by a variable C-terminal domain.[122–124] For the prokaryotic enzymes, the ATPase and TOPRIM

Structure and Function of DNA Topoisomerases 251

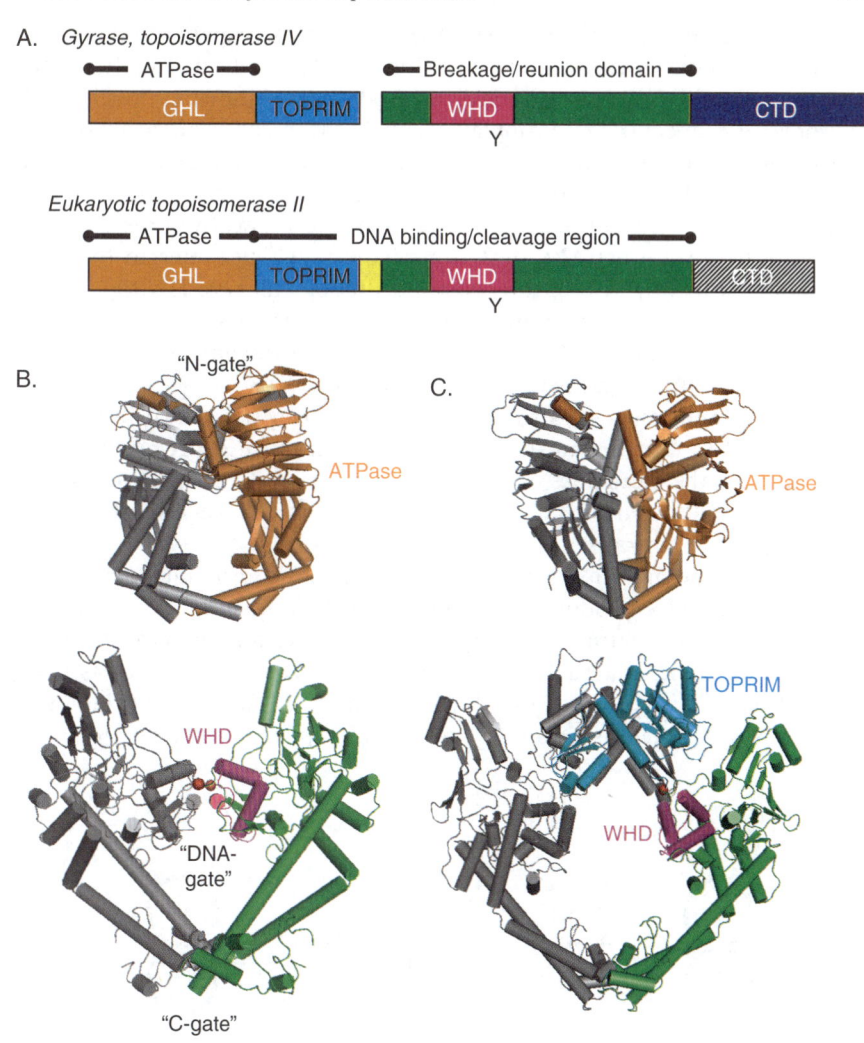

Figure 10.9 Type IIA topoisomerase domain arrangement and structures. (*A*) Schematic of domain organization. (*B*) (Upper): ATPase domain of *E. coli* GyrB (PDB code 1EI1); (lower): DNA breakage/reunion domain of *E. coli* GyrA (PDB code 1AB4). (*C*) (Upper): ATPase domain of *S. cerevisiae* topoisomerase II (PDB code 1PVG); (lower) DNA binding and cleavage core of *S. cerevisiae* topoisomerase II (PDB code 1BGW).

domains reside in one subunit, GyrB (gyrase) or ParE (topo IV), while the primary DNA binding region lies in the other (GyrA or ParC).

The finding that type IIA topos alter Lk by individual increments of ±2 led to early suggestions that these enzymes, like type IA topos, act as enzyme bridges that catalyze the passage of one DNA segment through another.[22] In contrast to bacterial topo I, however, the decatenation capability of type IIA topos, along with biochemical studies (Figure 10.2), indicated early on that both strands of DNA are transiently cleaved and separated to facilitate the transport of a second DNA duplex through the break,[112] a property that imbues type IIA topos with the capacity to separate tangled chromosomes prior to cell division. In this manner, a positive-handed DNA crossover in plectonemically supercoiled DNA can be converted into a negative crossover, and *vice versa*, to effect a net ΔLk of 2, a mechanism at times referred to as sign inversion[113] (Figure 10.10A). Interestingly, although all type IIA topoisomerases share this duplex passage mechanism, only gyrase has proven capable of negatively supercoiling DNA, a property that arises from an ability to wrap up to 120 bp of DNA around itself[125–128] (Figure 10.10B).

Type IIA topoisomerase structure and mechanism both have been extensively studied to date. From the biochemical side, much effort has gone into understanding the mechanics of DNA cleavage, separation, and transport, as well as how ATP turnover is coupled to these dynamic processes. Like type IB topos, type IIA enzymes are also targeted by various clinically important chemotherapeutic and antibacterial agents, such as fluoroquinolones (gyrase and topo IV) and epipodophyllotoxins (topo II), attracting a great deal of attention from the pharmaceutical and biomedical sectors. Emergent findings indicate that, rather than acting in isolation, type IIA enzymes can associate with a diverse array of exogenous factors involved in fundamental biological processes such as meiosis, apoptosis, and DNA replication and repair.

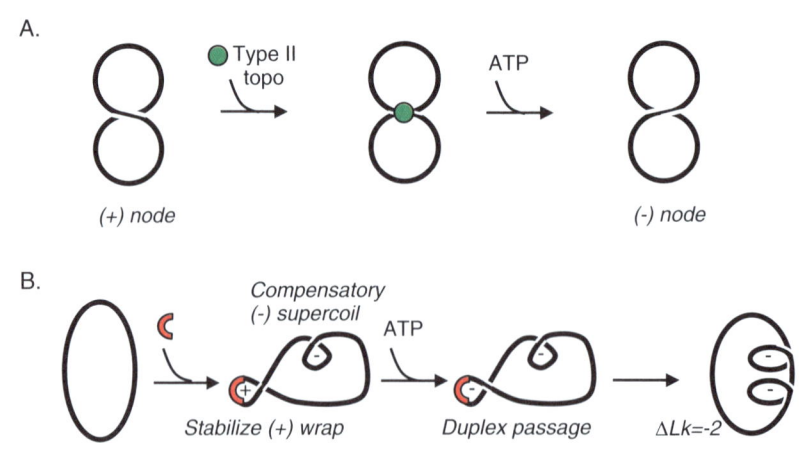

Figure 10.10 Linking number changes engendered by type II topoisomerase action. (*A*) Sign inversion mechanism of a type II topoisomerases can convert a (+) handed DNA crossover into a (−) handed crossover, or *vice versa*, leading to a net |ΔLk| of 2. (*B*) Gyrase stabilizes a positive DNA wrap, which in turn leads to the formation a compensatory (−) supercoil. Upon catalyzing one round of DNA transport, the Lk of DNA is altered by −2.

10.5.2 Structural Organization

Although no full-length type IIA topoisomerase has been imaged at high resolution, structures of individual domains have been important for understanding catalytic mechanism and substrate specificity. These models, together with biochemical studies, have revealed that there are three separable protein/protein interfaces in the holoenzyme.[129–133] One resides between the ATPase domains (the N-terminal or "N-gate", Figure 10.9B and C, upper), while the other two are located in the DNA binding and cleavage region of the enzyme (the "DNA-gate" and C-terminal or "C-gate", Figure 10.9B and 9C, lower).

The manner and timing by which the type IIA topoisomerase interfaces open and close to coordinate duplex transport has been the focus of much interest. Early biochemical studies showed that the non-hydrolyzable ATP analog, AMPPNP, can trap the enzyme on closed, circular DNA.[134] DNA captured in this manner cannot be released by high salt, but can dissociate upon linearization of the duplex by a restriction enzyme, indicating that that protein forms a topological clamp around nucleic acid segments when bound to nucleotide.[135] Subsequent studies have shown that binding of AMPPNP is sufficient to catalyze the passage and release of a second DNA segment through the first, even though the clamp remains closed and the enzyme is blocked from turning over.[136] Together, these findings suggested that the type IIA topoisomerase interfaces open and close in a prescribed manner to capture a duplex at one end of the enzyme and shuttle it out through the other, a reaction referred to as a "two-gate" mechanism.[136]

10.5.2.1 ATPase Domain

Key to the mechanism and control of strand passage is the binding and hydrolysis of ATP, and the subsequent release of ADP and inorganic phosphate. For type IIA topoisomerases, the ATPase domain consists not just of a GHKL fold but also a C-terminal $\alpha\beta$ region that participates in catalysis through the contribution of a conserved lysine to the active site. This domain, which is also linked to the metal-binding TOPRIM fold, has been termed a "transducer" element, because it is likely to communicate conformational signals initiated by ATP turnover to the rest of the enzyme. This domain is conserved in type II topoisomerases, Hsp90, and MutL, which together comprise a "GHL" subfamily of the GHKL ATPases. The structure of a type IIA ATPase region was first determined for *E. coli* DNA gyrase in the presence of AMPPNP;[133] this model, along with biochemical studies on the isolated ATPase region,[137] revealed that ATP binding induces dimerization of a pair of GHKL domains (Figure 10.9B, upper). The GyrB structure also showed that the ATPase dimer is further stabilized by a domain-swapping event between the extreme N-termini of dimer-related GHKL elements. Interestingly, the dimerized gyrase ATPase domains surround a large, ~ 20 Å diameter interior, hole, suggesting that these elements associate with DNA.

The type IIA topoisomerase ATPase domain has since been imaged from several other bacterial and eukaryotic enzymes (Figure 10.9C, upper).[138–140] These structures likewise demonstrate that nucleotide induces dimer formation, and further stabilizes a particular orientation between the GHKL and transducer domains to create a functional active site. Biochemical analyses and structure-guided mutagenesis studies have used these models to show that ATP hydrolysis proceeds by an acid–base mechanism,[137] that ATP binding and subunit dimerization are highly coupled,[141–143] and that the interior surface of the transducer domain likely interacts with DNA.[144–146] Structural data have also revealed that the interior hole in the eukaryotic ATPase dimer is significantly smaller than that seen for prokaryotic proteins,[139] and that the GHKL and transducer domains can alter their relative conformation in different nucleotide-bound states.[140] The contributions of these differences and dynamics to type IIA topoisomerase mechanism is still under investigation.

10.5.2.2 DNA Breakage/Reunion Domain and the DNA Binding/Cleavage Core

Additional insights into the mechanics of DNA recognition and duplex separation have come from structures of the DNA-breakage/reunion domains of S. cerevisiae topo II and E. coli gyrase.[129,131] Both models reveal a dimeric, heart-shaped architecture for this region (Figures 10.9B and C, lower). In each protomer, the active site tyrosine for DNA scission lies on the "wing" element of a WHD, which is in turn embedded within a mixed αβ region that forms a large electrostatic groove on the surface of the protein. These features, along with protein footprinting and mutational studies,[147,148] indicate that the groove likely serves as the primary site for DNA binding and cleavage. Extending from the C-terminus of this region is a long coiled-coil, which is capped by a small domain at its tip that forms the primary dimer interface (the C-gate) for the region.

Despite their similar tertiary structures, the E. coli and yeast models exhibit dramatically different conformational states. For E. coli gyrase, the WHD of each monomer make direct contacts with its partner on the opposite protomer, creating a 25 Å diameter interior hole.[131] In contrast, the WHDs of the yeast model are separated by nearly 30 Å, and the interaction between the WHDs is replaced by dimerization of the TOPRIM fold.[129] Other structures from S. cerevisiae topo II and E. coli ParC (the topo IV DNA binding/cleavage subunit) corroborate these movements, and reveal yet another distinct set of conformations for this region.[149,150] Together, these structures can be viewed as intermediate configurations that probably reflect the mechanical flexions of type IIA topoisomerases as they open and close a cleaved DNA segment, and indicate that the WHDs comprise an element (the DNA-gate) for regulating DNA movements within the protein.

The isolated breakage/reunion domain can bind, but not cleave, DNA on its own.[151,152] To carry out duplex scission, the TOPRIM fold is also required.[152] Biochemical studies indicate that, as for type IA topos, the TOPRIM domain is a divalent metal-binding center that collaborates with the active site tyrosine to form a competent DNA binding/cleavage core.[153,154] Comparison of the two yeast DNA binding/cleavage structures reveals that the TOPRIM domains are mobile with respect to the breakage/reunion domain.[129,150] Mutagenesis studies of catalytic amino acids in the TOPRIM and WHD indicate that the two domains act in *trans* with another,[155] such that the TOPRIM of one subunit partners with the active site tyrosine of its neighboring subunit.

Thus far, only electron microscopy has been able to provide any direct insights into how the ATPase domains link to the DNA binding and cleavage core.[156–158] Studies of full-length and truncated yeast topo II suggest that, in the absence of ATP, the ATPase regions extend out and away from the core, much like a pair of jaws. These images also show that the conformation of the enzyme can change dramatically, taking on a much more compact shape in the presence of DNA and/or nucleotides. Still other EM studies have shown that type IIA topos preferentially bind bent sequences over unbent regions[159] and favor associating with regions of high DNA curvature, such as the tips of plectonemically supercoiled DNA circles and DNA crossovers.[160,161] Despite these direct observations, the extent to which all type IIA topos bend DNA has come under some debate. Although the ability of gyrase to bend and wrap DNA is well established,[128,162,163] more recent biochemical data on topo II and topo IV have produced conflicting results.[160,164]

10.5.3 Duplex DNA Transport Mechanism

Since the discovery of gyrase, an extensive number of biochemical and structural studies on various orthologs have contributed to the development of a molecular framework for type IIA topoisomerase function (Figure 10.11).[129,136,165,166] The enzyme is thought first to bind one DNA duplex, termed the "Gate-" or G-segment, in its interior by associations with the DNA breakage/reunion domain. The footprint of this interaction for the core catalytic domain is fairly broad, measuring 25–30 bp,[167–170] and protein•DNA contacts are relatively sequence non-specific.[171–174] Next, a second DNA duplex, the "Transport-" or T-segment binds to the topoisomerase, priming the enzyme for strand passage. The binding of ATP to the ATPase domains triggers their dimerization, stably capturing the T-segment, while concomitantly stimulating cleavage of the G-segment by the

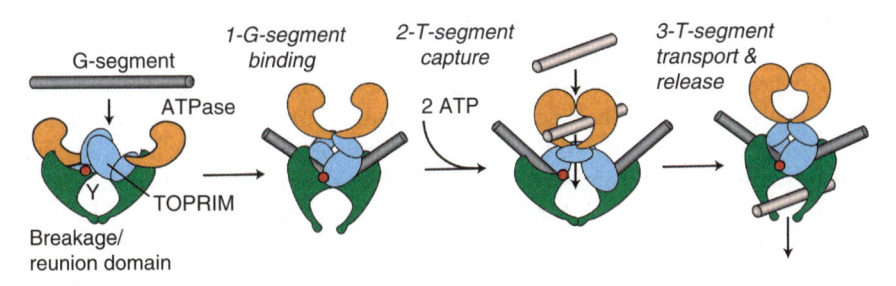

Figure 10.11 Duplex transport mechanism for type IIA topoisomerases.

catalytic tyrosine residues. Cleavage results in the formation of two 5′ phosphotyrosine linkages, staggered 4 bp apart,[175,176] and requires divalent metals[177–179] that are likely supplied by the TOPRIM fold.[32,154,180] DNA cleavage also permits opening of the G-segment and the subsequent passage of the T-segment through the break. Following transport, the G-segment is resealed, and the T-segment released. In support of the "two gate" mechanism of action, crosslinking studies have indicated that the T-segment exits through an interface opposite the ATPase domains, most likely the dimerization interface for the DNA breakage/reunion domain.[132] The order by which various subunit interfaces open and close also has been probed, indicating that the ATPase domains close before the G-segment is opened, and that the G-segment is resealed before T-segment expulsion.[132,135,136,166,181,182]

The relative contributions of ATP binding and hydrolysis to the DNA transport process have been difficult to fully ascertain. Mutations that block ATP binding abrogate type IIA topoisomerase activity,[183] but a heterodimeric mutant containing a single functional ATPase site can still support strand passage, albeit weakly.[184–186] Non-hydrolyzable ATP analogs block enzyme turnover and are unable to support processive relaxation of DNA supercoils,[135,187] but the initial binding event of such nucleotides does support a single round of T-segment transport and release.[136,187] Pre-steady kinetic analyses indicate that ATP binding is rapid and cooperative,[188] but that hydrolysis occurs asynchronously between active sites.[187] This asymmetry in active site firing appears to be tied to phosphate release from one ATPase domain, which serves as the rate-limiting step of the reaction and is coupled to T-segment transport.[189] Determining precisely how these ATP-dependent events control long-range chemo-mechanical movements in the rest of enzyme remains an important avenue of future inquiry.

10.5.3.1 Type IIA Topoisomerase Paralogs: Role of the C-terminal Domain in Modulating Duplex Transport

The minimal structure of a functional type IIA topoisomerase consists of the ATPase, TOPRIM and DNA breakage/reunion domains. Indeed, some viral type IIA topos, such as those from the *Chlorella* virus, consist only of these three elements.[190] The DNA breakage/reunion domains of type IIA topos from cellular organisms, however, are nearly always appended with a C-terminal domain (CTD). This region further differs between prokaryotic and eukaryotic type IIA homologs, and does not appear conserved between the two branches.

The first insights into the role of the CTD came from studies of DNA gyrase, which showed that removal of the domain did not eliminate DNA transport, but instead specifically abolished the enzyme's ability to negatively supercoil DNA.[125] Subsequent structural and biochemical studies of the isolated gyrase CTD show that it adopts a novel six-bladed all β-fold, termed a β-pinwheel, which can bend DNA by up to 180° and stabilize a chiral DNA wrap (Figure 10.12).[191,192] Based on studies of CTDs from several different gyrase and topo IV homologs, as well as

Structure and Function of DNA Topoisomerases 257

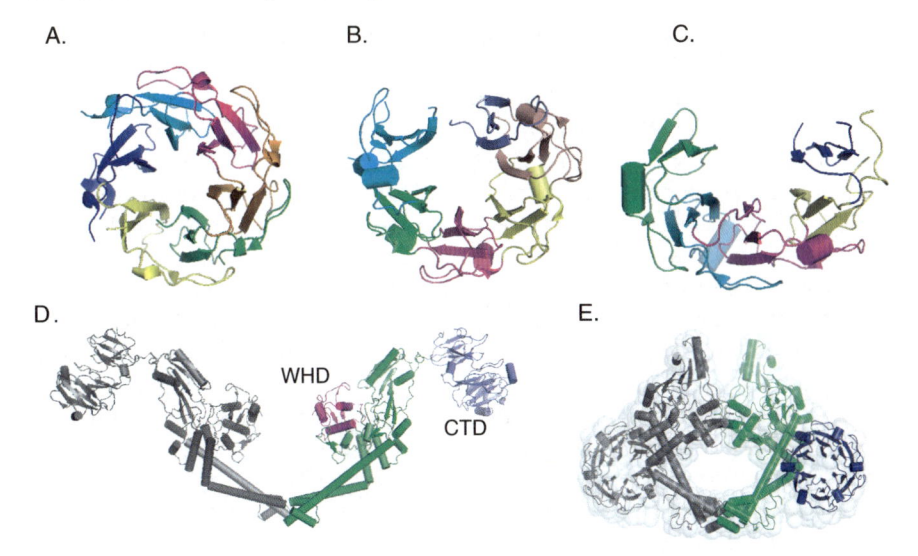

Figure 10.12 C-terminal domains (*A–C*) of type IIA topoisomerases of (*A*) *B. burgdorferi* DNA gyrase (PDB code 1SUU). (*B*) *E. coli* DNA gyrase (PDB code 1ZI0). (*C*) *E. coli* topo IV (PDB code 1ZVT). (*D*) DNA-breakage/reunion domain (ParC) of topo IV (PDB code 1ZVU). (*E*) Model for full-length GyrA derived from small-angle X-ray scattering data. (Courtesy of A. Maxwell.)

phylogenetic surveys, it now appears that there are a number of different structural permutations of the bacterial type IIA topoisomerase CTD, which include closed and open pinwheels,[193] and which contain anywhere from 1 to 8 blades (Figure 10.12).[149,194] Interestingly, blade number and amino acid sequence conservation for the CTD is quite good between gyrases, highlighted in particular by a signature sequence motif termed the "GyrA-box",[195] but breaks down significantly for topo IV orthologs. Several studies have shown that the GyrA box is essential for proper gyrase function.[195,196]

Why are bacterial CTDs so diverse? Part of the answer appears to derive from the fact that DNA gyrase and topoisomerase IV have significantly different substrate specificities and activities (Figure 10.2). DNA gyrase is the only enzyme that can actively negatively supercoil DNA, yet for a type IIA topo it is a remarkably poor decatenase.[53,197,198] By contrast, both bulk biochemical and single molecule measurements have shown that topo IV is extremely robust in its ability to relax positively-supercoiled DNA and unlink catenated DNA circles,[199–201] primarily due an ability to preferentially recognize and act on positive-handed DNA crossovers.[202] Just as removal of the gyrase CTD converts the enzyme into a standard type IIA topoisomerase,[125] deletion of the CTD in topo IV likewise eliminates its substrate selectivity.[149] The precise molecular basis for the control of bacterial type IIA topo function through the CTD is still uncertain, but appears to be linked to the ability of the CTD to bind *vs.* wrap DNA, and the relative position of the CTD with respect to the DNA breakage/reunion domain[149,203] (Figure 10.12D, E). Future work will be needed to tease apart CTD function, and ascertain why different bacteria have modified its structure and biochemical properties so dramatically, particularly for topo IV.

In eukaryotes, the role of the CTD is even less well understood. Secondary structure analysis indicates that for topo II, the CTD is relatively disordered. Various studies have shown that the CTD can be important for nuclear localization, and that the region is also a rich target for phosphorylation by serine/threonine protein kinases, but the exact purpose of this modification is unclear.[122,204,205] Studies to date also have not shown that topo II exhibits any preference for relaxing positively *vs.* negatively supercoiled DNAs, nor for decatenating linked plasmids.[206,207] The exception to this trend is human topo IIα, one of two type IIA topoisomerase paralogs, which does show an enhanced ability to relax positively-supercoiled DNAs.[208] Interestingly, this property is not exhibited by topo IIβ, but can be conferred on topo IIβ by swapping its CTD for the one borne by topo IIα.[208] These findings suggest that there may be higher levels of mechanistic regulation for eukaryotic type IIA enzymes awaiting discovery.

10.5.4 Physiological Specialization of Type IIA Topoisomerases

The duplex DNA passage mechanism of type IIA topoisomerases is used for several purposes by the cell. In many bacteria, particularly those that do not contain a type IB topoisomerase, DNA gyrase counteracts the activity of bacterial topo I to regulate supercoiling homeostasis.[51,209] For eukaryotes, DNA relaxation can be easily handled without consuming chemical energy (in the form of ATP) by their endogenous type IB enzymes,[103] although recent studies suggest that topo II may be the principal relaxase for chromatin, and that type IB enzymes may be partitioned to regions of histone-free DNA such as those found around sites of transcription.[210] The ability of topo IV to rapidly and selectively remove positive supercoils and catenanes has led to the proposal that it is primarily used to support DNA replication.[53,118,197,211]

It is also becoming clear that type IIA topoisomerases do not act as solo players in the cell. Topo IV associates directly with the chromosome partitioning motor FtsK, again supporting a role for its function in DNA synthesis and partitioning.[212] In eukaryotes, topo II has been found to associate with 14-3-3ε,[213] condensin,[214] the BRCT-domain protein TopBP1,[215,216] the Orc2 subunit of the Origin Recognition Complex,[217] the Lim15/Dmc1 recombinase, and Rad51;[218] in the latter instance, the interaction between the recombinase and topoisomerase may facilitate pairing of homologous chromosomes during meiosis. Moreover, there is a growing link between DNA breaks engendered by topo II and cellular or damage responses. For instance, double-stranded breaks caused by topoisomerase II have been associated with regulated transcription.[219] Topo II appears to be subject to control by SUMO-lation,[220,221] which regulates its function in centromeric cohesion and also plays a role in how the cell responds to drug-induced, topo II-mediated DNA lesions.[220,222,223] Recent work indicates that Tdp1, a protein used to hydrolyze 3′ tyrosine-DNA lesions produced by type

IB topos,[224] can also act on 5′ phosphotyrosine adducts produced topo II.[9] Altogether, this area of topoisomerase research is in its infancy, but growing rapidly.

10.6 Type IIB Topoisomerases

10.6.1 Overview

While type IIA topoisomerases have been found in all bacteria and eukaryotes, the first whole genome sequence of an archaeal organism (*M. jannaschii*) failed to reveal any obvious homologs of these proteins.[225] This finding was surprising, as a heterodimeric, ATP-dependent type II topoisomerase activity had been isolated previously from *S. sulfolobus*, a thermophilic archaeon.[226] The solution to this problem came with the discovery that the DNA cleavage subunit of this topoisomerase was not homologous to known type IIA topoisomerases, but to Spo11, a protein that generates double-strand DNA breaks to initiate meiotic recombination.[227,228] This distinction led to the classification of the archaeal enzyme as topoisomerase VI, and further demarcated topo VI as the founding member of a new group of enzymes termed type IIB topoisomerases. Since its initial isolation, topo VI has been found to be the principal type II topoisomerase of all archaeal organisms (although a few archaeons do retain DNA gyrase). Topo VI orthologs also have also been identified in plants, where they are required for endoreplication, a process by which a cell's genomic content is amplified in the absence of cell division to promote cell growth.[229,230]

10.6.2 Structure

Despite the lack of overall sequence similarity between type IIA and IIB topos, topo VI turns out to be an interesting amalgam of folds found in its type IIA counterparts (Figure 10.13A). For example, the DNA binding and cleavage subunit contains a metal-binding TOPRIM fold, along with a WHD that bears the active site tyrosine.[28,89,227,231] The ATP-binding subunit likewise contains a GHKL and transducer domain, as well as a helix-two-turn helix (H2TH) fold of unknown function that is not found in other topos.[206,232] Interestingly, the relative organization of these elements is quite distinct from that of type IIA topos; for example, the WHD domain lies N-terminal to the TOPRIM fold, while the GHL ATPase and TOPRIM modules reside on different subunits (Figure 10.13A).

The individual subunits of topo VI also appear to lack the large internal holes seen in the ATPase and DNA binding and cleavage regions of DNA gyrase and topo II (compare Figures 10.9 and 10.13).[129,131,133,139,231,232] Indeed, the structure of the isolated topo VI DNA-cleavage subunit reveals that the protein actually forms a "U"-shaped dimer, with a 20 Å wide, positively-charged channel running the length of the molecule.[231] The active site tyrosine of each protomer lies next to the TOPRIM fold of its partner subunit, forming

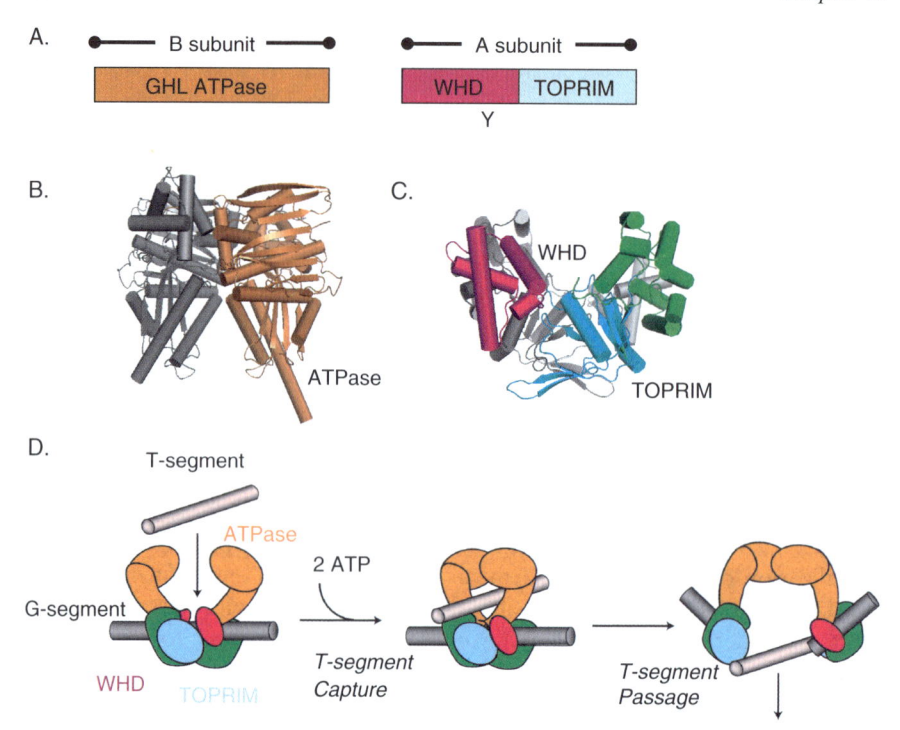

Figure 10.13 Type IIB topoisomerase structure and mechanism. (*A*) Domain organization of subunits. (*B*) ATPase domain of *S. shibatae* topoisomerase VI B subunit (PDB code 1MVS). (*C*) DNA cleavage domain of topoisomerase VI A subunit (PDB code 1D3Y). (*D*) Proposed type IIB topoisomerase mechanism.

two walls of the channel; the floor of this region is constructed from the dimer interface between the two subunits.

10.6.3 Mechanism

Although a full-length structure of topo VI has yet to be determined, pull-down data indicate that the C-terminus of the ATPase subunit binds to the N-terminus of the DNA cleavage protomer.[226] Structural and mutagenesis analyses further suggest that the channel in the DNA binding and cleavage subunits serves as the primary interaction site for DNA.[233–235] Studies thus far have shown that ATP binding dimerizes the GHKL domains of topo VI,[232] just as with type IIA topos, and that this event is required for DNA cleavage and strand passage.[234] DNA cleavage is dependent on divalent metals,[227] and occurs by the formation of 5′ phosphotyrosine intermediates at sites of scission staggered 2 bp apart;[234] unlike type IA and IIA topos, Mg^{2+} ions have been observed bound directly to the TOPRIM fold of topo VI.[231]

The overall similarities in ATPase domain structure and function, together with structural data indicating that the cleavage DNA will sit on a dimer interface, strongly suggest that type IIB topos, like their type IIA cousins, will employ a two-gate mechanism for strand passage[231,232] (Figure 10.13D). In this regard, a G-segment is first thought to bind to the DNA cleavage subunits, followed by binding of a T-segment to the holoenzyme. ATP binding triggers B-subunit dimerization, helping to stably capture the bound T-segment; DNA cleavage, which is strongly coupled to ATP binding,[233] likely takes place concomitantly. Once cut, the G-segment would be separated by opening of the DNA cleavage subunits, generating a gap to permit passage and expulsion of the T-segment. Interestingly, structural studies have delineated the full ATPase cycle of topo VI, providing evidence that a major conformational change between the GHKL and transducer domains takes place during the transition from an ADP•Pi to ADP-bound state.[236] This finding suggests that, as with topo II, phosphate release may be tied to structural rearrangements used to help facilitate G-segment opening, and T-segment transport though the DNA cleavage subunit interface.

10.7 Conclusions

Topoisomerases are critical enzymes for the cell. The various DNA strand manipulations catalyzed by these multifunctional proteins are essential for the formation and release of superhelical strain, and for resolving DNA tangles that arise from replication and repair processes. A wealth of structural, biochemical, and single-molecule studies have been instrumental in elucidating the physical basis for substrate recognition and topoisomerase mechanism. In addition to providing fundamental biophysical insights into the chemo-mechanical workings of complex molecular machines, studies of DNA topoisomerases also have long-standing implications for the development of antibacterial and antitumorogenic therapies. Despite these advances, however, there remain significant gaps in our understanding of topoisomerase biology, particularly with respect to the conformational dynamics accessed during various topological transactions, and regarding how topoisomerases are regulated in the cell. These outstanding issues, coupled with the ever-growing need for new pharmaceuticals, are signs that topoisomerases will continue to be a rich research field for the foreseeable future.

Acknowledgements

The authors thank members of the Berger lab and the general topoisomerase community for a great deal of stimulating discussion over the years. This work has been supported by the National Cancer Institute of the NIH (CA077373).

References

1. J.D. Watson and F.H. Crick, *Cold Spring Harb. Symp. Quant. Biol.*, 1953, **18**, 123.
2. J.C. Wang, *J. Mol. Biol.*, 1971, **55**, 523.

3. M. Gellert, K. Mizuuchi, M.H. O'Dea and H.A. Nash, *Proc. Natl. Acad. Sci. U.S.A.*, 1976, **73**, 3872.
4. J.J. Champoux and R. Dulbecco, *Proc. Natl. Acad. Sci. U.S.A.*, 1972, **69**, 143.
5. A.D. Bates and A. Maxwell, *DNA Topology*, 2nd edn., Oxford University Press, New York, 2005.
6. F.H. Crick, *Proc. Natl. Acad. Sci. U.S.A.*, 1976, **73**, 2639.
7. J.J. Pouliot, K.C. Yao, C.A. Robertson and H.A. Nash, *Science*, 1999, **286**, 552.
8. S.W. Yang, A.B. Burgin, Jr., B.N. Huizenga, C.A. Robertson, K.C. Yao and H.A. Nash, *Proc. Natl. Acad. Sci. U.S.A.*, 1996, **93**, 11534.
9. K.C. Nitiss, M. Malik, X. He, S.W. White and J.L. Nitiss, *Proc. Natl. Acad. Sci. U.S.A.*, 2006, **103**, 8953.
10. T.J. Cheng, P.G. Rey, T. Poon and C.C. Kan, *Eur. J. Biochem.*, 2002, **269**, 3697.
11. W.A. Denny, *Expert Opin. Emerg. Drugs*, 2004, **9**, 105.
12. Y. Pommier, *Nat. Rev. Cancer*, 2006, **6**, 789.
13. (a) E.L. Baldwin, N. Osheroff *Curr. Med. Chem. Anticancer Agents*, 2005, **5**, 363; (b) "Quinoloue Antimicrobial Agents," 3rd Ed., D.C. Hooper and E. Rubinstein (ed.), ASM Press, Washington, DC, 2003.
14. R.E. Depew, L.F. Liu and J.C. Wang, *J. Biol. Chem.*, 1978, **253**, 511.
15. D.E. Pulleyblank, M. Shure, D. Tang, J. Vinograd and H.P. Vosberg, *Proc. Natl. Acad. Sci. U.S.A.*, 1975, **72**, 4280.
16. P.L. Domanico and Y.C. Tse-Dinh, *J. Inorg. Biochem.*, 1991, **42**, 87.
17. Y.C. Tse-Dinh, *J. Biol. Chem.*, 1986, **261**, 10931.
18. C.X. Zhu, C.J. Roche and Y.C. Tse-Dinh, *J. Biol. Chem.*, 1997, **272**, 16206.
19. C.X. Zhu and Y.C. Tse-Dinh, *J. Biol. Chem.*, 2000, **275**, 5318.
20. Y.-C. Tse, K. Kirkegaard and J.C. Wang, *J. Biol. Chem.*, 1980, **255**, 5560.
21. Y. Tse and J.C. Wang, *Cell*, 1980, **22**, 269.
22. P.O. Brown and N.R. Cozzarelli, *Proc. Natl. Acad. Sci. U.S.A.*, 1981, **78**, 843.
23. A. Changela, R.J. DiGate and A. Mondragon, *Nature*, 2001, **411**, 1077.
24. C.D. Lima, J.C. Wang and A. Mondragon, *Nature*, 1994, **367**, 138.
25. N.H. Dekker, T. Viard, C.B. de La Tour, M. Duguet, D. Bensimon and V. Croquette, *J. Mol. Biol.*, 2003, **329**, 271.
26. Z. Li, A. Mondragon and R.J. DiGate, *Mol. Cell*, 2001, **7**, 301.
27. A. Mondragon and R. DiGate, *Struct. Fold Des.*, 1999, **7**, 1373.
28. L. Aravind, D.D. Leipe and E.V. Koonin, *Nucleic Acids Res.*, 1998, **26**, 4205.
29. K.S. Gajiwala and S.K. Burley, *Curr. Opin. Struct. Biol.*, 2000, **10**, 110.
30. J.L. Huffman and R.G. Brennan, *Curr. Opin. Struct. Biol.*, 2002, **12**, 98.
31. D.B. McKay and T.A. Steitz, *Nature*, 1981, **290**, 744.
32. J.M. Berger, D. Fass, J.C. Wang and S.C. Harrison, *Proc. Natl. Acad. Sci. U.S.A.*, 1998, **95**, 7876.

33. M.D. Been, R.R. Burgess and J.J. Champoux, *Nucleic Acids Res.*, 1984, **12**, 3097.
34. B. Thomsen, S. Mollerup, B.J. Bonven, R. Frank, H. Blocker, O.F. Nielsen and O. Westergaard, *EMBO J.*, 1987, **6**, 1817.
35. K.A. Edwards, B.D. Halligan, J.L. Davis, N.L. Nivera and L.F. Liu, *Nucleic Acids Res.*, 1982, **10**, 2565.
36. A. Changela, R.J. Digate and A. Mondragon, *J. Mol. Biol.*, 2007, **368**, 105.
37. H. Feinberg, C.D. Lima and A. Mondragon, *Nat. Struct. Biol.*, 1999 **6**, 918.
38. L. Yu, C.X. Zhu, Y.C. Tse-Dinh and S.W. Fesik, *Biochemistry*, 1995, **34**, 7622.
39. L. Yu, C.X. Zhu, Y.C. Tse-Dinh and S.W. Fesik, *Biochemistry*, 1996, **35**, 9661.
40. N.V. Grishin, *J. Mol. Biol.*, 2000, **299**, 1165.
41. G. Hansen, A. Harrenga, B. Wieland, D. Schomburg and P. Reinemer, *J. Mol. Biol.*, 2006, **358**, 1328.
42. Y.-C. Tse-Dinh and R.K. Beran-Steed, *J. Biol. Chem.*, 1988, **263**, 15857.
43. A. Ahumada and Y.C. Tse-Dinh, *Biochem. Biophys. Res. Commun.*, 1998, **251**, 509.
44. A. Ahumada and Y.C. Tse-Dinh, *BMC Biochem.*, 2002, **3**, 13.
45. K.S. Srivenugopal, D. Lockshon and D.R. Morris, *Biochemistry*, 1984, **23**, 1899.
46. M. Duguet, M.C. Serre and C. Bouthier de La Tour, *J. Mol. Biol.*, 2006, **359**, 805.
47. Y. Wang, A.S. Lynch, S.J. Chen and J.C. Wang, *J. Biol. Chem.*, 2002, **277**, 1203.
48. J.W. Wallis, G. Chrebet, G. Brodsky, M. Rolfe and R. Rothstein, *Cell*, 1989, **58**, 409.
49. R.J. DiGate and K.J. Marians, *J. Biol. Chem.*, 1988, **263**, 13366.
50. H. Hiasa and K.J. Marians, *J. Biol. Chem.*, 1994, **269**, 32655.
51. G.J. Pruss, S.H. Manes and K. Drlica, *Cell*, 1982, **31**, 35.
52. A.M. Bailis, L. Arthur and R. Rothstein, *Mol. Cell. Biol.*, 1992, **12**, 4988.
53. H. Hiasa, R.J. DiGate and K.J. Marians, *J. Biol. Chem.*, 1994, **269**, 2093.
54. P. Nurse, C. Levine, H. Hassing and K.J. Marians, *J. Biol. Chem.*, 2003, **278**, 8653.
55. T.J. Oakley, A. Goodwin, R.K. Chakraverty and I.D. Hickson, *DNA Repair (Amst)*, 2002, **1**, 463.
56. M. Duguet, *J. Cell Sci.*, 1997, **110**(Pt 12), 1345.
57. F.G. Harmon, R.J. DiGate and S.C. Kowalczykowski, *Mol. Cell*, 1999 **3**, 611.
58. J.L. Plank, J. Wu and T.S. Hsieh, *Proc. Natl. Acad. Sci. U.S.A.*, 2006, **103**, 11118.
59. L. Wu and I.D. Hickson, *Nucleic Acids Res.*, 2002, **30**, 4823.
60. M. Seki, T. Nakagawa, T. Seki, G. Kato, S. Tada, Y. Takahashi, A. Yoshimura, T. Kobayashi, A. Aoki, M. Otsuki, F.A. Habermann, H. Tanabe, Y. Ishii and T. Enomoto, *Mol. Cell. Biol.*, 2006, **26**, 6299.

61. A. Ui, M. Seki, H. Ogiwara, R. Onodera, S. Fukushige, F. Onoda and T. Enomoto, *DNA Repair (Amst)*, 2005, **4**, 191.
62. L. Wu and I.D. Hickson, *Nature*, 2003, **426**, 870.
63. H.J. Tsai, W.H. Huang, T.K. Li, Y.L. Tsai, K.J. Wu, S.F. Tseng and S.C. Teng, *J. Biol. Chem.*, 2006, **281**, 13717.
64. K.Y. Kwan and J.C. Wang, *Proc. Natl. Acad. Sci. U.S.A.*, 2001, **98**, 5717.
65. J.A. Cobb and L. Bjergbaek, *Nucleic Acids Res.*, 2006, **34**, 4106.
66. M.P. Killoran and J.L. Keck, *Nucleic Acids Res.*, 2006, **34**, 4098.
67. S. Sharma, K.M. Doherty and R.M. Brosh, Jr., *Biochem. J.*, 2006 **398**, 319.
68. A. Kikuchi and K. Asai, *Nature*, 1984, **309**, 677.
69. C. Bouthier de la Tour, C. Portemer, R. Huber, P. Forterre and M. Duguet, *J. Bacteriol.*, 1991, **173**, 3921.
70. C. Bouthier de la Tour, C. Portemer, M. Nadal, K.O. Stetter, P. Forterre and M. Duguet, *J. Bacteriol.*, 1990, **172**, 6803.
71. F. Confalonieri, C. Elie, M. Nadal, C. de La Tour, P. Forterre and M. Duguet, *Proc. Natl. Acad. Sci. U.S.A.*, 1993, **90**, 4753.
72. A.C. Declais, J. Marsault, F. Confalonieri, C.B. de La Tour and M. Duguet, *J. Biol. Chem.*, 2000, **275**, 19498.
73. A.C. Rodriguez and D. Stock, *EMBO J.*, 2002, **21**, 418.
74. A.C. Rodriguez, *Biochemistry*, 2003, **42**, 5993.
75. T.S. Hsieh and C. Capp, *J. Biol. Chem.*, 2005, **280**, 20467.
76. T.S. Hsieh and J.L. Plank, *J. Biol. Chem.*, 2006, **281**, 5640.
77. J.J. Champoux, *Proc. Natl. Acad. Sci. U.S.A.*, 1977, **74**, 3800.
78. M.D. Been and J.J. Champoux, *Nucleic Acids Res.*, 1980, **8**, 6129.
79. M.D. Been and J.J. Champoux, *Proc. Natl. Acad. Sci. U.S.A.*, 1981, **78**, 2883.
80. B.D. Halligan, J.L. Davis, K.A. Edwards and L.F. Liu, *J. Biol. Chem.*, 1982, **257**, 3995.
81. W.S. Dynan, J.J. Jendrisak, D.A. Hager and R.R. Burgess, *J. Biol. Chem.*, 1981, **256**, 5860.
82. R. Shaffer and P. Traktman, *J. Biol. Chem.*, 1987, **262**, 9309.
83. P. Caron and J.C. Wang, in *Advances in Pharmacology*, ed. L.F. Liu, Academic Press, San Diego, 1994, pp. 271.
84. C. Cheng, P. Kussie, N. Pavletich and S. Shuman, *Cell*, 1998, **92**, 841.
85. M.R. Redinbo, L. Stewart, P. Kuhn, J.J. Champoux and W.G. Hol, *Science*, 1998, **279**, 1504.
86. L. Stewart, M.R. Redinbo, X. Qiu, W.G. Hol and J.J. Champoux, *Science*, 1998, **279**, 1534.
87. J.C. Wang, *Annu. Rev. Biochem.*, 1996, **65**, 635.
88. B.O. Krogh and S. Shuman, *Proc. Natl. Acad. Sci. U.S.A.*, 2002, **99**, 1853.
89. N.V. Grishin, *Nucleic Acids Res.*, 2000, **28**, 2229.
90. W.K. Gillette, S. Rhee, J.L. Rosner and R.G. Martin, *Mol. Microbiol.*, 2000, **35**, 1582.
91. S. Shuman and J. Prescott, *J. Biol. Chem.*, 1990, **265**, 17826.
92. A. Sharma, R. Hanai and A. Mondragon, *Structure*, 1994, **2**, 767.

93. K. Perry, Y. Hwang, F.D. Bushman and G.D. Van Duyne, *Mol. Cell*, 2006, **23**, 343.
94. J.F. Carey, S.J. Schultz, L. Sisson, T.G. Fazzio and J.J. Champoux, *Proc. Natl. Acad. Sci. U.S.A.*, 2003, **100**, 5640.
95. M.H. Woo, C. Losasso, H. Guo, L. Pattarello, P. Benedetti and M.A. Bjornsti, *Proc. Natl. Acad. Sci. U.S.A.*, 2003, **100**, 13767.
96. A. Patel, S. Shuman and A. Mondragon, *J. Biol. Chem.*, 2006, **281**, 6030.
97. D.R. Davies, A. Mushtaq, H. Interthal, J.J. Champoux and W.G. Hol, *J. Mol. Biol.*, 2006, **357**, 1202.
98. B.O. Krogh and S. Shuman, *Mol. Cell*, 2000, **5**, 1035.
99. B.O. Krogh and S. Shuman, *J. Biol. Chem.*, 2002, **277**, 5711.
100. L. Tian, C.D. Claeboe, S.M. Hecht and S. Shuman, *Structure*, 2004 **12**, 31.
101. R. Nagarajan, K. Kwon, B. Nawrot, W.J. Stec and J.T. Stivers, *Biochemistry*, 2005, **44**, 11476.
102. L. Tian, C.D. Claeboe, S.M. Hecht and S. Shuman, *Mol. Cell*, 2003 **12**, 199.
103. D.S. Horowitz and J.C. Wang, *J. Mol. Biol.*, 1984, **173**, 75.
104. J.T. Stivers, T.K. Harris and A.S. Mildvan, *Biochemistry*, 1997, **36**, 5212.
105. M.R. Redinbo, L. Stewart, J.J. Champoux and W.G. Hol, *J. Mol. Biol.*, 1999, **292**, 685.
106. D.A. Koster, V. Croquette, C. Dekker, S. Shuman and N.H. Dekker, *Nature*, 2005, **434**, 671.
107. A.I. Slesarev, K.O. Stetter, J.A. Lake, M. Gellert, R. Krah and S.A. Kozyavkin, *Nature*, 1993, **364**, 735.
108. G.I. Belova, R. Prasad, S.A. Kozyavkin, J.A. Lake, S.H. Wilson and A.I. Slesarev, *Proc. Natl. Acad. Sci. U.S.A.*, 2001, **98**, 6015.
109. S.A. Kozyavkin, A.V. Pushkin, F.A. Eiserling, K.O. Stetter, J.A. Lake and A.I. Slesarev, *J. Biol. Chem.*, 1995, **270**, 13593.
110. B. Taneja, A. Patel, A. Slesarev and A. Mondragon, *EMBO J.*, 2006 **25**, 398.
111. P. Forterre, *Trends Biotechnol.*, 2006, **24**, 245.
112. L.F. Liu, C.C. Liu and B.M. Alberts, *Cell*, 1980, **19**, 697.
113. P.O. Brown and N.R. Cozzarelli, *Science*, 1979, **206**, 1081.
114. R. Lynn, G. Giaever, S.L. Swanberg and J.C. Wang, *Science*, 1986 **233**, 647.
115. T. Goto and J.C. Wang, *J. Biol. Chem.*, 1982, **257**, 5866.
116. K.G. Miller, L.F. Liu and P.T. Englund, *J. Biol. Chem.*, 1981, **256**, 9334.
117. L. Klevan and J.C. Wang, *Biochemistry*, 1980, **19**, 5229.
118. J. Kato, Y. Nishimura, R. Imamura, H. Niki, S. Hiraga and H. Suzuki, *Cell*, 1990, **63**, 393.
119. L.F. Liu, C.C. Liu and B.M. Alberts, *Nature*, 1979, **281**, 456.
120. M.K. Wall, L.A. Mitchenall and A. Maxwell, *Proc. Natl. Acad. Sci. U.S.A.*, 2004, **101**, 7821.
121. R. Dutta and M. Inouye, *Trends Biochem. Sci.*, 2000, **25**, 24.

122. N. Adachi, M. Miyaike, S. Kato, R. Kanamaru, H. Koyama and A. Kikuchi, *Nucleic Acids Res.*, 1997, **25**, 3135.
123. P.R. Caron, P. Watt and J.C. Wang, *Mol. Cell Biol.*, 1994, **14**, 3197.
124. D.S. Horowitz and J.C. Wang, *J. Biol. Chem.*, 1987, **262**, 5339.
125. S.C. Kampranis and A. Maxwell, *Proc. Natl. Acad. Sci. U.S.A.*, 1996, **93**, 14416.
126. P.O. Brown, C.L. Peebles and N.R. Cozzarelli, *Proc. Natl. Acad. Sci. U.S.A.*, 1979, **76**, 6110.
127. L.F. Liu and J.C. Wang, *Proc. Natl. Acad. Sci. U.S.A.*, 1978, **75**, 2098.
128. L.F. Liu and J.C. Wang, *Cell*, 1978, **15**, 979.
129. J.M. Berger, S.J. Gamblin, S.C. Harrison and J.C. Wang, *Nature*, 1996, **379**, 225.
130. A. Maxwell, *Nat. Struct. Biol.*, 1996, **3**, 109.
131. J.H. Morais Cabral, A.P. Jackson, C.V. Smith, N. Shikotra, A. Maxwell and R.C. Liddington, *Nature*, 1997, **388**, 903.
132. J. Roca, J.M. Berger, S.C. Harrison and J.C. Wang, *Proc. Natl. Acad. Sci. U.S.A.*, 1996, **93**, 4057.
133. D.B. Wigley, G.J. Davies, E.J. Dodson, A. Maxwell and G. Dodson, *Nature*, 1991, **351**, 624.
134. N. Osheroff, *J. Biol. Chem.*, 1986, **261**, 9944.
135. J. Roca and J.C. Wang, *Cell*, 1992, **71**, 833.
136. J. Roca and J.C. Wang, *Cell*, 1994, **77**, 609.
137. A.P. Jackson and A. Maxwell, *Proc. Natl. Acad. Sci. U.S.A.*, 1993, **90**, 11232.
138. S. Bellon, J.D. Parsons, Y. Wei, K. Hayakawa, L.L. Swenson, P.S. Charifson, J.A. Lippke, R. Aldape and C.H. Gross, *Antimicrob. Agents Chemother.*, 2004, **48**, 1856.
139. S. Classen, S. Olland and J.M. Berger, *Proc. Natl. Acad. Sci. U.S.A.*, 2003, **100**, 10629.
140. H. Wei, A.J. Ruthenburg, S.K. Bechis and G.L. Verdine, *J. Biol. Chem.*, 2005, **280**, 37041.
141. A.P. Jackson, A. Maxwell and D.B. Wigley, *J. Mol. Biol.*, 1991, **217**, 15.
142. J.E. Lindsley and J.C. Wang, *Nature*, 1993, **361**, 749.
143. S. Olland and J.C. Wang, *J. Biol. Chem.*, 1999, **274**, 21688.
144. T.R. Hammonds and A. Maxwell, *J. Biol. Chem.*, 1997, **272**, 32696.
145. V.H. Oestergaard, L. Bjergbaek, C. Skouboe, L. Giangiacomo, B.R. Knudsen and A.H. Andersen, *J. Biol. Chem.*, 2004, **279**, 1684.
146. A.P. Tingey and A. Maxwell, *Nucleic Acids Res.*, 1996, **24**, 4868.
147. R. Hanai and J.C. Wang, *Proc. Natl. Acad. Sci. U.S.A.*, 1994, **91**, 11904.
148. W. Li and J.C. Wang, *J. Biol. Chem.*, 1997, **272**, 31190.
149. K.D. Corbett, A.J. Schoeffler, N.D. Thomsen and J.M. Berger, *J. Mol. Biol.*, 2005, **351**, 545.
150. D. Fass, C.E. Bogden and J.M. Berger, *Nat. Struct. Biol.*, 1999, **6**, 322.
151. M. Gellert, L.M. Fisher and M.H. O'Dea, *Proc. Natl. Acad. Sci. U.S.A.*, 1979, **76**, 6289.

152. A. Sugino, N.P. Higgins and N.R. Cozzarelli, *Nucleic Acids Res.*, 1980, **8**, 3865.
153. C.G. Noble and A. Maxwell, *J. Mol. Biol.*, 2002, **318**, 361.
154. K.L. West, E.L. Meczes, R. Thorn, R.M. Turnbull, R. Marshall and C.A. Austin, *Biochemistry*, 2000, **39**, 1223.
155. Q. Liu and J.C. Wang, *Proc. Natl. Acad. Sci. U.S.A.*, 1999, **96**, 881.
156. P. Benedetti, A. Silvestri, P. Fiorani and J.C. Wang, *J. Biol. Chem.*, 1997, **272**, 12132.
157. T. Kirchhausen, J.C. Wang and S.C. Harrison, *Cell*, 1985, **41**, 933.
158. P. Schultz, S. Olland, P. Oudet and R. Hancock, *Proc. Natl. Acad. Sci. U.S.A.*, 1996, **93**, 5936.
159. M.T. Howard, M.P. Lee, T.S. Hsieh and J.D. Griffith, *J. Mol. Biol.*, 1991, **217**, 53.
160. A.V. Vologodskii, W. Zhang, V.V. Rybenkov, A.A. Podtelezhnikov, D. Subramanian, J.D. Griffith and N.R. Cozzarelli, *Proc. Natl. Acad. Sci. U.S.A.*, 2001, **98**, 3045.
161. E.L. Zechiedrich and N. Osheroff, *EMBO J.*, 1990, **9**, 4555.
162. K. Kirkegaard and J.C. Wang, *Cell*, 1981, **23**, 721.
163. R.J. Reece and A. Maxwell, *Nucleic Acids Res.*, 1991, **19**, 1399.
164. S. Trigueros, J. Salceda, I. Bermudez, X. Fernandez and J. Roca, *J. Mol. Biol.*, 2004, **335**, 723.
165. G. Orphanides and A. Maxwell, *Nucleic Acids Res.*, 1994, **22**, 1567.
166. N.L. Williams and A. Maxwell, *Biochemistry*, 1999, **38**, 13502.
167. M.P. Lee, M. Sander and T. Hsieh, *J. Biol. Chem.*, 1989, **264**, 21779.
168. H. Peng and K.J. Marians, *J. Biol. Chem.*, 1995, **270**, 25286.
169. K. Lund, A.H. Andersen, K. Christiansen, J.Q. Svejstrup and O. Westergaard, *J. Biol. Chem.*, 1990, **265**, 13856.
170. B. Thomsen, C. Bendixen, K. Lund, A.H. Andersen, B.S. Sorensen and O. Westergaard, *J. Mol. Biol.*, 1990, **215**, 237.
171. A.H. Andersen, K. Christiansen, E.L. Zechiedrich, P.S. Jensen, N. Osheroff and O. Westergaard, *Biochemistry*, 1989, **28**, 6237.
172. S.J. Froelich-Ammon, K.C. Gale and N. Osheroff, *J. Biol. Chem.*, 1994, **269**, 7719.
173. M. Sander and T.S. Hsieh, *Nucleic Acids Res.*, 1985, **13**, 1057.
174. A. Udvardy and P. Schedl, *Mol. Cell Biol.*, 1991, **11**, 4973.
175. L.F. Liu, T.C. Rowe, L. Yang, K.M. Tewey and G.L. Chen, *J. Biol. Chem.*, 1983, **258**, 15365.
176. M. Sander and T. Hsieh, *J. Biol. Chem.*, 1983, **258**, 8421.
177. T. Goto, P. Laipis and J.C. Wang, *J. Biol. Chem.*, 1984, **259**, 10422.
178. N. Osheroff, *Biochemistry*, 1987, **26**, 6402.
179. N. Osheroff and E.L. Zechiedrich, *Biochemistry*, 1987, **26**, 4303.
180. C. Sissi, E. Marangon, A. Chemello, C.G. Noble, A. Maxwell and M. Palumbo, *J. Mol. Biol.*, 2005, **353**, 1152.
181. N.L. Williams, A.J. Howells and A. Maxwell, *J. Mol. Biol.*, 2001, **306**, 969.
182. N.L. Williams and A. Maxwell, *Biochemistry*, 1999, **38**, 14157.
183. J.K. Tamura and M. Gellert, *J. Biol. Chem.*, 1990, **265**, 21342.

184. S.C. Kampranis and A. Maxwell, *J. Biol. Chem.*, 1998, **273**, 26305.
185. J.E. Lindsley and J.C. Wang, *J. Biol. Chem.*, 1993, **268**, 8096.
186. C. Skouboe, L. Bjergbaek, V.H. Oestergaard, M.K. Larsen, B.R. Knudsen and A.H. Andersen, *J. Biol. Chem.*, 2003, **278**, 5768.
187. C.L. Baird, T.T. Harkins, S.K. Morris and J.E. Lindsley, *Proc. Natl. Acad. Sci. U.S.A.*, 1999, **96**, 13685.
188. T.T. Harkins, T.J. Lewis and J.E. Lindsley, *Biochemistry*, 1998, **37**, 7299.
189. C.L. Baird, M.S. Gordon, D.M. Andrenyak, J.F. Marecek and J.E. Lindsley, *J. Biol. Chem.*, 2001, **276**, 27893.
190. O.V. Lavrukhin, J.M. Fortune, T.G. Wood, D.E. Burbank, J.L. Van Etten, N. Osheroff and R.S. Lloyd, *J. Biol. Chem.*, 2000, **275**, 6915.
191. K.D. Corbett, R.K. Shultzaberger and J.M. Berger, *Proc. Natl. Acad. Sci. U.S.A.*, 2004, **101**, 7293.
192. A.J. Ruthenburg, D.M. Graybosch, J.C. Huetsch and G.L. Verdine, *J. Biol. Chem.*, 2005, **280**, 26177.
193. T.J. Hsieh, L. Farh, W.M. Huang and N.L. Chan, *J. Biol. Chem.*, 2004, **279**, 55587.
194. W.M. Huang, *Annu. Rev. Genet.*, 1996, **30**, 79.
195. D. Ward and A. Newton, *Mol. Microbiol.*, 1997, **26**, 897.
196. V.M. Kramlinger and H. Hiasa, *J. Biol. Chem.*, 2006, **281**, 3738.
197. D.E. Adams, E.M. Shekhtman, E.L. Zechiedrich, M.B. Schmid and N.R. Cozzarelli, *Cell*, 1992, **71**, 277.
198. C. Ullsperger and N.R. Cozzarelli, *J. Biol. Chem.*, 1996, **271**, 31549.
199. R.W. Deibler, S. Rahmati and E.L. Zechiedrich, *Genes Dev.*, 2001**15**, 748.
200. H. Hiasa and K.J. Marians, *J. Biol. Chem.*, 1996, **271**, 21529.
201. E.L. Zechiedrich, A.B. Khodursky and N.R. Cozzarelli, *Genes Dev.*, 1997, **11**, 2580.
202. M.D. Stone, Z. Bryant, N.J. Crisona, S.B. Smith, A. Vologodskii, C. Bustamante and N.R. Cozzarelli, *Proc. Natl. Acad. Sci. U.S.A.*, 2003, **100**, 8654.
203. L. Costenaro, J.G. Grossmann, C. Ebel and A. Maxwell, *Structure*, 2005, **13**, 287.
204. J.S. Dickey and N. Osheroff, *Biochemistry*, 2005, **44**, 11546.
205. A. Sakaguchi, T. Akashi and A. Kikuchi, *Biochem. Biophys. Res. Commun.*, 2001, **283**, 876.
206. J. Roca and J.C. Wang, *Genes Cells*, 1996, **1**, 17.
207. G. Charvin, D. Bensimon and V. Croquette, *Proc. Natl. Acad. Sci. U.S.A.*, 2003, **100**, 9820.
208. A.K. McClendon, J.S. Dickey and N. Osheroff, *Biochemistry*, 2006, **45**, 11674.
209. E. Goldstein and K. Drlica, *Proc. Natl. Acad. Sci. U.S.A.*, 1984, **81**, 4046.
210. J. Salceda, X. Fernandez and J. Roca, *EMBO J.*, 2006, **25**, 2575.
211. E.L. Zechiedrich and N.R. Cozzarelli, *Genes Dev.*, 1995, **9**, 2859.
212. O. Espeli, C. Lee and K.J. Marians, *J. Biol. Chem.*, 2003, **278**, 44639.
213. E.U. Kurz, K.B. Leader, D.J. Kroll, M. Clark and F. Gieseler, *J. Biol. Chem.*, 2000, **275**, 13948.

214. O. Cuvier and T. Hirano, *J. Cell Biol.*, 2003, **160**, 645.
215. K. Yamane and T. Tsuruo, *Oncogene*, 1999, **18**, 5194.
216. K. Yamane, X. Wu and J. Chen, *Mol. Cell Biol.*, 2002, **22**, 555.
217. G. Abdurashidova, S. Radulescu, O. Sandoval, S. Zahariev, M.B. Danailov, A. Demidovich, L. Santamaria, G. Biamonti, S. Riva and A. Falaschi, *EMBO J.*, 2007, **26**, 998.
218. K. Iwabata, A. Koshiyama, T. Yamaguchi, H. Sugawara, F.N. Hamada, S.H. Namekawa, S. Ishii, T. Ishizaki, H. Chiku, T. Nara and K. Sakaguchi, *Nucleic Acids Res.*, 2005, **33**, 5809.
219. B.G. Ju, V.V. Lunyak, V. Perissi, I. Garcia-Bassets, D.W. Rose, C.K. Glass and M.G. Rosenfeld, *Science*, 2006, **312**, 1798.
220. Y. Mao, S.D. Desai and L.F. Liu, *J. Biol. Chem.*, 2000, **275**, 26066.
221. Y. Takahashi, V. Yong-Gonzalez, Y. Kikuchi and A. Strunnikov, *Genetics*, 2006, **172**, 783.
222. R.Y. Huang, D. Kowalski, H. Minderman, N. Gandhi and E.S. Johnson, *Cancer Res.*, 2007, **67**, 765.
223. S. Isik, K. Sano, K. Tsutsui, M. Seki, T. Enomoto, H. Saitoh and K. Tsutsui, *FEBS Lett.*, 2003, **546**, 374.
224. H. Interthal, J.J. Pouliot and J.J. Champoux, *Proc. Natl. Acad. Sci. U.S.A.*, 2001, **98**, 12009.
225. C.J. Bult, O. White, G.J. Olsen, L. Zhou, R.D. Fleischmann, G.G. Sutton, J.A. Blake, L.M. FitzGerald, R.A. Clayton, J.D. Gocayne, A.R. Kerlavage, B.A. Dougherty, J.F. Tomb, M.D. Adams, C.I. Reich, R. Overbeek, E.F. Kirkness, K.G. Weinstock, J.M. Merrick, A. Glodek, J.L. Scott, N.S. Geoghagen and J.C. Venter, *Science*, 1996, **273**, 1058.
226. A. Bergerat, D. Gadelle and P. Forterre, *J. Biol. Chem.*, 1994, **269**, 27663.
227. A. Bergerat, B. de Massy, D. Gadelle, P.C. Varoutas, A. Nicolas and P. Forterre, *Nature*, 1997, **386**, 414.
228. S. Keeney, C.N. Giroux and N. Kleckner, *Cell*, 1997, **88**, 375.
229. F. Hartung, K.J. Angelis, A. Meister, I. Schubert, M. Melzer and H. Puchta, *Curr. Biol.*, 2002, **12**, 1787.
230. K. Sugimoto-Shirasu, N.J. Stacey, J. Corsar, K. Roberts and M.C. McCann, *Curr. Biol.*, 2002, **12**, 1782.
231. M.D. Nichols, K. DeAngelis, J.L. Keck and J.M. Berger, *EMBO J.*, 1999, **18**, 6177.
232. K.D. Corbett and J.M. Berger, *EMBO J.*, 2003, **22**, 151.
233. C. Buhler, D. Gadelle, P. Forterre, J.C. Wang and A. Bergerat, *Nucleic Acids Res.*, 1998, **26**, 5157.
234. C. Buhler, J.H. Lebbink, C. Bocs, R. Ladenstein and P. Forterre, *J. Biol. Chem.*, 2001, **276**, 37215.
235. R.L. Diaz, A.D. Alcid, J.M. Berger and S. Keeney, *Mol. Cell Biol.*, 2002, **22**, 1106.
236. K.D. Corbett and J.M. Berger, *Structure*, 2005, **13**, 873.
237. B. Taneja, B. Schnurr, A. Sleserev, J.F. Marko and A. Mondragon, *Proc. Natl. Acad. Sci. U.S.A.*, 2007, **104**, 14670.

A. Schematic view of a transposon

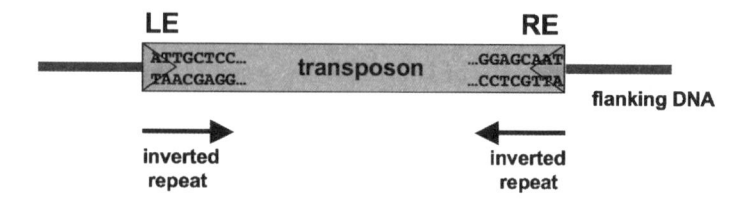

B. Classes of prokaryotic transposons

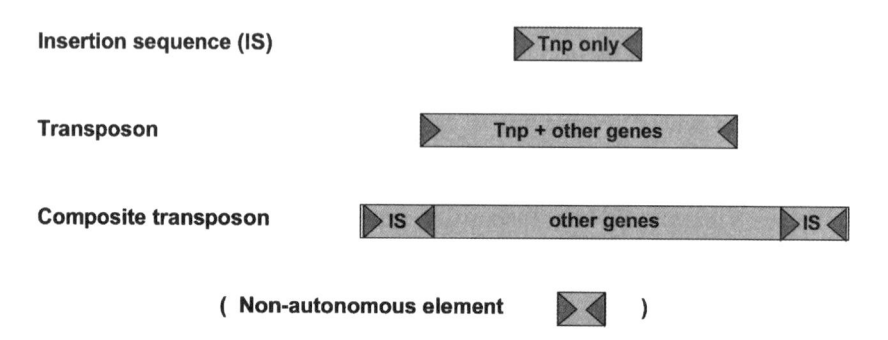

Insertion sequence (IS) — Tnp only

Transposon — Tnp + other genes

Composite transposon — IS ... other genes ... IS

(Non-autonomous element)

C. Examples of transposon end organization (not to scale)

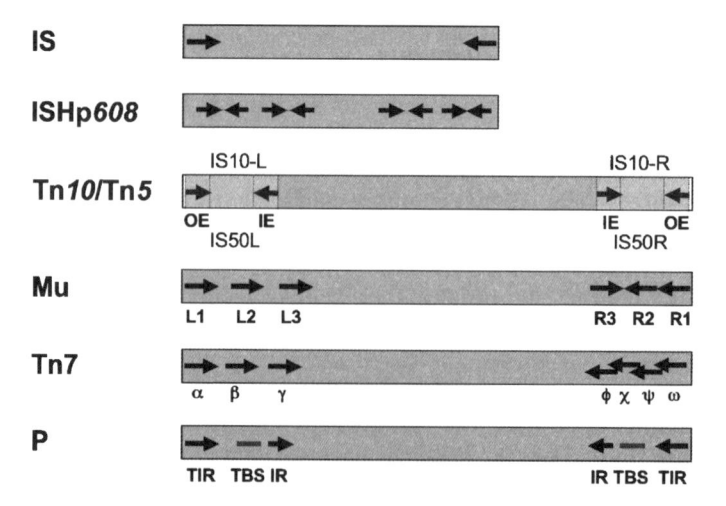

IS

ISHp*608*

Tn*10*/Tn*5*
  IS10-L   IS10-R
  OE IE   IE OE
  IS50L   IS50R

Mu
  L1 L2 L3   R3 R2 R1

Tn7
  α β γ   φ χ ψ ω

P
  TIR TBS IR   IR TBS TIR

Figure 11.1 Transposon organization. (*A*) Schematic view of a DNA transposon. At each transposon end, there are inverted repeats that are often symbolized by triangles or arrows. The genomic DNA on either side of the transposon is often called flanking DNA. (*B*) Types of prokaryotic transposons. (*C*) Variation in transposon end structures (not drawn to scale). The characteristic inverted (or direct) repeats for the various transposons discussed in the text are shown with arrows.

Simple transposons do not have embedded ISs but they can also encode for several proteins. For example, the 14-kb Tn7 transposon encodes not only the five proteins associated with its mobility and regulation but also three different antibiotic resistance determinants and an apparently inactive tyrosine recombinase.[9] Even larger than Tn7 is the 37-kb genome of the bacteriophage Mu that carries 56 genes[10] and is itself a highly regulated transposon.[11] Upon infection, the Mu transposase can catalyze the integration of the bacteriophage DNA into the genomes of many enteric bacteria.

11.1.1.2 Eukaryotic Elements

There is also great variation in the types of transposable elements found in eukaryotes. The known eukaryotic transposons can be divided into about a dozen superfamilies on the basis of sequence homology of their transposase genes and apparent transposition mechanism.[12] The prokaryotic terminology to classify them is not used and, instead, eukaryotic elements are usually referred to using their unique name or the name of the transposon superfamily to which they belong. Thus, the literature of these elements is enlivened by discussions of (for example) Harbingers, hobos, and Frog Princes. Using the convention for genes and gene products, the transposon name is often italicized whereas the transposase encoded by the transposon is not.

The genomes of many eukaryotes are littered with sequences that are transposon-derived. For example, about 45% of the human genome has been identified as having arisen from mobile elements.[13] Most of these elements are related to transposons that use an RNA intermediate, while sequences that are related to DNA transposition account for about 2–3% of the human genome. In many eukaryotic genomes, most of these sequences are no longer mobile because they have accumulated point mutations or deletions that render them immobile (or at least immobile when they are left to their own devices). Therefore, there are autonomous, non-autonomous, and immobile transposons. Autonomous transposons can move on their own as they still encode all the functions necessary for mobility. Non-autonomous ones have lost their ability to move on their own, but can be mobilized if a compatible transposase is available. Thus, *Ds* is a non-autonomous transposon because the *Ac* transposase can mobilize it. Immobile transposons are out of luck.

Given the great danger that untamed transposons present to their hosts, most exist in low copy numbers and/or their transposition is highly regulated. For example, transposition activity can be downregulated by methylation of transposon-containing DNA regions,[14] a mechanism known as silencing. In some organisms, DNA transposition is downregulated by RNA interference (RNAi).[15] Certain transposons have their own idiosyncratic regulation mechanisms. For example, the prokaryotic Tn5 transposon expresses an N-terminally truncated transposase that is a transdominant negative regulator of its own transposition.[16] Similarly, truncated and internally deleted versions of the *Drosophila* P element transposase repress transposition.[17]

Eukaryotic DNA transposons vary in length and are similar to prokaryotic ISs in the sense that they carry only those genes needed for transposition. The most widely distributed eukaryotic DNA transposon family is the Tc1/*mariner* superfamily, named after two of its well-studied members.[18] These transposons are generally 1300–2400 bp long and are found in protozoa, nematodes, arthropods, fish and humans; most are no longer active. However, it has been possible to reactivate a few by reversing inactivating mutations in the transposase gene. This is how the *Sleeping Beauty* transposon was created: careful examination and alignment of inactive sequences from salmonid fish species suggested which bases should be changed to reactivate the transposase.[19] Reconstructed, active *Sleeping Beauty* has proven to be a valuable tool for transgenesis and cancer gene identification as it can transpose in a range of human and other vertebrate cell types.[20]

Another widely distributed group of eukaryotic DNA transposons is the *hAT* superfamily which is named after three related transposons (*h*obo from *Drosophila melanogaster*, *A*c from maize, and *T*am3 from the snapdragon).[21–23] This family is widespread among higher eukaryotes yet completely absent from prokaryotes. The human genome harbors several *Charlie* elements that belong to the *hAT* family.[13] While *hAT* elements do not appear to directly benefit their hosts, they may have contributed to the evolution of the adaptive immune system, as the RAG1 component of the V(D)J recombinase shares a mechanistic and quite possibly a structural relationship with *hAT* transposases.[24–28]

We have not attempted to provide an exhaustive review of DNA transposons in the various kingdoms in life but rather to offer a sense of the immense diversity that exists. Clearly, some strategy is needed to sort them out.

11.2 Transposases and DNA

It is a general requirement of transposition that a transposase must be able to locate its transposon; in other words, a transposon must contain DNA sequences that are recognized and bound by its transposase. The transposon DNA that is recognized by a transposase is usually located at or very near to the ends of the transposon. This should not be a surprise as the transposase has to catalyze the cleavage of the phosphodiester DNA backbone at the ends of the transposon in order to liberate the transposon from its donor site. The two transposon ends are customarily designated the Left End (LE) and the Right End (RE) (Figure 11.1A). By convention, the LE is proximal to the transposase promoter with transcription proceeding from left to right.

11.2.1 Ends of Transposons

Most transposons contain inverted repeat sequences at their ends. These are similar sequences that are found in the opposite orientation at the LE and RE (Figure 11.1A). Schematically, they are often represented as arrows or triangles, and are sometimes called inverted terminal repeats (ITRs) or terminal inverted repeats (TIRs). The binding sites for the transposase are usually 10–40 bp long and are contained within or very near to these inverted repeats. The specific transposase binding sites usually do not extend to the very end of the transposon, a sensible arrangement that allows the transposase active site access to the transposon end where the cleavage reactions occur.

The sequences of the repeats at the two transposon ends need not be identical, in which case they are sometimes known as imperfect repeats. Some transposons have multiple inverted repeats at each end that can be arranged, depending on the system, as a mix of inverted and direct repeats and – again – their organization can differ at the transposon ends. Figure 11.1(C) shows some of the variation present in characterized systems. These organizational and sequence differences often have profound functional importance as they make it possible for the transposase to differentiate between the two transposon ends.

The simplest DNA binding site structures – a single inverted repeat at each end – are found in ISs. For prokaryotic compound transposons, the situation is more complicated. Here, inverted repeats border each of the two ISs, so the entire transposon has four of them. For instance, Tn*10* has nearly perfect 23 bp inverted repeats located at the ends of the two 1.3-kb IS*10*s. The situation is even more complex in the case of bacteriophage Mu. It has three 30 bp sequences in each end that are similar to each other and which share a 22 bp consensus sequence that is the binding site for its transposase.[29] As shown in Figure 11.1(C), on the LE, the repeat sequences are separated by intervening DNA and all point in the same direction. On the RE, the repeat sequences abut each other but the innermost repeat points in the opposite direction from the other two. The Mu transposase, known as MuA, can bind to all of these six repeats, although with variable affinity. The Tn*7* transposon has a similarly complicated structure with 30 bp nearly perfect inverted repeats at both LE and RE, three at the LE and four overlapping repeats at the RE. The TnsB protein, which forms part of the Tn*7* transposase, binds 22 bp sequences found within these repeats.[30,31] TnsB's partner in the transposase, TnsA, does not bind measurably to DNA on its own.

Eukaryotic DNA transposons also display variations in the organization of their terminal sequences. The Tc1/*mariner* and *hAT* transposons contain single inverted repeats at their ends. While some Tc1/*mariner* inverted terminal repeats are quite short (*e.g.*, those of *Mos1* are 28 bp), the inverted repeats of the Tc3 transposon are 462 bp and each contains two binding sites for the Tc3 transposase. Within the Tc1/*mariner* superfamily, there can either be one or two transposase binding sites per end; in the latter case, they can be either direct or inverted repeats. Some *hAT* transposon ends also feature subterminal regions extending for several hundred bp that contain multiple short repeat motifs whose role in transposition is still unclear.

Another well-characterized eukaryotic DNA transposon is the *Drosophila* P element, representing its self-named superfamily, a 2.9-kb transposon with 31 bp terminal inverted repeats (TIRs) at each end that are essential for transposition (Figure 11.1C).[17] About 100 bp in from each end is another short inverted repeat (IR), unrelated in sequence to the TIR, that acts as a transpositional enhancer.

The P element transposase does not bind to any of these. Rather, a specific 10 bp transposase binding site (TBS) is located between each TIR and IR. To initiate transposition, the transposase makes a staggered cut into the 31 bp TIRs to ultimately generate a "cut-out" transposon with 17 bp 3'-overhangs.

11.2.2 Chemistry of DNA Transposition

It is useful to differentiate between transposons based on how their transposases catalyze the necessary DNA cleavage and joining reactions. The major mechanistic dividing line is at the chemical step of phosphodiester bond cleavage. There are two possibilities, depending on whether a covalent intermediate is formed between the cleaved DNA strand and the transposase. Most ISs, compound transposons, and the vast majority of eukaryotic transposons have transposases whose catalytic domains are all structurally related. These are known as DDE transposases, and they do not form a covalent intermediate with DNA. Among those transposons that proceed through a covalent intermediate, the requirement for an active site amino acid that acts as the nucleophile of the cleavage reaction leads directly to their classification as either tyrosine (Y) or serine (S) transposases. These two classes of enzymes are very different with no topological or mechanistic relationship between them. Two other groups of transposases also use tyrosine as the nucleophile to cleave DNA strands, the Y1 and the Y2 transposases; these are structurally related to each other but not to Y transposases.

11.2.3 DDE Transposases

Historically, the first DNA transposase solved crystallographically was the catalytic core domain of MuA.[32] The fold that was observed, an $\alpha\beta$ fold centered around a mixed five-stranded β-sheet, was first seen in the crystal structure of the nuclease RNase H (see Chapter 13).[33,34] Like MuA, all DDE transposases use a water molecule as the nucleophile for the cleavage reaction and are assisted by two divalent metal ions, Mg^{2+}, that are coordinated by three acidic residues. In most cases, these are two aspartic acid residues and one glutamic acid (hence DDE), although some members of the *mariner* family use a DDD motif. These catalytic residues are always located on the same topological elements of the RNase H fold. The RNase H fold is also seen in the catalytic domains of retroviral integrases[35] and, very likely, the RAG1 component of the V(D)J recombinase.[36,37] This similarity is perhaps not surprising as retroviral integration and V(D)J recombination are mechanistically related to DNA transposition.

DDE transposases are modular molecules composed of multiple domains, typically with site-specific DNA binding activity and catalytic activity separated between distinct domains. It is common that the site-specific DNA binding domain (or domains) are upstream from the catalytic domain. Figure 11.2 shows a comparison of the domain organization of a few characterized DDE transposases. The DDE catalytic cores of those transposases whose three-dimensional structures have been solved to date are shown in Figure 11.3.

DNA Transposases 277

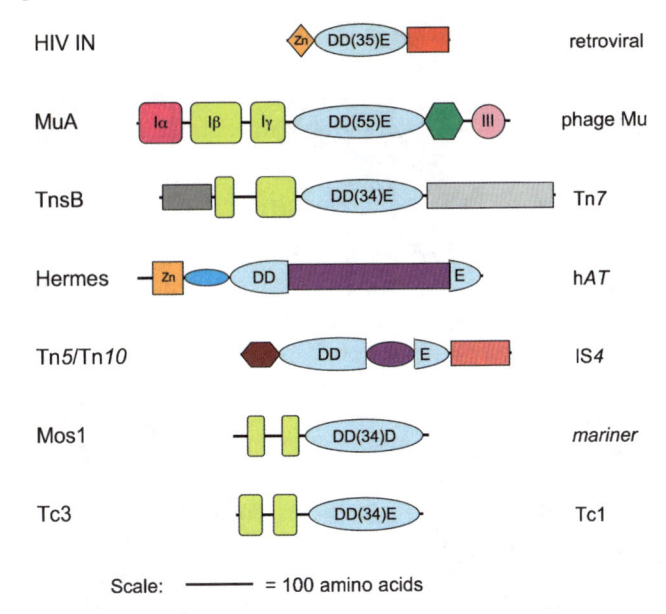

Figure 11.2 Domain organization of DDE transposases discussed in the text. Structurally unrelated domains are indicated with different shapes and colors. The defining DDE domain is shown as a light blue oval; interruptions of the DDE domain by an intervening domain are shown in purple. The yellow boxes represent demonstrated or predicted helix-turn-helix DNA binding domains.

11.2.4 Cut-and-paste, Copy-in, and Copy-out Transposition

There are two major ways how transposition can be catalyzed by DDE transposases. Using the terminology of editing documents, there is "cut-and-paste" and "copy-in" transposition. Cut-and-paste transposition is the simplest: the transposon is cut out from donor DNA and subsequently "pasted in" or integrated into the target DNA. In copy-in, or replicative, transposition, the transposition system collaborates with the cell's replication machinery to create a new copy of the transposon at the target site while leaving the original transposon intact at its original location.

Apart from myriad mechanistic differences that these two pathways imply, one significant issue relates to the number of cleavage reactions that occur at the transposon ends. In cut-and-paste transposition, the transposase must introduce double strand breaks at both transposon ends. In replicative transposition, this is not needed as the reaction goes through a stage called a Shapiro intermediate[38] in which only one DNA strand is cut at each transposon end.

11.2.4.1 *Mechanism of DNA Cleavage and Strand Transfer*

Figure 11.4 shows a summary of the chemical reaction steps of DDE transposition. The first step, a cleavage reaction, proceeds by an S_N2 in-line nucleophilic

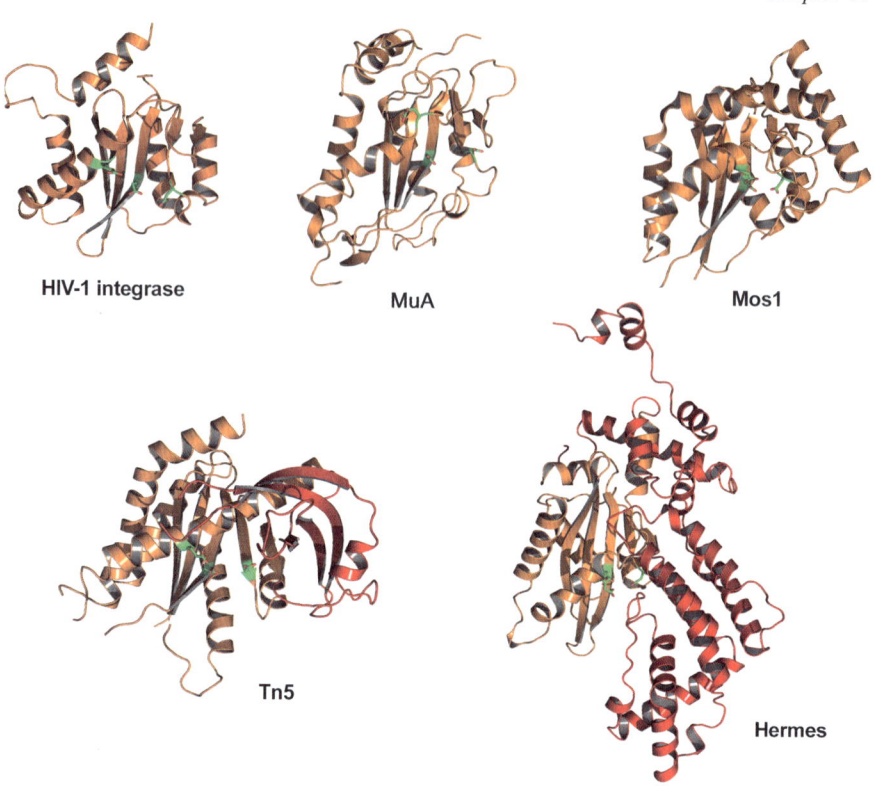

Figure 11.3 Aligned catalytic cores of structurally characterized DDE and DDD transposases. The three active site residues are shown in green with their carboxylate oxygen atoms in red. The domains inserted into the RNase H fold of the Tn5 transposase and Hermes are shown in dark red.

attack of an "activated" water (shown as an ⁻OH ion) on a phosphodiester bond at the transposon end. Cleavage always results in a free 3′ OH group, but its location depends on which DNA strand contained the scissile phosphate (Figure 11.4A). Subsequently, a 3′ OH group is the nucleophile that attacks the target DNA strand in a transesterification reaction that is also known as strand transfer (Figure 11.4B). When this occurs, a 3′ OH group is again liberated but now on the target DNA. Therefore, strand transfer is not really a reversal of the cleavage reaction but rather a "replay" of it, with the substantial difference that the attacking nucleophile is an activated water molecule during cleavage but a 3′ OH group of a DNA strand during strand transfer. In the prokaryotic systems that have been studied, a single active site catalyzes all the sequential nucleophilic attacks involving each end.[39,40] The nucleophilic groups used for cleavage and strand transfer are activated for catalysis using a two divalent metal ion mechanism similar to that proposed by Beese and Steitz for the exonuclease domain of the Klenow fragment of DNA polymerase.[41]

During strand transfer, the two terminal 3′ OH groups at the transposon ends created as a result of cleavage typically act in concert to attack opposite

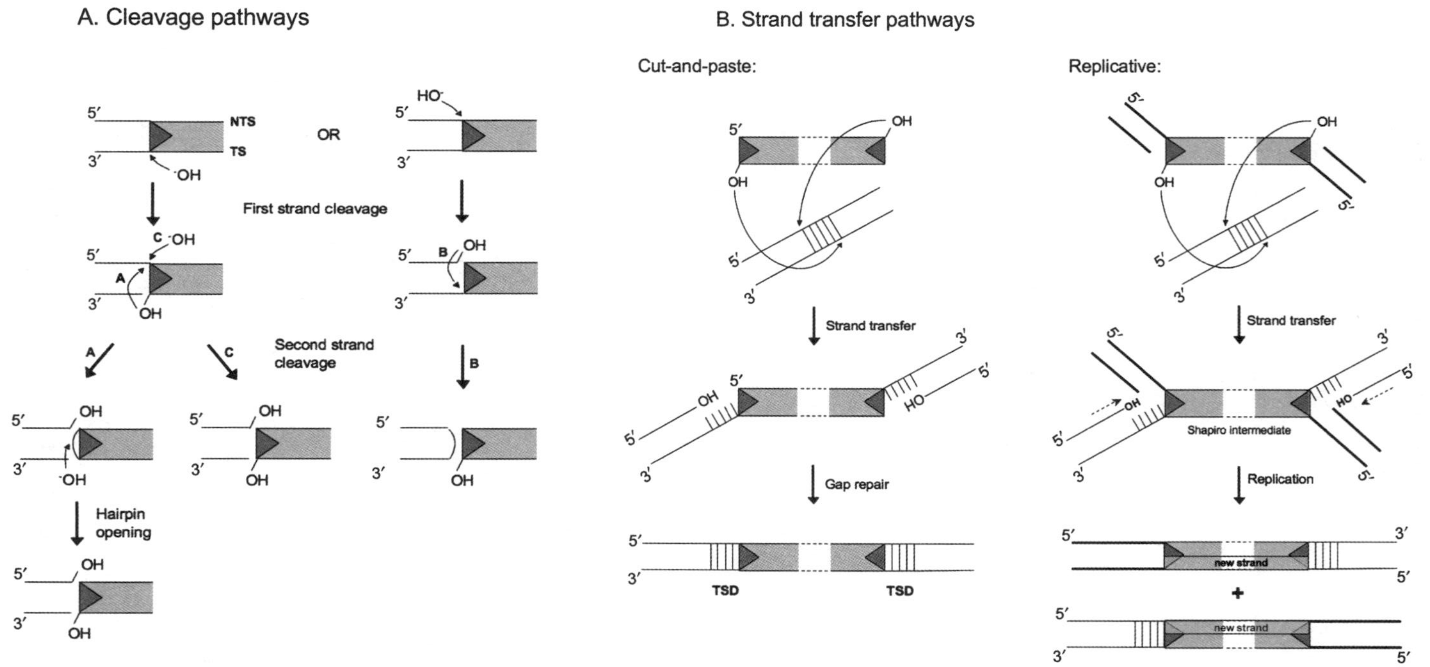

DNA Transposases 279

Figure 11.4 Chemical reactions of DDE transposition. (*A*) Cleavage pathways. (Left): Initial cleavage on the top strand (A pathway) results in hairpin formation on the transposon end. This is seen during Tn5 and Tn10 transposition. In the C pathway, a second cleavage reaction on the top strand – as by Tn7 – results in an excised transposon. (Right): If the first cleavage reaction occurs on the top strand, a 3′ OH group is liberated on the flanking DNA. In the B pathway, a hairpin is formed on the flanking DNA when the 3′ OH formed in the first step attacks the bottom strand, as during hAT transposition. (*B*) Strand transfer pathways. (Left): Nucleophilic attack by two 3′ OH groups at the transposon ends on a target DNA results in strand transfer. The gapped intermediate is repaired by host cell enzymes, leading to target site duplications (TSD). (Right): Schematic mechanism for replicative transposition, as carried out by bacteriophage Mu.

strands of the target DNA a few bps apart, where the exact number of bps varies from system to system (usually 2–9 bp). This results in a gapped intermediate that is subsequently processed by host cell DNA repair pathways. The final product, therefore, contains characteristic target site duplications (TSD) as a consequence of this offset attack (Figure 11.4B, left). Target site duplications bordering transposable elements in genomes (or elements that once upon a time transposed) are typical signatures of transposons and are one way in which they are identified during genome sequencing.

Transposition reactions do not use high energy co-factors for cleavage and strand transfer (with the possible exception of P element transposition[42]); therefore, any energy used must be paid by the energy of the bonds that are broken and by the binding energy of the complexes that form. While a cleavage reaction does decrease the number of phosphodiester bonds, the subsequent strand transfer step re-establishes them and it is assumed that the chemical driving force facilitating strand transfer is the binding energy of the final product complex that now includes target DNA.

11.2.4.2 Second Strand Cleavage by Cut-and-paste Transposases

Cut-and-paste transposases must cut both strands of DNA at each transposon end so the transposon can be liberated from the donor site. The second strand cleavage step can be accomplished several different ways, depending on the system. One pathway uses the 3′ OH group liberated by the initial cleavage to attack the opposite strand, thereby forming a hairpinned intermediate. If the polarity of the cleavage in the first step is such that the 3′ OH of the transposon DNA has been liberated, the hairpin is formed on the transposon end (pathway A in Figure 11.4A); this sets up a requirement for a subsequent chemical step in which the hairpin is opened, again catalyzed by the transposase using an activated water molecule as the nucleophile. Many ISs and the compound transposons Tn5 and Tn10 act in this way.

When the initial cleavage reaction liberates a 3′ OH end on the flanking donor DNA, then the subsequent attack of this group on the opposite strand yields a DNA hairpin on the flanking DNA (pathway B in Figure 11.4A). This pathway has been observed so far only for the *hAT* elements[24] and during the related process of V(D)J recombination. Since the fate of the flanking DNA is not important from the point of view of the *hAT* transposase, it does not appear to catalyze hairpin opening, and the host cell is left to deal with the consequences of transposition.

Another way to achieve second strand cleavage is to employ another transposon-encoded enzyme. Tn7, for example, encodes a restriction enzyme-like endonuclease, TnsA, to do just this (pathway C in Figure 11.4A).[9,43] Other DDE transposases mix, match, and modify some of these steps to transpose by different pathways. For example, the IS*3* family of transposases (such as the IS*911* transposase[44]) cleave only at one transposon end, and the resulting 3′ OH group attacks the opposite end to form a single-stranded circular intermediate. This is replicated *in vivo* to form a double-stranded transposon circle,[45] and another round of cleavage reactions introduces the 3' OH groups necessary for strand transfer. This pathway is known as "copy-out" transposition.[4]

11.2.4.3 Replicative Transposition

Replicative transposition, carried out by well-characterized systems such as Tn3[46] and phage Mu,[11] bypasses second strand cleavage altogether. After the first strand is cleaved to liberate 3' OH groups at the transposon ends, subsequent attack of these two 3' OH groups on opposite strands of target DNA results in a Shapiro intermediate (Figure 11.4B, right). The two 3' OH groups thus liberated on the target DNA can now be used by the host cell replication machinery, using each transposon strand as a template. The final result is a cointegrate in which two copies of the transposon and the donor DNA in between them are integrated into target DNA. In the case of Tn*3*, the two copies can be subsequently separated by a transposon-encoded dedicated site-specific recombination system.

A spectacular demonstration of the close relationship between cut-and-paste and replicative transposition was the experiment in which the introduction of a single inactivating point mutation in the TnsA endonuclease converted Tn7, a normally cut-and-paste transposon, into a replicative element.[47] As cleavage of the second strand by TnsA was blocked, the cut-and-paste pathway became unavailable. However, TnsB, the other component of the transposase, was able to carry out all the necessary DNA binding, cleavage, and strand transfer steps required for replicative transposition.

11.2.5 Transposition Happens in Context

Whether a DDE transposon mobilizes DNA by a cut-and-paste or a replicative pathway, the key to controlling the orderly events of the reaction is a nucleoprotein complex known as a synaptic complex or "transpososome". A transpososome typically consists of a multimer of the transposase, the transposon ends, and – on occasion – cellular proteins that are recruited to assist transposition. Understanding how transpososomes assemble and how they behave once assembled is central to understand how transposition happens and is regulated.[48] "Understanding a transpososome" can be re-stated as understanding the three-dimensional relationship between its components, the changes in their configuration, and the timing of these changes during the transposition reaction.

To date, there are only a very few experimental, three-dimensional structures of transpososome components bound to DNA. This relative paucity is due the immense difficulties encountered in assembling these complexes *in vitro* in quantities needed for structure determination experiments. The inherent ability of these complexes to undergo the conformational changes that are needed along the recombinational pathways implies built-in flexibility. This can cause even further difficulties as three-dimensional structural methods rely on ensemble averages requiring large numbers of nearly identical assemblies to be present during the experiment.

11.3 Tn5 Transposase: The Minimum Necessary

The first three-dimensional transpososome structure to be determined was that of the post-cleavage synaptic complex of an IS transposase, Tn5. Tn5 is a compound transposon bordered by two IS*50* elements and is a member of the IS*4* family. The crystal structure (Figure 11.5A) contains a dimer of the 476-amino acid transposase (with two hypertransposing point mutations) bound to two blunt-ended 20 bp oligonucleotides that represent the outer inverted repeats (OE or Outer End; Figure 11.1C) of the ISs.[49]

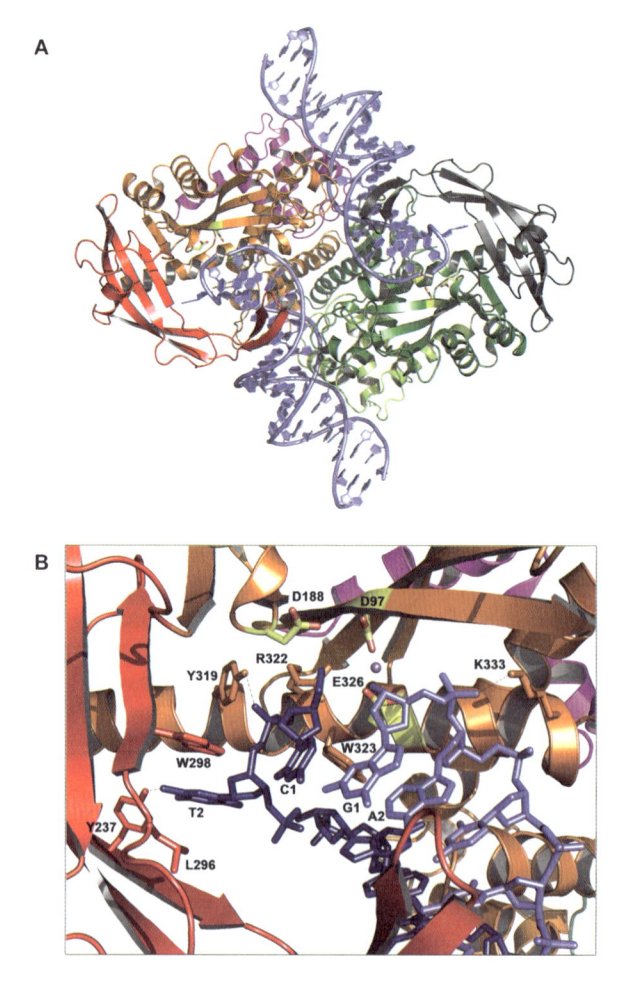

Figure 11.5 Tn5 transpososome. (*A*) Two transposase monomers, shown in red and green shades, bind two DNA molecules (blue) to form a synaptic complex.[49] (*B*) Details of the interaction between the tip of the Tn5 transposon and its transposase. The horizontal α-helix in the middle of the figure contains the residues of the YREK motif.

11.3.1 Overview of the Tn5 Transposition Pathway

To understand the significance of the structural details of the Tn5 transposase-DNA complex, we must define the state on the transposition pathway that the structure represents. The progression of DNA intermediates along the Tn5 reaction pathway is shown in pathway A of Figure 11.4(A). After synaptic complex assembly, Tn5 transposase cuts one of the DNA strands precisely at the transposon end, creating a free 3' OH group. The cut strand is called the "transferred strand", or TS. This new 3' OH group of the TS then attacks the other strand, the "non-transferred strand" (NTS), precisely at the transposon end to form a tight DNA hairpin. In the next step, the hairpin is opened to regenerate the 3' OH group on the TS. The solved structure represents the configuration of the complex immediately following this hairpin opening step.

Like transposition catalyzed by MuA and the Tn10 transposase,[39,40] data suggests that all the chemical steps of Tn5 transposition are carried out by the same active site and that the TS does not leave the active site after its initial binding. Instead, the various scissile phosphates come and go, and the major difference between the steps is the nature of the nucleophile. Thus, the Tn5 transposase active site must be able to activate both water molecules and DNA terminal 3' OH groups for attack on DNA strands.

11.3.2 Structure of a Tn5 Transposase Dimer Bound to DNA

The Tn5 transpososome structure reveals several key aspects of Tn5 transposition. Although the transposase is a monomer when not bound to DNA, OE binding results in dimerization. In the crystal structure, two OE oligonucleotides and two transposase subunits are interwoven to form the transpososome in which each OE oligonucleotide contacts both monomers. The two DNA molecules are antiparallel with the ends representing the opened hairpins converging on one face of the dimer. Although there is a small C-terminal dimerization domain, it appears that protein-DNA interactions are more important in dimer stabilization than protein-protein interactions between monomers.

As shown schematically in Figure 11.2, the Tn5 transposase is a three-domain protein. The N-terminal site-specific DNA binding domain is responsible for recognizing positions 7 to 13 of the OE (counting from position 1, the cleavage site at the end of the transposon). However, the cleavage site of this OE is located within the active site of the second monomer. This observed arrangement is consistent with "*trans*" cleavage activity.[50] In other words, the N-terminal DNA binding domain binds the OE in *cis* but the catalytic subunit acts on it in *trans*. This provides an important regulatory mechanism that ensures that cleavage can occur only after the transposase has located and synapsed the two OE of Tn5, thereby preventing wanton DNA cleavage.

Initial recognition of the OE is likely driven by the *cis* interactions between the OE and the N-terminal domain of the transposase. The α-helix between Ser-55 and Asn-66 is crucial as it lies deeply in the major groove of the OE about one helical turn away from the cleavage site. The interactions between the OE and the transposase close to the cleavage site (positions 1–6) are all in *trans*, contributed by the other two domains (Figure 11.5B). The large number of intricate interactions between DNA and the transposase achieves two things. First, they appropriately position the transposon DNA in the active site so that the cleavage and joining steps happen at the right place. Second, they promote the formation of the hairpin intermediate by stabilizing a thymine residue of the NTS at the 2nd position (T2) in a flipped-out, extrahelical position. As DNA hairpins contain several unpaired bases at their tips, the tight hairpin seen here simply could not happen if T2 remained base-paired to its partner, A2, of the TS. The transposase stabilizes the DNA structure at the tip of the OE by sandwiching the flipped-out T2 between Leu-296 and Trp-298 and providing a H-bond from Tyr-237 to O4 of T2.

The three transposase residues that interact with T2 are part of a 90-residue insertion (shown in purple in Figure 11.2, red in Figure 11.3) into the catalytic DDE domain that extends the β-sheet of the standard RNase H-like fold by four strands. Clearly, one of the functions of this insertion is to stabilize the hairpin intermediate. Another feature of the insertion is a long β hairpin that clamps down on the major groove of the DNA close to the tip of the OE, making numerous interactions both with bases and the phosphate backbone (Figure 11.5A).

The other key structural element of the transposase involved in *trans* interactions with the DNA is the α-helix just downstream of the inserted domain that harbors the catalytically essential E of the DDE motif, Glu-326. This helix contains a Y(2)R(3)E(6)K motif (consisting of Tyr-319, Arg-322, Glu-326, and Lys-333) that is conserved in the IS*4* family. The first two residues of the YREK motif are intimately involved in hairpin stabilization (Figure 11.5B) as they bind the phosphate group between the first and second bases of the NTS (C1 and T2). Clamping this phosphate in place allows C1 of the NTS to stack on *top* of G1 of the TS instead of forming a Watson–Crick basepair. This helps to push T2 into the flipped-out position. Another residue on the same helix as the YREK residues, Trp-323, is important for hairpin formation: it is inserted into the widened minor groove so that it is less than 4 Å away from A2 of the TS, thus helping A2 to unpair from T2. Trp-323 promotes the flipping-out of T2 by disrupting base stacking and helping to bend the DNA.[51] The net effect of these interactions is that, at the tip of the OE, four bases are unpaired and the DNA axis is bent by 41° around the *cis* binding N-terminal domain.

Although the structure of the Tn5 synaptic complex does not directly address strand transfer, the 3′ OH groups of the two TSs are 41 Å apart, which is slightly larger than expected if they are to insert into opposite strands of target DNA with a 9 bp stagger. The structure and the arrangement of basic residues suggest where target DNA might bind, but it is also clear that some conformational changes have to take place for target DNA binding and strand transfer to occur.

11.4 Bacteriophage Mu

While the Tn5 transposase is one of the simplest DDE transposases, the transposition system of phage Mu is probably one of the most highly regulated and sophisticated. Therefore, it is especially interesting to consider it in comparison with that of Tn5.

11.4.1 Organization of the Mu Genome Ends

The DNA structure at the ends of the Mu genome is asymmetric (Figure 11.1C) with three sparsely spaced inverted repeats at the LE, designated L1–L3, and three closely spaced repeats at the RE, R1–R3.[29] Additional binding sites elsewhere in the genome for MuA and for host cell proteins required for transposition hint at the complex requirements for transpososome assembly. For example, the 80 bp between the two leftmost repeats (L1 and L2) contain a binding site for the host cell protein HU which can introduce a sharp bend in the DNA and bring L1 and L2 close to each other. One kb away from the LE is the internal activating sequence, or IAS, whose presence increases the efficiency of Mu transposition 100-fold *in vivo*,[11] and is essential in the assembly of the authentic transpososome. Within the IAS are several binding sites for MuA and one for IHF, another host DNA-bending protein. IHF binding, like that of HU, causes a large bend in DNA, resulting in a near U-turn (see Chapter 8).[52] Binding and bending brings distant DNA sequences close together so that they can be bridged by an oligomer of MuA.

11.4.2 Domain Structure of MuA

The elaborate Mu DNA end site structure is echoed in the structure of the MuA transposase, which is 663 residues long and composed of six domains (Figure 11.2). Although the structure of full-length MuA is unavailable, structures of five of its domains have been determined either by X-ray crystallography or by NMR (Figure 11.6A).

Starting at the N-terminus, MuA has a small Iα domain (red in Figure 11.6A) containing a winged-helix motif[53] to bind the IAS, followed by two larger domains (yellow and green), Iβ and Iγ,[54,55] that bind to MuA binding sites at each transposon end. Iβ and Iγ together bind to the 22 bp binding sites in two adjacent major grooves using their respective recognition helices of a helix-turn-helix (HTH) motif and interacting with the intervening minor groove (see Chapter 3 for an introduction to these DNA binding motifs).

Following these DNA-binding modules is the DDE catalytic domain, which is intimately associated at its C-terminus with a small, β-barrel non-specific DNA binding domain; together these form a large, 303 residue unit (blue in Figure 11.6A).[32] Unlike Tn5, there is no insertion domain between the fifth β-strand and the helix containing the catalytic E residue, Glu-392. This reflects the lack of a hairpinned intermediate as phage Mu transposases by the replicative pathway. Instead, a fairly mobile loop, which probably becomes ordered upon DNA binding, leads to the helix that contains Glu-392. Interestingly, a similar situation has been seen with HIV-1 integrase,[35] the enzyme that catalyzes the insertion of the viral DNA into the chromosomes of infected cells. Second strand processing is not required by HIV-1 since the DNA copy of the retroviral genome is generated with blunt ends, and they need only to be

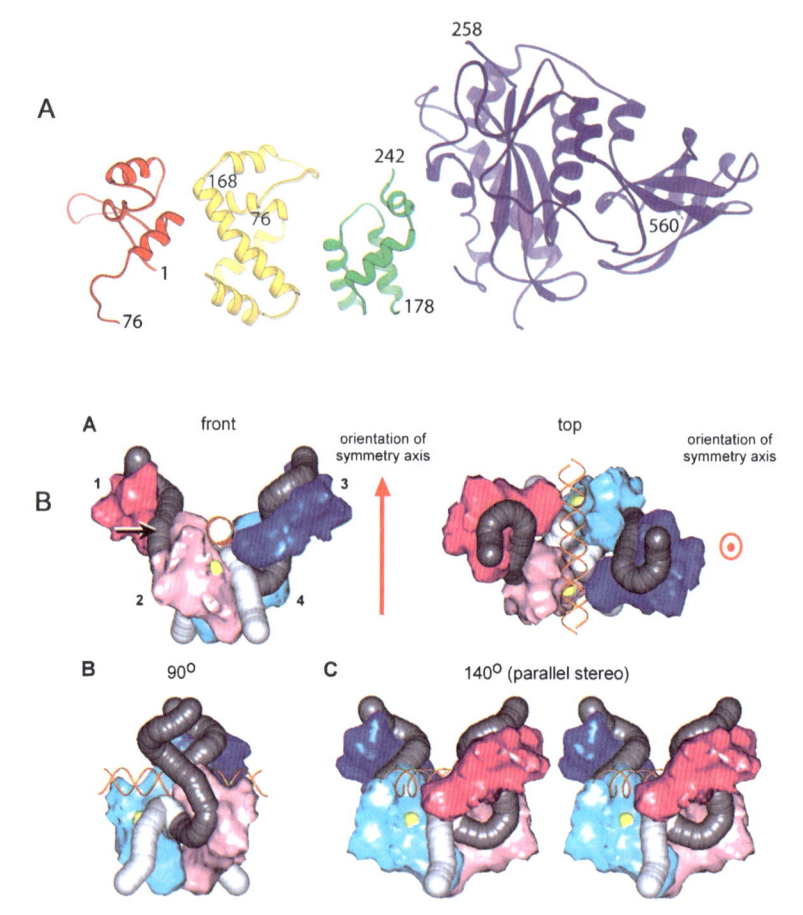

Figure 11.6 MuA transpososome. (*A*) Structures of individual domains of MuA that have been determined. (Reprinted with permission from ref. 89.) (*B*) Model of the Mu transpososome in different orientations. Four MuA molecules are shown in shades of pink and blue with the path of the transposon DNA in dark gray. Active sites are colored yellow, modeled flanking DNA in light gray, and modeled target DNA in orange. (Reprinted with permission from ref. 58.)

minimally trimmed (also by integrase) to create the 3′ OH groups needed for strand transfer. Like MuA, HIV-1 integrase lacks an insertion domain and, in some of the available structures, has only a disordered loop leading into the helix containing the catalytic E.

The as-yet structurally uncharacterized C-terminal region of MuA is involved in protein-protein interactions with other proteins that are regulators of transposition. One of these is MuB, a non-specific DNA binding protein with ATPase activity whose interaction with target DNA is needed for efficient transposition. Its role is to deliver target DNA to a pre-assembled MuA oligomer holding the transposon ends.

11.4.3 Putting Mu Ends and MuA together: The Mu Transpososome

Similarly to the Tn5 transposase, MuA is a monomer in solution and oligomerizes only after binding to the ends of the Mu genome. However, as transposition proceeds, there are four distinct tetrameric complexes with somewhat different configurations and of increasing stability. Each complex type contains the two Mu ends while the last also contains the target DNA. This last complex is the most stable, and an ATP-dependent protease, ClpXP, is needed to disassemble it and release the transposition reaction products.[56,57] The sequence of these complexes forms an ordered pathway for transposition.

Direct experimental structural information, obtained through electron microscopy (EM) at 35 Å resolution, exists on the MuA cleaved donor complex (CDC).[58] This is the complex in which TS cleavage has occurred but before target DNA has bound. For structure determination, the complex was assembled using 50 bp oligonucleotides containing the closely spaced R1 and R2 repeats such that the TS ends precisely at the authentic cleavage site whereas the NTS extends 4 bp into host DNA. Although Mu transposition *in vivo* requires two different ends (LE and RE), the use of identical oligonucleotides is common in structural biology due to the obvious complication of a heterogeneous mix of complexes associated with using different ones. This simplification certainly introduces a degree of artificiality. The good news is that MuA is active on the type of RE oligonucleotides used for the structural studies,[59] suggesting that a great deal of truth has been captured in this experimental structure.

While the relatively low resolution of the EM structure (shown in various orientations in Figure 11.6B) does not permit a detailed description of the MuA-DNA interactions, the availability of high-resolution structures of five MuA domains allowed them to be docked into the EM images, giving a decent idea of what is located where. Two of the most important features revealed by the structure of the Tn5 synaptic complex are recapitulated in the MuA transpososome model. First, the tetrameric complex is largely held together by protein-DNA interactions and, second, transposon end cleavage occurs in *trans*. Thus, a structural explanation is provided for the known biochemical result that a MuA monomer bound to the LE cleaves and catalyzes strand transfer of the RE, and *vice versa*.[60,61]

The overall shape of the modeled complex resembles the letter "V". Just like the letter, the complex has two-fold symmetry relating two MuAs and bound DNA alongside one side of the V to the others that lie alongside the other side. Each DNA molecule (shown in dark gray in Figure 11.6B) winds through the two MuAs that form each side of the V such that each DNA molecule interacts mainly with only these two MuAs. At the bottom of the V, two MuAs are close to each other (shown in light pink and blue); these are the ones that are bound to the two R1 repeats and whose active sites are close to the very end of the DNAs where they are well-positioned for *trans* cleavage. The two MuAs at the top of the V (darker pink and blue) only minimally interact with each other, and whereas their DNA binding domains, Iβ and Iγ, are engaged with the DNA at each R2 their active sites are not used. This "oversupply" of active sites might be a common feature of highly regulated transposition systems.

Despite the sharp 80° bends in the DNA molecules as they pass by the Iβ and Iγ domains of each of the two MuAs at the bottom of the V, the DNA molecules are not wrapped around each other. One very attractive feature of the observed V shape is the saddle in the middle, naturally suggesting where target DNA (shown in orange in Figure 11.6B) might bind. This would bring target DNA close to the active sites of the two MuAs at the bottom of the V and therefore close to the 3′ OH ends. Another consequence of imagining that target DNA binds in the saddle is that this would place the C-termini of the two MuA molecules that are on the top of the V close to target DNA. As these are precisely the MuA regions that interact with MuB, the protein responsible for delivering target DNA to the CDC, their location makes a lot of sense.

It is unclear how a Mu transpososome assembled on authentic LE and REs, facilitated by IAS, HU, and IHF, would differ from the transpososome imaged by EM. Considering that features such as DNA-dependent assembly and *trans* catalysis appears to be faithfully displayed, the probability of fundamental differences is small. Nevertheless, recent experiments probing the topology of transpososome-bound DNA suggest that some differences are likely.[62,63]

11.5 Tc1/*mariner* Transposases

The Tc1/*mariner* superfamily is one of the most widespread DNA transposition systems among eukaryotes. Although the terminal repeat structure of Tc1/*mariner* elements varies considerably,[18] the domain organization of the associated transposases is less divergent. The rule, followed by Tn5 and MuA, that the site-specific DNA binding domains responsible for end recognition are upstream of an RNase H-like catalytic domain, is upheld. However, the DNA-binding domains are present in a dipartite arrangement in which two small N-terminal domains are connected by a flexible linker. To date, the three-dimensional structures of portions of two Tc1/*mariner* transposases have been determined: the N-terminal DNA binding domains of the *C. elegans* Tc3 transposase bound to ITR DNA,[64] and the catalytic domain of the *Mos1* transposase[65] from an element isolated from *Drosophila mauritiana*.

All Tc1/*mariner* superfamily members move *via* the cut-and-paste pathway, and the transposase is the only protein needed for transposition.[66–68] The ability to introduce double strand breaks at the ends of the transposon without relying on either hairpinned intermediates or another nuclease makes this superfamily unique among DNA transposons. At each end, cuts are introduced with a 2 or 3 bp stagger in which the TS is cut precisely at the end of the transposon whereas the NTS is cut two or three bases into the transposon. A general feature of Tc1/*mariner* transposition is a preference for insertion into TA target sequences with a 2 bp stagger, resulting in characteristic TA target site duplications.

11.5.1 Tc3 N-terminal Domains and DNA

In the crystal structure of the Tn3 transposase-DNA complex (Figure 11.7), two sequential binding sites within a 26 bp oligonucleotide representing the Tc3 ITR are bound by two small, 42 residue HTH domains (in green) that are separated by an 18 amino acid linker (in yellow).[64] Both domains bind in the major groove about a turn and a half (14 bp) away from each other. The linker connecting them runs parallel to the DNA and follows the minor groove between the two phosphodiester backbones. This mode of DNA binding is similar to that of AT-hook motifs.[69] As a consequence of this minor groove binding, both HTH motifs are bound to the same DNA molecule, which is bent about 30°. Although, in principle, an 18 residue linker could allow the two HTH domains to bind independently to two DNA molecules and promote the synapse of the two transposon ends, apparently this is not the case.

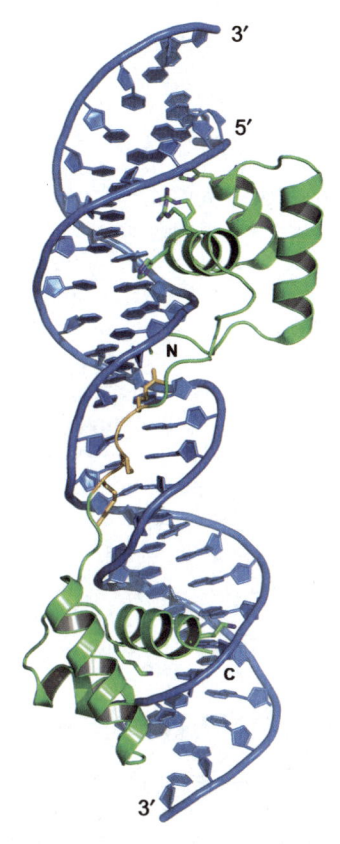

Figure 11.7 Tc3 N-terminal domains and inverted terminal repeat DNA. Two helix-turn-helix domains (green), connected by a 18 amino acid linker (yellow), bind to the major groove of the Tc3 ITR.[64]

11.5.2 *Mos1* Transposase

The three-dimensional structure of the C-terminal catalytic domain of the *Mos1* transposase (Figure 11.3, top right) shows as expected an RNase H-like fold but of the DDD variety.[65] The DDD motif is characteristic of *mariner* elements such as *Mos1*, whereas Tc1 elements have the more usual DDE arrangement.[70] The structural result that the *Mos1* transposase does not possess an insertion domain or a YREK motif is consistent with the introduction of double strand breaks at the transposon ends without a hairpinned intermediate.

11.5.3 What is the Active Assembly?

Although the active oligomeric state of Tc1/*mariner* transposases is not known, Tc1/*mariner* transpososomes contain at least two transposase monomers, thereby supplying one active site per end. If only a dimer is present, after the first strand cleavage (which in the case of *Mos1* occurs on the NTS[71]), each active site would have to release the cleaved DNA strand to allow in the TS for the second cleavage. Such acrobatics might be expected to require substantial conformational changes. On the other hand, cleaving the NTS first is reasonable, given that the TS with its terminal 3′ OH has to end up in the active site for the subsequent strand transfer reaction. In principle, a higher order multimer such as a tetramer could supply four active sites in which each would be responsible for a single cleavage reaction.

In the Tc3 transposase crystal structure, a dimer observed between two DNA-bound HTH domains suggests how a synaptic complex might assemble on two transposon ends. There is also evidence that the N-terminal part of the *Mos1* transposase is important for multimerization, and that the *Mos1* transposase and Sleeping Beauty form tetramers.[72–74] Nevertheless, one model that avoids the need for a tetramer[65] suggests that NTS cleavage at each transposon end might happen in the context of a monomer and is an uncoordinated event. Subsequent TS cleavage would then occur only in a synaptic assembly containing a transposase dimer in which both NTS had been cleaved and the TS now occupies the active sites.

Although resolution of this issue awaits, it is certainly possible that, since the number of transposase binding sites per transposon end varies within the superfamily, different members of the Tc1/*mariner* family use synaptic assemblies with differing numbers of transposase molecules.

11.6 *hAT* Elements: A First Glimpse

Another large superfamily of eukaryotic DDE transposases is the *hAT* superfamily. Transposition by one family member, *Hermes* from *Musca domestica*,[23] has been reconstituted *in vitro*.[24] The crystal structure of an N-terminally truncated form of Hermes[25] (Figure 11.8) provides some insight into what may be generalizable mechanistic themes of *hAT* transposition.

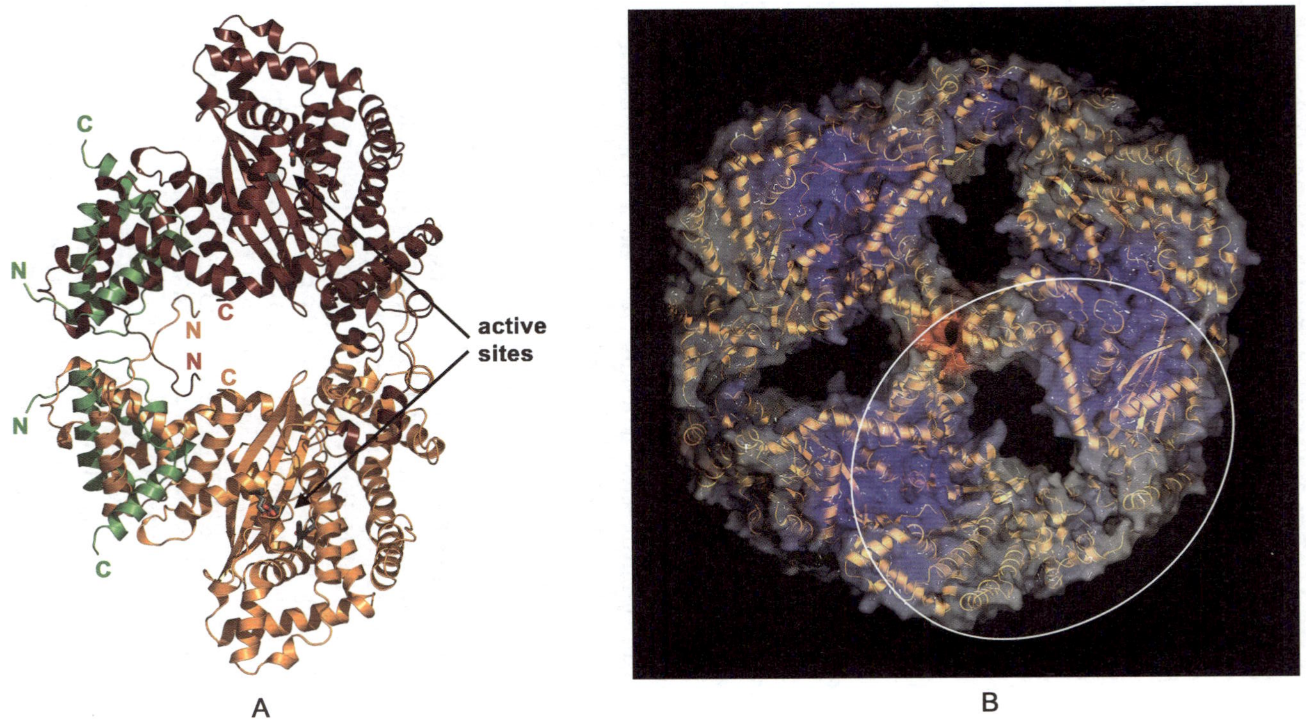

Figure 11.8 Hermes, a *hAT* transposase. (*A*) In the crystallographically observed assembly,[25] two subunits (shown in orange and red) were bound to two identical small domains (residues 79–162, shown in green). This heterotetramer arises from proteolytic cleavage of the biologically relevant hexamer during expression in *E. coli*. Active sites, indicated with arrows, are ~70 Å apart. (*B*) Modeled hexameric assembly of Hermes. The calculated electrostatic surface shows extensive patches of positive charge (shown in blue), suggesting where DNA substrates might bind. The crystallographically observed heterotetramer is circled.

Hermes moves *via* a cut-and-paste mechanism using hairpinned intermediates.[24] As seen in Tc1/*mariner* transposition, the first cleavage occurs on the NTS but, in this case, one base into flanking donor DNA from the transposon end. The newly created 3′ OH on the flank then attacks the TS precisely at the transposon end to create a hairpin on flanking donor DNA (pathway B in Figure 11.4A), reminiscent of the hairpinned intermediates formed during V(D)J recombination.[28] The "signal ends" of V(D)J recombination are analogous to the cut-out *hAT* transposon, and the hairpins formed on the V(D)J "coding ends" are similar to the hairpins on flanking donor DNA. It appears that neither *hAT* transposases nor the V(D)J recombinase can open the hairpins whose formation they catalyze.

Upon hairpin formation, a new 3′ OH group is formed on the TS at the transposon end, which is subsequently used to attack target DNA. *hAT* transposition is characterized by 8 bp target site duplications, reflecting the spacing of the staggered cuts that result from strand transfer.

11.6.1 Domain Organization of Hermes

The *Hermes* transposase is 612 amino acids long and composed of four domains (Figure 11.2).[25] The first N-terminal 78 residues were not part of the crystallized Hermes fragment, but based on the primary sequence this region appears to contain a BED domain[75] that binds a structural Zn^{2+} ion and is predicted to bind DNA non-specifically.

A second domain located between residues 87 and 150 is most likely responsible for specific recognition of the transposase binding site located in the *Hermes* 17 bp TIRs. This domain forms an unusual three-dimensional structure (shown in green + red and green + yellow in Figure 11.8A) that consists of a completely intertwined and very stable dimer in which the polypeptide folds around a copy of itself to form a tightly packed hydrophobic core. Dimer formation through this domain seems absolutely mandatory.

Following these two domains is the DDE catalytic domain with its expected RNase H-like fold. Not surprisingly, given the hairpinned intermediate, it has a large insertion domain (red in Figure 11.3) between the 5th β-strand of the central sheet and the helix that contains the catalytic E residue. The structure of the insertion domain is unique among currently known protein structures and bears no resemblance to the insertion seen in the Tn*5* transposase. The Hermes insertion domain contains a tryptophan residue conserved among *hAT* transposases, Trp-319, that projects towards the active site. By analogy to Trp-298 or Trp-323 of the Tn*5* transposase, it is very likely that Trp-319 plays a crucial role in assisting hairpin formation on the donor end by either stacking against a flipped-out base or pushing the base out of duplex DNA. Similarly, Trp residues in RAG1, also located in a predicted insertion domain, have been shown to play a role in DNA hairpin stabilization.[26,27]

Biophysical and electron microscopy data yielded the surprising result that the oligomeric state of the N-terminally truncated *Hermes* transposase in solution and in the absence of DNA is a hexamer.[25] These assemblies exhibit specific DNA binding to the *Hermes* TIR (Hickman and Dyda, unpublished observations) and are active for *in vitro* hairpin formation and strand transfer.[24] Although only a heterotetramer was seen in the crystal structure, a hexamer can be readily modeled based on the two crystallographically observed dimerization interfaces (Figure 11.8B). Pleasingly, in this model, the hexamer possesses surfaces with positive electrostatic potential that are contiguous with the surface of the DNA binding domain.

The observation of oligomers of Hermes prior to DNA binding suggests that one of the key regulatory features seen in prokaryotic DDE transposases, *i.e.*, that oligomerization occurs only in the context of a synaptic complex containing the two transposon ends, might not be a necessary property of eukaryotic DNA transposases. The suggestion that a hexamer is the active assembly for *Hermes* transposition again raises the question of the number of required active sites. The modeled hexameric arrangement brings two active sites close together and the use of two active sites per end appears structurally reasonable. If only one active site is used per transposon end, then it must somehow coordinate the release of the NTS after cleavage and hairpin formation, and the capture of the TS for strand transfer.

11.7 Transposases that Form Covalent Phosphotyrosine or Phosphoserine Intermediates

Although most DNA transposases have a catalytic domain containing a DDE (or DDD) motif, others employ a catalytic mechanism that involves a covalent reaction intermediate between the transposase and DNA. There are four recognized structural classes of these transposases (Y, S, Y1, and Y2)[4,76] and each class has close structural relatives among enzymes that are involved either in site-specific recombination (Y and S) or in the initiation of rolling circle replication (Y1 and Y2). For the most part, these structural relationships have been deduced from primary and secondary sequence analysis, as only two transposases from the Y1 family have been structurally characterized to date.[76,77]

11.7.1 Y and S Transposases

Y and S transposases produce closed dsDNA circle intermediate forms of their associated transposons through a common mechanistic pathway that is a modification of the cut-and-paste pathway.[5] To make this dsDNA circle, Y and S transposases catalyze a recombination event between two sites at the end of the transposon in which all four DNA strands are cut and rejoined to create a new site at the newly formed junction of the circle. In the second target joining step, this new site is recombined with a suitable site on target DNA; again, four strands are cut and subsequently rejoined.

Although there is no direct structural information on any of these transposases, biochemical evidence indicates that the broad features of DNA recognition and rearrangement are similar to those of their relatives that carry out conservative site-specific recombination. It is assumed that Y transposases act similarly to tyrosine recombinases such as lambda integrase, Cre, and Flp (Chapter 12), and S transposases are likely to be similar to serine recombinases such as γδ resolvase.[78]

Known Y transposases are almost always associated with conjugative transposons, where the transposon also encodes for proteins that promote cell-to-cell plasmid transfer. This is a very effective way to horizontally spread antibiotic resistance determinants and to establish resistant bacterial strains.[79] For example, the recently determined genome sequence of the highly virulent bacterium, *Clostridium difficile*, a multiantibiotic-resistant "superbug" found in hospitals, revealed that its genome is packed with various conjugative transposons.[80]

11.7.2 Y2 Transposases

Y2 transposases are so named because their catalytic subunits have two tyrosines close to each other that are required for transposition. Structurally, they belong to a vast superfamily of nucleases[81] sometimes referred to as Rolling Circle Replication (RCR) initiators or members of the HUH superfamily. In addition to catalytic tyrosine residues, they bear a catalytically essential histidine-hydrophobic residue-histidine (HUH) motif that provides two His ligands that coordinate a required cofactor divalent metal ion (such as Mg^{2+}). This contrasts with Y and S transposases, which do not require metal ions. HUH nucleases are also involved in the initiation of conjugative plasmid transfer, in which case they are called relaxases.

HUH nucleases nick DNA by forming a covalent phosphotyrosine linkage with the 5′ phosphate at the cleavage site, liberating the 3′ OH. This polarity of cleavage is opposite to that of Y transposases and tyrosine recombinases which release the 5′ OH at the cleavage site. Since the cellular replication machinery can only use a 3′ OH group as a primer to extend DNA, the action of HUH nucleases is often followed by DNA replication whereas this is not the case for tyrosine recombinases.

Y2 transposases are of the "copy-in" type, meaning that the transposon is recreated at the donor site. ssDNA transposon circles are generated during Y2 transposition, supporting the structural and mechanistic link to RCR initiators. However, the details of the reactions catalyzed by Y2 transposases have not yet been completely established. It is known that in the case of the well-characterized IS*91* transposon[82] its transposase cleaves a single strand of DNA by forming a phosphotyrosine linkage and releasing the 3′ OH on the flanking DNA as expected for an HUH nuclease. This cleavage is both strand- and site-specific, and is followed by replication starting from the free 3′ OH. Transposon strand displacement follows, presumably assisted by a cellular 5′ to 3′ helicase, leading to single-stranded transposon circles. Figure 11.9 shows the possible steps along this pathway. However, alternative mechanisms for Y2 transposition have been proposed.[83]

Although there is no direct structural data for a Y2 transposase, the structures of protein-DNA complexes of related HUH nucleases have provided

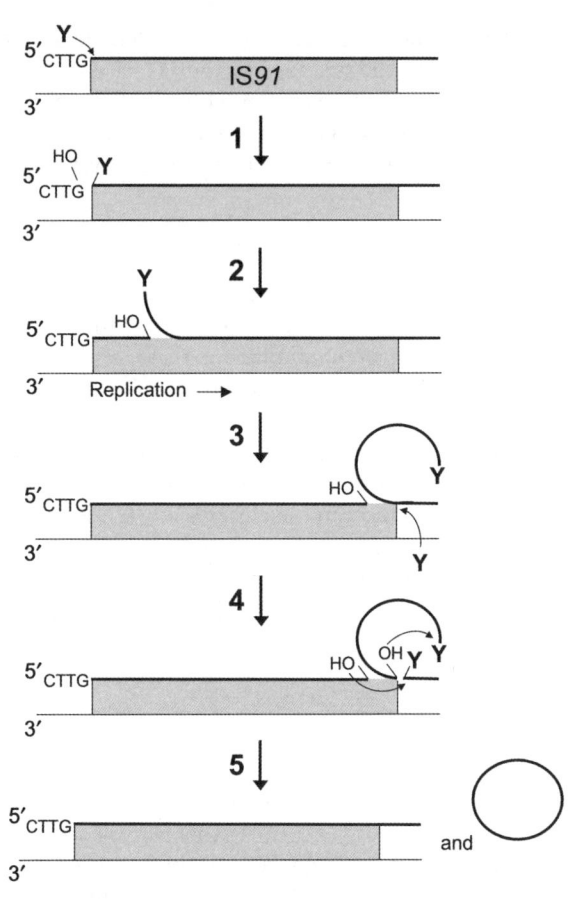

Figure 11.9 A model for the production of single-stranded transposon circles during Y2 transposition. Steps include two sequential strand-specific cleavage reactions (steps 1 and 4), each of which results in a 3' OH and a 5' phosphotyrosine covalent intermediate, and two nucleophilic attacks (step 5) that seal the top strand of the transposon and release the displaced single-stranded transposon circle.

insights into comparable steps along the transposition pathway. For example, important features of Y2 transposon ends are stretches of inverted repeats or palindromes with the ability to form DNA hairpins. Similar hairpin-forming sequences are found at *oriT* (origin of transfer) sites that are binding sites for relaxases. The structure of the TrwC relaxase domain from plasmid R388 bound to a 25-mer *oriT* sequence that contains a hairpin[84] suggests how Y2 transposases might recognize their transposon ends. In the structure, TrwC specifically recognizes two DNA regions: four Watson–Crick base pairs in the stem of the hairpin and the eight nucleotide ssDNA leading to the *nic* site that winds its way from the bottom of the hairpin into the HUH active site.

Insight into the mechanism of cleavage by Y2 transposases comes from the structure of the F plasmid TraI relaxase domain bound to *oriT* ssDNA.[85] This shows the arrangement of the active site close to the scissile phosphate, with the catalytic metal ion bound to the residues of the HUH motif. The geometry is such that the phosphate is perfectly positioned for a nucleophilic attack by one of the active site tyrosines. Given the presumed structural similarity, it is likely that Y2 transposases carry out chemistry in a very similar way.

11.7.3 Y1 Transposases

Some transposases of the IS*200* family also have an identifiable HUH motif and a conserved tyrosine residue, so the assumption that they were likely to resemble Y2 transposases was understandable. However, structural studies on the transposase of ISHp*608*[76] and a related IS*200*-like transposase[77] revealed some important variations on the standard organization of HUH proteins. Interestingly, the ISHp*608* transposon family is directly responsible for the emergence of an antibiotic resistant strain of *Helicobacter pylori* in which a mini-IS*605* element, a non-autonomous IS*200*-like sequence, has transposed into the nitroreductase gene.[86] The clinically used antibiotic, metronidazole, is a pro-drug that is activated by nitroreductase, and this is an example of how transposition by an IS that does not carry resistance genes can give rise to new, antibiotic-resistant strains.

11.7.3.1 ISHp608 TnpA

ISHp*608* is 1.8 kb long and contains two ORFs; the first encodes for the transposase, TnpA, while the product of the second ORF is not needed for transposition, at least in *E. coli*.[87] How ISHp*608* transposes is still an open question. Transposition has been analyzed *in vitro* and most steps, including target capture and insertion, can be reproduced using single-stranded oligonucleotides.[88] It is likely that the transposase acts together with a helicase (similarly to Y2 transposases) and in a way that might be closely coupled to DNA replication. TnpA is the smallest known functional transposase, and its relative simplicity might be a reflection of a heavy reliance on cellular proteins. It is also possible that TnpA works by using ssDNA byproducts of DNA repair or recombination. This would be a reasonable strategy for controlling transposition frequency, as suitable structures that TnpA could bind would only arise infrequently.

The ISHp*608* transposon ends contain an intricate arrangement of multiple short repeated sequences that are shown schematically in Figure 11.1(C). The repeat sequences at each end are, in fact, imperfect palindromes that can form imperfect DNA hairpins. One set of palindromic sequences is recessed 20 bp from the LE and only 9 bp from the RE. ISHp*608* always inserts just 3′ from a TTAC sequence, and transposition proceeds without target site duplication. Although dsDNA circular forms of the transposon have been found during ISHp*608* transposition,[88] it is not clear that these are necessarily part of the pathway.

The crystal structures of TnpA alone and bound to a 22-mer hairpin formed by the palindrome at the transposon RE (Figure 11.10) reveal that, whereas

DNA Transposases 297

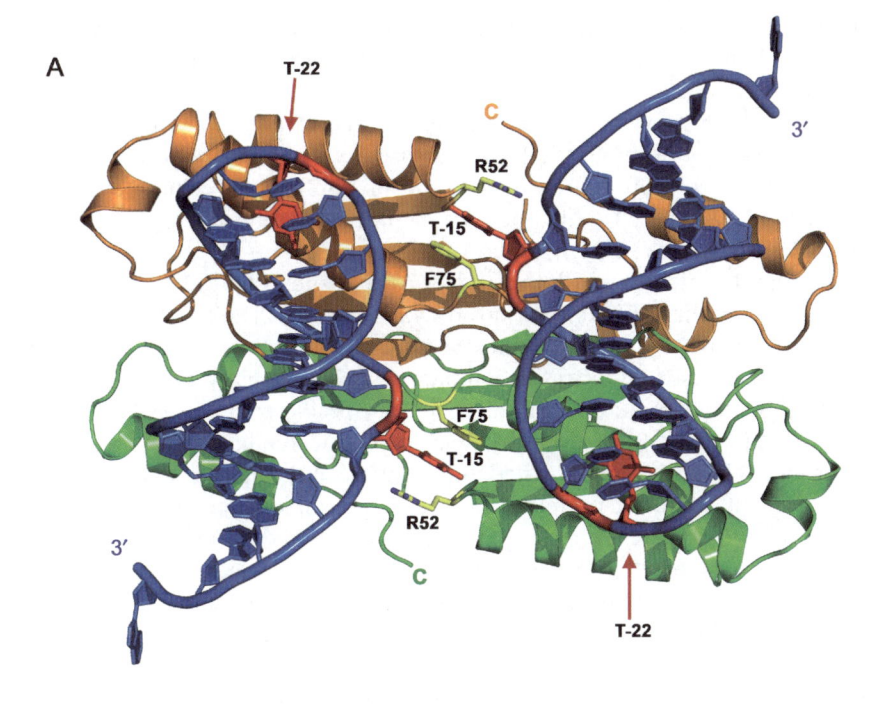

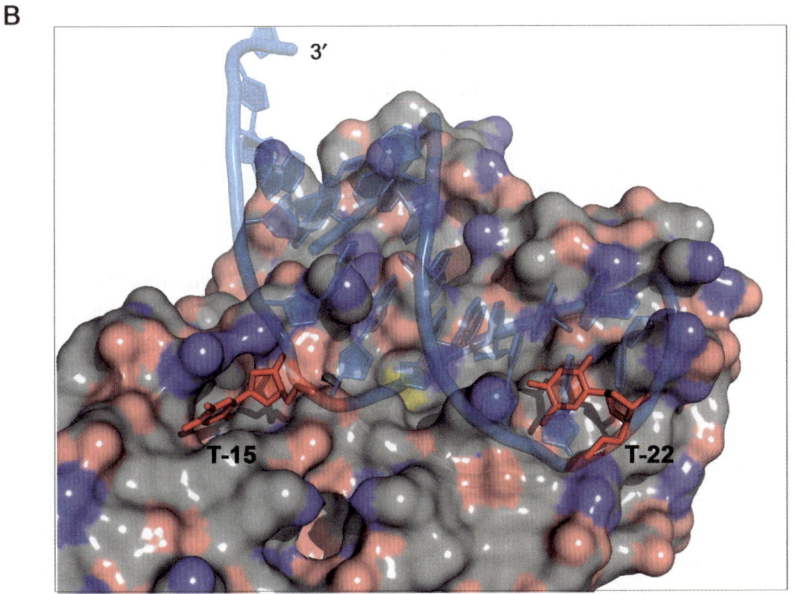

Figure 11.10 TnpA from IS*Hp608*. (*A*) Two TnpA monomers shown in orange and green bind two DNA hairpins (blue) that are formed by imperfect palindromes located in subterminal regions of the IS ends.[76] The bases shown in red, T-15 which is flipped out from the axis of the DNA and T-22 located at the hairpin tip, are specifically recognized by TnpA. (*B*) Surface representation of TnpA showing two pockets in the transposase surface that recognize T-15 and T-22. The colors represent atom type: N in blue, O in pink, C in gray, and S in yellow.

TnpA is indeed a member of the HUH superfamily, it forms an obligatory dimer with a shared active site in which the tyrosine nucleophile of one monomer crosses over into the vicinity of the HUH motif of the other monomer.[76] This contrasts with the other structurally characterized HUH nucleases (and presumably Y2 transposases) which are monomers, and is reminiscent of the active site organization of Flp, a site-specific recombinase (Chapter 12). As there is only one tyrosine in the neighborhood of each HUH motif, TnpA has been designated a Y1 transposase.

The complex of TnpA with the 22-mer hairpin shows that DNA recognition is distinct from that of Y and S transposases or recombinases. TnpA consists of a single domain without separate site-specific DNA binding or catalytic domains. In the structure of the complex, two 22-mer hairpins are bound on one face of the TnpA dimer (Figure 11.10). Specific recognition by TnpA is mediated through an unpaired thymine base, T-15 (counting from the transposon end at base 1), that is located in the middle of the hairpin stem and which is flipped out from the double helix into a pocket on the surface of TnpA (Figure 11.10B). There it forms an array of H-bonds and is sandwiched between the side chains of Phe-75 and Arg-52. A second point of specific DNA recognition is at the tip of the hairpin, where there are two unpaired thymines. One, T-22, is flipped inward where it points toward the bottom of the hairpin and forms an H-bond to a guanine base in the stem. It sits close to the protein surface snugly in a pocket. The rest of the hairpin is recognized non-specifically through contacts to phosphate groups.

The structure reveals that the features of the hairpin that are important for recognition are precisely those that deviate from a regular Watson–Crick base-paired structure; in other words, the imperfections. This gives TnpA an obvious way to discriminate between hairpins formed on opposite DNA strands. Clearly, a double-stranded palindromic sequence can form DNA hairpins on both strands, generating a cruciform-like structure. If this occurs, the sequence of the hairpin formed by the top strand differs from that formed by the bottom strand. Thus, in the case of ISHp*608*'s RE hairpin, the equivalent of the flipped out T-15 of the top strand is a flipped out A-15 on the bottom strand hairpin. The crystal structure shows that an A base could not be accommodated in the observed binding site on the TnpA surface. Similarly, the equivalents of two T bases at the tip of the top strand RE hairpin are A bases on the bottom hairpin; again, an A base would not fit the TnpA binding pocket for T-22. It appears that, to initiate transposition, TnpA cleaves the top strand of the transposon,[88] so ensuring recognition of only the top strand hairpins makes sense. The observed dimer with two bound DNA hairpins (Figure 11.10A) suggests how TnpA can bring the ends of ISHp*608* into close proximity. Such proximity is required to create both the closed transposon circles and sealed donor DNA ends that are observed during transposition.

Clearly, we are just beginning to understand the intricacies of molecular systems that mobilize DNA and their regulation. Nevertheless, it is certain that the application of structural methods, combined with clever biochemical experiments, is one of the most effective ways to move forward.

Acknowledgements

This work was funded by the Intramural Program of the National Institutes of Health. We are grateful to Drs Michael Chandler and Kiyoshi Mizuuchi for critical reading of the manuscript.

References

1. N.C. Comfort, *J. Hist. Biol.*, 1999, **32**, 133.
2. B. McClintock, *Proc. Natl. Acad. Sci. U.S.A.*, 1950, **36**, 344.
3. D.J. Finnegan, *Curr. Opin. Cell Biol.*, 1990, **2**, 471.
4. M.J. Curcio and K.M. Derbyshire, *Nat. Rev. Mol. Cell Biol.*, 2003, **4**, 865.
5. F. Cornet and M. Chandler, in *Microbial Evolution: Gene Establishment, Survival and Exchange*, ed. R.V. Miller and M.J. Day, ASM Press, Washington DC, 2004, p. 36.
6. *Mobile DNA II*, ed. N.L. Craig, R. Craigie, M. Gellert, and A.M. Lambowitz, ASM Press, Washington DC, 2002.
7. J. Mahillon and M. Chandler, *Microbiol. Mol. Biol. Rev.*, 1998, **62**, 725.
8. D.B. Haniford, *Crit. Rev. Biochem. Mol. Biol.*, 2006, **41**, 407.
9. J.E. Peters and N.L. Craig, *Nat. Rev. Mol. Cell Biol.*, 2001, **2**, 806.
10. G.J. Morgan, G.F. Hatfull, S. Casjens and R.W. Hendrix, *J. Mol. Biol.*, 2002, **317**, 337.
11. K. Mizuuchi, *Annu. Rev. Biochem.*, 1992, **61**, 1011.
12. V.V. Kapitonov and J. Jurka, *DNA Cell Biol.*, 2004, **23**, 311.
13. E.S. Lander et al., *Nature*, 2001, **409**, 860.
14. J.A. Yoder, C.P. Walsh and T.H. Bestor, *Trends Genet.*, 1997, **13**, 335.
15. N.L. Vastenhouw and R.H.A. Plasterk, *Trends Genet.*, 2004, **20**, 314.
16. W.S. Reznikoff, in *Mobile DNA II*, ed. N.L. Craig, R. Craigie, M. Gellert and A.M. Lambowitz, ASM Press, Washington DC, 2002, p. 403.
17. D.C. Rio, in *Mobile DNA II*, ed. N.L. Craig, R. Craigie, M. Gellert and A.M. Lambowitz, ASM Press, Washington DC, 2002, p. 484.
18. R.H.A. Plasterk, Z. Izsvák and Z. Ivics, *Trends Genet.*, 1999, **15**, 326.
19. Z. Ivics, P.B. Hackett, R.H. Plasterk and Z. Izsvák, *Cell*, 1997, **91**, 501.
20. Z. Ivics, C.D. Kaufman, H. Zayed, C. Miskey, O. Walisko and Z. Izsvák, *Curr. Issues Mol. Biol.*, 2004, **6**, 43.
21. E. Rubin, G. Lithwick and A.A. Levy, *Genetics*, 2001, **158**, 949.
22. B.R. Calvi, T.J. Hong, S.D. Findley and W.M. Gelbart, *Cell*, 1991, **66**, 465.
23. W.D. Warren, P.W. Atkinson and D.A. O'Brochta, *Genet. Res.*, 1994, **64**, 87.
24. L. Zhou, R. Mitra, P.W. Atkinson, A.B. Hickman, F. Dyda and N.L. Craig, *Nature*, 2004, **432**, 995.
25. A.B. Hickman, Z.N. Perez, L. Zhou, P. Musingarimi, R. Ghirlando, J.E. Hinshaw, N.L. Craig and F. Dyda, *Nat. Struct. Mol. Biol.*, 2005, **12**, 715.
26. C.P. Lu, H. Sandoval, V.L. Brandt, P.A. Rice and D.B. Roth, *Nat. Struct. Mol. Biol.*, 2006, **13**, 1010.
27. G.J. Grundy, J.E. Hesse and M. Gellert, *Proc. Natl. Acad. Sci. U.S.A.*, 2007, **104**, 3078.

28. M. Gellert, *Annu. Rev. Biochem.*, 2002, **71**, 101.
29. R. Craigie, M. Mizuuchi and K. Mizuuchi, *Cell*, 1984, **39**, 387.
30. L.K. Arciszewska and N.L. Craig, *Nucleic Acids Res.*, 1991, **19**, 5021.
31. L.K. Arciszewska, R.L. McKown and N.L. Craig, *J. Biol. Chem.*, 1991, **266**, 21736.
32. P. Rice and K. Mizuuchi, *Cell*, 1995, **82**, 209.
33. K. Katayanagi, M. Miyagawa, M. Matsushima, M. Ishikawa, S. Kanaya, M. Ikehara, T. Matsuzaki and K. Morikawa, *Nature*, 1990, **347**, 306.
34. W. Yang, W.A. Hendrickson, R.J. Crouch and Y. Satow, *Science*, 1990, **249**, 1398.
35. F. Dyda, A.B. Hickman, T.M. Jenkins, A. Engelman, R. Craigie and D.R. Davies, *Science*, 1994, **266**, 1981.
36. M.A. Landree, J.A. Wibbenmeyer and D.B. Roth, *Genes Dev.*, 1999, **13**, 3059.
37. D.R. Kim, Y. Dai, C.L. Mundy, W. Yang and M.A. Oettinger, *Genes Dev.*, 1999, **13**, 3070.
38. J.A. Shapiro, *Proc. Natl. Acad. Sci. U.S.A.*, 1979, **76**, 1933.
39. S. Bolland and N. Kleckner, *Cell*, 1996, **84**, 223.
40. A.K. Kennedy, D.B. Haniford and K. Mizuuchi, *Cell*, 2000, **101**, 295.
41. L.S. Beese and T.A. Steitz, *EMBO J.*, 1991, **10**, 25.
42. P.D. Kaufman and D.C. Rio, *Cell*, 1992, **69**, 27.
43. A.B. Hickman, Y. Li, S.V. Mathew, E.W. May, N.L. Craig and F. Dyda, *Mol. Cell*, 2000, **5**, 1025.
44. P. Rousseau, C. Normand, C. Loot, C. Turlan, R. Alazard, G. Duval-Valentin and M. Chandler, in *Mobile DNA II*, ed. N.L. Craig, R. Craigie, M. Gellert and A.M. Lambowitz, ASM Press, Washington DC, 2002, p. 367.
45. G. Duval-Valentin, B. Marty-Cointin and M. Chandler, *EMBO J.*, 2004, **23**, 3897.
46. N.D.F. Grindley, in *Mobile DNA II*, ed. N.L. Craig, R. Craigie, M. Gellert and A.M. Lambowitz, ASM Press, Washington DC, 2002, p. 272.
47. E.W. May and N.L. Craig, *Science*, 1996, **272**, 401.
48. E. Gueguen, P. Rousseau, G. Duval-Valentin and M. Chandler, *Trends Microbiol.*, 2005, **13**, 543.
49. D.R. Davies, I.Y. Goryshin, W.S. Reznikoff and I. Rayment, *Science*, 2000, **289**, 77.
50. T.A. Naumann and W.S. Reznikoff, *Proc. Natl. Acad. Sci. U.S.A.*, 2000, **97**, 8944.
51. J. Bischerour and R. Chalmers, *Nucleic Acids Res.*, 2007, **35**, 2584.
52. P.A. Rice, S.W. Yang, K. Mizuuchi and H.A. Nash, *Cell*, 1996, **87**, 1295.
53. R.T. Clubb, J.G. Omichinski, H. Savilahti, K. Mizuuchi, A.M. Gronenborn and G.M. Clore, *Structure*, 1994, **2**, 1041.
54. S. Schumacher, R.T. Clubb, M. Cai, K. Mizuuchi, G.M. Clore and A.M. Gronenborn, *EMBO J.*, 1997, **16**, 7532.
55. R.T. Clubb, S. Schumacher, K. Mizuuchi, A.M. Gronenborn and G.M. Clore, *J. Mol. Biol.*, 1997, **273**, 19.

56. I. Levchenko, L. Luo and T.A. Baker, *Genes Dev.*, 1995, **9**, 2399.
57. H. Nakai, V. Doseeva and J.M. Jones, *Proc. Natl. Acad. Sci. U.S.A.*, 2001, **98**, 8247.
58. J.F. Yuan, D.R. Beniac, G. Chaconas and F.P. Ottensmeyer, *Genes Dev.*, 2005, **19**, 840.
59. H. Savilahti, P.A. Rice and K. Mizuuchi, *EMBO J.*, 1995, **14**, 4893.
60. H. Savilahti and K. Mizuuchi, *Cell*, 1996, **85**, 271.
61. H. Aldaz, E. Schuster and T.A. Baker, *Cell*, 1996, **85**, 257.
62. S. Pathania, M. Jayaram and R.M. Harshey, *Cell*, 2002, **109**, 425.
63. R.M. Harshey and M. Jayaram, *Crit. Rev. Biochem. Mol. Biol.*, 2006, **41**, 387.
64. S. Watkins, G. van Pouderoyen and T.K. Sixma, *Nucleic Acids Res.*, 2004, **32**, 4306.
65. J.M. Richardson, A. Dawson, N. O'Hagan, P. Taylor, D.J. Finnegan and M.D. Walkinshaw, *EMBO J.*, 2006, **25**, 1324.
66. D.J. Lampe, M.E.A. Churchill and H.M. Robertson, *EMBO J.*, 1996, **15**, 5470.
67. J.C. Vos, I. DeBaere and R.H.A. Plasterk, *Genes Dev.*, 1996, **10**, 755.
68. L.R.O. Tosi and S.M. Beverley, *Nucleic Acids Res.*, 2000, **28**, 784.
69. L. Aravind and D. Landsman, *Nucleic Acids Res.*, 1998, **26**, 4413.
70. H.M. Robertson, *J. Insect Physiol.*, 1995, **41**, 99.
71. A. Dawson and D.J. Finnegan, *Mol. Cell*, 2003, **11**, 225.
72. L. Zhang, A. Dawson and D.J. Finnegan, *Nucleic Acids Res.*, 2001, **29**, 3566.
73. C. Augé-Gouillou, B. Brillet, S. Germon, M.H. Hamelin and Y. Bigot, *J. Mol. Biol.*, 2005, **351**, 117.
74. Z. Izsvák, D. Khare, J. Behlke, U. Heinemann, R.H. Plasterk and Z. Ivics, *J. Biol. Chem.*, 2002, **277**, 34581.
75. L. Aravind, *Trends Biochem. Sci.*, 2000, **25**, 421.
76. D.R. Ronning, C. Guynet, B. Ton-Hoang, Z.N. Perez, R. Ghirlando, M. Chandler and F. Dyda, *Mol. Cell*, 2005, **20**, 143.
77. H.H. Lee, J.Y. Yoon, H.S. Kim, J.Y. Kang, K.H. Kim, D.J. Kim, J.Y. Ha, B. Mikami, H.J. Yoon and S.W. Suh, *J. Biol. Chem.*, 2006, **281**, 4261.
78. W. Li, S. Kamtekar, Y. Xiong, G.J. Sarkis, N.D.F. Grindley and T.A. Steitz, *Science*, 2005, **309**, 1210.
79. G. Whittle, N.B. Shoemaker and A.A. Salyers, *Cell. Mol. Life Sci.*, 2002, **59**, 2044.
80. M. Sebaihia et al., *Nat. Genet.*, 2006, **38**, 779.
81. E.V. Koonin and T.V. Ilyina, *Biosystems*, 1993, **30**, 241.
82. M.P. Garcillán-Barcia, I. Bernales, M.V. Mendiola and F. de la Cruz, in *Mobile DNA II*, ed. N.L. Craig, R. Craigie, M. Gellert and A.M. Lambowitz, ASM Press, Washington DC, 2002, p. 891.
83. N. Tavakoli, A. Comanducci, H.M. Dodd, M. Lett, B. Albiger and P. Bennett, *Plasmid*, 2000, **44**, 66.
84. A. Guasch, M. Lucas, G. Moncalián, M. Cabezas, R. Pérez-Luque, F.X. Gomis-Rüth, F. de la Cruz and M. Coll, *Nat. Struct. Biol.*, 2003, **10**, 1002.

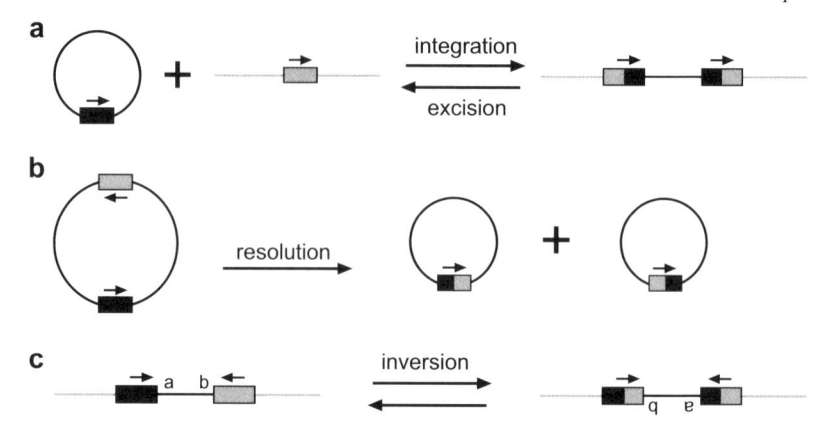

Figure 12.1 Examples of site-specific recombination. Specific recombination sequences are indicated by light and dark boxes and arrows indicate the relative orientation of the sites. (*a*) Integration of a circular DNA molecule into a specific site on a second molecule. The reverse reaction excises the integrated DNA as a circle. (*b*) The resolution reaction is a special case of the excision pathway shown in (a). In the excision/resolution reactions, the recombination sequences are arranged in a head-to-tail orientation (direct repeats). (*c*) When the sites are oriented in a head-to-head fashion (inverted repeats), the outcome of site-specific recombination is inversion of the DNA located between the sites.

number of the yeast 2μ plasmid. Table 12.1 gives examples of well-studied enzyme systems that carry out these reactions.

Several families of enzymes carry out biological functions similar to those listed in Table 12.1. The retroviral integrases, for example, catalyze the integration of viral DNA into a host's genome[3] and the Rag1 and Rag2 recombinases catalyze the excision of DNA segments during immunoglobulin and T-cell receptor gene rearrangements.[4] Although both have "site-specific" elements and both are recombinases they use mechanisms that are more closely related to transposition (see Chapter 11) than to what is generally referred to as site-specific recombination. This chapter is restricted to discussion of "conservative site-specific recombinases", which are those enzymes that form transient phosphotyrosine or phosphoserine linkages during recombination. These systems have the characteristic feature that they do not involve any net hydrolysis or synthesis of phosphodiester linkages during the course of the reaction.

Essentially all of the conservative site-specific recombinases that have been identified to date fall into one of two distinct families: the tyrosine recombinases or the serine recombinases. The two families are named after the protein nucleophile responsible for formation of covalent recombinase-DNA intermediates during the exchange of DNA strands. The tyrosine recombinases (YRs) and serine recombinases (SRs) appear to be unrelated at the level of primary sequence and proteins from the two families have quite different three-

Site-specific Recombinases

Table 12.1 Functions of site-specific recombinases.

Function	Examples[a]	Family[b]
Bacteriophage integration & excision	Bacteriophage λ-integrase[5,38]	YR
	Bacteriophage φC31-integrase[83]	SR
Chromosome/plasmid dimer resolution	Bacteriophage P1 Cre recombinase[29,93]	YR
	E. coli XerCD recombinases[47,94]	YR
	S. aureus Sin recombinase[82]	SR
Cointegrate resolution during transposition	Transposon Tn3-resolvase[64,95]	SR
	Transposon γδ-resolvase[7,64]	SR
DNA inversion gene expression switch	*S. typhimurium* Hin invertase[74]	SR
	E. coli FimB/FimE recombinase[96]	YR
Plasmid copy amplification	*S. cerevisiae* Flp recombinase[24,97]	YR
Integron gene cassette integration/excision	*Vibrio cholerae* IntI4 recombinase[14,55]	YR
DNA excision during developmental gene expression	*B. subtilis* SpoIVCA[98]	SR
Integration & excision of conjugative transposons	Transposon Tn916 integrase[13,99]	YR

[a] Only a subset of characterized site-specific recombinases is listed as examples.
[b] YR = tyrosine recombinase, SR = serine recombinase.

dimensional structures. As discussed throughout this chapter, the YRs and SRs also use very different mechanisms to carry out recombination. Given the lack of similarity at the sequence, structural, and mechanistic levels, one might expect that the two recombinase families would have evolved to become specialized for distinct types of DNA rearrangements. As shown in Table 12.1, this is clearly not the case. There is nearly a complete overlap in function between the two families of site-specific recombinases.

An interesting example of the functional overlap between the YR and SR families is the bacteriophage integrase enzymes. The integrase from bacteriophage λ is the grandfather of the YRs; indeed, this group of recombinases was previously referred to as the λ-integrase family.[5] As expected, numerous λ-like integrases have been identified in other bacteriophages and some have been characterized biochemically. It is now known, however, that the SRs also include a growing number of identified "serine integrases" that carry out integration and excision reactions similar to their YR counterparts.[6]

As in many other nucleic acid processing systems, structural biology has made important contributions to understanding the tyrosine and serine recombinases. Table 12.2 summarizes the YR and SR systems where experimental structures have been reported. Both the YRs and SRs are multi-domain proteins and most have been refractory to crystallization in the absence of DNA substrates or are too poorly behaved in solution to be readily studied using NMR methods. Structures of isolated domains and DNA-binding

Table 12.2 Representative structures of site-specific recombinases and recombinase/DNA complexes.

Protein	Source	Family[a]	Structure	PDB code[b]
λ-integrase	Bacteriophage λ	YR	Pre-cleavage tetramer (a.a. 75-356)[43]	1Z19
			Covalent protein-DNA intermediate[43]	1Z1B
			Holliday junction intermediate[43]	1Z1G
			Integrase/half-site complex[42]	1P7D
			Catalytic domain[41]	1AE9
			Arm-binding domain[39]	1KJK
Cre recombinase	Bacteriophage P1	YR	Pre-cleavage complex[32]	4CRX
			Covalent protein-DNA intermediate[18]	1CRX
			Holliday junction intermediate[35]	1XO0
			Cre/3-way junction complex[100]	1F44
Flp recombinase	Yeast 2μ plasmid	YR	Holliday junction intermediate[36]	1FLO
IntI4 integrase	*Vibrio cholerae*	YR	Pre-cleavage synaptic complex[57]	2A3V
XerD recombinase	*E. coli*	YR	Unliganded protein[53]	1AOP
HP1 integrase	Bacteriophage HP1	YR	Catalytic domain[101]	1AIH
Tn916 integrase	Transposon Tn916	YR	Arm-binding domain/DNA complex[102]	1B69
γδ-resolvase	Transposon γδ	SR	Resolvase dimer/*res* site I complex[70]	1GDT
			Resolvase/DNA cleaved complex[71]	1ZR4
			Unliganded resolvase[69]	2RSL
			Activated resolvase tetramer[68]	2GM5
			Catalytic domain[67]	1HX7
			DNA-binding domain[66]	1RES
Hin invertase	*S. typhimurium*	SR	DNA-binding domain/DNA complex[103]	1HCR

[a] YR = tyrosine recombinases, SR = serine recombinases.
[b] Accession code for the Protein Data Bank. A single representative accession code is given in cases where multiple structures have been reported for a given protein or protein-DNA complex. Structures where coordinates have not been deposited in the PDB are not included.

domain/DNA complexes have provided useful insights, as have the structures of a small number of unliganded proteins. However, the most revealing data have come from structures of recombinase-DNA complexes that represent "snapshots" of stages in the recombination pathway. In some cases (*e.g.*, Cre,

Flp, λ-integrase, and γδ-resolvase), several independent structures have been determined for one or more of the reaction intermediates.

In addition to revealing the overall architectures of tyrosine and serine recombinase-DNA assemblies, the structural models represented in Table 12.2 have also made important contributions to understanding the phosphoryl transfer chemistry that forms the basis for the cleavage and ligation steps of the site-specific recombination pathways. This is true both at the level of catalysis within the enzyme active sites and with respect to the mechanism of strand exchange within the assembled recombination complex. In this chapter, I summarize what is known about the structures and mechanisms of action of the tyrosine and serine recombinases, with an emphasis on systems where structural biology has played an important role. I also provide some background on additional systems where structural data is not yet available, but would provide valuable new insights.

The mechanistic aspects of YR and SR site-specific recombination have been discussed in depth in a recent review.[7] Many of the individual systems have also been reviewed in detail in *Mobile DNA II*, a comprehensive text covering many aspects of site-specific recombination and transposition.[8]

12.2 Tyrosine Recombinases

Although the DNA sequences required for site-specific recombination by YR family members can be complex, all YRs utilize a core recombination site where the strand exchange reactions are catalyzed. These sites share a similar organization, with two recombinase binding elements (RBEs) arranged with approximate dyad symmetry flanking a central crossover region (Figure 12.2(a)). The RBEs vary in size up to ~13 bp and the RBE sequences are specific for each different recombinase. Most YR systems utilize a single recombinase enzyme that binds to the RBEs (*i.e.*, the recombinases labeled A–D in Figure 12.2a are identical) and the RBEs are similar or identical in sequence. In some systems, however, two different recombinases bind to the core site and the corresponding RBEs differ in sequence. The size of the crossover region also varies from ~6 to 8 bp among the biochemically characterized YR systems.

Tyrosine recombinase family members share a common mechanism of strand exchange that involves formation of covalent 3′-phosphotyrosine linkages between the recombinase enzyme and the DNA substrate. These linkages are then targets for ligation by 5′-hydroxyl groups that differ from the ones released as a result of the initial cleavage events. The YRs use this strategy to exchange one pair of DNA strands between two associated core sites to form a four-way Holliday junction (HJ) intermediate. The HJ intermediate is then a substrate for a second round of cleavage and ligation reactions, but this time the complementary strands are exchanged to produce recombinant products. Figure 12.2 illustrates the reaction schematically.

A key feature of the YR mechanism is that strand exchange is a stepwise process, where the two strand exchange steps are linked by the central

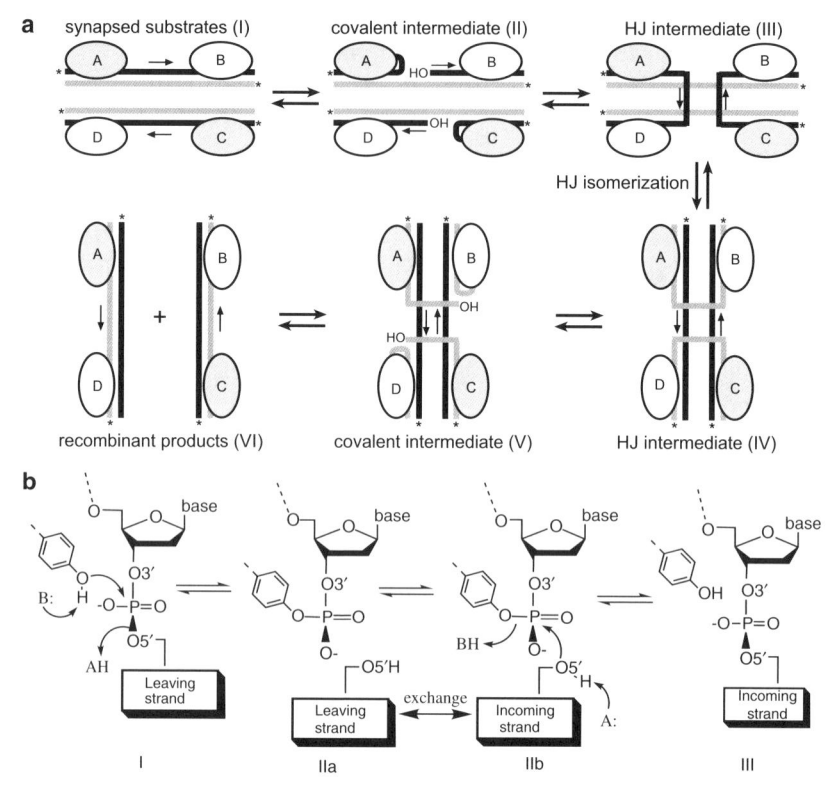

Figure 12.2 Mechanism of tyrosine recombinase site-specific recombination. (*a*) Schematic of the reaction pathway. Only the core recombination sites are shown. Each core site is composed of two recombinase binding elements (RBEs) arranged as inverted repeats surrounding a central crossover region where strand exchange takes place. Upon synapsis (association) of two sites to form a synaptic complex (I), two of the recombinase subunits are active for cleavage (A and C). Cleavage by a conserved tyrosine residue results in formation of a covalent intermediate (II) in which the recombinase subunits form 3′-phosphotyrosine linkages to the dark strands of the DNA substrates. Strand exchange and ligation by the 5′-hydroxyl group from the partner substrate generates a four-way Holliday junction (HJ) intermediate (III). The HJ intermediate isomerizes to adopt a configuration in which recombinases B and D are now active to catalyze exchange of the light DNA strands to generate recombinant products in the bottom half of the pathway, using the same cleavage/ligation chemistry used in the top half. In some YR systems, recombinases A and B are identical; in others, they are different proteins. In all cases, however, neighboring subunits are stereochemically distinct and the core recombination complex has approximate local twofold symmetry. (*b*) Close-up of the active sites for intermediates I–III in part (*a*), illustrating the phosphoryl transfer steps involved in YR strand exchange. Transition states are not shown. Two states are shown for the covalent intermediate; IIa is the cleavage product shown in part (*a*) and IIb is the result of strand exchange prior to chemical ligation.

HJ intermediate. The overall architecture of the YR mechanism implies that an antiparallel alignment (or an alignment skewed towards antiparallel) of sites is favored and that the reaction pathway passes through intermediates with approximate twofold symmetry (12.2(a)).

In most of the well-studied YR systems there is a strict requirement for identical crossover sequences between the recombining sites. This requirement has been explained by a strand exchange mechanism referred to as "strand swapping".[9] For the 5'-hydroxyl group of one DNA substrate to attack the phosphotyrosine linkage of the other substrate, base-pairs must be melted from the complementary strand and then annealed to the partner substrate's complementary strand.[10–12] The exact number of base-pairs melted and annealed during strand transfer is not known for all of the well-studied systems, but is most likely to be ~ 3 bp. If mismatches are encountered during the strand swap, ligation is inhibited and strand exchange is reversed in favor of re-forming the initial substrates. While this scenario describes the behavior of YR family members such as λ-integrase, Cre, and Flp, there are also YRs that contain mismatched crossover regions. These systems violate the identical crossover sequence rule, yet they still manage to exchange strands.[13,14]

The phosphoryl transfer reaction used by YR enzymes is essentially the same reaction that is catalyzed by type I topoisomerases (see Chapter 10). The primary difference is that in the topoisomerase reaction the attacking 5'-hydroxyl group is the same one that was displaced in the original cleavage event. It is now well established that the tyrosine recombinases and type IB topoisomerases (TopIBs) share a conserved catalytic domain, active site, and mechanism of transesterification.[15] The phosphoryl transfer catalytic steps are summarized in Figure 12.2(b). For both the type IB topoisomerases and the YRs, general acid catalysis to assist the 5'-OH leaving group is crucial for the initial cleavage event. Mutation of the conserved lysine that acts as general acid results in YR and TopIB enzymes with greatly reduced activity.[16,17] Acid/base catalysis at the tyrosine nucleophile appears to be less important. Although a conserved histidine residue has been identified as a likely candidate for this role,[18,19] substitutions of this residue in Cre and Flp are still active[19,20] and the TopIB enzymes have no obvious protein residue to carry out this function.

12.2.1 Cre and Flp Recombinases

The Cre and Flp recombinases are often referred to as "simple" members of the YR family of enzymes because they can carry out site-specific recombination on simple substrates without the need of accessory factors or auxiliary DNA sequences. The minimal requirements for recombination are DNA substrates containing the core recombination sequence (a 34-bp *loxP* site for Cre and a 34-bp *frt*-derived site for Flp) and the recombinase enzymes themselves. Cre and Flp will also recombine DNA substrates in various topological contexts, which greatly simplifies investigation and application of the *in vitro* reaction. Both systems have been extensively studied biochemically and in many ways have become the prototype YR family members.[21–24] Cre and Flp are also widely used in genome engineering applications and as tools for *in vitro* and *in vivo* cloning procedures.[25–28]

The natural function of Cre recombinase is to maintain the bacteriophage P1 chromosome in a monomeric state.[29] The lysogenic state of phage P1 is a unit copy episome that rarely integrates into the *E. coli* chromosome. Flp recombinase mediates copy number amplification for the *S. cerevisiae* 2μ circle in the event that segregation defects lead to plasmid reduction in daughter cells.[24] Cre therefore functions as a resolvase (Figure 12.1b) and Flp uses both the resolution and inversion functions (Figure 12.1c) to generate multiple plasmid copies from a single initiation of replication.[30] Neither functions well as an integrase, since the product of integration is the natural substrate for these enzymes and the intramolecular excision reaction is strongly favored.

Despite extensive efforts, both Cre and Flp have proven resistant to attempts to determine the structures of individual domains or of the isolated proteins. All structural data for these systems have therefore come from crystal structures of recombinase-DNA complexes. Cre and Flp are both two-domain proteins (Figure 12.3). The N-terminal "core-binding" or CB domains have distinct folds, but the C-terminal domains share the catalytic domain core structure that is common to all YR enzymes. Flp has additional C-terminal sequences that further elaborate the catalytic domain. Both proteins have at least one structural element that is exchanged in a cyclic manner with neighboring subunits in the recombinase tetramer that is assembled at the start of recombination (Figure 12.2a). For Cre, a helix located at the C-terminus forms a cyclic crosslink that plays a major role in stabilizing the synaptic complex. For Flp, both an inter-domain helix and a short helix containing the catalytic tyrosine are delivered to adjacent subunits. In both cases, these quaternary interactions are involved in regulating the activity of recombinase subunits during recombination.[21,31]

An interesting difference between the two systems concerns the source of the tyrosine nucleophile. Cre cleaves its core half-site using an active site tyrosine from the same subunit bound to that half-site (*cis*-cleavage). Flp forms a shared active site, where the tyrosine nucleophile is donated in *trans* from an adjacent subunit. The *trans*-cleavage variation appears to be unique to the yeast branch of YR enzymes that are most closely related to Flp recombinase. The cartoon mechanism in Figure 12.2(a) can be readily modified to accommodate the Flp-like YR systems by changing the covalent protein-DNA linkages from A and C to B and D, respectively, in the top half of the pathway and from B and D to A and C in the bottom half, without moving the subunits from their current positions.

Our mechanistic understanding of YR site-specific recombination has been strongly influenced by the structural biology of Cre-*loxP* and Flp-*frt* complexes. In the Cre system, it has been possible to determine crystal structures representing each of the reaction intermediates shown in Figure 12.2(a). In each case, multiple independent structures have been reported. Indeed, schemes such as the one shown in Figure 12.2(a) are the result of interpreting these structural

Site-specific Recombinases 311

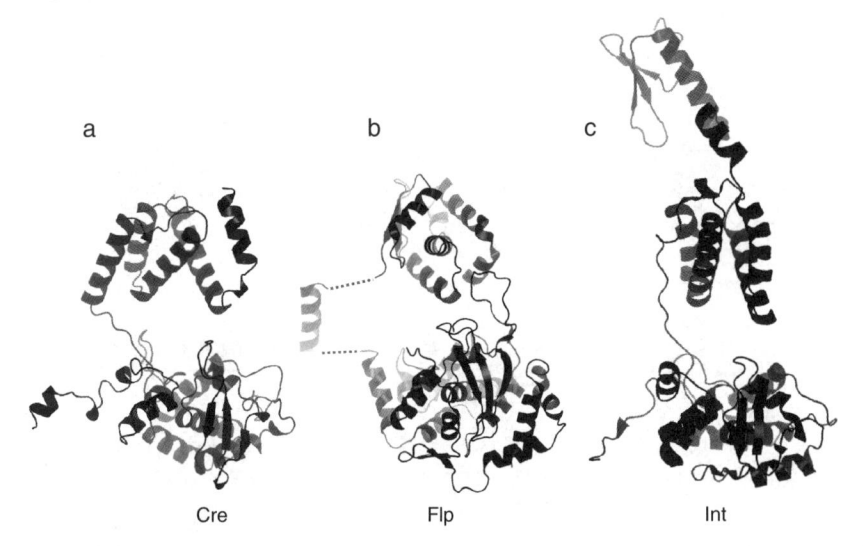

Figure 12.3 Domain structures of tyrosine recombinases, as observed in the DNA complexes listed in Table 12.2. (*a*) Cre recombinase has N-terminal core-binding (CB) and C-terminal catalytic domains, and a C-terminal helix that interacts with neighboring subunits during recombination (PDB entry 1CRX). (*b*) Flp recombinase has CB and catalytic domains, and an inter-domain helix that interacts with neighboring subunits (1FLO). (*c*) In addition to CB and catalytic domains, λ-integrase has an N-terminal arm-binding domain that binds to accessory sequences outside of the core recombination site (1Z1B). λ-Integrase also has a C-terminal segment that interacts with neighboring subunits analogous to the C-terminal helix in Cre. These and other structures in this chapter were drawn using PyMol.[92]

models in the context of years of biochemical studies in the Cre, Flp, XerCD, and λ-integrase systems.

Figure 12.4(a) shows the Cre-HJ complex structure. The pre-cleavage synaptic complex[32,33] (intermediate I) and covalent intermediate[18,33] (intermediate II) complexes are nearly superimposable on the HJ intermediate[20,34,35] (III); only the DNA connectivity is changed in the central part of the complex. Presumably, the lack of large changes in quaternary structure between these intermediates (representing the top half of the pathway in Figure 12.2a) and the enclosed nature of the strand exchange region is what allowed many of these structures to be determined in nearly isomorphous crystal forms.

The mechanistic implications of these observations are also important. During synapsis of *loxP* sites to form a two-fold symmetric, antiparallel alignment, Cre bends the substrates to form a framework in which strand exchange can take place in the absence of any large changes in quaternary structure. Isomerization of the HJ intermediate most likely does involve modest structural changes.[34] This can best be described as a subtle scissoring motion of

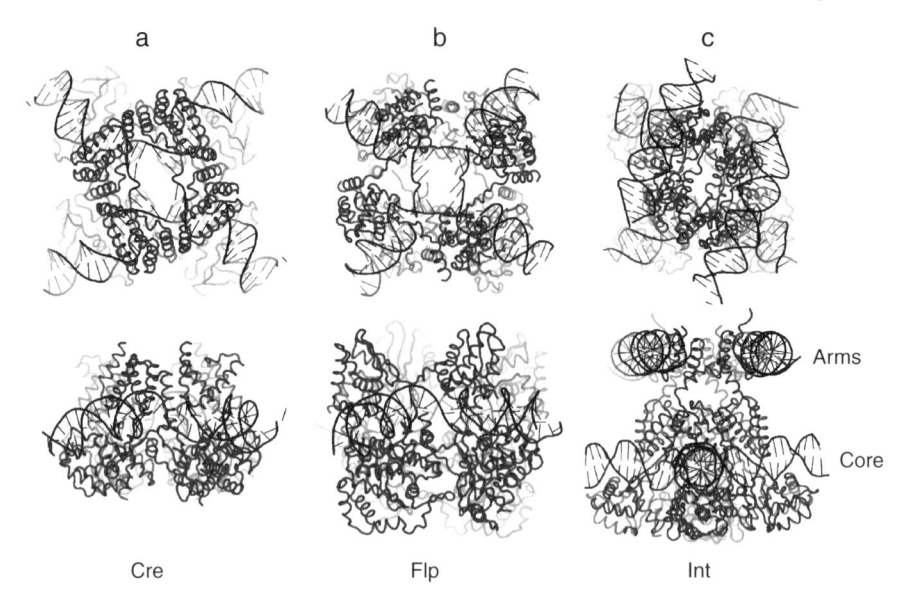

Figure 12.4 Structures of tyrosine recombinase-Holliday junction complexes. Top panels are views from the CB-domain side of the complex (protein N-termini) and the minor groove faces of the junctions. Bottom panels show an orthogonal view, generated by 90° rotation about a horizontal axis, with CB-domains on the top and catalytic domains on the bottom. (*a*) Cre-HJ complex (PDB entry 1XO0); (*b*) Flp-HJ complex (1FLO); (*c*) λ-integrase-HJ complex, with arm-binding domains and arm/DNA duplexes present (1Z1G).

the junction arms to convert from one two-fold symmetric form into the alternative one (exaggerated in the Figure 12.2a cartoon). Note that the HJ intermediate shown in Figure 12.4(a) is nearly four-fold symmetric. The difference between alternative isomers is therefore small, but functionally important.

In the Flp system, structures have been described for the HJ intermediate of the recombination pathway,[36,37] but not for the pre-cleavage synaptic complex. Interestingly, the HJ complex structures have a 7-bp crossover sequence, rather than the 8-bp crossover that is found in the *frt* site. Since Flp will recombine substrates with the 7-bp crossover, these HJ complexes still represent informative snapshots of the Flp reaction. In particular, the Flp-HJ complexes have revealed in detail the differences between the *trans*-cleaving YR systems and the *cis*-cleaving systems represented by Cre.[31]

The Flp-HJ complex structure (Figure 12.4b) has a similar overall architecture to the Cre-HJ complex, particularly with respect to the arrangement of DNA arms. In both cases, the complexes are nearly, but not exactly, fourfold symmetric and the DNA arms are nearly coplanar. Both complexes show a small amount of curvature in the DNA substrates, creating a concave surface on the catalytic domain side of the HJ, where the major grooves of the DNA

Site-specific Recombinases 313

converge. As expected, the nature of inter-domain interactions differs considerably in the Cre and Flp systems, both in the CB-domains and in the catalytic domains.

12.2.2 λ-Integrase and XerCD Recombinases

The minimal nature of the Cre and Flp systems implies that they would be ill-suited for biological functions that demand a high level of regulation of the forward and reverse reactions. The bacteriophage λ-integrase (λ-int) system is the paradigm for such regulated activity.[5,38] Integration of bacteriophage λ into the *E. coli* chromosome involves site-specific recombination between phage *attP* and bacterial *attB* sites to form *attL* and *attR* sites flanking the integrated phage (Figure 12.5). While *attB* is a minimal core site similar to *loxP* in the Cre system, the ~230-bp *attP* site contains several regulatory sequences in the "arms" that flank the core site. Some of these regulatory sequences are binding sites for accessory proteins that are responsible for dictating whether λ-int

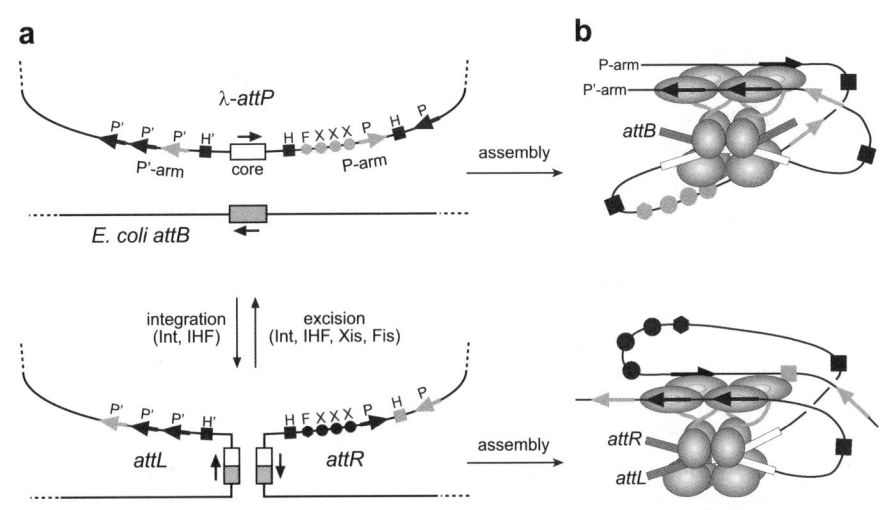

Figure 12.5 Site-specific recombination by λ-integrase. (*a*) Integration reaction between *attP* and *attB* to produce *attL* and *attR* sites and the excision reaction between *attL* and *attR* to produce *attP* and *attB*. IHF is required for integration. IHF and Xis are required for excision and Fis further stimulates the excision reaction. Arm domain binding sites are labeled P and P′, IHF binding sites H, Xis binding sites X, and Fis binding sites F. Black symbols indicate accessory sites that are used in the integration reaction (top) or excision reaction (bottom) and gray symbols indicate sites that are not used. (*b*) Cartoon models of the assembled complexes poised to carry out integration (top) and excision (bottom), based on recent structural and biochemical results.[43,45,46] Note that excision is not simply the reverse of integration; an entirely different complex is formed.

should carry out the integration or excision reaction. Others are binding sites for the arm-binding domains of λ-int.

λ-Integrase is an example of a bipartite DNA-binding protein. In addition to the CB and catalytic domains observed in Cre and Flp, λ-int has an N-terminal arm-binding domain (Figure 12.3c) whose structure has been studied by NMR.[39] The integrase protein therefore binds simultaneously to core sites and to sites in the flanking arms (labeled P and P' in Figure 12.5). The *E. coli* Integration Host Factor (IHF) protein promotes the integration reaction by forming sharp bends in the P and P' arms that allow assembly of a specific integration complex. When the phage-encoded Xis protein is present, formation of a specific excision complex is promoted and integration is inhibited. The *E. coli* protein Fis further stimulates excision by binding to *attR* and stimulating cooperative assembly of a short Xis-DNA filament.[40]

Decades of research have been devoted to understanding the integration and excision reactions of bacteriophage λ. In this context, contributions from structural biology have been relatively recent, but nonetheless quite informative. In addition to NMR studies of the arm-binding domain, crystal structures of the catalytic domain and an integrase-half-site DNA complex have been described.[41,42] In fact, the λ-int and HP1 integrase catalytic domain structures were the first reported for the YR family. Most recently, crystal structures of tetrameric λ-int/DNA complexes representing reaction intermediates were described.[43] Two of these (the covalent intermediate and HJ intermediate) also contain duplex DNA segments with arm-binding sequences (Figure 12.4c). A third structure represents a pre-cleavage synaptic complex without the arm-binding domains or arm DNA present.

Most of the features relating to the core-binding and catalytic domains of λ-int are similar to those described earlier for Cre and Flp. The full-length integrase/DNA complexes, although determined at low resolution (3.8 and 4.5 Å for the covalent intermediate and HJ intermediate, respectively) were unprecedented achievements and revealed a domain-swapped architecture for the arm-binding domains that was not anticipated from earlier work. As shown in cartoon form in Figure 12.5, the arm domains of λ-int are exchanged in a cyclic fashion in the integrase tetramer such that the arm domain of a given subunit "sits" on the CB-domain of an adjacent subunit. Identification of which integrase domains (four arm-binding and four core-binding) bind to which sites (four core half-sites and three of the available arm sites) is therefore a complex biochemical and structural issue. Recent studies using fluorescence resonance energy transfer (FRET) experiments to map interatomic distances in protein/DNA assemblies containing λ-int and accessory proteins have also been quite informative in establishing the architectures shown in Figure 12.5(b).[44–46]

Detailed structural models of the λ-int/DNA assemblies specific for integration and excision have been proposed based on the full-length integrase/DNA crystal structures.[43] These are shown in cartoon form in Figure 12.5(b). In both cases, it has been established that a specific three out of the five possible arm-binding sites are used and it is also known which IHF-binding sites are required for the two reactions.[5] At the time that these models were constructed, however, the stoichiometry and likely structure of the Fis-Xis region of *attR* had not yet been established. More recent biochemical and structural data for these accessory proteins and sites have led to additional constraints on the system and have allowed modeling of this region of *attR*.[40,46] These new data appear to be inconsistent with the structural model proposed for the excision complex. It seems plausible that contributions from intrinsic DNA-bending, perhaps coupled with the inherent flexibility of IHF and HU-like DNA bends, could lead to a model that is stereochemically consistent with all of the available structural and biochemical data, but this remains to be seen.

The XerCD recombinases are a second example of a well-studied YR system where the forward and reverse reactions are carefully regulated *in vivo*. This is also the prototype for a two-recombinase system, where the recombination pathway is inherently asymmetric even on simple core recombination sites. Xer recombination functions to maintain circular replicons in a monomeric state to promote efficient segregation upon cell division.[47] Essentially all eubacteria have XerC and XerD homologs, which resolve chromosome dimers (generated, for example, during repair of stalled replication forks) to monomers. *Escherichia coli* Xer recombination also resolves ColE1 and pSC101 plasmid multimers to monomers, using a regulatory mechanism that is quite different than that used in maintenance of the bacterial chromosome.[48]

On the *E. coli* chromosome, XerCD recombinases act on 28-bp *dif* sites.[47] Like the Cre and Flp systems, XerC and XerD will bind and cleave *dif* sites. However, XerCD will not execute a complete recombination reaction *in vitro*. *In vivo*, the cell division FtsK protein is required for recombination at *dif*. FtsK is a hexameric DNA-translocase that is tethered to the membrane at the septum during the process of cell division. In an early model for this regulatory process, strand exchange catalyzed by XerC (where XerD is the "inactive subunit" in Figure 12.2a) occurs in the absence of accessory proteins, but HJ isomerization and resolution by XerD-catalyzed strand exchange requires activation by FtsK.[49] More recent models for Xer regulation propose that FtsK facilitates initial strand exchange by XerD, which then leads to efficient HJ resolution to form products by XerC.[50] Although structures of FtsK domains have recently been reported,[51,52] the only structural data for the XerCD recombinases thus far is for the unliganded XerD protein.[53]

Xer-catalyzed resolution of plasmid dimers has also been studied in some detail. The recombination sites on ColE1 and pSC101 plasmids are called *cer* and *psi*, respectively. In addition to the core recombination sequences, *cer* and *psi* contain binding sites for accessory proteins that are required for recombination to occur. The *cer* site, for example, contains binding sites for the arginine repressor (ArgR) and leucine aminopeptidase (PepA). A specific nucleoprotein complex formed between two *cer* sites arranged as direct repeats in the same molecule (*i.e.*, in a plasmid dimer) results in juxtaposition of the core recombination sites and facilitates XerC-catalyzed strand exchange to form a HJ intermediate.[54] Cellular factors are then responsible for resolution of the HJ intermediate to plasmid monomers. Interestingly, the core recombination site in *cer* has an 8-bp crossover region, compared to the 6-bp crossover in *dif*. The 2-bp insertion into the core site of *cer* causes it to become completely dependent upon formation of accessory sequences to undergo strand exchange. The result is a "topological filter" that ensures that Xer recombination at *cer* is exclusively intramolecular.[48]

12.2.3 Integron Integrases

Integrons are mobile DNA modules present in many bacterial genomes.[14,55] They are composed of an integrase gene (*intI*), an integration site (*attI*), and a tandem array of integron cassettes, each of which contains the coding sequence of a gene to be transferred among a bacterial population (*e.g.*, genes coding for antibiotic resistance or virulence factors). The cassettes also contain recombination sequences, termed *attC*, which allow the genes to be excised from an integron as circles (*via attC* x *attC* recombination) and to be integrated into other integrons (*via attC* x *attI* recombination).

The integron integrases are members of the YR family, but their recombination sequences do not resembles those of other well-studied systems. The *attC* sites that flank integron cassette genes, for example, vary in length from ~50 to ~150 bp and do not contain readily identifiable core recombination sequences. In fact, the IntI enzyme does not even bind to duplex *attC* sites, yet it catalyzes excisive recombination between these sites *in vivo*. The answer to this paradox is that IntI binds to single-stranded *attC* sites that have folded into unusual duplex structures containing extrahelical bases. These folded sites still bear only little resemblance to the core recombination sites used in other YR systems[56] (Figure 12.6a). IntI-mediated exchange of one specific pair of strands between these sites then generates a single-stranded integron cassette (Figure 12.6b). The single-stranded *attC* sites are presumably generated during processes such as replication or conjugation.

The crystal structure of an IntI enzyme from *Vibrio cholerae* bound to a single-stranded *attC* DNA site in a pre-cleavage synaptic complex has shed considerable light on this atypical YR system[57] (Figure 12.6c,d). There are few direct contacts to DNA bases in the major groove of the *attC* duplex regions of the structure. Instead, most of the specificity appears to be derived from recognition of the extrahelical bases. One of the flipped-out residues (a thymidine) is buried *in cis* to a bound recombinase subunit. An extrahelical guanine base is bound *in trans* in a deep pocket formed by the catalytic domain of the integrase subunit located across the synaptic interface. Together, the recognition of extrahelical bases provides an intriguing explanation for how *attC* sites with little sequence similarity can be specifically recognized and efficiently recombined.

The unusual IntI/*attC* structure also explains the required strand exchange specificity. Exchange of only one of the two folded *attC* strands will give rise to the desired ss-integron cassette. An alternative way of stating the issue is that antiparallel synapsis of the folded *attC* sites could occur in two ways that differ

Site-specific Recombinases

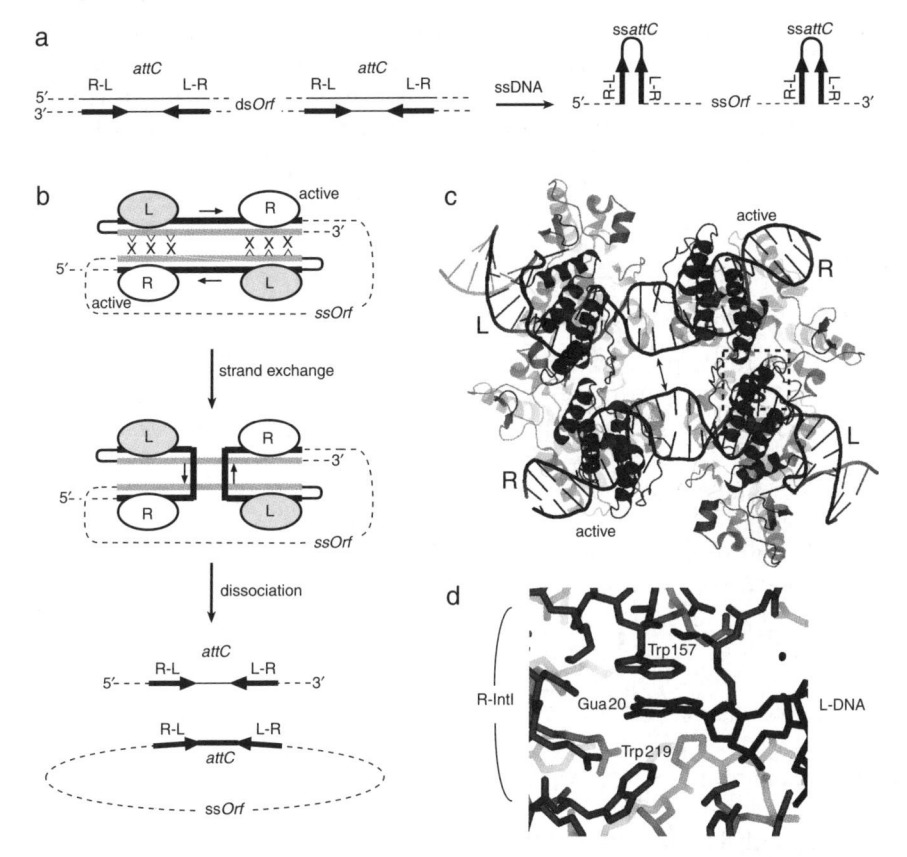

Figure 12.6 Excision of integron cassettes by integron integrases. (*a*) Single-stranded DNA is generated from an integron cassette (*e.g.*, during replication or conjugation). Only one of the generated strands is shown. The two *attC* sites in the ss-cassette fold into core recombination sites containing left (L) and right (R) half-sites. The left half-sites contain extrahelical bases. (*b*) IntI binds to and synapses the folded *attC* sites. A unique complex forms in which the subunits bound opposite to the half-sites containing extrahelical bases are active for cleavage of a specific pair of strands. Strand exchange yields the circular ss-cassette. (*c*) Structure of a *Vibrio cholerae* IntI/*attC* synaptic complex. Active subunits are indicated and the strands about to be exchanged are indicated with arrows. (*d*) Close-up of the region boxed in (c), showing recognition of the extrahelical guanine, viewed from an orthogonal direction.

by the bend direction of the sites and by the strand that will be exchanged. The IntI/*attC* interface apparently uses recognition of extrahelical bases to establish a unique synaptic configuration that is committed to exchange of the correct strand.

An outstanding question in this fascinating YR system concerns the integration reaction between *attC* and *attI* sites. The overall architecture of this reaction will be different than that observed for the *attC* x *attC* complex, since IntI will presumably bind *attI* as standard duplex DNA and will not have extrahelical bases to recognize. Given that the IntI protein strongly resembles Cre and XerD in its two-domain organization and in its general mode of DNA binding as a two-domain clamp, it would not be surprising to find that IntI is also capable of specific binding in a more traditional sense, as observed in the Cre, Flp, and λ-int systems.

12.2.4 Other Tyrosine Recombinases

Two other groups of tyrosine recombinases have been studied in some detail and are obvious targets for structural investigations. The first group includes the conjugative transposons typified by Tn916 and the CTnDOT transposons found in Bacteroides species[13] (see Chapter 11 for a discussion of transposition). These mobile genetic elements move by a three-step process: (i) excision of the transposon from the host by site-specific recombination, (ii) transfer to an alternative host by conjugation, and (iii) integration into the new host by site-specific recombination. Steps (i) and (iii) are mediated by an integrase enzyme that shares many features with λ-integrase. An interesting mechanistic puzzle in this group of YR enzymes is that the reactions are only partly site-specific. Unlike λ-integrase, there is little preference for host sites in the integration reaction and, as a consequence, both the integration and excision reactions involve formation of heteroduplex DNA in the crossover region.

The second family of enzymes is the telomere resolvases.[58] These enzymes (only four have been identified to date) are required for the replication of linear genomes with hairpin ends. They function by carrying out sequential trans-esterification reactions at specific sequences in the replicated ends of the chromosome, resulting in formation of monomeric chromosomes with hairpin ends. The catalytic domains, active sites, and mechanisms of phosphoryl transfer are in many ways similar to the YR family enzymes, but there are also some interesting differences.[59,60] It remains to be seen whether the telomere resolvases should be classified in the YR family; structural data will surely be informative in this regard.

12.3 Serine Recombinases

The core recombination sites for the serine recombinase (SR) family follow a similar modular organization as described for the YR systems. Two recombinase binding elements are arranged surrounding a central crossover region, often as inverted repeats. The crossover sequence, defined here as those residues that are exchanged between sites, is generally only 2 bp for the SRs. In all characterized systems, the core half-sites are bound by the same type of recombinase protein; there are no XerCD equivalents known for the SR family. An interesting difference between the YR and SR families at the level of recombination sites is in the nature of accessory sequences. Some SR systems use additional copies of elements resembling the core sites as accessory sequences, often with an alternative spacing between the RBEs. The recombinase proteins thus play a dual role in these systems, acting both to directly catalyze strand exchange (at the core sites) and to serve an architectural/regulatory role at accessory sites.

The serine recombinases exchange DNA strands between substrates using a mechanism that differs in nearly every conceivable way from that utilized by the tyrosine recombinases.[7] The basic reaction is presented in cartoon form in Figure 12.7(a). Upon synapsis of the core sites in an approximately parallel orientation, all four of the recombinase subunits cleave the DNA substrates with conserved serine side chains to form covalent 5'-phosphoserine linkages. These linkages are then targets for ligation by 3'-hydroxyl groups derived from the corresponding strands on the partner substrate. Rather than utilizing a stepwise mechanism where one pair of strands is cleaved and exchanged at a time, the SR enzymes use a concerted mechanism in which staggered double-strand breaks are formed in each substrate and then all four ligations occur together to generate products.

One of the most interesting features of the SR strand exchange mechanism has also been a puzzling and extensively debated issue. Once the DNA substrates have been cleaved, how do pairs of duplex arms exchange positions to allow for product formation? In the YR family enzymes, the sites are brought into proximity at the start of the process and single-stranded segments are swapped between partners with only small changes in the overall structure of the synaptic complex. It has been known for some time that this is unlikely to be the case for the SRs, based on biochemical and topological data.[61] The currently favored model for this process is illustrated in Figure 12.7(a) and discussed in more detail in the next section. Two subunits in the tetramer of covalently linked recombinase subunits rotate 180° about a symmetry axis parallel to the recombining substrates, re-creating a functional tetramer in which ligation can take place. As a consequence of this rotation, the DNA strands are crossed, changing the substrate's linking number (for intramolecular recombination). In cases where the DNA substrates are negatively supercoiled, the direction of subunit rotation is such that supercoiling is relaxed in the recombinant products.

An interesting aspect of the SR mechanism is the question of whether strand rotation can continue for additional rounds of 180° rotations. Like most of the YR family systems, the SRs require identical sequences in their crossover regions between recombining sites to efficiently form recombinant products. If non-identical 3'-overhanging sequences (2 bp long) are encountered after the initial rotation, the reaction does indeed proceed to a full 360° circle where starting substrates are restored by ligation. This event can be monitored by the change in substrate topology.[62,63]

Figure 12.7(b) illustrates a close-up of the enzyme active sites, showing phosphoryl transfer. The transesterification steps mirror those used by the YR

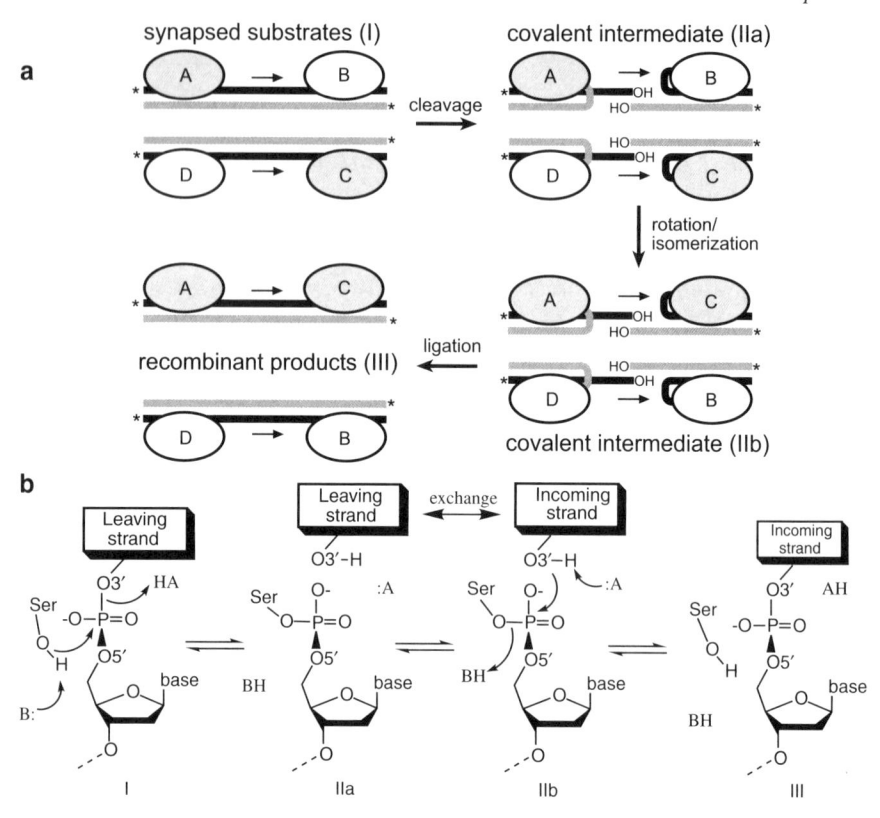

Figure 12.7 Site-specific recombination by the serine recombinases. (*a*) Schematic of the SR mechanism, based on the most favored current model.[7] Only the core sites where strand exchange occurs are shown. Two recombinases bind (as a dimer) to each core site and the sites are associated to form a roughly parallel synaptic complex (I). All four subunits (A–D) cleave the substrates to form 5′-phosphoserine linkages, generating free 3′-hydroxyl groups (IIa). The B and C subunit/DNA complexes rotate by 180° about a horizontal axis, resulting in a positional exchange, where subunit A is now partnered with C and subunit D with B (IIb). Re-ligation generates recombinant products (III). (*b*) Close-up of the active sites for the reaction intermediates shown in (a), illustrating the likely phosphoryl transfer catalytic cycle.

enzymes, except that a serine nucleophile is used and the phosphate linkages are reversed from 3′-phosphotyrosine to 5′-phosphoserine. Less is known about the nature of catalysis in the SR systems, but a conserved group of active site residues have been identified. Three conserved arginine and one conserved aspartic acid residue appear to be responsible for catalysis, although their specific roles are not yet known. Additional residues not yet identified as important for catalysis might, possibly, also play a role.

12.3.1 Tn3 and γδ-Resolvases

The resolvase enzymes from γδ and Tn3 transposons have been extensively studied and a great deal of our current knowledge about the SR family comes from this work. Both enzymes catalyze the resolution of a transposition intermediate, called a cointegrate, in which the entire transposon donor has inserted into the target site flanked by duplications of the transposon.[64] The sites of recombination, called *res*, are arranged as direct repeats in the cointegrate and site-specific recombination between the sites results in excision of the donor DNA, leaving a copy of the transposon in the target.

Although work in the Tn3 resolvase system has resulted in several important advances in understanding the SRs, the structural biology of the SR family has been almost completely dominated by the γδ-resolvase system. As indicated in Table 12.2, structural data have been reported for isolated domains,[65–67] unliganded protein in two different oligomeric states,[68,69] a protein dimer bound to a core recombination site[70] and tetrameric covalent resolvase/DNA complexes.[68,71]

Figure 12.8(a) shows the domain structure of γδ-resolvase as observed in the DNA-bound form. An N-terminal catalytic domain of ~100 residues is connected to a C-terminal helix-turn-helix DNA-binding domain through a long α-helix, termed the E-helix. Resolvase enzymes (and other SRs) are dimeric in solution and the E-helix forms a substantial fraction of the dimerization interface within the catalytic domain. The conserved serine nucleophile and active site are located at the end of the first β-strand in the catalytic domain.

The crystal structure of γδ-resolvase bound to a core recombination site (*res* site I) was an important step in understanding the basic architectural features likely to be present in the full recombination assembly.[70] In this structure (Figure 12.8b), the resolvase catalytic domain dimer straddles the minor groove at the center of the site and the E-helices track along the minor grooves into the two half-sites. Helix-E leads to a linker segment that also binds in the minor groove until crossing the phosphate backbone and connecting to a DNA-binding domain (DBD) bound in the major groove on the opposite face of the duplex relative to the catalytic domain dimer. A puzzling feature of this structure was that the phosphates expected to be cleaved in the normal recombination reaction were located far from the active sites of the nearest recombinase subunits. Conformational changes were therefore predicted to occur during formation of the pre-cleavage synaptic complex.[70]

A major advance in understanding the structural and mechanistic aspects of this system came when the structure of a tetrameric resolvase-DNA complex was reported.[71] The protein used in this study was an "activated" resolvase, where several residues had been mutated to produce an enzyme that was capable of recombining core recombination sites without the need for accessory sequences or proteins. The DNA half-sites used in forming the complex had been cleaved by the activated resolvase subunits, resulting in a tetrameric assembly in which each resolvase is covalently linked to its bound DNA half-site (Figure 12.8c,d).

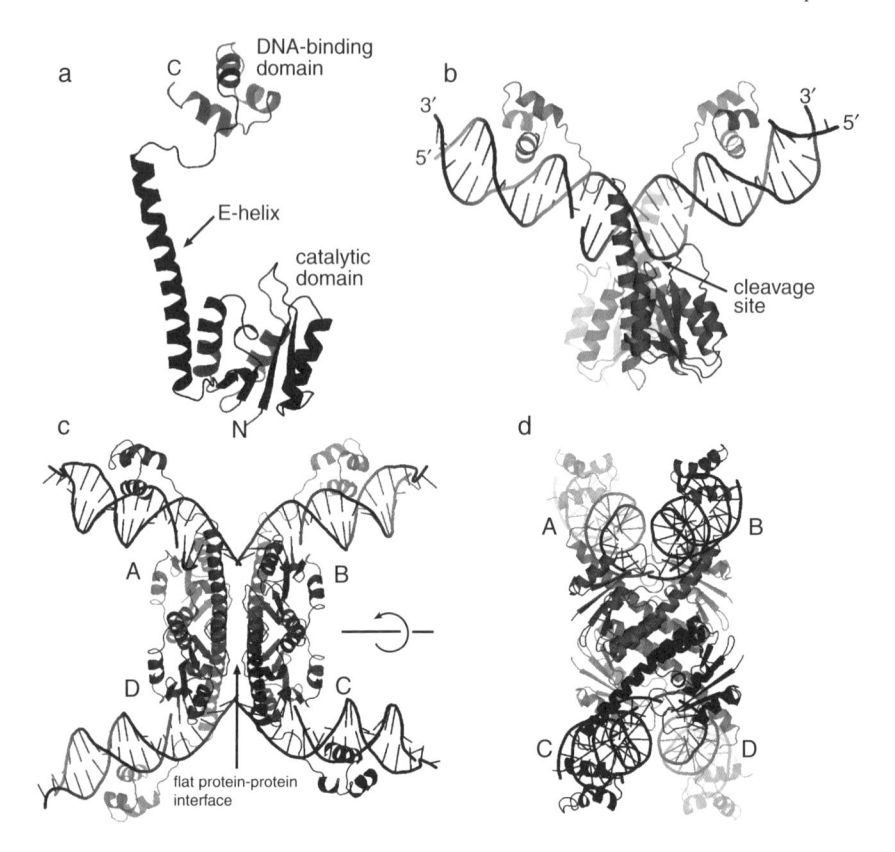

Figure 12.8 Structural models of γδ-resolvase. (*a*) Domain structure of the resolvase protein. The structure shown is half of the resolvase dimer bound to DNA, as shown in (*b*). Although the structure of resolvase has been determined in the unliganded form, the region C-terminal to the catalytic domain is disordered. (*b*) Structure of γδ-resolvase bound to a core recombination site (referred to as *res* site I) (PDB code 1GDT). The location of one of the phosphates to be cleaved during recombination is indicated. (*c*) Structure of an activated γδ-resolvase mutant/DNA tetramer (1ZR4). The four resolvase subunits have cleaved their duplex half-sites to form covalent 5′-phosphoserine linkages. The relatively flat interface formed between A-D and B-C halves of the complex is indicated. The expected subunit rotation based on the mechanism in Figure 12.7(a) is also indicated. It has been proposed that the flat interface could facilitate this rotation. (*d*) Orthogonal view of the structure shown in (c), viewed down the subunit rotation axis.

The most remarkable finding from this work was that the catalytic domains had undergone significant changes in both tertiary and quaternary structure, effectively remodeling the original dimer interface (as shown in Figure 12.8b) in favor of two new dimer interfaces in a complex with overall D_2 symmetry. The remodeled dimer interface was found to be nearly featureless relative to the interdigitated interfaces normally observed between protein subunits in oligomeric proteins (Figure 12.8c). This "flat" interface strongly suggested a mechanism by which the two subunits on one side of the interface could exchange places *via* a simple rotation of 180°, without requiring a dissociation of the recombinase-DNA tetramer. Simple calculations and modeling supported this idea.

The subunit rotation model for SR recombination had been discussed for some time prior to the resolvase tetramer crystal structure,[61] but lacked a structural foundation that could explain how the protein subunit interfaces could be remodeled in such a dramatic fashion. Additional data supporting the subunit rotation model come from crosslinking experiments in the resolvase[71] and Hin recombinase[72] systems. One aspect of the resolvase tetramer structure that is not yet understood is the location of the active sites relative to the 3'-hydroxyl ends that will be involved in the subsequent ligation. Like the resolvase dimer/DNA complex shown in Figure 12.8(b), the distances are sufficiently large (>10 Å) that significant conformational changes need to take place to allow ligation to occur. Perhaps the tetramer crystal structure represents one of several intermediate states on the pathway between cleavage, rotation, and ligation.

An intriguing architectural and mechanistic issue in the SR systems concerns the disposition of protein subunits *vs.* DNA substrates during the recombination pathway. In the YR systems, it is clear that the recombinases conspire to bring the sites of strand exchange close together in the center of the recombination complex. The structures in Figure 12.8(c, d) imply that the SR enzymes do exactly the opposite; the DNA is on the outside of the complex. This model had also been suggested well before supporting structural data was available.[7] In addition to the crystal structures, solution scattering of Tn3 resolvase/DNA synaptic complexes has provided independent structural support for the "DNA on the outside" architecture of SR recombination.[73]

12.3.2 Hin and Gin Recombinases

Of the serine recombinases that promote DNA inversion reactions, the Hin and Gin recombinases are the most well-studied.[74] Hin recombinase inverts a promoter-containing segment to cause switching of the surface flagellar proteins of *Salmonella* (a process referred to as phase variation) and Gin recombinase inverts a segment that contains alternative versions of genes coding for bacteriophage Mu tail fibers. The two recombinases are closely related and, based on sequence comparisons, have domain structures identical to γδ-resolvase. Indeed, important contributions to understanding the mechanism of SR recombination have also come from biochemical and topological studies in the Hin and Gin systems.

Neither the Hin or Gin system has been amenable to structural analysis of the full-length proteins or protein/DNA complexes, but the Hin DNA-binding domain/DNA interaction has been extensively studied on a structural and biochemical level.[75] Interestingly, the Hin-DBD/DNA interface includes two well-ordered water molecules that mediate interactions between protein and DNA. One of these waters plays a crucial role in the specific recognition of the core recombination sites by Hin.

12.3.3 Regulation by Accessory Sites

The SR mechanism cartoon shown in Figure 12.7(a) describes the current view of what occurs within the core recombination sites for the resolvase and invertase enzymes. Like the λ-integrase system, however, the core sites are only part of the story. The Tn3 and γδ-resolvase enzymes, for example, will not efficiently synapse, cleave, or recombine their core recombination sites when present in DNA substrates that do not contain the accessory sequences in the correct relative positions. In these systems, the complete *res* site actually contains three binding sites for resolvase dimers, termed site I, site II, and site III. Site I serves as the core recombination site where strand exchange occurs (shown schematically in Figure 12.7a) and sites II and III serve as the accessory sites.[7]

Interestingly, the spacing between the 12-bp resolvase RBEs differs for the three dimer binding sites; 4 for site I, 10 for site II, and 1 for site III. Clearly, the resolvase dimer must have sufficient structural plasticity to accommodate the quite different half-site spacings. The flexible linker between helix E and the autonomous DNA-binding domains in the γδ-resolvase/DNA complex structures is likely to be responsible for this adaptability, and is reminiscent of the Gal4 and steroid/nuclear receptor families of transcription factor DNA-binding domains (see Chapter 3).

The role of the resolvase accessory sequences is similar to that described earlier for Xer recombination at plasmid *cer* and *psi* sites. A cartoon model for formation and resolution of the complete synaptic complex is shown in Figure 12.9(a). The resolvase dimers bound at sites II and III appear to be entirely responsible for formation of the synaptic complex and will do so even in the absence of site I.[76] As a result of this accessory site-mediated synapsis, the two *res* core recombination sites are presented to one another in a manner that is optimal for catalysis. Direct contacts between resolvase dimers bound at accessory sites and the dimers bound at site I are also thought to play an important role in activation of catalysis.[77] The reason for requiring formation of this complex assembly is to ensure that only intramolecular recombination between *res* sites arranged as direct repeats on a supercoiled substrate will result in recombinant products. Other permutations of these requirements will block recombination, effectively preventing unwanted integration and inversion rearrangements.

Although there is currently no experimental structure available for the model shown in Figure 12.9(a), topological and biochemical data are consistent with the architecture shown. Indeed, the essential nature of this "topological filter" was first established for the Tn3 system.[61] Inter-subunit contacts identified in crystal structures of γδ-resolvase[69] have inspired biochemical experiments leading to several plausible three-dimensional models for this complex assembly as well.[69,78,79] Note that Figure 12.9(a) implies that the simplest products of

Site-specific Recombinases 325

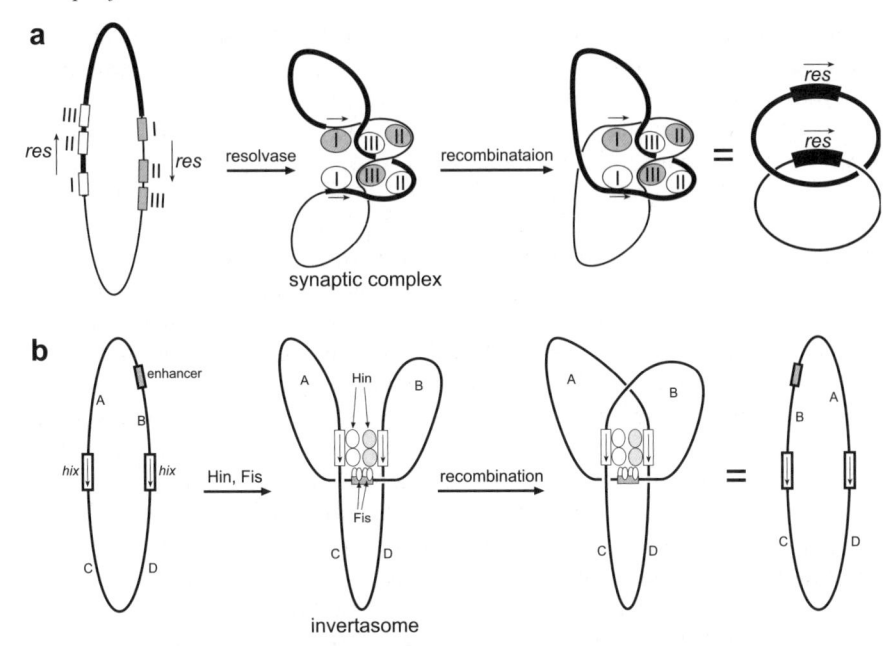

Figure 12.9 Roles of accessory sites in serine recombinase site-specific recombination. (*a*) Formation of a synaptic complex and recombination between *res* sites by Tn3 and γδ-resolvase. *Res* sites (composed of sites I, II, and III) are arranged as direct repeats in a supercoiled substrate. Resolvase dimers bound to sites II and III are responsible for forming an intramolecular resolution-specific synaptic complex that allows site I bound dimers to catalyze strand exchange. Three negative nodes are trapped during formation of the synaptic complex and one is removed during strand exchange to generate catenated circular products. Supercoils in the DNA are not shown. (*b*) Formation of the invertasome, a synaptic complex containing Hin (or Gin) dimers bound to *hix* sites arranged as inverted repeats in a supercoiled substrate. The enhancer is bound by two dimers of Fis. Hin synapses the sites regardless of whether Fis is present, but Fis is required for Hin to catalyze strand exchange. The DNA between *hix* sites is inverted as a consequence of recombination. The A, B, and C/D domains are supercoiled in the normal Hin substrates and the invertasome forms at a local branch site.

recombination between directly repeated *res* sites are two singly linked catenated circles. Three negative nodes are trapped in the process of synapsis, but one is removed as a consequence of strand exchange.

The Hin and Gin SR systems also require an accessory site for efficient recombination, but the situation is quite different in these inversion systems.[74] The recombination sites (called *hixL and hixR* in the Hin system) consist of only the core recombination sequences, which are similar in structure to *res* site I in the resolvase systems. The accessory site (called the enhancer) contains recognition sequences for two dimers of the Fis protein (which also stimulates the λ-integrase excision reaction). A cartoon of the synaptic complex formed in the Hin inversion reaction, termed the invertasome, is shown in Figure 12.9(b). The specific location of the enhancer is not crucial, as long as it is present in the same DNA molecule.

Interestingly, Hin will synapse *hix* sites under various conditions that do not require the Fis protein.[80] However, Hin will only catalyze strand exchange when Fis is present in the context of a correctly assembled invertasome, thereby preventing integration and excision reactions from occurring. Fis dimers have been demonstrated to contact the recombinase proteins in this reaction and this interaction is thought to be responsible for activation of the recombinase subunits for catalysis.[81] As in the $\gamma\delta$ and Tn3-resolvase systems, mutated versions of Hin are available that will catalyze recombination on simple substrates in the absence of Fis, greatly facilitating biochemical analyses.[72]

The Sin recombinase, representative of a group of resolvases from *Staphylococcus* plasmids, has a similar domain structure as the $\gamma\delta$ and Tn3 resolvases and carries out a similar reaction between directly repeated sites.[82] In the Sin system, however, the *res* sites contain only site I and site II, and site II has an unexpected head-to-tail arrangements of RBEs. Recombination in this system also requires a DNA-bending protein related to *E. coli* HU. Together, the recombinase and HU-like accessory protein are thought to generate a synaptic complex analogous to that shown in Figure 12.9(a) for $\gamma\delta$ and Tn3 resolvase. The structure of a Sin/DNA complex has recently been determined (P.A. Rice, personal communication).

12.3.4 Large Serine Recombinases

The resolvase and invertase systems discussed in the previous sections share a common domain structure where a N-terminal catalytic domain is linked to a small C-terminal DNA-binding domain *via* helix-E and a flexible linker. A subfamily of SR enzymes termed the "large serine recombinses" (LSRs), or the "serine integrases", also share the highly conserved catalytic domain and the extended E-helix.[6] In place of the HTH domain at the C-terminus, however, these proteins have large domains or perhaps multiple domains that share some sequence similarity with one another. The sizes of the LSRs identified thus far vary from ~450 residues to the smallest examples to over 700 residues for the largest. In comparison, $\gamma\delta$-resolvase is only 182 residues in size.

The best characterized LSR systems are the integrase proteins from the *Streptomyces* bacteriophage ϕC31[83] and mycobacteriophage Bxb1.[84] As illustrated in Table 12.1, LSR functions are not restricted to bacteriophage integration and excision. Examples have been identified with biological functions similar to many of the non-integrase YR and small SR systems as well. In the ϕC31 and Bxb1 systems, the recombination sites required for integration (*attP* and *attB*) and excision (*attL* and *attR*) have also been well-studied. In contrast to the λ-integrase system, where *attP* is quite complex and accessory sites and accessory proteins play a crucial role in regulating recombination, the LSR *attP* sites are only ~50 bp in length and appear to include only the core recombination sites. The *attB* sites are somewhat shorter, and in both cases the sites contain RBEs arranged as approximate inverted repeats surrounding a 2-bp crossover sequence.

Perhaps the most intriguing aspect of the serine integrases is that they will recombine DNA substrates containing simple *attP* and *attB* sites *in vitro* and in mammalian cells, with no topological requirements such as supercoiling.[85] The minimal requirements for core recombination sites and recombinase protein are reminiscent of the Cre and Flp systems from the YR family. Indeed, several LSR enzymes have also been useful in genetic engineering of eukaryotic cells and show promise for applications in human gene therapy.[86,87] In contrast to the Cre and Flp systems, however, the serine integrases will not recombine substrates containing the product sites of integration, *attL* and *attR*. In fact, the serine integrases will not recombine any other combination of sites (such as *attP* x *attP*, *attB* x *attB*, *etc.*). In the TP-901 and Bxb1 systems, Xis proteins have been identified that appear to be responsible for promoting *attL* x *attR* recombination *in vivo*,[88,89] so it seems likely that the other integrases also have corresponding accessory factors.

There are currently no structural models for LSR family members, but this is clearly an area that will benefit tremendously from structural insights. In particular, sequence-dependent conformations in the synaptic complexes formed between recombining sites are likely to be responsible for allowing or disallowing recombination between a given pair of sites and the nature of this specificity will be difficult to dissect using biochemical experiments alone.

12.4 Summary

One of the most interesting aspects of the structural biology of site-specific recombinases that has emerged from work over the past 10–15 years comes from a comparison of the YR and SR systems. It was clear from early sequence and biochemical analyses that there were some fundamental differences between the two families and structural models have provided an opportunity to investigate these systems at atomic resolution. It is intriguing that two very different structural and mechanistic frameworks have evolved to orchestrate essentially the same array of DNA arrangements and corresponding biological functions. As outlined in this chapter, the YR and SR systems could not be more different both in terms of structure and in mechanism, yet in some cases they carry out identical functions – often in a highly regulated manner.

The initial structures of individual proteins and protein domains in the YR and SR systems provided long-awaited information about overall folds, active site composition, and possible architectures of active protein-DNA assemblies. Structures of protein-DNA complexes have provided even more informative platforms to test mechanistic hypotheses. It clearly has not been the case that an initial structure in any of the systems that have been amenable to crystallographic analysis was sufficient, when combined with biochemical studies, to explain all aspects of site-specific recombination. Instead, structural and biochemical approaches have worked hand-in-hand to advance functional and mechanistic models in nearly every case.

Given the diversity present in each of the site-specific recombinase families, it is also clear that a detailed structural and mechanistic understanding of one system will not necessarily provide the answers for all systems. Cre, Flp and λ-int are examples. Although three-dimensional structures of reaction intermediates were first reported for the Cre system, many aspects of the Flp and λ-int systems could not be inferred from this work. It is also true that detailed structural information for a single intermediate in the recombination pathway has not necessarily provided the answers to questions concerning the other intermediates. For example, in the γδ-resolvase system, the structure of the pre-cleavage synaptic complex is still unknown and cannot be readily inferred from that of the tetrameric cleaved intermediate due to the complex conformational changes that are presumed to take place.

The structural biology of site-specific recombination is obviously far from complete. Important work remains to be done in several systems. A higher order complex structure for λ-int that includes accessory sites and accessory proteins would be a challenging, yet most welcome, addition to understanding the YR family integrases. The relaxed requirement for sequence identity in the crossover sites of CTnDOT-related integrases[90] also remains an important unanswered puzzle in the YR family. In the SR family, additional structures of γδ-resolvase reaction intermediates, as well as accessory site complexes, would also provide much-needed data. The LSR enzymes remain an entirely open field, starved of structural models to explain existing biochemical data and to inspire new experiments.

Although most of the structural data for the YR and SR systems discussed in this chapter come from crystallographic studies, single molecule approaches have also begun to be applied to these systems. For example, a complete λ-int recombination reaction has recently been visualized using single molecule light microscopy, revealing that the reaction is remarkably efficient once synapsis occurs.[91] Kinetic analysis of the reaction also indicates that a rate-limiting step exists following formation of the synaptic complex and prior to formation of the HJ intermediate. It seems likely that other systems will also greatly benefit from these types of studies.

References

1. H.A. Nash, in *Escherichia coli and Salmonella: Cellular and Molecular Biology*, (ed.) F.C.E.A. Neidhardt, ASM Press, Washington DC, 1996.
2. W.M. Stark, M.R. Boocock and D.J. Sherratt, *Trends Genet.*, 1992, **8**, 432.
3. R. Craigie, in *Retroviral DNA Integration*, ed. N.L. Craig, R. Craigie, M. Gellert and A.M. Lambowitz, Washington DC, 2002.
4. J.M. Jones and M. Gellert, *Immunol. Rev.*, 2004, **200**, 233.

5. M.A. Azaro and A. Landy, in *Mobile DNA II*, (ed.) N.L. Craig, R. Craigie, M. Gellert and A.M. Lambowitz, ASM Press, Washington DC, 2002, p. 118–148.
6. M.C. Smith and H.M. Thorpe, *Mol. Microbiol.*, 2002, **44**, 299.
7. N.D. Grindley, K.L. Whiteson and P.A. Rice, *Annu. Rev. Biochem.*, 2006, **75**, 567.
8. N.l. Craig, R. Craigie, M. Gellert and A.M. Lambowitz, *Mobile DNA II*, ASM Press, Washington DC, 2002.
9. S. Nunes-Düby, M.A. Azaro and A. Landy, *Curr. Biol.*, 1995, **5**, 139.
10. X.D. Zhu, G. Pan, K. Luetke and P.D. Sadowski, *J. Biol. Chem.*, 1995, **270**, 11646.
11. J. Lee and M. Jayaram, *J. Biol. Chem.*, 1995, **270**, 4042.
12. A.B. Burgin and H.A. Nash, *Curr. Biol.*, 1995, **5**, 1312.
13. G. Churchward, in *Conjugative Transposons and Related Mobile Elements*, ed. N.L. Craig, R. Craigie, M. Gellert and A.M. Lambowitz, Washington, DC, 2002.
14. G.D. Recchia and D.J. Sherratt, in *Gene Acquisition in Bacteria by Integron-mediated Site-specific Recombination*, ed. N.L. Craig, R. Craigie, M. Gellert and A.M. Lambowitz, Washington DC, 2002.
15. C. Cheng, P. Kussie, N. Pavletich and S. Shuman, *Cell*, 1998, **92**, 841.
16. K. Ghosh, C.K. Lau, K. Gupta and G.D. Van Duyne, *Nat. Chem. Biol.*, 2005, **1**, 275.
17. B.O. Krogh and S. Shuman, *Mol. Cell*, 2000, **5**, 1035.
18. F. Guo, D.N. Gopaul and G.D. Van Duyne, *Nature*, 1997, **389**, 40.
19. K.L. Whiteson, Y. Chen, N. Chopra, A.C. Raymond and P.A. Rice, *Chem. Biol.*, 2007, **14**, 121.
20. S.S. Martin, E. Pulido, V.C. Chu, T.S. Lechner and E.P. Baldwin, *J. Mol. Biol.*, 2002, **319**, 107.
21. G.D. Van Duyne, *Annu. Rev. Biophys. Biomol. Struct.*, 2001, **30**, 87.
22. G.D. Van Duyne, in *Mobile DNA II*, (ed.) N.L. Craig, R. Craigie, M. Gellert and A.M. Lambowitz, ASM Press, Washington DC, 2002, p. 93–117.
23. M. Jayaram, I. Grainge and G. Tribble, in *Mobile DNA II*, (ed.) N.L. Craig, R. Craigie, M. Gellert and A.M. Lambowitz, ASM Press, Washington DC, 2002, p. 192–218.
24. P.D. Sadowski, *Progr. Nucleic Acid Res. Mol. Biol.*, 1995, **51**, 53.
25. B. Sauer, in *Mobile DNA II*, (ed.) N.L. Craig, R. Craigie, M. Gellert and A.M. Lambowitz, ASM Press, Washington DC, 2002, p. 38–58.
26. Y. Le and B. Sauer, *Mol. Biotechnol.*, 2001, **17**, 269.
27. A.L. Garcia-Otin and F. Guillou, *Front Biosci.*, 2006, **11**, 1108.
28. C.S. Branda and S.M. Dymecki, *Dev. Cell*, 2004, **6**, 7.
29. N. Sternberg, D. Hamilton, S. Austin, M. Yarmolinsky and R. Hoess, *Cold Spring Harbor Symp. Quant. Biol.*, 1981, **1**, 297.
30. A.B. Futcher, *J. Theor. Biol.*, 1986, **119**, 197.
31. Y. Chen and P.A. Rice, *Annu. Rev. Biophys. Biomol. Struct.*, 2003, **32**, 135.
32. F. Guo, D.N. Gopaul and G.D. Van Duyne, *Proc. Natl. Acad. Sci. U.S.A.*, 1999, **96**, 7143.

33. E. Ennifar, J.E. Meyer, F. Buchholz, A.F. Stewart and D. Suck, *Nucleic Acids Res.*, 2003, **31**, 5449.
34. D.N. Gopaul, F. Guo and G.D. Van Duyne, *EMBO J.*, 1998, **17**, 4175.
35. K. Ghosh, C.K. Lau, F. Guo, A.M. Segall and G.D. Van Duyne, *J. Biol. Chem.*, 2005, **280**, 8290.
36. Y. Chen, U. Narendra, L.E. Iype, M.M. Cox and P.A. Rice, *Mol. Cell*, 2000, **6**, 885.
37. A.B. Conway, Y. Chen and P.A. Rice, *J. Mol. Biol.*, 2003, **326**, 425.
38. A. Landy, *Annu. Rev. Biochem.*, 1989, **58**, 913.
39. J.M. Wojciak, D. Sarkar, A. Landy and R.T. Clubb, *Proc. Natl. Acad. Sci. U.S.A.*, 2002, **99**, 3434.
40. C.V. Papagiannis, M.D. Sam, M.A. Abbani, D. Yoo, D. Cascio, R.T. Clubb and R.C. Johnson, *J. Mol. Biol.*, 2007, **367**, 328.
41. H.J. Kwon, R. Tirumalai, A. Landy and T. Ellenberger, *Science*, 1997, **276**, 126.
42. H. Aihara, H.J. Kwon, S.E. Nunes-Duby, A. Landy and T. Ellenberger, *Mol. Cell*, 2003, **12**, 187.
43. T. Biswas, H. Aihara, M. Radman-Livaja, D. Filman, A. Landy and T. Ellenberger, *Nature*, 2005, **435**, 1059.
44. M. Radman-Livaja, C. Shaw, M. Azaro, T. Biswas, T. Ellenberger and A. Landy, *Mol. Cell*, 2003, **11**, 783.
45. M. Radman-Livaja, T. Biswas, D. Mierke and A. Landy, *Proc. Natl. Acad. Sci. U.S.A.*, 2005, **102**, 3913.
46. X. Sun, D.F. Mierke, T. Biswas, S.Y. Lee, A. Landy and M. Radman-Livaja, *Mol. Cell*, 2006, **24**, 569.
47. D.J. Sherratt, B. Soballe, F.X. Barre, S. Filipe, I. Lau, T. Massey and J. Yates, *Philos. Trans. R. Soc. London, B*, 2004, **359**, 61.
48. S.D. Colloms, J. Bath and D.J. Sherratt, *Cell*, 1997, **88**, 855.
49. F.X. Barre, M. Aroyo, S.D. Colloms, A. Helfrich, F. Cornet and D.J. Sherratt, *Genes Dev.*, 2000, **14**, 2976.
50. L. Aussel, F.X. Barre, M. Aroyo, A. Stasiak, A.Z. Stasiak and D. Sherratt, *Cell*, 2002, **108**, 195.
51. V. Sivanathan, M.D. Allen, C. de Bekker, R. Baker, L.K. Arciszewska, S.M. Freund, M. Bycroft, J. Lowe and D.J. Sherratt, *Nat. Struct. Mol. Biol.*, 2006, **13**, 965.
52. T.H. Massey, C.P. Mercogliano, J. Yates, D.J. Sherratt and J. Lowe, *Mol. Cell*, 2006, **23**, 457.
53. H.S. Subramanya, L.K. Arciszewska, R.A. Baker, L.E. Bird, D.J. Sherratt and D.B. Wigley, *EMBO J.*, 1997, **16**, 5178.
54. R. McCulloch, L.W. Coggins, S.D. Colloms and D.J. Sherratt, *EMBO J.*, 1994, **13**, 1844.
55. D. Mazel, *Nat. Rev. Microbiol.*, 2006, **4**, 608.
56. M. Bouvier, G. Demarre and D. Mazel, *EMBO J.*, 2005, **24**, 4356.
57. D. MacDonald, G. Demarre, M. Bouvier, D. Mazel and D.N. Gopaul, *Nature*, 2006, **440**, 1157.
58. G. Chaconas, *Mol. Microbiol.*, 2005, **58**, 625.

59. J. Deneke, A.B. Burgin, S.L. Wilson and G. Chaconas, *J. Biol. Chem.*, 2004, **279**, 53699.
60. J. Deneke, G. Ziegelin, R. Lurz and E. Lanka, *Proc. Natl. Acad. Sci. U.S.A.*, 2000, **97**, 7721.
61. W.M. Stark, D.J. Sherratt and M.R. Boocock, *Cell*, 1989, **58**, 779.
62. W.M. Stark, N.D. Grindley, G.F. Hatfull and M.R. Boocock, *EMBO J.*, 1991, **10**, 3541.
63. K.A. Heichman, I.P. Moskowitz and R.C. Johnson, *Genes Dev.*, 1991, **5**, 1622.
64. N.D. Grindley, in *Mobile DNA II*, (ed.) N.L. Craig, R. Craigie, M. Gellert and A.M. Lambowitz, ASM Press, Washington DC, 2002, p. 272–302.
65. B. Pan, M.W. Maciejewski, A. Marintchev and G.P. Mullen, *J. Mol. Biol.*, 2001, **310**, 1089.
66. T. Liu, E.F. DeRose and G.P. Mullen, *Protein Sci.*, 1994, **3**, 1286.
67. P.A. Rice and T.A. Steitz, *Structure*, 1994, **2**, 371.
68. S. Kamtekar, R.S. Ho, M.J. Cocco, W. Li, S.V. Wenwieser, M.R. Boocock, N.D. Grindley and T.A. Steitz, *Proc. Natl. Acad. Sci. U.S.A.*, 2006, **103**, 10642.
69. P.A. Rice and T.A. Steitz, *EMBO J.*, 1994, **13**, 1514.
70. W. Yang and T.A. Steitz, *Cell*, 1995, **82**, 193.
71. W. Li, S. Kamtekar, Y. Xiong, G.J. Sarkis, N.D. Grindley and T.A. Steitz, *Science*, 2005, **309**, 1210.
72. G. Dhar, E.R. Sanders and R.C. Johnson, *Cell*, 2004, **119**, 33.
73. M. Nollmann, J. He, O. Byron and W.M. Stark, *Mol. Cell*, 2004, **16**, 127.
74. R.C. Johnson, in *Mobile DNA II*, (ed.) N.L. Craig, R. Craigie, M. Gellert and A.M. Lambowitz, ASM Press, Washington DC, 2002, p. 230–271.
75. T.K. Chiu, C. Sohn, R.E. Dickerson and R.C. Johnson, *EMBO J.*, 2002, **21**, 801.
76. M.A. Watson, M.R. Boocock and W.M. Stark, *J. Mol. Biol.*, 1996, **257**, 317.
77. L.L. Murley and N.D. Grindley, *Cell*, 1998, **95**, 553.
78. G.J. Sarkis, L.L. Murley, A.E. Leschziner, M.R. Boocock, W.M. Stark and N.D. Grindley, *Mol. Cell*, 2001, **8**, 623.
79. A.E. Leschziner and N.D. Grindley, *Mol. Cell*, 2003, **12**, 775.
80. K.A. Heichman and R.C. Johnson, *Science*, 1990, **249**, 511.
81. S.K. Merickel, M.J. Haykinson and R.C. Johnson, *Genes Dev.*, 1998, **12**, 2803.
82. S.J. Rowland, W.M. Stark and M.R. Boocock, *Mol. Microbiol.*, 2002, **44**, 607.
83. H.M. Thorpe and M.C. Smith, *Proc. Natl. Acad. Sci. U.S.A.*, 1998, **95**, 5505.
84. A.I. Kim, P. Ghosh, M.A. Aaron, L.A. Bibb, S. Jain and G.F. Hatfull, *Mol. Microbiol.*, 2003, **50**, 463.
85. A.C. Groth and M.P. Calos, *J. Mol. Biol.*, 2004, **335**, 667.
86. A. Keravala, A.C. Groth, S. Jarrahian, B. Thyagarajan, J.J. Hoyt, P.J. Kirby and M.P. Calos, *Mol. Genet. Genomics*, 2006, **276**, 135.

87. M.P. Calos, *Curr. Gene Ther.*, 2006, **6**, 633.
88. P. Ghosh, L.R. Wasil and G.F. Hatfull, *PLoS Biol.*, 2006, **4**, e186.
89. A. Breuner, L. Brondsted and K. Hammer, *J. Bacteriol.*, 1999, **181**, 7291.
90. K. Malanowska, A.A. Salyers and J.F. Gardner, *Mol. Microbiol.*, 2006, **60**, 1228.
91. J.P. Mumm, A. Landy and J. Gelles, *EMBO J.*, 2006, **25**, 4586.
92. W. Delano, *The PyMOL Molecular Graphics System*, DeLano Scientific, San Carlos, CA, 2002.
93. S. Austin, M. Ziese and N. Sternberg, *Cell*, 1981, **25**, 729.
94. F.X. Barre and D.J. Sherratt, in *Mobile DNA II*, (ed.) N.L. Craig, R. Craigie, M. Gellert and A.M. Lambowitz, ASM Press, Washington DC, 2002, p. 149–161.
95. W.M. Stark, M.R. Boocock and D.J. Sherratt, *Trends Genet.*, 1989, **5**, 304.
96. P. Klemm, *EMBO J.*, 1986, **5**, 1389.
97. M. Jayaram, I. Grainge and G. Tribble, in *Mobile DNA II*, (ed.) N.L. Craig, R. Craigie, M. Gellert and A.M. Lambowitz, ASM Press, Washington DC, 2002, p. 192–218.
98. B. Kunkel, R. Losick and P. Stragier, *Genes Dev.*, 1990, **4**, 525.
99. J.R. Scott and G.G. Churchward, *Annu. Rev. Microbiol.*, 1995, **49**, 367.
100. K.C. Woods, S.S. Martin, V.C. Chu and E.P. Baldwin, *J. Mol. Biol.*, 2001, **313**, 49.
101. A.B. Hickman, S. Waninger, J.J. Scocca and F. Dyda, *Cell*, 1997, **89**, 227.
102. J.M. Wojciak, K.M. Connolly and R.T. Clubb, *Nat. Struct. Biol.*, 1999, **6**, 366.
103. J.A. Feng, R.C. Johnson and R.E. Dickerson, *Science*, 1994, **263**, 348.

CHAPTER 13
DNA Nucleases

NANCY C. HORTON

Department of Biochemistry and Molecular Biophysics, University of Arizona, 1041 E. Lowell Street, Tucson, AZ 85721, Arizona, USA

13.1 Introduction

This chapter reviews structural studies of DNA nucleases that cleave the backbone of DNA at the P–O3' bond, utilizing water as a nucleophile (with one exception) and leave 5'phosphate and 3'OH ends. The three-dimensional crystal structures of over 90 different nucleases have been solved to date. As biologically important enzymes, nucleases are found across all species, and as such represent a wide variety of folds, functions, and mechanisms of DNA cleavage. Rather than describe each individually, the nucleases have been grouped such that commonalities within a particular group could be presented instead. The grouping could have been made based on biological function, such as DNA repair, recombination, immunity, defense, or apoptosis. Alternatively, it could have been by substrate preference and mode of cutting, *e.g.*, single *versus* double-stranded DNA, sequence specific *versus* non-sequence specific, or endonucleolytic *versus* exonucleolytic. Finally, the DNA nucleases could also have been grouped according to features of their catalytic mechanism, such as the type and the number of divalent cations used. Instead, the nucleases reviewed here have been grouped by fold. This choice was made since different nucleases with the same fold will often have other characteristics in common, such as DNA cleavage mechanism, DNA binding, and sometimes, although clearly not always, function.

13.2 Summaries by Fold

The definition of fold that has been used here is that defined by SCOP,[1] and 13 different folds have been identified among the DNA nucleases (Table 13.1). These fold groupings vary widely with respect to the number of members of each fold, from 35 in the case of the Restriction Endonuclease-like Fold, to only 1 in the DHH Phosphodiester Fold, Metallodependent Phosphatase Fold, and the Phospholipase C-like/P1 Nuclease Fold. When more than one member of a particular fold is found, the DALI server[2] was used to align them, and ribbon and topology diagrams (made with TOPS[3]) have been shaded to emphasize the elements of the fold common within the group. For each fold group, a brief description of the members, common fold, oligomerization states, and biological functions is given. A proposed catalytic mechanism based on those described in the reports of the structures is presented, along with notes on variations of the mechanistic models found in the literature. Finally, the mode of DNA binding and recognition is described briefly.

13.2.1 Restriction Endonuclease-like Fold

This group contains the greatest number of structurally characterized DNA nucleases, (33 at the time of writing, Table 13.1), including most of the type II restriction endonucleases (REs), several DNA repair endonucleases, two Holliday junction resolvases, and a transposase. All of these RE-like fold nucleases require divalent cations, with Mg(II) being the likely *in vivo* cofactor, and all bind and cleave (at least one strand) of duplex DNA.

The common core that defines this fold consists of five beta strands flanked by two alpha helices, with active site placement as shown by spheres or crosses in Figure 13.1(A) and (B), respectively. The degree of conservation of the secondary structural elements, as assessed by the ability of DALI to produce an alignment, is shown by shading (darkest = most conserved, lightest least conserved) mapped onto the structure of Hjc, a Holliday junction resolvase.[4] This non-sequence specific nuclease possesses fewer elaborations on the basic fold. Such elaborations occur to perform specific functions such as DNA recognition (sequence specific or structural), and oligomerization.

Many members of the Restriction Endonuclease-like Fold (RE-like Fold) cleave both strands of duplex DNA, and form dimers to position one active site on each of the DNA strands. However, the mode of oligomerization is not dictated by the core fold, and is remarkably variable. Among the subgroup comprising the canonical type II REs, most are dimers but some are tetramers. These enzymes possess elaborations on the RE-like fold that connect sequence specific recognition to DNA cleavage. The cleavage mode, in other words, whether they leave blunt or sticky ends, affects substructures responsible for different dimerization modes, which are necessary to position the active sites of the two monomers of the dimer appropriately on the duplex DNA. Dali alignments roughly identify two groups: the EcoRI group, which leave overhanging nucleotides or sticky ends on the 5′ends of the cleaved DNA

DNA Nucleases 335

Table 13.1 Protein structures reviewed with corresponding references and PDB codes.

Structure	PDB file
Restriction endonuclease-like fold	
HincII	1DC6, 1TW8, 1TX3, 1XHU, 1XHV, 2AUD, 2GIE, 2GIG, 2GIH, 2GII, 2GIJ
EcoRV	1AZO, 1B94-7, 1BGB, 1BSS, 1BSU, 1BUA, 1EO3, 1EO4, 1EON, 1EOO, 1EOP, 1RV5, 1RVA, 1RVB, 1RVC, 1RVE, 1STX, 1SUZ, 1SX5, 1SX8, 2BOD, 2BOE, 2GE5, 2RVE, 4RVE
EcoRI	1CKQ, 1CL8, 1ERI, 1QC9, 1QPS, 1QRH, 1QRI
PvuII	1H56, 1EYU, 1F0O, 1KOZ, 1PVI, 2PVI, 3PVI
BamHI	1BAM, 1BHM, 1ESG, 2BAM, 3BAM
BglII	1D2I, 1DFM, 1ES8
BglI	1DMU
MunI	1D02
Cfr10I	1CFR
BsoB	1DC1
NaeI	1EV7, 1IAW
NgoMIV	1FIU
FokI	1FOK, 2FOK
Bse634I	1KNV
EcoRII	1NA6
MspI	1SA3, 1YFI
BstYI	1SDO, 1VRR
EcoO109I	1WTD, 1WTE
HinPI	1YNM, 2FCK, 2FKH, 2FL3, 2FLC
SfiI	2EZV
Ecl18kI	2FQZ, 2GB7
Human ERCC1	2A1I
A. Pernix XPF	2BGW, 2BHN
Hef from *P. furiosus* (XPF homolog)	1J22-5, 1X2I
MutH	1AZO, 2AOQ, 2AOR
Vsr	1CW0, 1ODG, 1VSR
T7 endonuclease I	1MOD, 1MOI, 1FZR
Lambda exonuclease	1AVQ
TN7 transposon catalytic component TnsA	1F1Z
Archaeal Holliday junction resolvase Hjc	1GEF, 1IPI, 1HH1
Hje	1OB8-9
N.BspD6I	2EWF
SdaI	2IXS
RNase H-like fold	
WRN-exo	2FBY, 2FBV, 2FBX, 2FC0, 2FBT, 2E6L, 2E6M, 1VK0
TREX 1 $3' \rightarrow 5'$ exonuclease	2IOC, 2O4I, 2O4G, 2OA8
TREX 2 $3' \rightarrow 5'$ exonuclease	1Y97
Exo I	1FXX
Epsilon subunit of *E. coli* DNA pol III	2IDO, 1J53, 1J54

Table 13.1 (*continued*)

Structure	PDB file
RB69 DNA polymerase	1CLQ, 1Q9X, 1RV2, 1WAF, 1WAJ, 2ATQ
T7 DNA polymerase	1SKR, 1SKS, 1SKW, 1SL0, 1SL1, 1SL2, 1T7P, 1T8E, 1TK0, 1TK5, 1TKD, 2X9M, 1X9S, 1X9W, 1ZYQ, 2AJQ,
E. coli DNA polymerase I Klenow fragment	1D8Y, 1D9D, 1D9F, 1D9H, 1DPI, 1KFD, 1KFS, 1KLN, 1KRP, 1KSP, 1QSL, 2KZM, 2KFZ, 2KZM, 2KZZ
T4 DNA polymerase	1NOY, 1NOZ
Archaebacterial *D. Tok* DNA polymerase	1D5A, 1QQC
Thermostable type B DNA polymerase	1TGO
Bacillus DNA polymerase I	1XWL, 2BDP, 3BDP, 4BDP
Phi29 DNA pol	2EX3, 1XHX
RuvC resolvase	1HJR, 1KCF
Taq polymerase 3′→5′exonuclease domain	1TAQ
Herpes virus DNA polymerase	2GV9
Archaeal KOD1 DNA polymerase	1WN7
Homing endonuclease-like fold	
I-CreI	1AF5, 1G9Y, 1BP7, 1G9Z, 1N3E, 1NEF, 1T9I, 1T9J, 1U0C, 1U0D, 2I3Q, 2I3P
I-DmoI	1B24
PI-SceI	1DFA, 1GPP, 1LWS, 1LWT, 1VDE, 1EF0
I-SceI	1R7M
PI-PfuI	1DQ3
I-MsoI	1M5X
I-DreI (I-CreI/I-DmoI chimera)	1MOW
I-AniI	1P8K
PI-TkoII	2CW7, 2CW8
I-Tsp061I	2DCH
I-CeuI	2EX5
His-Me finger endonuclease fold	
I-PpoI	1A73-4, 1CYQ, 1CZ0, 1EVW, 1EVX, 1IPP
Serratia (SM) Nuclease	1QL0, 1G8T, 1QAE, 1SMN
Periplasmic endo Vvn	1OUO, 1OUP
Nuclease A from *Anabaena*	1ZM8, 2O3B
T4 endo VII	1E7D, 1E7L, 1EN7
Vibrio cholerae endo I	2G7E, 2G7F
I-Hmu	1U3E
Caspase activated DNase (CAD)	1V0D
Colicin E9	1FR2, 1EMV, 1BXI, 1V13-5, 1AYI, 1FSJ
Colicin E7	1UJZ, 1MZ8, 1PT3, 1ZNS, 1MO8, 7CEI
SAM domain-like/PIN domain-like fold	
Methanococcus jannaschii Flap endo FEN-1	1A76, 1U7B

DNA Nucleases

Table 13.1 (*continued*)

Structure	PDB file
FEN-1 from *P. furiosus, A. fulgidus*, and *S. sulfataricus*	1B43, 1RXV, 1RXW, 1RXZ, 2IZO
T5 5′-flap exonuclease	1EXN, 1XO1, 1UT5, 1UT8
Taq pol 5′ → 3′ exo	1TAQ
T4 RNaseH	1TFR
DNase I-like fold	
H. ducreyi cytolethal distending toxin	1SR4
APE-1	1DE8, 1DE9, 1DEW, 1HD7, 1E9N, 1ISI, 1BIX
DnaseI	1DNK, 2A40, 2DNJ, 3DNI, 1ATN, 2A3Z, 2A42
LINE-1	1VYB
TRAS1-EN	1WDU
Exonuclease III	1AKO
Phospholipase C-like/P1 nuclease fold	
P1 nuclease	1AK0
Phospholipase D-like/nuclease fold	
BfiI	2C1L
Nuc	1BYR
TIM beta/alpha barrel fold	
Endonuclease IV	1QTW, 1QUM, 1XP3
TatD Dnase	1J6O, 1XWY
DHH phosphoesterases fold	
RecJ exonuclease	1IR6
GIY-YIG endonuclease fold	
UvrC (NT domain)	1YD0, 1YD1, 1YD2, 1YD3, 1YD4, 1YD5, 1YD6
I-TevI (catalytic domain)	1MK0, 1LN0 (DNA binding domain = 1I3J, 1T2T)
Metallo-dependent phosphatases fold	
Mre11-3	1II7, 1S8E
***Bacillus chorismate* mutase-like fold**	
APE1501 putative endonuclease from *A. pernix*	2CWJ
RusA Holliday junction resolvase	1Q8R

(BsoBI, EcoRI, MunI, BstYI, BamHI, BglII, NgoMIV, Cfr10I, Bse634I, EcoRII, Eco0109I, Ecl18kI, FokI), and EcoRV group, which leave either blunt or 3′ sticky ends (EcoRV, HincII, PvuII, BglI, SdaI, NaeI SfiI), although they are actually better grouped into six based on topology.[5] Figure 13.1(C) shows the dimeric structure of PvuII,[6] which places the two active sites across the duplex to leave blunt ends. The Holliday junction resolvases also form dimers, but recognize the structure of a four-way ("Holliday") junction and make two

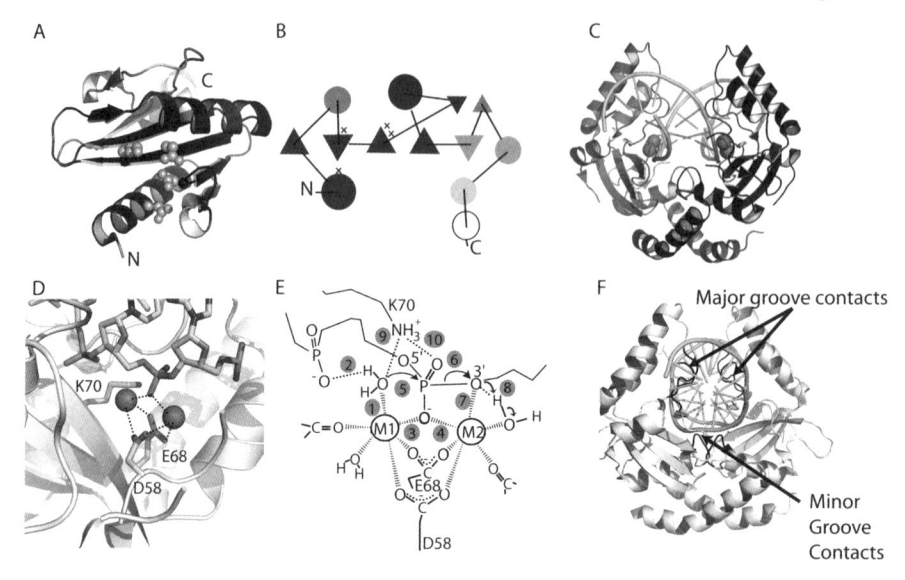

Figure 13.1 Restriction endonuclease-like fold. (*A*) Ribbon diagram of one monomer of Hjc Resolvase (PDB code 1GEF) shaded by common features within fold family (darkest = most conserved, light = least conserved). Active site residues shown as spheres. (*B*) Topology diagram shaded by common features within fold family as in (*A*). The position of active site residues shown as crosses. (*C*) Dimeric structure of PvuII bound to DNA (PDB code 1F0O), chain A, gray, chain B, black, DNA, light gray. Ca(II) ions shown as spheres. Bound DNA shown as cartoon. (*D*) Active site of PvuII (PDB code 1F0O). Ca(II) shown as spheres. Ca(II)-ligand interactions shown as dashed lines. (*E*) A proposed mechanism for DNA cleavage by PvuII. See text for details. (*F*) Protein-DNA interfaces in PvuII (PDB code 1F0O). Protein loops interacting with DNA shown as darker ribbons.

cleavages at the appropriate locations to allow separation of the DNA into two duplexes (Chapter 6).

The type IIS noncanonical REs, such as FokI,[7] contain a separate domain for sequence specific DNA binding, tethered to a nonspecific nuclease of the RE-like fold. Other type II restriction endonucleases form tetramers, and bind to two duplexes of DNA (Type IIE: NaeI and EcoRII, Type IIF: NgoMIV, Cfr10I, SfiI, Bse634I, Ecl18kI). The type IIE enzymes cleave only one of the bound duplexes, but the IIF enzymes cleave both. The structures of NaeI (ref. 8) and EcoRII (ref. 9) bound to their target DNA sequences show how this is accomplished. The protein chain folds into two different domains that combine to form a symmetric dimer with two DNA binding clefts. Both clefts bind a copy of the target DNA sequence specifically; however, only the domain with the restriction-endonuclease-like fold is capable of cleaving DNA. These proteins have been proposed to show an evolutionary link between endonucleases and recombinases/topoisomerases.

Several family members function in DNA repair. Lambda exonuclease is a 5'→3' exonuclease that binds double-stranded DNA and degrades a single strand in the 5'→3' direction. It has a toroidal structure of three identical subunits, which may be responsible for its processivity.[10] A different DNA repair protein, the structure-specific nuclease XPF, has two domains: a RE-like nuclease domain and a DNA binding head containing two copies of a non-sequence-specific helix-hairpin-helix (HhH) motif (also seen in topo V; see Chapter 10). It forms a dimer, and interacts with DNA using both HhH domains and one nuclease domain to make an endonucleolytic cleavage in one strand of double-stranded DNA 5' to a single-stranded region.[11] XPF acts in the nucleotide excision repair pathway of eukaryotes and some archaea. Additional DNA repair nucleases possessing the RE-like fold are MutH[12,13] and Vsr.[14–17] These are monomeric nucleases; however, MutH, which recognizes hemimethylated GATC sequences, acts in concert with other members of the mismatch repair pathway. Vsr recognizes TG mismatches and makes a single cleavage 5' to the T.

The type II REs, of which most of this class consists, have the general active site motif of $DX_n(D/E)XK$.[18] The two acidic groups serve to ligate one or two divalent cations (depending on the enzyme in question), while the lysine residue has a less clear role in the cleavage mechanism, yet is important as mutagenesis leads to drastic reductions in DNA cleavage rates.[18] Figure 13.1(D) shows groups in one active site of the homodimer PvuII bound to DNA and Ca(II) ions.[6] Figure 13.1(E) shows a proposed DNA cleavage mechanism based on this structure. Though all DNA nucleases in this fold family utilize Mg(II), they typically will also function, although with lower activity, using Co(II) or Mn(II), and are inhibited by Ca(II). Many RE structures have been determined in the presence of Ca(II) to stall the reaction prior to DNA cleavage, and it is thought that the ions bind in roughly the same sites as Mg(II). The reason for inhibition by Ca(II) is not known; however, it may be related to the larger ion size leading to slight but mechanistically important mispositioning of active site groups, the increased number of potential ligands [6 for Mg(II), 6–8 for Ca(II)], the greater potential for Mg(II) to ionize ligated water molecules, or to the inability of the ions to be positioned close enough to each other to stabilize the pentacovalent transition state (or intermediate).[19–21] In some cases, the structure with an active metal ion such as Mg(II) or Mn(II) has also been determined, usually with cleaved DNA, or with inactive mutants or noncleavable DNA analogs. The positions and even number of metal ions of Ca(II), Mg(II), Mn(II) found bound in an active site are not always identical even with the same protein [e.g., a single Ca(II) binds per active site in HincII for Ca(II),[22] but at least two Mn(II) bind to the same protein active site[23]], which may be one reason that several different mechanistic proposals have been made for DNA cleavage. Alternatively, some differences may be real.

The most common mechanistic proposal for DNA cleavage by the RE-like fold nucleases utilizes two Mg(II) ions (Figure 13.1E). The gray circled numbers show different features of the mechanism. First, the function of the M1 Mg(II) ion is to orient a nucleophilic water molecule and to activate it by lowering its pK_a and to produce hydroxide (a better nucleophile) (1). In some cases a neighboring phosphate has been implicated as assisting in handling one of the two protons from the nucleophilic water[24,25] (2). This Mg(II) ion also ligates a non-esterified oxygen of the scissile phosphate (3), which could serve to reduce the electron density on the phosphorus atom, making it more electrophilic and susceptible to nucleophilic attack, as well as stabilize the additional negative charge on the transition state that results from attack by the hydroxide nucleophile. A similar function could also be performed by the ligation of the second metal ion, M2, to the same non-esterified phosphate oxygen (4). Attack of the phosphate by the nucleophile (5) affords a pentacovalent transition state (or intermediate), which decomposes into the product with cleavage of the P–O3' bond (6). The leaving group, the O3', has a relatively high pK_a and is stabilized by direct ligation to M2 (7). Prior to or during dissociation it is protonated, potentially from an M2 ligated water (8). The third member of the active site motif, K70, could perform several possible functions, including activation of the nucleophile (10), and orienting the nucleophile, and stabilizing the transition state (9). Proposed mechanistic models that differ from this as presented include the absence of one or the other metal ion (*e.g.*, M2 is absent in mechanistic models for BglII, and EcoRI), absence of the interaction with the neighboring phosphate, and identity and therefore role of the residue in the lysine's position (which is variable: *e.g.*, it is glutamine in BstYI, glutamate in BamHI and asparagine in BglII). The second metal ion, M2, when present, is not always found ligated to the leaving O3' group, and therefore such an interaction is not always proposed (as in PvuII). Further, both metal ions are not always proposed to ligate the non-esterified oxygen of the scissile phosphate.[23,26–28] Notably, in addition to the possibility of active site group mispositioning resulting from using substitutions for Mg(II) [such as Ca(II)], and side-chain mutations or substrate analogues, which allow visualization of the stalled pre-reaction ground state structure, it is also recognized that the transition state structure differs from the ground state structure in some subtle but important way, which should be taken as a caveat in structure-based mechanistic models.

The sequence specific endonucleases make direct contacts to most of the bases in their recognition sequence, which can explain their specificity.[29] In some cases, indirect readout, involving other interactions not including direct contacts (such as DNA distortion), has been implicated as well[30–32] (see Chapter 4 for a general discussion of indirect readout). Figure 13.1(F) emphasizes loops of PvuII that interact with the DNA in both the major and minor grooves. The types of structures (loops, beta strands or alpha helices) and the exact contacts to the DNA are particular to each endonuclease; however, this general type of interaction is common to the canonical Type II REs. Some more unusual features of DNA interaction by members of this group include DNA intercalation, as found in the co-crystal structures of HincII,[31] HinPI,[33] and Vsr,[15] which may play a role in recognition. For Vsr, three aromatic side-chains are inserted into the DNA duplex, and appear to be part of the recognition of TG mismatches. One nuclease, Ecl18kI, actually flips out the two bases of a central base pair, then contacts the remaining bases pairs in a fashion very similar to NgoMIV, which recognizes the same site, but without the central base pair.[34]

13.2.2 RNaseH-like Fold

This fold is found with the next highest frequency in structurally characterized nucleases (17 to date). In addition to these DNA nucleases, this fold is also found in some RNaseH enzymes, retroviral integrases and transposases, which are not included here (see Chapter 11 for the latter).[35] All DNA nucleases of this fold appear to utilize two divalent cations in their reaction mechanism; however, their identity varies. In some cases both ions are thought to be Mg(II), but in others one Mg(II) and one Zn(II) are implicated. In addition, all but one (RuvC) are $3' \to 5'$ exonucleases.

The fold common to all members of this group is depicted by a monomer of TREX2 (ref. 36) in Figure 13.2(A) and (B). It consists of a mixed parallel and antiparallel beta sheet surrounded by alpha helices, two on one side of the sheet and three on the other. This fold is slightly more complex than the RE-like fold. Elaborations on the basic fold occur to perform functions from oligomerization to substrate binding. Many members are found covalently linked to or noncovalently associated with a polymerase (Figure 13.2C), while others (TREX1, TREX2, RuvC) form dimers.[36–38] The Werner exonuclease (WRN-exo) forms a hexameric ring.[39] ExoI is monomeric but contains an addition domain that appears to provide processivity to the enzyme, by encircling around the DNA substrate.[40]

The biological functions of members of this fold group vary, but in most cases they are associated with DNA polymerases and act in proofreading. Of the non-polymerase associated members included here, four (ExoI, TREX1, TREX2, WRN-exo) are still 3'-5' exonucleases but are involved in DNA recombination or repair,[36–37,39,40] and RuvC is an endonuclease that resolvase Holliday junctions.[38] None of these enzymes are strongly sequence specific; however, RuvC does preferentially cleave certain DNA sequences in the context of a Holliday junction.[38]

Members of this group can be divided, according to active site residues, into two subgroups: DnaQ-Y or DnaQ-H.[41] DnaQ is the name of the gene that codes for the epsilon subunit of E. coli DNA polymerase III. Common side-chains found in the active site include four acidic residues, which coordinate the usually observed two divalent cations, and the Y or H refers to a tyrosine in some cases, or a histidine in others, which interacts with the nucleophile. (Some members are vestigial and inactive, and the active site residues have been mutated, e.g., the exonuclease domains of Taq and Bacillus DNA polymerases (PDB codes 1TAQ and 1XWL).) Figure 13.2(D) shows active site interactions in the epsilon chain of E. coli DNA polymerase III bound to Ca(II) and dTMP, which is found in two different conformations in the crystal.[41] The two conformations inspired a mechanistic proposal, summarized in Figure 13.2(E), which is similar to that proposed for the exonuclease domain

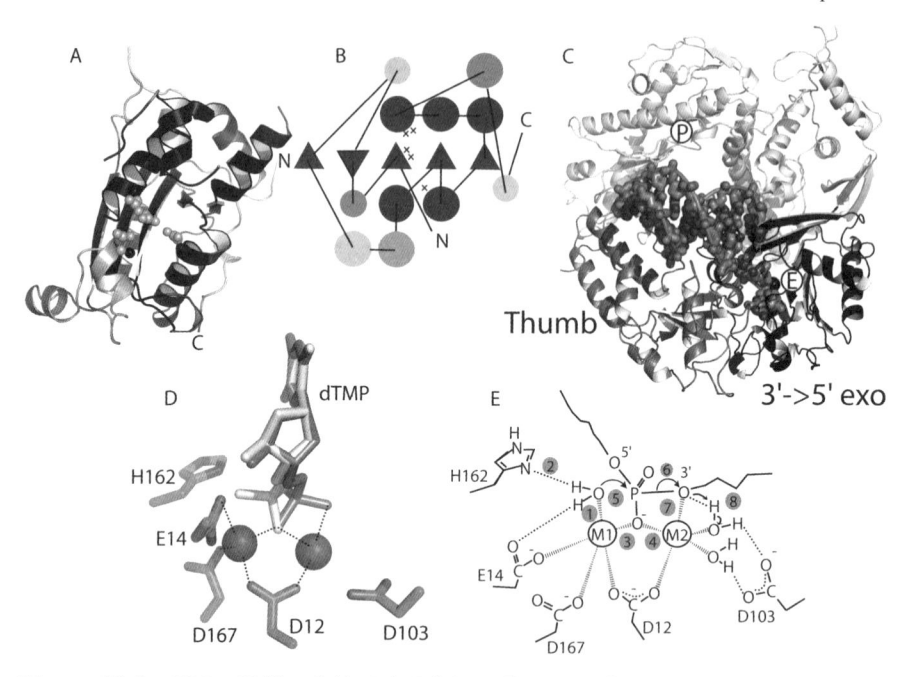

Figure 13.2 RNaseH-like fold. (*A*) Ribbon diagram of one monomer of TREX2 (PDB code 1Y97) shaded by common features within fold family (darkest = most conserved, light = least conserved). (*B*) Topology diagram shaded by common features within fold family as in (*A*). The positions of active site residues are shown as crosses. (*C*) Structure of RB69 DNA polymerase GP43 bound to DNA (PDB code 1CLQ), in white with the thumb domain (residues 729–903) in gray, and the 3′→5′ exo domain (residues 109–338) in black. Ca(II) ions shown as spheres. The exonuclease active site is marked with E and the polymerase active site marked with P. DNA is shown as sticks. (*D*) Active site of the epsilon chain of *E. coli* DNA polymerase III (PDB code 1J54) bound to dTMP in two conformations (white and gray). Ca(II) shown as spheres. Ca(II)-ligand interactions shown as dashed lines. (*E*) A proposed mechanism for DNA cleavage by the epsilon subunit of *E. coli* DNA polymerase III. See text for details.

of *E. coli* DNA polymerase I.[42,43] (Note that M1 and M2 correspond to sites A and B often cited in the literature.) Many features are similar to the mechanism proposed for the RE-like fold (Figure 13.1E), and the numbering scheme between the proposals has been preserved. Differences include the absence of the lysine and presence of, in this case, histidine, making it a member of the DnaQ-H subgroup, along with TREX1, TREX2, ExoI. This histidine is thought to interact with and orient the nucleophile (2). In DnaQ-Y members (which include the other exonucleases of this fold group) it is a tyrosine. (RuvC appears to have no comparable residue.) The histidine in the epsilon subunit has also been proposed to function in proton abstraction,[41] unlike the similarly positioned tyrosine of the *E. coli* of DNA polymerase I 3′-5′ exonuclease.[42,43]

The interaction of the nucleophile with the carboxylate of E14 has also been proposed to orient the nucleophile.[41] A similar function has also been proposed for the corresponding side-chain in the Klenow fragment.[42,43] Another difference between this and the mechanism of Figure 13.1(E) is the absence of an interaction with a neighboring phosphate group. The chemical identity of M1 may also vary; although typically assumed to be Mg(II), in three cases it has been argued to be Zn(II) (*E. coli* DNA polymerase I, T7 DNA polymerase, and T4 DNA polymerase).[44-46]

Two structures of polymerases exist with DNA bound at the 3'-5' exonuclease active site: RB69 DNA polymerase[47] and DNA polymerase I.[48] Figure 13.2(C) shows a ribbon of the structure of RB69 DNA polymerase bound to DNA, with the exonuclease and polymerase active sites marked by a circled E and P, respectively. The 3'-5' exonuclease domain is colored in black, and the thumb domain in gray. Comparisons with the apo structure, and with the structure of T7 DNA polymerase bound to DNA in polymerizing mode,[45] suggest that the thumb domain plays a critical role in shuffling the newly synthesized DNA into the exonuclease active site for editing.[47] In RB69 DNA polymerase structure, three base pairs of the duplex DNA are melted, with the 3' end of one strand in the exonuclease active site, such that the phosphate between the 3' and penultimate nucleotide interacts with a Ca(II) ion at the M2 site. An arginine side-chain, R260, blocks the template strand from entering, and the side-chain of F123 intercalates between the two 3' bases, stacking onto the penultimate one.[47]

13.2.3 Homing Endonuclease-like Fold

To date, eleven structures of this fold have been characterized. This group is also known by their conserved sequence LAGLIDADG, which can be found in an alpha helix that forms the dimerization interface. All utilize Mg(II) as a cofactor, function in homing, and sequence-specifically cleave both strands of duplex DNA.

The common elements of this group are emphasized by shading the monomeric structure and topology diagram of I-CreI[49] (Figure 13.3A, B). The most conserved region of the protein is closest to the active site, which is in fact the dimerization interface as well. Features more removed from the active site/dimerization interface appear less conserved. The long beta strands bind in the major groove of the DNA and make sequence specific contacts (Figure 13.3F). All family members are either homodimers (I-CreI, I-MsoI, I-CeuI) (Figure 13.3C), or pseudodimers (I-DmoI, PI-SceI, I-SceI, PI-PfuI, I-MsoI, I-AniI, PI-TkoII, I-Tsp061I), where both subunits are part of the same chain. The pseudodimeric enzymes are not constrained to recognizing symmetric DNA targets.

All members of this group are homing endonucleases responsible for cleaving DNA at particular sequences (although with quite a bit of sequence permissiveness). Repair of the break through homologous recombination introduces the intervening sequence coding for the endonuclease from an infected allele.[50]

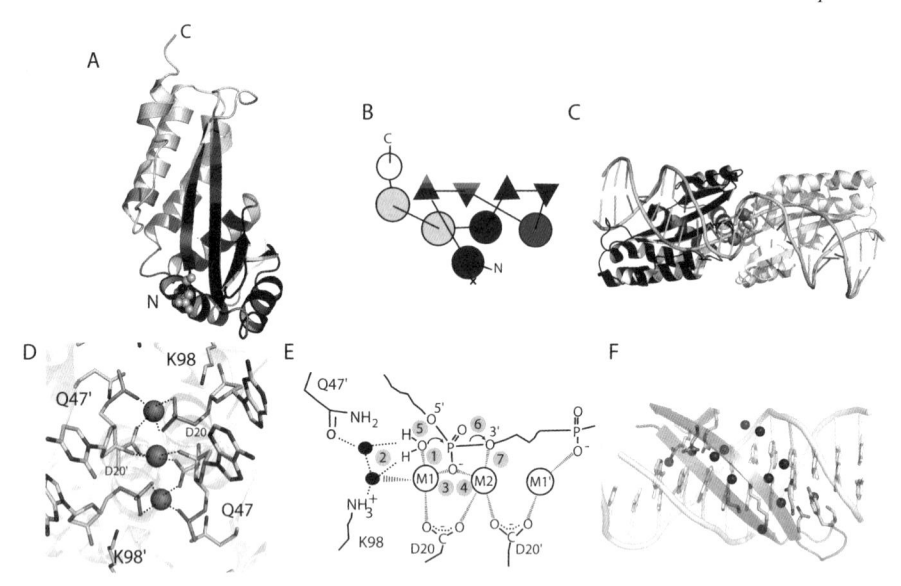

Figure 13.3 Homing endonuclease-like fold. (*A*) Ribbon diagram of one monomer of I-CreI (PDB code 1N3E) shaded by common features within fold family (darkest = most conserved, light = least conserved). Active site residue shown as spheres. (*B*) Topology diagram shaded by common features within fold family as in (*A*). The position of active site residues shown as crosses. (*C*) Dimeric structure of I-CreI (PDB code 1N3F), chain A, white, chain B, black. Ca(II) ions shown as spheres. Bound DNA shown as cartoon. (*D*) Active site of I-CreI (PDB code 1G9Y). Ca(II) shown as spheres. Ca(II)-ligand interactions shown as dashed lines. (*E*) A proposed mechanism for DNA cleavage by I-CreI. See text for details. (*F*) Protein-DNA interfaces in I-CreI (PDB code 1N3F). Water molecules at the interface shown as spheres.

Some members have an additional function of protein splicing (PI-SceI, PI-PfuI, PI-TkoII) and contain a corresponding additional domain (Hint domain). These intein homing endonucleases also contain other domains (DRR in PI-SceI, stirrup in PI-PfuI, domain III in PI-TkoII) with possible DNA binding functions.[51–53] The Hint domain of PI-SceI contains most of the DNA binding specificity.[51] The LAGLIDADG homing endonucleases are usually encoded in the mitochondrial or chloroplast genomes of single-cell eukaryotics.[54] The nuclease I-DreI is an artificial endonuclease created by fusing features of I-CreI and I-DmoI.[55]

Extensive investigation into the mechanism of DNA cleavage by I-CreI has been done using protein mutations, substrate substitutions, and Ca(II), Mn(II), and Mg(II) binding.[56] Work with I-MsoI, I-DreI, I-and SceI support the model proposed for I-CreI.[55,57,58] Active site interactions in one of the structures [wild type with cleaved DNA and Ca(II)] are shown in Figure 13.3(D). The proposed mechanism is summarized in Figure 13.1(E). Three metal ions, M1, M2, M1' are found near the scissile phosphate, although only two are directly involved in the cleavage of a single strand (M1 and M2). The third metal ion (M1'), is involved in the cleavage reaction at the opposite strand, and the middle metal ion (M2) is shared in the two active sites. Another important difference between the current and the previously described mechanisms (*i.e.*, for the RE-like fold, and for the RNaseH-like fold) is the absence of a side-chain directly interacting with the nucleophile. Instead, a network of water molecules is found that are hydrogen bonded to neighboring side-chains. Divalent cation binding sites have also been located in I-SceI,[58] and three Ca(II) ions are somewhat displaced relative to the three metal ion binding sites found in I-CreI. I-SceI is a monomer with a pseudo-twofold axis (a pseudodimer), and the two active sites are not identical. The position of one of the two unshared ions (*i.e.*, M1) overlaps better with that found in I-CreI, a finding that correlates with the faster cleavage of the DNA strand near this metal ion. An unusual feature of this group of nucleases is that only the acidic side-chains responsible for divalent cation binding are strictly conserved. Side-chains capable of hydrogen bonding line the active sites, but interact with solvent rather than with the DNA directly.[54]

All members of this group recognize and cleave specific sequences, or families of sequences, and with differing degrees of promiscuity. This type of recognition is flexible, presumably to tolerate polymorphisms in their homing site without increasing toxicity, which would occur with reduced endonuclease specificity.[54] The typical length of a homing site for the non-protein splicing homing endonuclease (not PI-SceI, PI-PfuI or PI-TkoII) is 18–22 bp, and the DNA is cut in both strands across the minor groove resulting in four base 3′ overhangs. That for the endonucleases with the protein splicing domain is longer, *e.g.*, 31 bp for PI-SceI. The long target site, in combination with "undersaturated" contacts, appears to be the strategy for flexible recognition. Water molecules also appear to play a prominent role at the protein-DNA interface, allowing for some degenerate recognition of DNA sequence (Figure 13.3F). Surprisingly, the symmetric homodimer I-CeuI prefers an asymmetric site over symmetric sites containing either half-site.[59] This asymmetric recognition of a symmetric dimer is accomplished by utilization of different side-chain conformations of residues involved in protein-DNA contacts, along with the use of water-mediated contacts to recognize different bases in each half site.[59] The DNA is also asymmetrically bent, where the bend is better accommodated by the asymmetric sequence than by either symmetric sequence.

13.2.4 His-Me Finger Endonucleases

This is the most divergent structural group of nucleases, having ten structurally characterized members (Table 13.1) that share only a single ββα motif (Figure 13.4). The motif contains the active site, which includes a divalent cation ("Me", meaning any one of several metal ions) ligated by residues that vary among the members, as well as a conserved histidine residue, which appears to act as a general base. The ββα motif bears some structural similarity to C2H2 zinc fingers;[60] however, it interacts with DNA very differently. Within

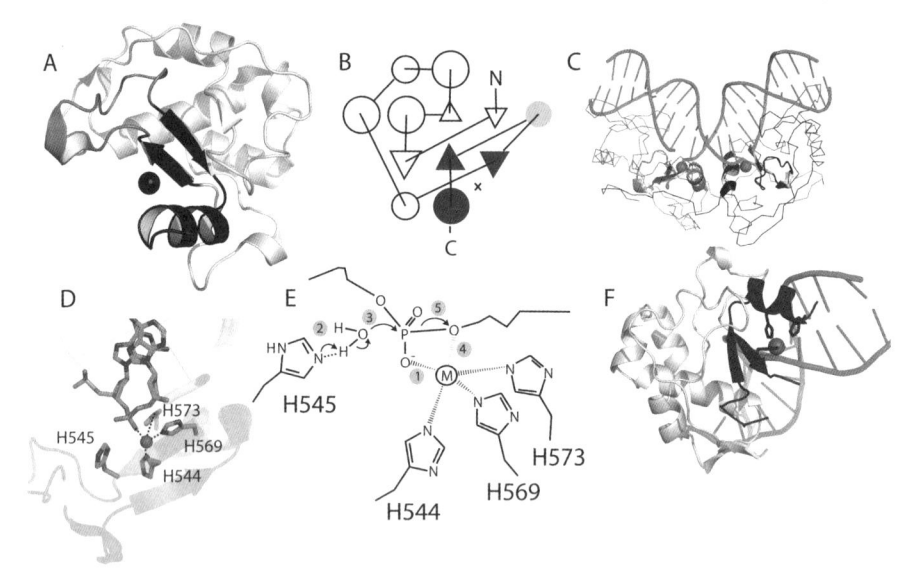

Figure 13.4 His-Me finger endonuclease fold. (*A*) Ribbon diagram of colicin E7 (PDB code 1ZNS) shaded by common features within fold family (darkest = most conserved, light = least conserved). Bound Zn(II) shown as a sphere. (*B*) Topology diagram shaded by common features within fold family as in A. The position of the bound Zn(II) shown as a cross. (*C*) Dimeric structure of I-PpoI (PDB code 1A73), chain A, white, chain B, black. Zn(II) ions shown as spheres. Bound DNA shown as cartoon. (*D*) Active site of colicin E7 (PDB code 1ZNS with the side-chain of H545 from 1PT3). Zn(II) shown as a sphere. Zn(II)-ligand interactions shown as dashed lines. (*E*) A proposed mechanism for DNA cleavage by colicin E7. See text for details. (*F*) Protein-DNA interfaces in colicin E7 (PDB code ZNS). ββα motif shown by darker ribbons.

the His-Me Finger Endonucleases are subgroups of more closely related nucleases: I-PpoI[61] and I-HmuI,[62] *Serratia marcescens* endonuclease[63] and NucA,[64] *Vibrio cholera* Endo I[65] and Vvn,[66] while colicin E7,[67] E9,[68] CAD,[69] and T4 endo VII[70] have no similarity to each other or to the other members.

All His-Me Finger Endonucleases can be subdivided into either the HNH-N or the HNH-H group, depending on the identity of one of the variable metal ion liganded residues (N or H), and named for the three conserved residues (HNH) that define the sequence motif. Only a single divalent cation is implicated in the catalytic mechanism of each enzyme; however, some use predominantly Mg(II), while others may use Zn(II).

All but three His-Me Finger Endonuclease Fold nucleases are monomers, and most cleave DNA nonspecifically. One of the dimeric nucleases, I-PpoI (Figure 13.4C), is a homing endonuclease that binds and cleaves both strands of duplex DNA sequence specifically. Another dimeric member, T4 endo VII, is a Holliday junction resolvase. Finally, CAD also functions as a dimer, and has a deep cleft for binding DNA made up of two long structures, one from either monomer. CAD functions in apoptosis, and is normally kept in an inactive state by an inhibitory protein that binds the monomer, thus preventing its dimerization.[69]

The group I homing endonucleases, I-HmuI and I-PpoI, share not only the ββα motif, but also structure extending by an additional 70 residues.[62] However, they belong to different classes of homing endonucleases that are identified by mode of action and conserved sequence motifs, designated before structural studies had showed the common features: I-PpoI is designated a His-Cys box type, and I-HmuI is an HNH type. I-HmuI contains a DNA binding domain that is very similar to that of I-TevI, although the nuclease domain of I-TevI is of a different fold (GIY-YIG Endonuclease Fold, see below). In both I-HmuI and I-TevI, this domain is responsible for sequence specific recognition, and the nuclease domain introduces the cleavage at a fixed distance away that is determined by the length of the protein linker between the two domains. I-PpoI, though having the ββα motif of the His-Me Finger Endonucleases in its active site, has additional features, including a zinc binding structural domain, placing it in the His-Cys Box class of group I homing endonucleases.[54] The dimeric nature of I-PpoI, along with the structural integration of its catalytic and DNA binding domains, is more reminiscent of type II REs, while the modular nature of I-HmuI is more similar to I-TevI.

Other functions performed by this group, in addition to homing, include the cleavage of Holliday junctions by T4 endo VII, which resolves branch points prior to packaging of the T4 DNA into the phage head.[70] The colicins E7 and E9 are toxins secreted by bacteria.[71] CAD functions in apoptosis and acts to degrade cellular DNA.[69] Vvn is a periplasmic protein that protects the bacterial cell from transformed DNA, and will also degrade RNA.[66] The functions of Endo I from *V. cholera*, Serratia nuclease, and Nuc A are thought to include moving through mucous, acting as bacteriocides, and/or creating dNMPs for nutritional purposes.[64–65,72]

The active site of colicin E7, an HNH-H subgroup member, bound to Zn(II) and DNA is shown in Figure 13.4(D), and a proposed cleavage mechanism in Figure 13.4(E).[62] All His-Me Finger Endonuclease Fold proteins position the ββα finger similarly, with the alpha helix in the minor groove. Three histidine residues ligate the metal ion, M, which ligates a non-esterified oxygen of the scissile phosphate (1), and in only some structures also ligates the leaving group (4). A histidine activates the nucleophile by acting as a general base (2), abstracting a proton and thereby creating hydroxide. Differences in the proposed cleavage mechanisms among members of this fold group occur mainly in the ligation of the metal ion. In HNH-N enzymes, one ligand is asparagine (N); specifically, I-HmuI has only one N and one D ligating to the Mg(II) ion, and I-PpoI has only a single N ligating a Mg(II) ion. Finally, although Mg(II) is the suspected cofactor in most of the nucleases of the His-Me Finger Endonuclease Fold, colicin E9 (an HNH-H enzyme) appears to be able to use Mg(II), Ca(II), Ni(II), or Co(II), but is inhibited by Zn(II),[73] while Zn(II) has been argued to be the cofactor for E7.[74,75]

13.2.5 SAM Domain-like/PIN Domain-like Fold

Five structurally characterized nucleases contain the combination of a SAM domain-like and a PIN domain-like fold (Table 13.1). The fold containing both SAM and PIN like domains has also been denoted FEN-1 like or FLAP endo like. All are 5'-3' exonucleases that utilize the divalent cations Mg(II) or Mn(II) and bind two such ions in cases examined. None of the available structures show binding of the DNA to the active site; however, binding of DNA to an adjacent site on the structure of FEN-1 inspired a model for an active protein-DNA complex.

The overall structure of the FEN-1 from *Methanococcus jannaschii*[76] is shown in Figure 13.5(A), and the topology diagram in Figure 13.5(B). All members of this class possess, with one exception, an arch with an opening

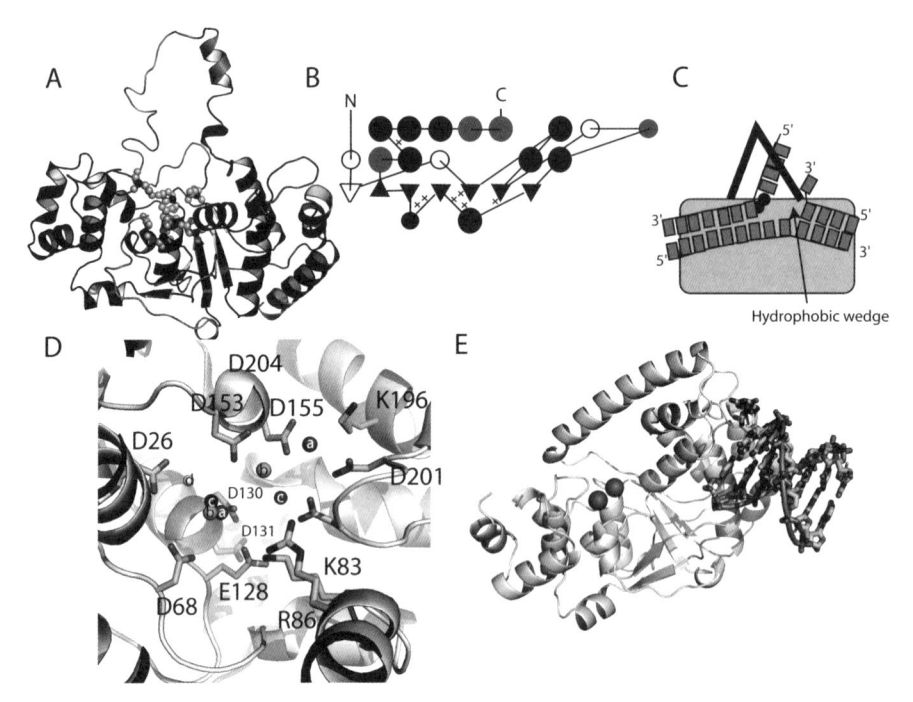

Figure 13.5 SAM-like/PIN-like fold. (*A*) Ribbon diagram of *Methanococcus jannasschii* FLAP endonuclease (PDB code 1A76) shaded by common features within fold family (darkest = most conserved, light = least conserved). Active site residues shown as spheres. (*B*) Topology diagram shaded by common features within fold family as in (A). The positions of active site residues are shown as crosses. (*C*) Cartoon of DNA binding proposed for FEN-1.[80] (*D*) Active site of T5 5'-exonuclease (PDB code 1UT5) with bound ions [Mn(II), labeled a] and ions from *Methanococcus jannasschii* FLAP endonuclease (PDB code 1A76) [Mn(II), labeled b], and T4 endonuclease (PDB code 1TFR) [Mg(II), labeled c] and *Taq* polymerase 5'-3' nuclease (PDB code 1TAQ) [Zn(II), labeled d]. Structural superpositions were done using all alpha carbons aligned by the DALI server.[2] (*E*) DNA bound structure of FEN-1 (PDB code 1UT5) with active site ions [Mn(II)] shown as spheres.

large enough for single-stranded DNA to thread through, but which is too small for duplex DNA. The exact topology of the arch region differs in the different proteins. The active site is found at the base of this arch. The member without an arch is the 5'→3' exonuclease of *Taq* polymerase. All other nucleases of this fold function as monomers; however, they may associate to other proteins such as PCNA, as FEN-1 does during DNA replication.

These 5'→3' exonucleases have varied functions. The FEN-1 enzymes are from archaea but have eukaryotic homologues. They are involved in DNA replication, where they cleave the nucleotide left by RNase H at the RNA-DNA junction of Okazaki fragments. In addition, they function in DNA repair by cleaving 5' flap overhangs, and also possess nick and gap exonuclease activity.[77] Okazaki fragment removal during DNA replication in *Thermus aquaticus* is performed by the 5'→3' exonuclease domain of its polymerase. This polymerase is homologous *E. coli* DNA polymerase I. However, in that case the 5'→3' exonuclease is absent from the structurally characterized Klenow fragment. Similar functions are performed by the T4 RNaseH and T5 5'-exonuclease, as the phages T4 and T5 express these proteins separate from their polymerases.

No structure of any member of this group has been solved with substrate nucleic acid bound at the active site, although soaking experiments have determined the sites of divalent cation binding for most. Figure 13.5(D) shows the positions of these ions after superimposition onto the structure of T5 5'-exonuclease (PDB code 1UT5), using the alpha carbon atoms of the corresponding active site residues shown. One of the two ions appears better matched in the structures (that on the left) than the other. Their positions are somewhat variable, perhaps due to the absence of DNA. Based on the observation of two bound Mn(II) ions in the 5'→3' exonuclease of *Taq* polymerase, a two metal ion mechanism was proposed[78] that is similar to that for the 3'→5' exonuclease activity of the Klenow fragment of *E. coli* DNA polymerase I (Figure 13.2E).[42] Metal binding by calorimetry and by crystallography, in conjunction with DNA cleavage assays of active site mutants T5 5'-exonuclease,[79] led to the proposal that only one of the two metal ions is required for the flap endonuclease activity; however, both are required for the 5'→3' exonuclease activity.

Only one structure has been determined in the presence of bound DNA – that of *Archaeoglobus fulgidus* FEN-1 (Figure 13.5E).[80] Though the DNA is not bound at the active site [spheres denote the bound Mn(II) ions in the active site], the structure, along with FRET studies of DNA binding and bending, inspired a model for binding of 5' flap DNA (Figure 13.5C).[80] The duplex binds FEN-1 with a 3' nucleotide pried away from the remainder of the duplex, and stacked onto the surface of the protein. A "wedge" consisting of several hydrophobic side-chains appears responsible for this effect. The model in Figure 13.5(C) shows how this would connect to the remainder of the flap DNA, allowing the 5' overhang to be threaded through the arch, and placing it over the active site (black spheres) to be cleaved.

13.2.6 DNase I-like Fold

Six structurally characterized nucleases contain this fold (Table 13.1). All cleave one strand of duplex DNA, either exonucleolytically or endonucleolytically. All require a divalent cation cofactor, most likely Mg(II). Most proposed mechanisms make use of a single metal ion in the active site; however, one report invokes a two metal ion mechanism. Biological function varies among the members, as does their specificity.

The overall common DNase I fold is shown in Figure 13.6(A,B), as exemplified by the structure of bovine pancreatic DNase I.[81] The fold consists of two mixed parallel-antiparallel beta sheets surrounded by alpha helices on both sides. The active site residues occur at one end of the two beta sheets and in nearby loops intervening (crosses, Figure 13.6B). The core beta strands are the most conserved, as are some, but not all, of the surrounding alpha helices. Inserts into loops near the active site confer different specificities among the members, *e.g.*, abasic site specificity in APE1 (Figure 13.6E). All act as monomers, although

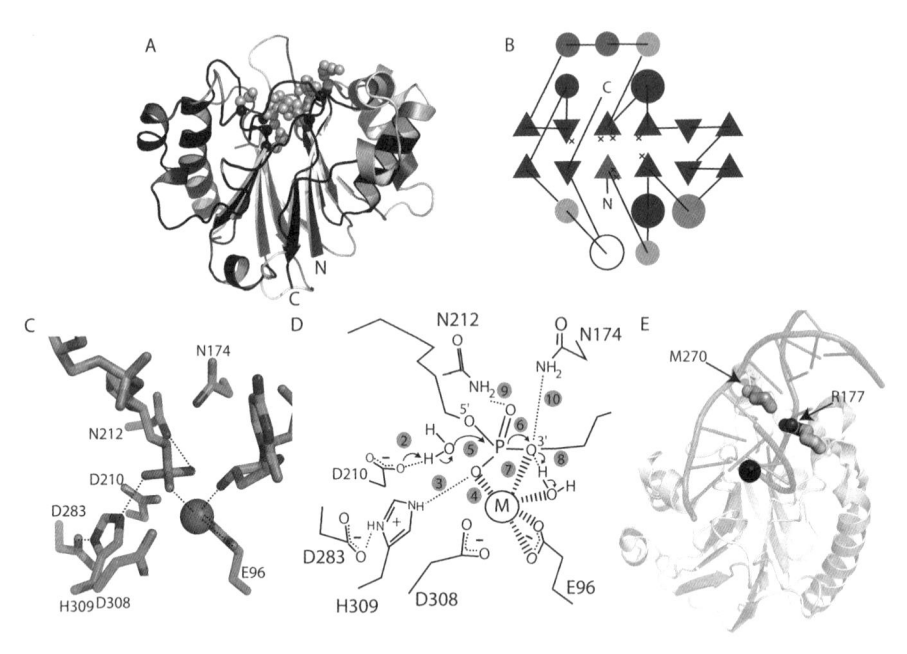

Figure 13.6 DNaseI-like fold. (*A*) Ribbon diagram of DNaseI (PDB code 1DNK) shaded by common features within fold family (darkest = most conserved, light = least conserved). Active site residues shown as spheres. (*B*) Topology diagram shaded by common features within fold family as in (A). The positions of active site residues are shown as crosses. (*C*) Active site of APE1 (PDB code 1DE9). Mn(II) shown as spheres. Mn(II)-ligand interactions and selected hydrogen bonds shown as dashed lines. (*D*) A proposed mechanism for DNA cleavage by APE1. See text for details. (*E*) Protein-DNA interface in APE1 (PDB code 1DE9). Intercalating residues shown as dark sticks and the Mn(II) ion is shown as a sphere.

DNA Nucleases

one, the cytolethal toxin, is found in a complex with other proteins prior to activation, which hides its active site from substrate DNA until it enters the target cell.

The biological functions of the members of this fold are varied and include cell killing (cytolethal toxin), DNA repair (APE-1, Exo III), transposition (LINE-1, TRAS1), or digestive degradation (DNase I). The cytolethal distending toxin is produced by the bacteria *Haemophilus ducreyi* and induces cell cycle arrest and apoptosis in eukaryotic cells.[82] It is produced as a holotoxin with the DNase I-like CdtB nuclease in a complex with two ricin-like lectin domains CdtA and CdtC, which block DNA cleavage activity and translocate CdtB into cells where it produces DNA lesions.[82] The two DNA repair enzymes, APE-1 and ExoIII have inserts into loops near their active sites that confer their particular specificities. Human APE-1 is an apurinic/apyrimidic (AP) nuclease, which cleaves DNA 5' of AP sites after damage-specific DNA glycosylases remove the damaged base.[83] Exo III has a similar function and is found in *E. coli*. In addition to cleaving at AP sites, this enzyme has $3' \to 5'$ exonuclease, 3'-phosphomonoesterase, and ribonuclease activities.[84] The LINE-1 and TRAS1 proteins are sequence specific nucleases that function as part of retrotransposon mobile elements in vertebrates, and are found as part of a larger protein that contains a reverse transcriptase.

Figure 13.6(C) shows the active site of human APE1 (PDB code 1DE9) with Mn(II). Dotted lines indicate important hydrogen bonds or ligations from the Mn(II) ion (shown as a sphere). The DNA is cleaved in this structure, and both the cleaved phosphate and the 3' oxygen are within ligation distance to the ion. Figure 13.6(D) shows a proposed reaction mechanism, in this case involving only one ion.[85] Many functions are similar to those described for the restriction endonuclease fold (Figure 13.1E), indicated at (4)–(8). The nucleophile in this case interacts with an aspartic acid residue, D210, postulated to have an elevated pK_a and act as a general base, accepting a proton from the nucleophile water (2). Nearby, the histidine, H309, has a pK_a potentially raised by neighboring D283, and interacts with the non-esterified oxygen of the scissile phosphate to potentially also stabilize the transition state and/or intermediate (3). The asparagine residues N212 and N174 hydrogen bond to the other non-esterified oxygen of the scissile phosphate and the leaving group respectively (9) and (10). These interactions, along with that from H309, serve to orient the scissile phosphate in the active site. In a variation of this mechanism, proposed much earlier[81] based on a the structure of DNase I with Ca(II) and thymidine 3'-5' diphosphate, the histidine (H309 in APE1) functions as the general base, rather than the aspartate (D210 in APE1). In addition, the leaving group, O3', is proposed to be protonated by H134. Finally, a two metal ion mechanism has been proposed[86] based on the observation of two Pb(II) ions bound in the active site, and the stimulation of Ca(II) on the Mg(II) dependent cleavage reaction. In this proposal the new metal ion would be ligated by H309, D210, N212, the nucleophilic water and a nonesterified oxygen of the scissile phosphate. This mechanism is very similar to those proposed for enzymes of the RE-like fold (Figure 13.1E) and the RNaseH-like fold (Figure 13.2E).

Two co-crystal structures of DNase I with DNA have been solved that help elucidate its sequence preferences.[87,88] Although DNaseI is a nonspecific nuclease, its activity is decreased with runs of A or T.[89,90] The two structures of DNaseI bound to different sequences show differences in DNA conformation. The enzyme induces structural perturbations in the DNA by stacking a tyrosine residue (Y76) onto a deoxyribose group, and by inserting an arginine into the minor groove. The DNA is bent away from the enzyme and the minor groove is widened. Therefore, the resistance to cleavage by sequences exhibiting narrower minor grooves, namely tracks of A, T, or alternating AT, could be due to steric hindrance of the binding of these groups into the narrower minor groove, coupled with the inability to simultaneously maintain optimal contacts to both phosphodiester backbones.

For APE1, the crystal structures with duplex DNA containing an abasic site, and also with or without Mn(II), have been described.[85,91] When Mn(II) is bound, the DNA is cleaved (Figure 13.6C) and the nucleotide at the abasic site is flipped out of the DNA helix (Figure 13.6E). The side-chain of M270 intercalates into the DNA duplex from the minor groove and packs against the orphan would-be base pairing partner of the abasic site, occupying the space where it would be in unkinked DNA. The side-chain of R177 intercalates into the DNA from the major groove and hydrogen bonds to the phosphate 3′ to the abasic site. These interactions stabilize the kinked DNA, allowing the abasic site to be engulfed and therefore specifically recognized, and both residues are found on loops added to the basic DNaseI-like fold particular only to the abasic recognizing endonucleases APE1 and Exo III. Surprisingly, mutagenesis studies of these two residues show that specific recognition and cleavage of AP sites is not significantly affected by their mutation to alanine.[85] They have therefore been proposed to act in strong binding to the cleaved product DNA, before handing it off to the next step in DNA repair.

13.2.7 Phospholipase C/P1 Nuclease Fold

Only a single DNA nuclease of this fold has been structurally characterized (Table 13.1), P1 nuclease from *Penicillium citrinum*.[92,93] The fold is named after phospholipase C enzymes that cleave phospholipid molecules to diacylglycerol and a phosphorylated lipid head group. In addition to the common fold, many aspects of their catalytic mechanisms appear similar, in particular, the use of three zinc ions.

P1 nuclease acts as a monomer and is secreted as a glycoprotein to cleave single-stranded DNA and RNA for catabolic purposes. It can also act as a phosphomonoesterase in removing 3′ terminal nucleotides. The fold (Figure 13.7A and B) is nearly all alpha helical with a small two stranded beta sheet. The active site is indicated by the three zinc ions (spheres in Figure 13.7A), and crosses marking the active site residues in Figure 13.7(B).

A close-up of the interactions in the active site is shown in Figure 13.7(D), where three zinc ions can be located. Structures have been solved with

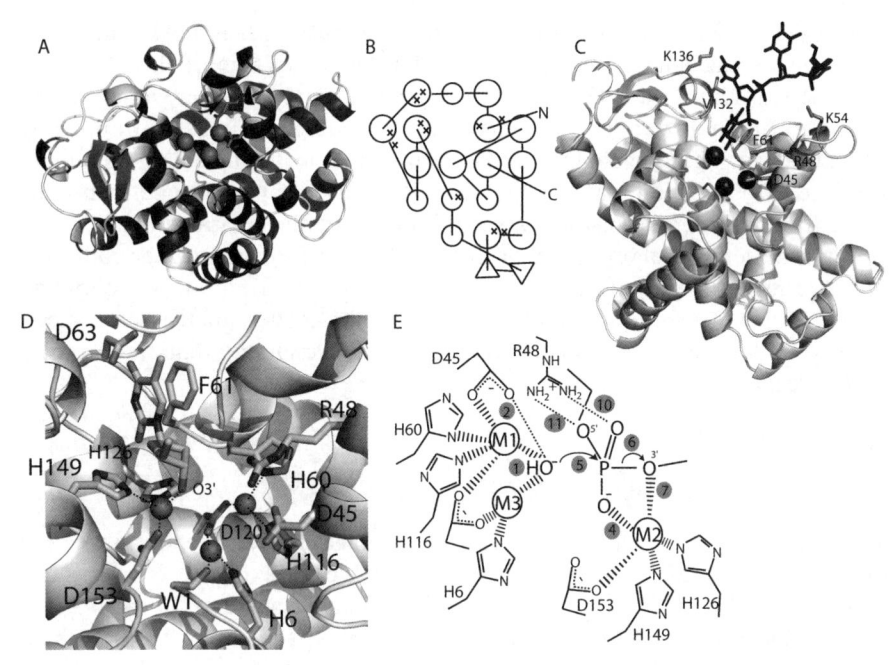

Figure 13.7 Phospholipase C-like fold. (*A*) Ribbon diagram of P1 nuclease (PDB code 1AK0). Bound Zn(II) ions shown as spheres. (*B*) Topology diagram of P1 nuclease (PDB code 1AK0). The positions of active site residues are shown as crosses. (*C*) Protein-DNA interface in P1 nuclease (PDB code 1AK0). (*D*) Active site of P1 nuclease (PDB code 1AK0). Zn(II) shown as spheres. Zn(II)-ligand interactions shown as dashed lines. (*E*) A proposed mechanism for DNA cleavage by P1 nuclease. See text for details.

single-stranded DNA containing non-cleavable phosphorothioate substitutions that are 2, 4 and 6 nucleotides in length. The structure deposited in the PDB (PDB code 1AK0) includes four nucleotides of the DNA where both non-esterified oxygen atoms at each phosphate are substituted with sulfur. The DNA binds to the surface of P1 nuclease (Figure 13.7C) with the 3′ base in the active site mimicking a product complex. The O3′ of the 3′ base ligates to the M2 (Figure 13.7E) Zn(II) ion. Its base is stacked onto the ring of F61 (Figure 13.7C and D), which is consistent with observations that the presence of the base 5′ to a phosphate is necessary for its cleavage by P1 nuclease,[94] and that the aromaticity, as tested by base analogues, is also important to activity.[95] An unexpected finding is the hydrogen bonding between the side-chain of D63 and this base, since the nuclease is non-sequence specific; however, hydrogen bonds can be formed with a protonated D63 to any base. The D63 side-chain is expected to be protonated under the acidic conditions that are optimal for P1 nuclease.

Though the phosphate part of the mechanism shown in Figure 13.7(E) is not present in the structure, a model can be proposed[93] based on the location of the

O3′ oxygen, where a water molecule ligated by the Zn(II) ions at M1 and M3 is activated for nucleophilic attack (1). In addition to ligation of the M1 Zn(II) ion, the side-chain of D45 is also within hydrogen bonding distance to the putative nucleophile and may serve in its orientation and also possibly activation (2). The side-chain of R48 orients the scissile phosphate and could stabilize the transition state (10). After attack by the nucleophile hydroxide (5), a pentacovalent transition state or intermediate is formed that inverts configuration by the breakage of the P–O3′ bond (6). M2 Ligation to a non-esterified oxygen of the scissile phosphate (4) orients the group and stabilizes the transition state/intermediate, while ligation of the same metal to the O3′ leaving group (7) stabilizes the negative charge that results from bond breakage.

13.2.8 Phospholipase D/Nuclease Fold

Two nucleases that have been structurally characterized show this fold (Table 13.1): Nuc,[96] a nonspecific nuclease (cleaving DNA, RNA, single stranded, double stranded) from *Salmonella typhimurium*, and a type IIS RE, BfiI,[97] which cleaves both strands of duplex DNA at a fixed distance from an asymmetric sequence. Remarkably, both enzymes are unusual for DNA nucleases in that they are metal independent, and may utilize two histidine residues in the DNA cleavage reaction, similar to the reaction on phospholipids by phospholipase D. Neither has been structurally characterized in complex with DNA.

Figure 13.8(A) shows the fold of one of the monomers of the Nuc dimer, with the topology diagram in Figure 13.8(B). For BfiI, an additional domain is found in each monomer that belongs to the DNA binding pseudobarrel domain endonuclease effector domain fold, similar to the N-terminal domain of EcoRII.[97] Figure 13.8(C) shows the dimeric structure of BfiI, with one monomer in white and one in black. The active site is formed by residues from each monomer at the center of the dimer (spheres, Figure 13.8C). The active site can cleave only a single strand of DNA at a time. The BfiI dimer binds to two recognition sites, sequence specifically, with one bound at each DNA binding domain. Kinetic studies show that each strand of DNA is cleaved sequentially, with a dissociation and reassociation occurring between cleavages.[98] The structure was solved in the absence of DNA, and a loop that connects the two domains lies on the top of the active site, presumably blocking DNA binding. The nuclease domain cleaves DNA nonspecifically in the absence of the DNA binding domains. Their role is thought to be to bring the nuclease domain near the correct DNA site. DNA binding likely causes a conformational change that removes the loop, allowing DNA access to the active site.[97]

A close-up of the active site of BfiI is shown in Figure 13.8(D). In the Nuc structure, a tungstate molecule, which is structurally similar to a phosphate molecule, binds the active site in a way that inspires the mechanism outlined in Figure 13.8(E).[96] The enzyme forms a covalent intermediate with the DNA using

DNA Nucleases 355

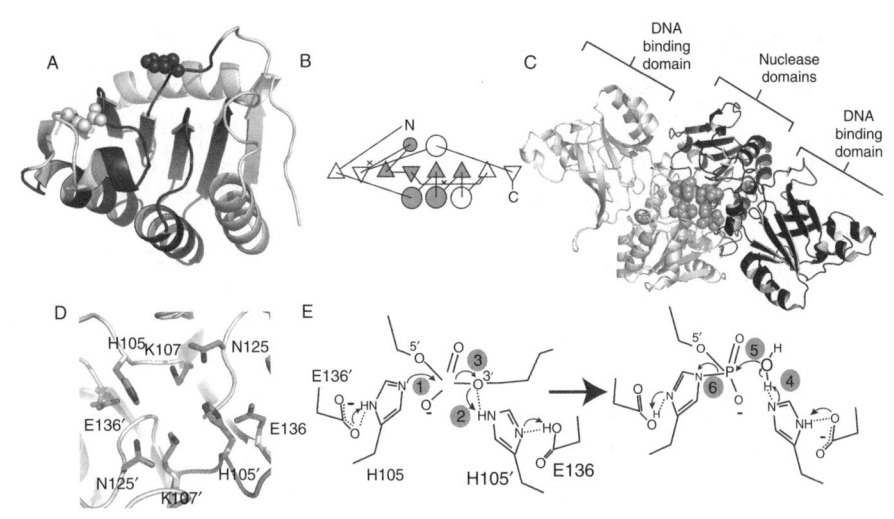

Figure 13.8 Phospholipase D-like fold. (*A*) Ribbon diagram of one monomer of Nuc (PDB code 1BYR). The core phospholipid D elements found in Nuc and BfiI are colored black. Active site residues shown as spheres. (*B*) Topology diagram of a monomer of Nuc. The positions of active site residues are shown as crosses. (*C*) Dimeric structure of BfiI (PDB code 2C1L), chain A, white, chain B, black. Active site residues shown as spheres. (*D*) Active site of BfiI (PDB code 2C1L). (*E*) A proposed mechanism for DNA cleavage by BfiI. See text for details.

a nitrogen atom from one of the histidine residues, H105, for nucleophilic attack on the phosphorus (1). The histidine from the other monomer in the dimer, H105′, is in a protonated state, and protonates the O3′ leaving group (2), as the P–O3′ bond breaks (3). In the second step, the same histidine, H105′, deprotonates a water molecule (4) to attack the phosphorus (5), allowing the N–P bond on the first histidine, H105, to break (6). The nearby glutamate residues, E136, from each monomer act as proton shuttles to and from the histidine residues.

13.2.9 TIM beta/alpha Barrel Fold

Two structurally characterized nucleases possess the TIM beta/alpha barrel, TatD and Endo IV (Table 13.1). Only the latter has been described in the literature at the time of writing, and has been solved in the presence of DNA and Zn(II).[99] It appears to use three Zn(II) ions much like the mechanism shown for P1 nuclease (Figure 13.7E), of the Phospholipid C/P1 Nuclease fold. However, the topology of the enzyme is totally different. Endo IV recognizes abasic (AP) sites and cleaves the phosphodiester backbone 5′ from the site, just as Exo III and APE1 do from the DNase I fold. It is found in both eubacteria and eukaryotes.

Figure 13.9(A) shows the overall fold of Endo IV (PDB code 1QUM), which is a beta barrel surrounded by alpha helices, with the active site residues at the

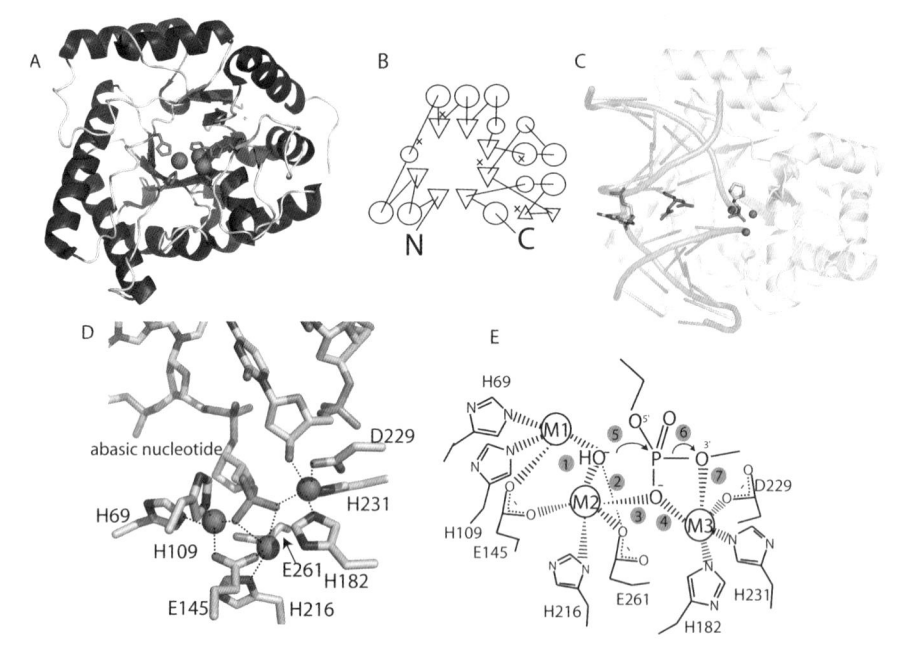

Figure 13.9 TIM alpha/beta barrel fold. (*A*) Ribbon diagram of Endo IV (PDB code 1QTW). (*B*) Topology diagram of Endo IV. The positions of active site residues are shown as crosses. (*C*) Cartoon representation of DNA bound Endo IV (PDB code 1QTW), emphasizing the intercalating arginine residue in two conformations as well as the abasic and orphan nucleotides (dark sticks). Bound Zn(II) ions shown as spheres. (*D*) Active site of Endo IV (PDB code 1QTW). Zn(II) shown as spheres. Zn(II)-ligand interactions shown as dashed lines. (*E*) A proposed mechanism for DNA cleavage by Endo IV. See text for details.

center. This TIM barrel fold is common among enzymes that perform diverse chemical reactions.[100,101] The enzyme functions as a monomer and binds three Zn(II) ions near the center of the barrel (spheres, Figure 13.9A). The ions are ligated to histidine, aspartate, and glutamate side-chains. AP site-containing duplex DNA in the structure binds to the surface of the barrel (Figure 13.9C) with the orphaned sugar buried in the active site. Three residues, L73, R37, and Y72, intercalate into the DNA duplex from the minor groove side, causing flipping out of the AP nucleotide, as well as its orphaned base pairing partner, and stack on the base pairs on either side of the AP site. Figure 13.9(C) shows the side-chain of R37 (dark sticks) modeled in two conformations, intercalated into the DNA.

The DNA was found cleaved at the phosphate 5′ to the AP site. The cleaved phosphate and the O3′ hydroxyl were found ligated to the Zn(II) ions in the active site (Figure 13.9D). The proposed reaction mechanism (Figure 13.9E) is very similar to that proposed for P1 nuclease (Figure 13.7E) with some small differences. The only significant difference is the ligation of the non-esterified

DNA Nucleases 357

oxygen of the scissile phosphate by two zinc ions, M2 and M3 (note numbering of metal ions is different in the different mechanisms). In addition, no arginine or lysine residue occurs nearby to perform the function of R48 in the P1 nuclease mechanism.

13.2.10 DHH Phosphoesterases Fold

RecJ from *T. thermophilus* is the only structurally characterized nuclease with this fold (Table 13.1), named for the conserved DHH active site residues (Figure 13.10).[102] It is a $5' \rightarrow 3'$ exonuclease of single-stranded DNA that is found in both prokaryotes and eukaryotes. It functions in DNA repair and recombination and requires Mg(II) as a cofactor.

Although the DNA-bound structure of RecJ is not known, the site of Mn(II) binding has been identified (Figure 13.10C). A cleft between the two domains appears to be a likely place for DNA binding, with the active site located just at

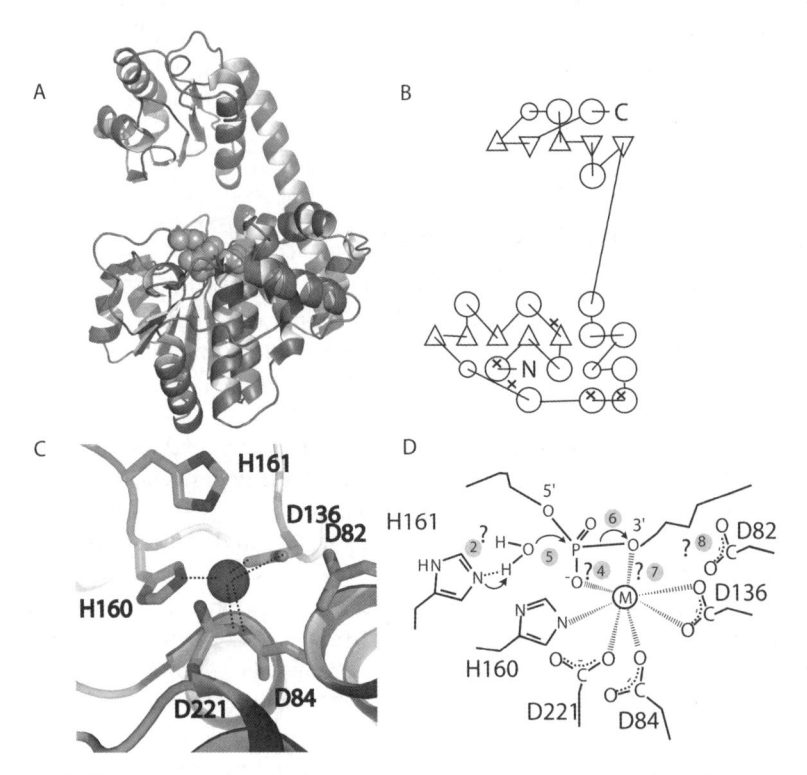

Figure 13.10 DHH phosphoesterase fold. (*A*) Ribbon diagram of RecJ exonuclease (PDB code 1IR6). (*B*) Topology diagram of RecJ exonuclease. The positions of active site residues are shown as crosses. (*C*) Active site of RecJ (PDB code 1IR6). The bound Mn(II) ions shown as a spheres. Mn(II)-ligand interactions shown as dashed lines. (*D*) A proposed mechanism for DNA cleavage by RecJ. See text for details.

the bottom. The cleft readily accommodates single-stranded DNA, but is too small for double-stranded DNA.

The single Mn(II) binds at the site of the conserved active site residues (Figure 13.10C) and ligates D136, D84, D221 and H160. Ligation to a nitrogen ligand such as H160 is unusual for Mg(II), but Mn(II) has a higher affinity for such ligands. Therefore, the ligation to Mg(II) may be slightly different than that seen for Mn(II).[102] The side-chains of H161 and D82 are nearby, and may also play important roles in the DNA cleavage mechanism. Finally, additional divalent cations may bind when substrate DNA is bound, as the DNA provides some of the ligands. One of several possible mechanisms is shown in Figure 13.10(D). Without bound DNA, interactions with it can only be speculated based on other better characterized mechanisms. The observed metal ion could ligate the leaving O3' group (7) and possibly a non-esterified oxygen of the scissile phosphate (4). One or both histidine side-chains could activate the nucleophile (2) for nucleophilic attack onto the phosphorus (5). Cleavage of the P–O3' bond would result from breakdown of the pentacovalent transition state or intermediate (6). The side-chain of D82 could have an important role such as protonating the leaving group (8). Alternatively, with a repositioning of the scissile phosphate, D82 could activate the nucleophilic water molecule, and H161 could then have a role in transition state and leaving group stabilization.

13.2.11 GIY-YIG Endonuclease Fold

Two members of this fold have been structurally characterized: the catalytic domains of the homing endonuclease I-TevI[103] and of the bacterial DNA repair protein UvrC[104] (Table 13.1). Both require Mg(II) for catalytic function, and contain the motif GIY followed by YIG with an intervening span of 9–10 residues.

Figure 13.11(A) shows the overall fold of the N-terminal domain of UvrC (PDB code 1YD1). The catalytic domain of I-TevI differs from that of UvrC in that the N terminal alpha helix is absent. The $GIYX_{9-10}YIG$ motif resides are found on the first and second beta strands (middle and right-hand strands in Figure 13.11B) in both structures.

I-TevI is a homing endonuclease, involved in the activity of mobile elements, where it cleaves DNA sequence specifically to allow homologous recombination, converting an uninfected allele into one possessing the mobile element. Unlike the LAGLIDADG family (Homing Endonuclease-like Fold), or the His-Cys Box homing endonucleases I-PpoI (His-Me Finger Endonuclease Fold) described above, but like I-HmuI, a HNH homing endonuclease (His-Me Finger Endonuclease Fold), I-TevI is quite modular with a separate domain making the sequence specific contacts,[105] fused to this nonspecific nuclease domain. UvrC is a member of the prokaryotic nucleotide excision repair pathway responsible for making the cuts in DNA 5' and 3' to the site of DNA damage. It cuts only the damaged strand, and is aided by the complex with UvrB and

DNA Nucleases 359

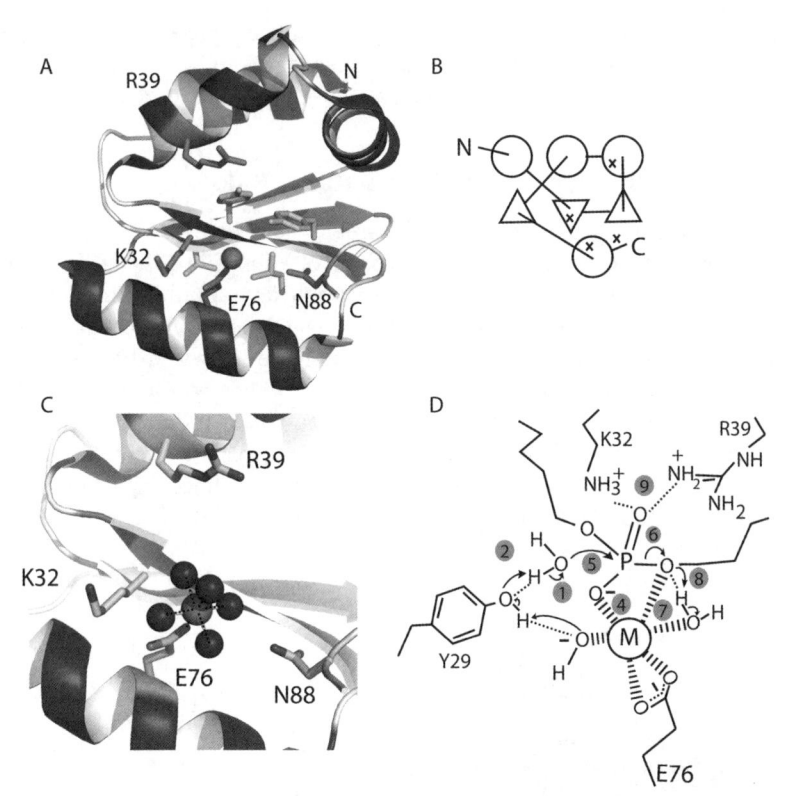

Figure 13.11 GIY-YIG homing endonuclease-like fold. (*A*) Ribbon diagram of the N terminal nuclease domain of UvrC (PDB code 1YD1). Active site residues shown as dark sticks. Side-chain residues of the GIY-YIG motif shown in light sticks. (*B*) Topology diagram of the N terminal nuclease domain of UvrC. The positions of active site residues are shown as crosses. (*C*) Active site of the N terminal nuclease domain of UvrC (PDB code 1YD1). The bound Mn(II) ion is shown as a lighter sphere, with ligated water molecules shown as darker spheres. (*D*) A proposed mechanism for DNA cleavage by UvrC. See text for details.

UvrA proteins for specificity. The domain described here is that responsible for the 3′ incision.

Though substrate DNA binding has not been structurally characterized, the binding of a Mg(II) to the UvrC domain has[104] (Figure 13.11C). The ion ligates to the side-chain of E76, with the remaining five ligands as water molecules. Based on this and mutagenic studies, a mechanistic model for DNA cleavage has been proposed (Figure 13.11D).[104] It utilizes a single metal ion (M) to ligate a non-esterified oxygen of the scissile phosphate (4) and the leaving group, O3′ (7). A metal ligated hydroxide abstracts the proton from the second of the two tyrosine residues of the GIY-YIG motif (1), which activates the tyrosine to act as a general base, activating the nucleophilic water (2). Formation of the penta-covalent transition state/intermediate results from nucleophilic attack upon the phosphorus (5), which is stabilized by the positively charged side-chains of K32 and R39 (9). After or during P–O3' bond breakage (6), and before dissociation of the cleaved DNA, a metal ligated molecule may protonate the O3' (8).

13.2.12 Metallo-dependent Phosphatases Fold

A single member of this fold has been characterized, *Pyrococcus furiosus* Mre11,[106] which in a complex with Rad50 is involved in DNA repair. It functions as a single-stranded DNA endonuclease as well as a double-stranded 3'→5' exonuclease. It binds and utilizes two Mn(II) ions in the DNA cleavage reaction.

The overall structure and topology of Mre11 are shown in Figure 13.12(A) and (B). The overall shape of the molecule is similar to RecJ (DHH Phosphoesterases Fold, Figure 13.10A), but the topology is unique.

The structure with Mn(II) and dAMP resembles a product complex (Figure 13.12C). The two Mn(II) are ligated by histidine, aspartate, and an asparagine residue (N84, not shown in Figure 13.12D). The Mn(II) at M1

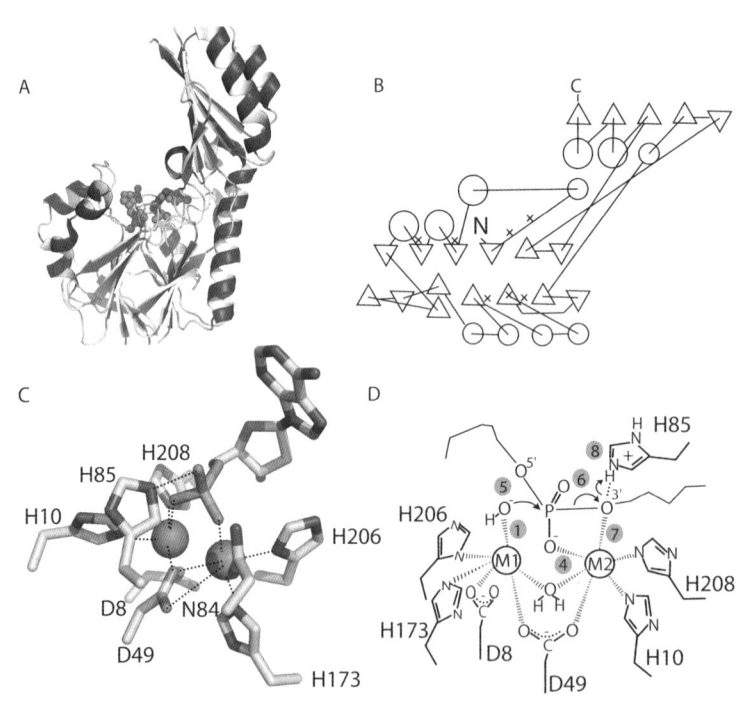

Figure 13.12 Metallo-dependent phosphatase fold. (*A*) Ribbon diagram of Mre11 (PDB code 1II7). Active site residues shown as spheres. (*B*) Topology diagram of Mre11. The positions of active site residues are shown as crosses. (*C*) Active site of Mre11 (PDB code 1II7). The bound Mn(II) ions shown as spheres. Mn(II)-ligand interactions shown as dashed lines. (*D*) A proposed mechanism for DNA cleavage by Mre11. See text for details.

DNA Nucleases 361

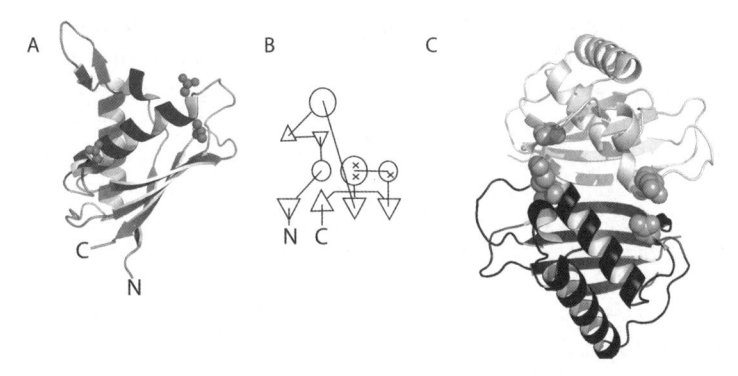

Figure 13.13 *Bacillus chorismate* mutase-like fold. (*A*) Ribbon diagram of one monomer of RusA (PDB code 1Q8R). Active site residues shown as spheres. (*B*) Topology diagram of RusA. The positions of active site residues are shown as crosses. (*C*) Dimeric structure of RusA (PDB code 1Q8R), chain A, white, chain B, black.

activates the nucleophile by reducing its pK_a to generate hydroxide (1). The Mn(II) at M2 ligates a non-esterified oxygen of the scissile phosphate (4) and the leaving group (7) to stabilize the transition state and the leaving group, respectively. Histidine 85 could protonate the leaving group (8). The structure also shows why Rad50 is required for some of the Mre11/Rad50 functions, since the DNA binding cleft requires that double-stranded DNA be deformed. Such a function could be performed by the ATPase Rad50, through conformational cycling concurrent with ATP binding, hydrolysis, and release.[106]

13.2.13 *Bacillus chorismate* Mutase-like Fold

Two nucleases (or putative nucleases) have been structurally characterized with this fold (Table 13.1): RusA Holliday junction resolvase,[107] and APE1501,[108] a putative endonuclease from *A. pernix*. Only the former has been described in full. It resolves Holliday junctions in *E. coli*, and is encoded by a cryptic prophage.

The overall structure of a monomer of RusA (PDB code 1Q8R) is shown in Figure 13.13(A), and the topology diagram in Figure 13.13(B). Active site residues (D70, D72 and D90) are highlighted as gray spheres or crosses in the two figures, respectively. These residues are distant from each other in the monomer, but come together from opposing chains in the dimer (Figure 13.13C), indicating the dimer formation is necessary for divalent cation binding and DNA cleavage. The enzyme requires Mg(II) to function. No structure with divalent cations or substrate DNA is available.

13.3 Conclusion

This survey found over 90 different structurally characterized DNA nucleases, possessing any of 13 different folds. Some folds are very well represented in this group, such as the Restriction Endonuclease-like Fold (35 DNA nucleases) and the RNase H-like Fold (17 DNA nucleases). Others, such as the Phospholipase C-like/P1 Nuclease, DHH Phosphoesterase, and the Metallo-dependent Phosphatases Folds, are found in only a single example. It is tempting to speculate that some folds are more versatile or effective than others, but the differences in the number of examples of each fold could merely reflect technical reasons, such as the ease of the structural studies with proteins from bacteria *versus* those from other sources. It remains to be seen what the true frequency of these folds among DNA nucleases is in Nature.

Within each fold group, it is possible to find DNA nucleases that perform very different biological functions. For example, within the RE-like Fold group, DNA nucleases that function as restriction endonucleases are common, but several DNA repair proteins and a Holliday junction resolvase are also found. Conversely, DNA nucleases that perform similar functions, such as recognizing and cleaving at abasic sites in DNA, can be found with very different folds: the DNase I fold of APE1 and ExoIII, and the TIM beta/alpha Barrel fold of Endo IV. Similarly, homing endonucleases are found in four fold groups, Holliday junction resolvases in four, and restriction endonucleases in two.

The proposed catalytic mechanisms for DNA nucleases of each fold have been presented, and in all but one case the enzymes make use of divalent cations. The number and identity of the ions can differ, both between folds, and even in some cases within the same fold group. Many use Mg(II), with either one or two ions per active site. Others use Zn(II), with either one or three ions per active site, and one enzyme, Mre11, appears to use two Mn(II) ions. Some enzymes of the RNase H-like Fold appear to use one Mg(II) and one Zn(II). The catalytic mechanisms proposed for DNA cleavage by DNA nucleases have many features in common. Two very different enzymes, with different folds, P1 nuclease and Endo IV, appear to use a similar mechanism with three zinc ions. When Mg(II) is a cofactor, Ca(II) generally inhibits the enzyme, although colicin (E9) is unusual in that Ca(II) can easily substitute for Mg(II) and confer nuclease activity. If the inhibition by Ca(II) in Mg(II) dependent reactions is the result of the inability of two Ca(II) ions to achieve the close spacing required in a two metal ion mechanism,[21,109] then it is clear why Ca(II) would inhibit mechanisms requiring two divalent cations (the RE-like Fold, RNase H-like Fold, Homing Endonuclease-like Fold, SAM domain-like/PIN domain-like Fold), but function in enzymes which use only one (E9 of the His-Me Finger Endonuclease Fold).

The mode of DNA binding reflects the specificity and function of each DNA nuclease. Many are non-sequence specific, and some are even nonspecific with respect to the type of sugar (they can cleave RNA or DNA), and interactions with the bases and sometimes sugar are minimal. Others are sequence, structure, or damage specific, and contain structural features that provide these functions. Most nucleases have the specificity determinants integrated into their fold, while others use a modular approach, such as some restriction and homing endonucleases, which have a nonspecific nuclease attached to a sequence specific DNA binding domain.

DNA Nucleases

References

1. A.G. Murzin, S.E. Brenner, T. Hubbard and C. Chothia, *J. Mol. Biol.*, 1995, **247**, 536.
2. L. Holm and C. Sander, *Nucleic Acids Res.*, 1998, **26**, 316.
3. D.R. Westhead, T.W. Slidel, T.P. Flores and J.M. Thornton, *Protein Sci.*, 1999, **8**, 897.
4. T. Nishino, K. Komori, D. Tsuchiya, Y. Ishino and K. Morikawa, *Structure*, 2001, **9**, 197.
5. M.Y. Niv, D.R. Ripoll, J.A. Vila, A. Liwo, E.S. Vanamee, A.K. Aggarwal, H. Weinstein and H.A. Scheraga, *Nucleic Acids Res.*, 2007, **35**, 2227.
6. J.R. Horton and X. Cheng, *J. Mol. Biol.*, 2000, **300**, 1049.
7. D.A. Wah, J. Bitinaite, I. Schildkraut and A.K. Aggarwal, *Proc. Natl. Acad. Sci. U.S.A.*, 1998, **95**, 10564.
8. Q. Huai, J.D. Colandene, Y. Chen, F. Luo, Y. Zhao, M.D. Topal and H. Ke, *EMBO J.*, 2000, **19**, 3110.
9. X.E. Zhou, Y. Wang, M. Reuter, M. Mucke, D.H. Kruger, E.J. Meehan and L. Chen, *J. Mol. Biol.*, 2004, **335**, 307.
10. R. Kovall and B.W. Matthews, *Science*, 1997, **277**, 1824.
11. M. Newman, J. Murray-Rust, J. Lally, J. Rudolf, A. Fadden, P.P. Knowles, M.F. White and N.Q. McDonald, *EMBO J.*, 2005, **24**, 895.
12. C. Ban and W. Yang, *EMBO J.*, 1998, **17**, 1526.
13. J.Y. Lee, J. Chang, N. Joseph, R. Ghirlando, D.N. Rao and W. Yang, *Mol. Cell*, 2005, **20**, 155.
14. K.A. Bunting, S.M. Roe, A. Headley, T. Brown, R. Savva and L.H. Pearl, *Nucleic Acids Res.*, 2003, **31**, 1633.
15. S.E. Tsutakawa, H. Jingami and K. Morikawa, *Cell*, 1999, **99**, 615.
16. S.E. Tsutakawa, T. Muto, T. Kawate, H. Jingami, N. Kunishima, M. Ariyoshi, D. Kohda, M. Nakagawa and K. Morikawa, *Mol. Cell*, 1999, **3**, 621.
17. S.E. Tsutakawa and K. Morikawa, *Nucleic Acids Res.*, 2001, **29**, 3775.
18. A. Pingoud and A. Jeltsch, *Nucleic Acids Res.*, 2001, **29**, 3705.
19. J.P. Glusker, *Adv. Protein Chem.*, 1991, **42**, 1.
20. M.E. Maguire and J.A. Cowan, *Biometals*, 2002, **15**, 203.
21. W. Yang, J.Y. Lee and M. Nowotny, *Mol. Cell*, 2006, **22**, 5.
22. C. Etzkorn and N.C. Horton, *Biochemistry*, 2004, **43**, 13256.
23. C. Etzkorn and N.C. Horton, *J. Mol. Biol.*, 2004, **343**, 833.
24. A. Jeltsch, J. Alves, H. Wolfes, G. Maass and A. Pingoud, *Proc. Natl. Acad. Sci. U.S.A.*, 1993, **90**, 8499.
25. A. Jeltsch, M. Pleckaityte, U. Selent, H. Wolfes, V. Siksnys and A. Pingoud, *Gene*, 1995, **157**, 157.
26. N.C. Horton and J.J. Perona, *Nat. Struct. Biol.*, 2001, **8**, 290.
27. A. Pingoud, *Restriction Endonucleases*, Springer-Verlag, Berlin, New York, 2004.
28. E.A. Galburt and B.L. Stoddard, *Biochemistry*, 2002, **41**, 13851.

29. N.M. Luscombe, R.A. Laskowski and J.M. Thornton, *Nucleic Acids Res.*, 2001, **29**, 2860.
30. D.R. Lesser, M.R. Kurpiewski and L. Jen-Jacobson, *Science*, 1990, **250**, 776.
31. N.C. Horton, L.F. Dorner and J.J. Perona, *Nat. Struct. Biol.*, 2002, **9**, 42.
32. A.M. Martin, M.D. Sam, N.O. Reich and J.J. Perona, *Nat. Struct. Biol.*, 1999, **6**, 269.
33. J.R. Horton, X. Zhang, R. Maunus, Z. Yang, G.G. Wilson, R.J. Roberts and X. Cheng, *Nucleic Acids Res.*, 2006, **34**, 939.
34. M. Bochtler, R.H. Szczepanowski, G. Tamulaitis, S. Grazulis, H. Czapinska, E. Manakova and V. Siksnys, *EMBO J.*, 2006, **25**, 2219.
35. D. Esposito and R. Craigie, *Adv. Virus Res.*, 1999, **52**, 319.
36. F.W. Perrino, S. Harvey, S. McMillin and T. Hollis, *J. Biol. Chem.*, 2005, **280**, 15212.
37. U. de Silva, S. Choudhury, S.L. Bailey, S. Harvey, F.W. Perrino and T. Hollis, *J. Biol. Chem.*, 2007, **282**, 10537.
38. M. Ariyoshi, D.G. Vassylyev, H. Iwasaki, H. Nakamura, H. Shinagawa and K. Morikawa, *Cell*, 1994, **78**, 1063.
39. J.J. Perry, S.M. Yannone, L.G. Holden, C. Hitomi, A. Asaithamby, S. Han, P.K. Cooper, D.J. Chen and J.A. Tainer, *Nat. Struct. Mol. Biol.*, 2006, **13**, 414.
40. W.A. Breyer and B.W. Matthews, *Nat. Struct. Biol.*, 2000, **7**, 1125.
41. S. Hamdan, P.D. Carr, S.E. Brown, D.L. Ollis and N.E. Dixon, *Structure*, 2002, **10**, 535.
42. L.S. Beese and T.A. Steitz, *EMBO J.*, 1991, **10**, 25.
43. V. Derbyshire, N.D. Grindley and C.M. Joyce, *EMBO J.*, 1991, **10**, 17.
44. C.A. Brautigam and T.A. Steitz, *J. Mol. Biol.*, 1998, **277**, 363.
45. S. Doublie, S. Tabor, A.M. Long, C.C. Richardson and T. Ellenberger, *Nature*, 1998, **391**, 251.
46. J. Wang, P. Yu, T.C. Lin, W.H. Konigsberg and T.A. Steitz, *Biochemistry*, 1996, **35**, 8110.
47. Y. Shamoo and T.A. Steitz, *Cell*, 1999, **99**, 155.
48. L.S. Beese, J.M. Friedman and T.A. Steitz, *Biochemistry*, 1993, **32**, 14095.
49. M.S. Jurica, R.J. Monnat Jr. and B.L. Stoddard, *Mol. Cell*, 1998, **2**, 469.
50. M. Belfort and P.S. Perlman, *J. Biol. Chem.*, 1995, **270**, 30237.
51. C.M. Moure, F.S. Gimble and F.A. Quiocho, *Nat. Struct. Biol.*, 2002, **9**, 764.
52. K. Ichiyanagi, Y. Ishino, M. Ariyoshi, K. Komori and K. Morikawa, *J. Mol. Biol.*, 2000, **300**, 889.
53. H. Matsumura, H. Takahashi, T. Inoue, T. Yamamoto, H. Hashimoto, M. Nishioka, S. Fujiwara, M. Takagi, T. Imanaka and Y. Kai, *Proteins*, 2006, **63**, 711.
54. B.L. Stoddard, *Q. Rev. Biophys.*, 2006, **38**, 49.
55. B.S. Chevalier, T. Kortemme, M.S. Chadsey, D. Baker, R.J. Monnat and B.L. Stoddard, *Mol. Cell*, 2002, **10**, 895.
56. B. Chevalier, D. Sussman, C. Otis, A.J. Noel, M. Turmel, C. Lemieux, K. Stephens, R.J. Monnat Jr. and B.L. Stoddard, *Biochemistry*, 2004, **43**, 14015.

57. B. Chevalier, M. Turmel, C. Lemieux, R.J. Monnat Jr. and B.L. Stoddard, *J. Mol. Biol.*, 2003, **329**, 253.
58. C.M. Moure, F.S. Gimble and F.A. Quiocho, *J. Mol. Biol.*, 2003, **334**, 685.
59. P.C. Spiegel, B. Chevalier, D. Sussman, M. Turmel, C. Lemieux and B.L. Stoddard, *Structure*, 2006, **14**, 869.
60. M.J. Sui, L.C. Tsai, K.C. Hsia, L.G. Doudeva, W.Y. Ku, G.W. Han and H.S. Yuan, *Protein Sci.*, 2002, **11**, 2947.
61. K.E. Flick, M.S. Jurica, R.J. Monnat Jr. and B.L. Stoddard, *Nature*, 1998, **394**, 96.
62. B.W. Shen, M. Landthaler, D.A. Shub and B.L. Stoddard, *J. Mol. Biol.*, 2004, **342**, 43.
63. M.D. Miller, J. Tanner, M. Alpaugh, M.J. Benedik and K.L. Krause, *Nat. Struct. Biol.*, 1994, **1**, 461.
64. M. Ghosh, G. Meiss, A. Pingoud, R.E. London and L.C. Pedersen, *J. Biol. Chem.*, 2005, **280**, 27990.
65. B. Altermark, A.O. Smalas, N.P. Willassen and R. Helland, *Acta Crystallogr. Sect. D: Biol. Crystallogr.*, 2006, **62**, 1387.
66. C.L. Li, L.I. Hor, Z.F. Chang, L.C. Tsai, W.Z. Yang and H.S. Yuan, *EMBO J.*, 2003, **22**, 4014.
67. T.P. Ko, C.C. Liao, W.Y. Ku, K.F. Chak and H.S. Yuan, *Structure*, 1999, **7**, 91.
68. M.J. Mate and C. Kleanthous, *J. Biol. Chem.*, 2004, **279**, 34763.
69. E.J. Woo, Y.G. Kim, M.S. Kim, W.D. Han, S. Shin, H. Robinson, S.Y. Park and B.H. Oh, *Mol. Cell*, 2004, **14**, 531.
70. H. Raaijmakers, O. Vix, I. Toro, S. Golz, B. Kemper and D. Suck, *EMBO J.*, 1999, **18**, 1447.
71. K.F. Chak, M.K. Safo, W.Y. Ku, S.Y. Hsieh and H.S. Yuan, *Proc. Natl. Acad. Sci. U.S.A.*, 1996, **93**, 6437.
72. M.D. Miller, J. Cai and K.L. Krause, *J. Mol. Biol.*, 1999, **288**, 975.
73. A.J. Pommer, R. Wallis, G.R. Moore, R. James and C. Kleanthous, *Biochem. J*, 1998, **334**(Pt 2), 387.
74. L.G. Doudeva, H. Huang, K.C. Hsia, Z. Shi, C.L. Li, Y. Shen, Y.S. Cheng and H.S. Yuan, *Protein Sci.*, 2006, **15**, 269.
75. W.Y. Ku, Y.W. Liu, Y.C. Hsu, C.C. Liao, P.H. Liang, H.S. Yuan and K.F. Chak, *Nucleic Acids Res.*, 2002, **30**, 1670.
76. K.Y. Hwang, K. Baek, H.Y. Kim and Y. Cho, *Nat. Struct. Biol.*, 1998, **5**, 707.
77. B. Shen, P. Singh, R. Liu, J. Qiu, L. Zheng, L.D. Finger and S. Alas, *Bioessays*, 2005, **27**, 717.
78. Y. Kim, S.H. Eom, J. Wang, D.S. Lee, S.W. Suh and T.A. Steitz, *Nature*, 1995, **376**, 612.
79. M. Feng, D. Patel, J.J. Dervan, T. Ceska, D. Suck, I. Haq and J.R. Sayers, *Nat. Struct. Mol. Biol.*, 2004, **11**, 450.
80. B.R. Chapados, D.J. Hosfield, S. Han, J. Qiu, B. Yelent, B. Shen and J.A. Tainer, *Cell*, 2004, **116**, 39.

81. D. Suck and C. Oefner, *Nature*, 1986, **321**, 620.
82. C.L. Pickett and C.A. Whitehouse, *Trends Microbiol.*, 1999, **7**, 292.
83. D.M. Wilson 3rd and L.H. Thompson, *Proc. Natl. Acad. Sci. U.S.A.*, 1997, **94**, 12754.
84. B. Demple and L. Harrison, *Annu. Rev. Biochem.*, 1994, **63**, 915.
85. C.D. Mol, T. Izumi, S. Mitra and J.A. Tainer, *Nature*, 2000, **403**, 451.
86. P.T. Beernink, B.W. Segelke, M.Z. Hadi, J.P. Erzberger and D.M. Wilson, 3rd and B. Rupp, *J. Mol. Biol.*, 2001, **307**, 1023.
87. A. Lahm and D. Suck, *J. Mol. Biol.*, 1991, **222**, 645.
88. S.A. Weston, A. Lahm and D. Suck, *J. Mol. Biol.*, 1992, **226**, 1237.
89. G.P. Lomonossoff, P.J. Butler and A. Klug, *J. Mol. Biol.*, 1981, **149**, 745.
90. H.R. Drew and A.A. Travers, *Cell*, 1984, **37**, 491.
91. C.D. Mol, D.J. Hosfield and J.A. Tainer, *Mutat. Res.*, 2000, **460**, 211.
92. A. Volbeda, A. Lahm, F. Sakiyama and D. Suck, *EMBO J.*, 1991, **10**, 1607.
93. C. Romier, R. Dominguez, A. Lahm, O. Dahl and D. Suck, *Proteins*, 1998, **32**, 414.
94. M. Weinfeld, M. Liuzzi and M.C. Paterson, *Nucleic Acids Res.*, 1989, **17**, 3735.
95. M. Weinfeld, K.J. Soderlind and G.W. Buchko, *Nucleic Acids Res.*, 1993, **21**, 621.
96. J.A. Stuckey and J.E. Dixon, *Nat. Struct. Biol.*, 1999, **6**, 278.
97. S. Grazulis, E. Manakova, M. Roessle, M. Bochtler, G. Tamulaitiene, R. Huber and V. Siksnys, *Proc. Natl. Acad. Sci. U.S.A.*, 2005, **102**, 15797.
98. G. Sasnauskas, S.E. Halford and V. Siksnys, *Proc. Natl. Acad. Sci. U.S.A.*, 2003, **100**, 6410.
99. D.J. Hosfield, Y. Guan, B.J. Haas, R.P. Cunningham and J.A. Tainer, *Cell*, 1999, **98**, 397.
100. G.K. Farber and G.A. Petsko, *Trends Biochem. Sci.*, 1990, **15**, 228.
101. D. Reardon and G.K. Farber, *FASEB J.*, 1995, **9**, 497.
102. A. Yamagata, Y. Kakuta, R. Masui and K. Fukuyama, *Proc. Natl. Acad. Sci. U.S.A.*, 2002, **99**, 5908.
103. P. Van Roey, L. Meehan, J.C. Kowalski, M. Belfort and V. Derbyshire, *Nat. Struct. Biol.*, 2002, **9**, 806.
104. J.J. Truglio, B. Rhau, D.L. Croteau, L. Wang, M. Skorvaga, E. Karakas, M.J. DellaVecchia, H. Wang, B. Van Houten and C. Kisker, *EMBO J.*, 2005, **24**, 885.
105. P. Van Roey, C.A. Waddling, K.M. Fox, M. Belfort and V. Derbyshire, *EMBO J.*, 2001, **20**, 3631.
106. K.P. Hopfner, A. Karcher, L. Craig, T.T. Woo, J.P. Carney and J.A. Tainer, *Cell*, 2001, **105**, 473.
107. J.B. Rafferty, E.L. Bolt, T.A. Muranova, S.E. Sedelnikova, P. Leonard, A. Pasquo, P.J. Baker, D.W. Rice, G.J. Sharples and R.G. Lloyd, *Structure*, 2003, **11**, 1557.
108. C. Takemoto-hori, K. Suetsugu-hanawa, K. Murayama, M. Shirouzu and S. Yokoyama, to be published.
109. M. Nowotny and W. Yang, *EMBO J.*, 2006, **25**, 1924.

CHAPTER 14
RNA-modifying Enzymes

ADRIAN R. FERRÉ-D'AMARÉ

Division of Basic Sciences, Fred Hutchinson Cancer Research Center, 1100 Fairview Avenue North, Seattle, WA 98109-1024, USA

14.1 Introduction: Scope of RNA Modification

In rapidly growing cells, most metabolic output is employed for the synthesis of ribosomal RNAs and proteins.[1] If synthesis and maturation of other non-coding RNAs, particularly tRNAs, are also considered, it becomes clear that RNA modification, which broadly speaking includes transcription, post-transcriptional modification and degradation, represents a central biochemical function of the cell. Consistent with this, genes encoding proteins that directly participate in RNA metabolism appear to comprise 3–11% of the open reading frames of the genomes of archaea, bacteria and eukarya.[2] In this chapter, we confine our attention to enzymes responsible for sequence-specific post-transcriptional modification of RNA nucleobases, thus excluding RNA polymerases, aminoacyl-tRNA synthases, nucleases, nucleolytic ribozymes, self-splicing introns and other catalysts that participate in RNA metabolism; those enzymes have been reviewed elsewhere.[3–5]

Over 100 chemically distinct modified nucleotides have been discovered in cellular RNAs. These range from relatively simple modifications such as dihydrouridine (**1**) or 7-methylguanosine (**2**), to the extremely elaborate base Y or wybutosine (**3**). As analytical techniques evolve and RNAs from a wider range of organisms are examined, it is likely that even more types of post-transcriptionally modified nucleotides will be discovered. An updated database of modifications is maintained by McCloskey and collaborators[6] (http://library.med.utah.edu/RNAmods).

RSC Biomolecular Sciences
Protein-Nucleic Acid Interactions: Structural Biology
Edited by Phoebe A. Rice and Carl C. Correll
© Royal Society of Chemistry 2008

(1) dihydrouridine   (2) 7-methylguanosine   (3) base Y

Although the extent of metabolic investment in their biosynthesis indicates that modified nucleotides must have important biochemical functions, in many instances these are only beginning to be understood. tRNAs appear to be the most densely modified cellular RNAs. It has long been established that post-transcriptional modifications of tRNAs at the anticodon loop modulate, and in some cases determine, interaction with their cognate mRNA codon(s). Some tRNA modifications (such as the dihydrouridine and the ribothymidine that give the D- and T-loops, respectively, their names) are present in nearly all tRNAs and are thought to stabilize the characteristic L shape of tRNAs. The function of the many other modifications, most of which are unique to a small set or a single tRNA in an organism, has remained mysterious.[7] Recent work by Uhlenbeck and co-workers suggests that the role of at least some of these idiosyncratic modifications is to fine-tune the different aminoacyl-tRNAs so that they have comparable affinities (and association and dissociation kinetics) for the various components of the translation machinery.[8]

The availability of whole-genome sequences has given new impetus to the discovery and characterization of enzymes responsible for post-transcriptional RNA modifications. Perhaps 50% of the enzymes that synthesize the known RNA modifications remain to be characterized.[9] Using a comparative genomics approach, de Crécy-Lagard and her colleagues discovered enzymes responsible for wyosine[10] (4), queuosine[11] (5), and dihydrouridine[12] (1) biosynthesis. Phizicky and co-workers have carried out genome-wide expression and purification of yeast proteins,[13] and used this tool to discover genes encoding dihydrouridine synthases,[14] the tRNA^{His} guanylyltransferase (a remarkable enzyme that functions as a *de facto* 3′ to 5′ RNA polymerase),[15] as well as a guanosine-7-methyltransferase[16] and a tRNA-specific 2′-O-methyltransferase.[17] By combining genome-wide reverse genetics and mass spectrometry, Suzuki and co-workers identified five genes required for 2-thiouridine (6) formation in bacteria in a technical *tour de force*.[18] The currently known RNA modification enzymes from *Escherichia coli* and *Saccharomyces cerevisiae*, their position in modification pathways, and their loci of action are compiled in the "Modomics" database[19] (http://genesilico.pl/modomics).

RNA-modifying Enzymes

(4) wyosine   (5) queuosine   (6) 2-thiouridine

The first structure of an RNA nucleobase-modifying enzyme bound to a substrate, that of the pseudouridine (Ψ) synthase TruB, was reported in 2001.[20] Since then, several structures of enzyme-RNA complexes that illustrate various remarkable facets of protein-nucleic acid recognition have been reported. Structural studies of RNA-modifying enzymes and their complexes with RNA can shed light on the mechanisms of substrate recognition and catalysis, the physiologic function of the enzymes and the modifications, and evolutionary relationships among the enzymes. In the following sections three enzymes that have been well characterized from a structural standpoint are discussed and their RNA recognition strategies are compared.

14.2 The tRNA Adenosine Deaminase TadA

Site-specific hydrolytic deamination of RNA is a widespread process that contributes importantly to gene expression. Deamination of adenine to yield inosine (7), and of cytosine to yield uridine results, at the mRNA level, in transition mutations. Cells employ this to expand the coding potential of mRNAs, and as a defense mechanism. Site-specific deamination of nucleobases of non-coding RNAs is also a frequent cellular process.[21]

(7) inosine

14.2.1 RNA Recognition by Loop Eversion

In bacteria, inosine is found at the wobble position (residue 34) of the anticodon of tRNA^{Arg2}. Inosine at this position allows degenerate pairing with A,

C or U, allowing a single tRNA to recognize multiple codons. TadA, the bacterial enzyme responsible for the deamination of the A34 residue of the precursor to this tRNA, functions as a homodimer. In yeast, a heterodimeric complex of which the catalytic subunit is orthologous to TadA catalyzes the equivalent transformation. Both the homodimeric bacterial and heterodimeric yeast enzymes are essential for viability.

In addition to full-length pre-tRNA^{Arg2}, TadA efficiently catalyzes deamination of A34 residues in small RNA hairpins with the sequence of the anticodon stem-loop (ASL) of its cognate substrate. Verdine and coworkers determined the structure of *Staphylococcus aureus* TadA bound to such an ASL RNA at 2.0 Å resolution.[22] The structure reveals a homodimeric protein that can simultaneously bind to two separate ASL RNAs (Figure 14.1A). The binding site for each RNA consists of residues from both protein protomers, but one subunit contributes the active site residues involved in catalyzing deamination in one RNA substrate. The ASL RNA approaches the protein in such a manner that the rest of the tRNA would project away from the enzyme, as expected from full functionality of the minimal ASL as a substrate.

The ASLs of tRNAs in isolation adopt the classic U-turn conformation, in which the backbone makes a sharp turn after residue 33, and residues 34–36 stack underneath the 3′ stack of the anticodon stem (Figure 14.2A). Crystal structures of the ribosome bound to mRNA and tRNA reveal that the ASLs of tRNAs adopt essentially the same conformation even when base pairing with their cognate codons in the decoding site of the ribosome.[23] In contrast, when bound to TadA, nucleotides 33–35 and 37 of the ASL are flipped out from the loop and bound in pockets in the protein that individually recognize the bases. In addition, the capping C•A base pair of the loop is directly read out by the protein. Of the flipped-out nucleotides, residue 34 (the site of deamination) is the most completely everted from the ASL. Although they do not introduce chemical modifications in the ASL, aminoacyl-tRNA synthetases such as GlnRS,[24] LysRS,[25] and ThrRS[26] have also been shown to employ eversion of ASL nucleotides and direct readout of individual nucleobases as their primary mode of sequence-specific RNA recognition.

14.2.2 Hydrolytic Deamination by a Zinc-activated Water

Sequence homology indicated that TadA has an active site similar to that of adenosine deaminases (which act on the isolated nucleotide, rather than on RNA). Those enzymes are inhibited by nebularine (**8**) which is observed in the hydrated form in co-crystal structures (Scheme 14.1).

Verdine and co-workers co-crystallized TadA with ASL RNAs in which residue A34 was replaced with nebularine.[22] Unexpectedly, nebularine behaves as a substrate analogue, rather than an intermediate analogue (Scheme 14.2). It thus appears that, compared to the active sites of adenosine deaminases, the TadA active site results in reduced nucleophilicity of the zinc-bound water. The structure confirms the presence of a catalytic zinc ion close to the everted

RNA-modifying Enzymes   371

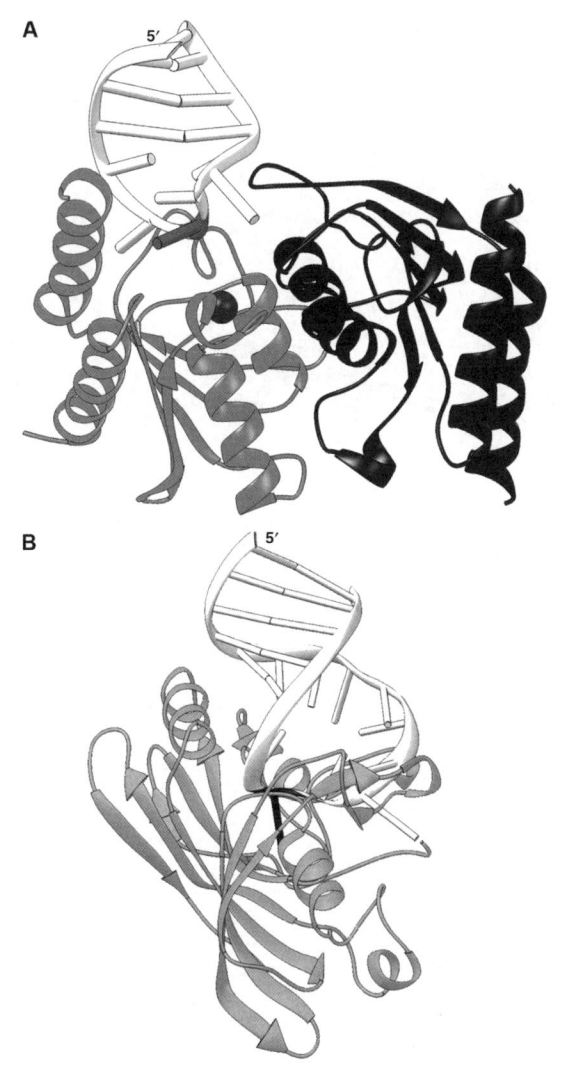

Figure 14.1 Crystal structures of two enzymes that catalyze post-transcriptional modification of nucleobases of the anti-codon stem-loop. (*A*) The adenosine deaminase TadA bound to an ASL RNA substrate.[22] RNA is shown as a white ribbon (top) with nucleobases represented as sticks (site of modification is denoted by the gray stick), the two subunits of the homodimeric enzyme in dark gray and black (left and right, respectively). Although each subunit binds independently to a separate RNA, only one RNA, and the catalytic zinc ion (sphere) bound to the corresponding catalytic protein subunit are shown for clarity. The nucleotide that is deaminated by TadA is shown as a gray stick. (*B*) The pseudouridine synthase RluA bound to an ASL RNA substrate.[28] RNA is in white with the site of modification in black; protein is in gray.

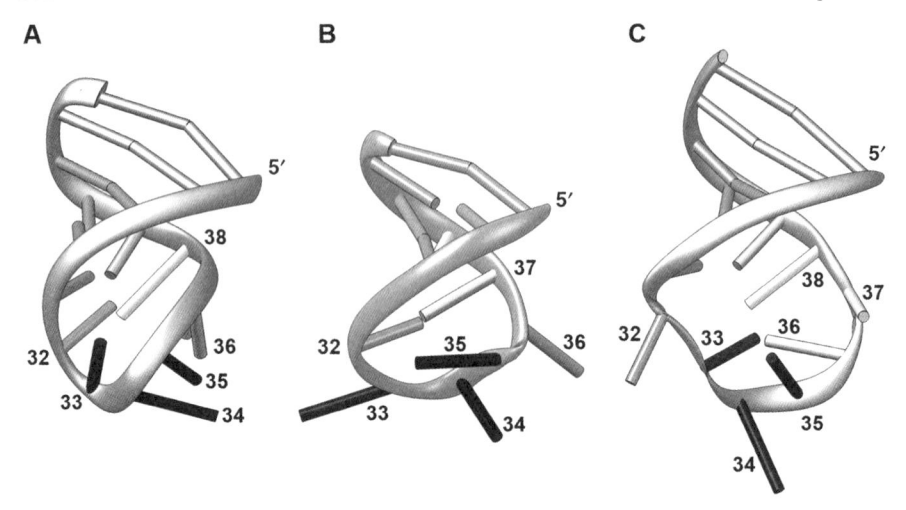

Figure 14.2 Comparison of the structure of ASL RNAs free and bound to two enzymes. (*A*) Schematic representation of the conformation of the free ASL of tRNA[Phe].[40] The bases of residues 32 through 38 are colored in shades of gray and numbered. (*B*) Conformation of the ASL bound to the hydrolytic deaminase TadA.[22] The RNA is colored using the same scheme as in panel (A). The site of modification is residue 34. (*C*) Conformation of the ASL bound to the pseudouridine synthase RluA.[28] The site of modification is residue 32.

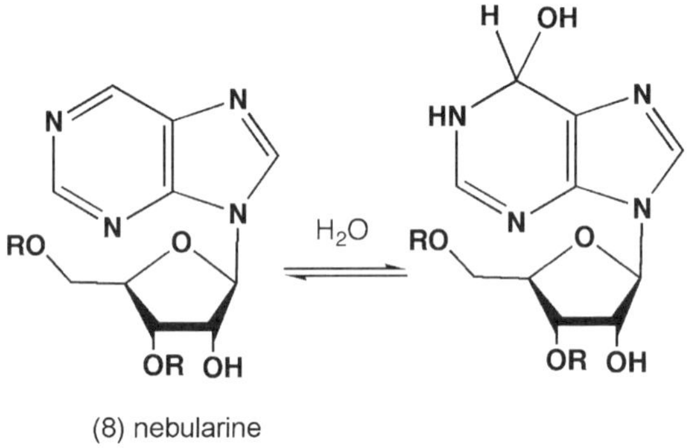

(8) nebularine

Scheme 14.1

nucleobase 34, and indicates that glutamate residue 55 probably participates in catalysis as well.

14.3 The tRNA Pseudouridine Synthase RluA

Pseudouridine (Ψ, **9**), the C5-glycoside isomer of uridine, is the most abundant post-transcriptional modification of cellular RNAs. Many sites of pseudouridylation are highly conserved across phylogeny. Thus, all elongator tRNAs

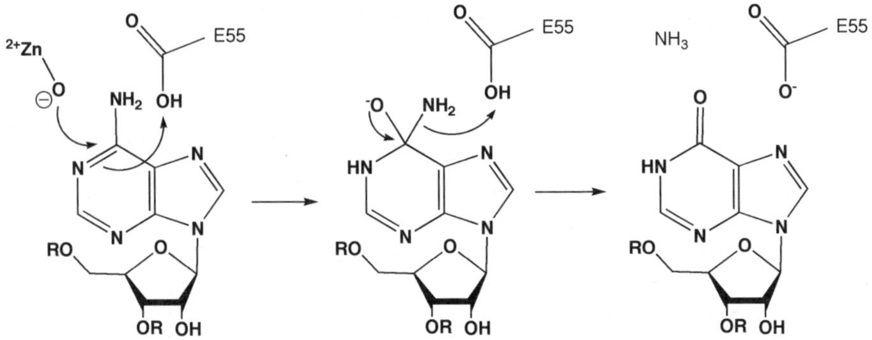

RNA-modifying Enzymes

Scheme 14.2

have Ψ55, Ψ residues are present in most 16S and 23S rRNAs, and all spliceosomal snRNAs appear to have Ψ in homologous positions. The Watson–Crick face of Ψ is identical to that of uridine; therefore, Ψ base pairs with A. Depending on its sequence and structural context, Ψ has been shown to stabilize codon–anticodon pairing, to stabilize the structure of the anti-codon stem-loop of certain tRNAs, and to favor a reactive conformation of the branch-point adenosine in the duplex formed between an intron and the U2 snRNA.

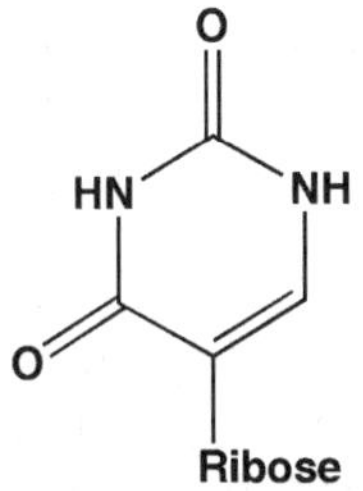

(9) pseudouridine

All known Ψ synthases can be classified into five families based on their sequence. Structure determination of members of the five families demonstrated that, despite minimal sequence similarity, all Ψ synthases share a catalytic domain with a common core fold. Thus, all Ψ synthases appear to be descended from a common molecular ancestor. Moreover, their distribution in all phyla implies that enzymes belonging to the five families were already present in the last common ancestor. The conserved Ψ synthase core is extended by peripheral structural elements and accessory domains that presumably confer on each enzyme its particular substrate specificity. The Ψ synthases of the Cbf5/dyskerin subfamily achieve specificity not by extensions of their polypeptide chain, but by associating with several accessory proteins and a guide RNA. The latter have conserved structural elements that recruit the Ψ synthase, and segments that base-pair with sequences flanking the site of pseudouridylation in their substrate RNAs (reviewed in ref. 27).

14.3.1 RNA Recognition through Protein-induced Base-pairing

RluA is a single-polypeptide *E. coli* Ψ synthase with dual substrate specificity. It is responsible for Ψ32 in several tRNAs and also for Ψ746 in 23S rRNA. Because the sites of modification in all its substrates share a common flanking sequence, it was expected that RluA would directly read out the sequence of its substrate RNA in a manner analogous to that employed for substrate recognition by TadA and other sequence-specific RNA-binding proteins.

The 2.05 Å resolution structure of RluA bound to a substrate ASL RNA (Figure 14.1B) shows that the enzyme adopts the canonical Ψ synthase fold, and that the substrate RNA binds in the deep active site cleft that bisects the protein.[28] Comparison of the structure of the ASL free and bound to RluA (Figures 14.2A, C) reveals that, rather than simply everting the nucleotides of the anticodon loop and making sequence-specific contacts with the exposed nucleobases, RluA enforces a new base-pairing scheme on the ASL, and indirectly probes its RNA sequence. The co-crystal structure, as well as the results of structure-guided mutagenesis, indicate that RluA indirectly recognizes its substrates by testing their ability to adopt the rearranged secondary structure. The site of modification (U32) is, as expected, flipped out from the helical stack, and projects deeply into the active site of RluA. This recognition strategy is very different from that employed by TadA (above). It is also very different from that employed by the Ψ synthase TruB. The structure of the latter enzyme bound to its substrate, a T-stem-loop (TSL) RNA, revealed a recognition strategy primarily based on shape complementarity with the characteristic structure (and underlying conserved sequence) of a T-loop. Shape complementarity is combined with direct readout of the identity of a cytosine that is universally conserved in TruB substrates, and is flipped out of the TSL structure when the RNA is bound by the Ψ synthase.[20] The structures of the Ψ synthases RluA and TruB bound to their respective substrates show how structurally similar enzymes with a highly-conserved core domain can recognize their respective substrates using completely different strategies.

14.3.2 In-line Displacement or Michael Addition?

No cofactors or tightly bound metal ions are employed by Ψ synthases. Analyses of mutant Ψ synthases belonging to all five families have demonstrated the presence of a catalytically essential aspartate residue. Thus, by analogy to glycosylases, a reaction mechanism was proposed in which the aspartate would disconnect the uracil base at the site of isomerization through a nucleophilic in-line attack on the anomeric carbon of the ribose. The disconnected nucleobase would then somehow rotate and reattach to the ribose through its C5 position (Scheme 14.3).[29]

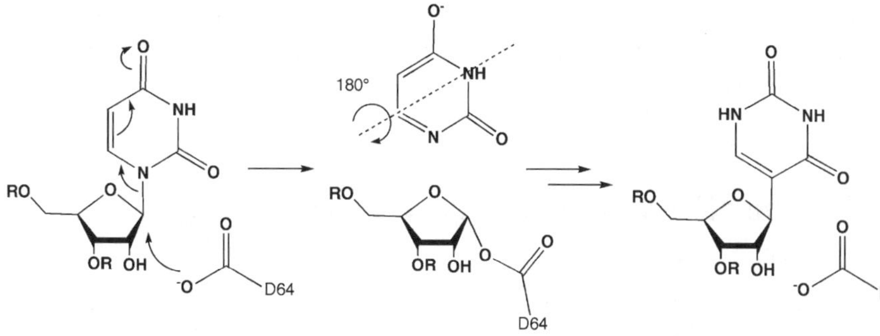

Scheme 14.3

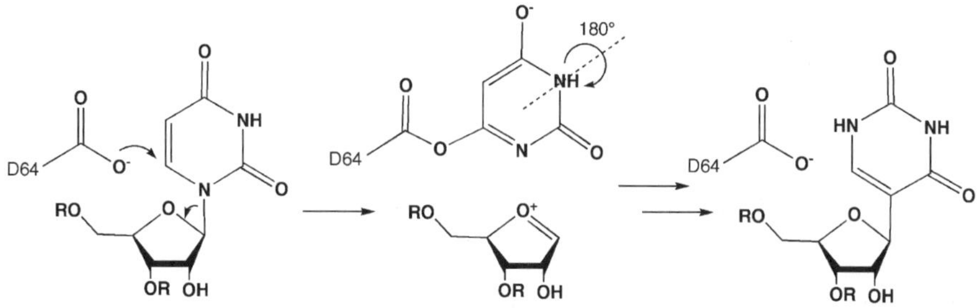

Scheme 14.4

The catalytic aspartate of the Ψ synthase TruA forms a covalent adduct with substrate RNAs in which the uracil at the site of isomerization is replaced with a 5-fluorouracil. Based on this observation, Santi and co-workers proposed an alternate mechanism in which the catalytic aspartate makes a Michael-type addition to the C6 position of uracil, leading to disconnection of the base, followed by rotation of the base (which would remain tethered by an ester linkage to the catalytic aspartate) and reattachment (Scheme 14.4).[30]

TruB and RluA were co-crystallized with RNAs in which their respective target uridines were replaced with 5-fluorouridine. Structure determination of TruB revealed that this Ψ synthase is not trapped by such a modified RNA.[20] Rather, TruB efficiently isomerizes 5-fluorouridine into 5-fluoropseudouridine.[31] RluA does form a stable covalent adduct with 5-fluorouridine-containing ASL RNAs. However, this adduct was found to be highly X-ray sensitive, precluding its structural characterization.[28] Nonetheless, the co-crystal structures of TruB and RluA indicate that if an in-line displacement mechanism were to operate the ribose at the site of isomerization would have to undergo a large conformational change. Thus, at present, the structures support either the Michael addition mechanism, or an S_N1-type substitution on the anomeric carbon. Comparison of Ψ synthase structures shows that the active sites of these enzymes are primarily hydrophobic, with only three conserved amino acids: the catalytic aspartate, a basic residue that makes a buried salt bridge with the aspartate, and an aromatic residue (either phenylalanine or tyrosine). How such a small set of catalytic residues can carry out the isomerization of uridine remains a perplexing mechanistic question.

14.4 The tRNA Archaeosine Transglycosylase ArcTGT

Archaeosine (**10**) is a hypermodified derivative of 7-deazaguanosine. This modification [and the related queuosine (**5**) of bacteria] is introduced in multiple steps. First, a tRNA-guanine-transglycosylase (TGT) replaces a guanine base at the site of modification with a 7-deaza purine base [PreQ$_0$ (**11**, Scheme 14.5), in the case of archaeosine]. Then, other enzymes modify the attached PreQ$_0$ to yield the final hypermodified base (in eukarya, pre-formed queuosine is introduced in a single step by a TGT that is related to the archaeal and bacterial enzymes).

14.4.1 RNA Recognition by Tertiary Structure Rearrangement

Yokoyama and co-workers determined the structure of the archaeosine TGT from *Pyrococcus horikoshii* complexed with tRNA^{Val} at 3.3 Å resolution.[32] ArcTGT is a dimeric, multi-domain enzyme. The enzymatic activity resides in a TIM barrel. The crystal structure of the dimeric enzyme bound simultaneously to two independent substrate RNAs (Figure 14.3) shows how the acceptor stem is bound in a cleft formed between the catalytic and C-terminal domains, and the anticodon stem is bound by the C-terminal domains (one of which, the PUA domain,[33,34] has a fold that is present in accessory domains of many unrelated RNA-modifying enzymes[35]).

Binding to ArcTGT causes a dramatic rearrangement of the tertiary structure of tRNA. ArcTGT transglycosylates residue 15, which in the native L-shaped structure of tRNA (Figure 14.4A) is part of the D-loop, and lies deeply buried in the core of the molecule, forming tertiary interactions. To access its site of

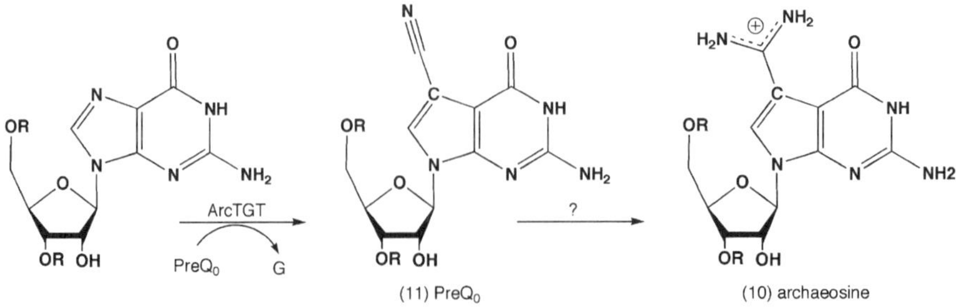

Scheme 14.5

RNA-modifying Enzymes 377

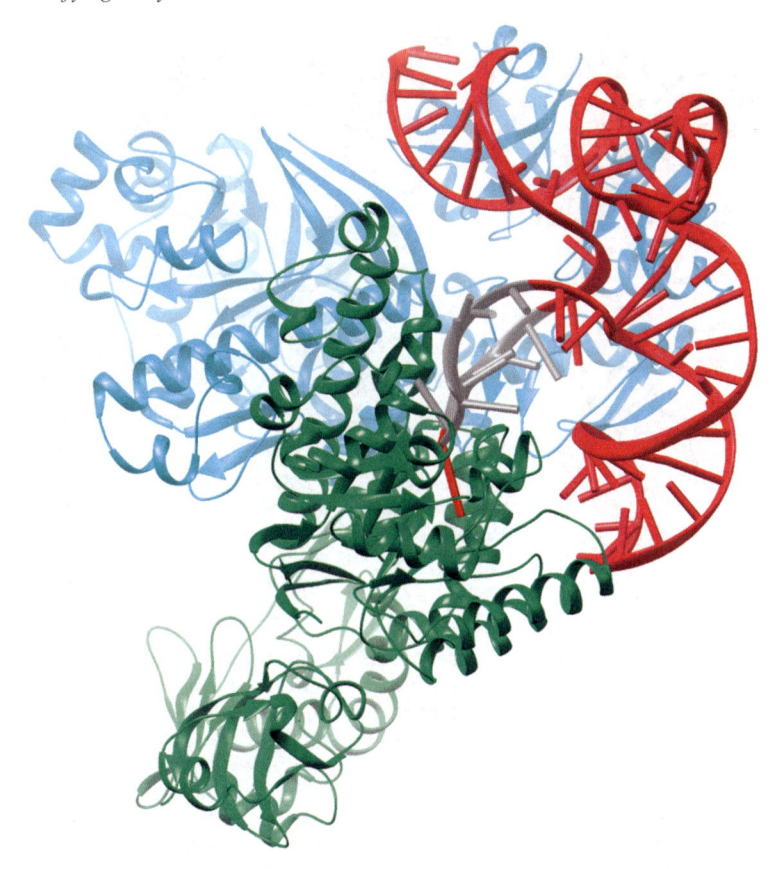

Figure 14.3 Crystal structure of the archaeosine transglycosylate ArcTGT.[32] The two protomers forming the dimeric enzyme are shown in different shades of blue. Only one of the two independently-bound tRNA molecules is shown. The tRNA is colored yellow, except for the site of modification (red) and surrounding nucleotides in gray. The enzymatic activity resides in a canonical TIM barrel fold.

modification, ArcTGT completely unfolds the D-loop of tRNA (Figure 14.4B). In the process, a new helical stem is formed between several residues formerly part of the D-loop and those in the variable, or V-stem. This DV stem is compatible with the sequences of several tRNAs, prompting Yokoyama and colleagues to suggest that this reorganized form of tRNA, which they term the λ-form, is adopted by tRNA when in complex with the large number of enzymes that need to access residues that are buried in the native L-form of tRNA. The rearranged λ-form tRNA has an intact, folded TSL. This, and modeling experiments, suggest that TruB, for instance, could act simultaneously on a tRNA being modified by ArcTGT. This is consistent with genetic evidence that several RNA modifying enzymes may act in concert, and that cells may have higher-order assemblies of RNA modifying enzymes.[32]

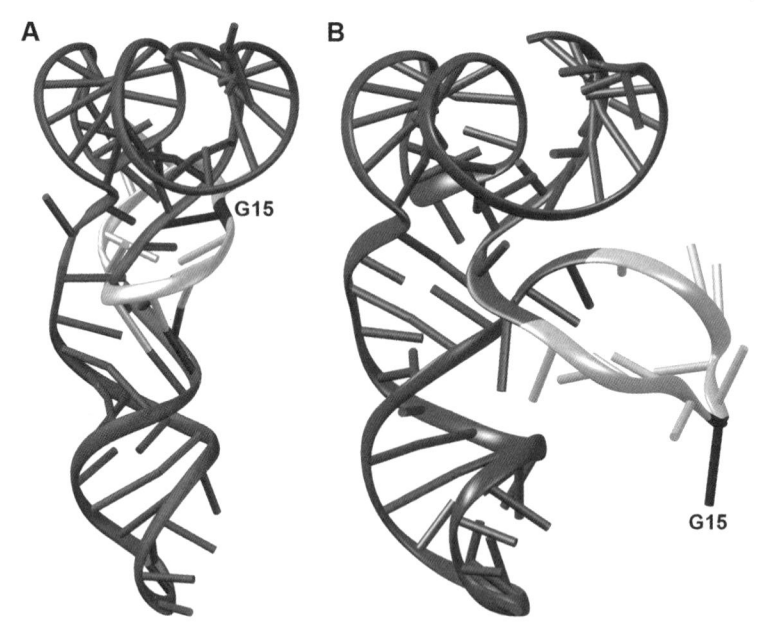

Figure 14.4 Comparison of the structures of tRNA free and bound to ArcTGT. The site of modification (residue G15) is shown and labeled. Nucleotides 10–21 are in gray. (*A*) Structure of natively folded tRNA in its characteristic L-shape. The acceptor stem projects toward the viewer in the upper right side. The T-loop lies at upper left, and makes extensive tertiary contacts with the D-loop (gray and black). The anticodon is at the bottom of the panel. (*B*) Structure of the radically rearranged tRNA bound to ArcTGT. In this "λ-form" the D-loop is completely everted from the tRNA, and a new "DV" helix stacks above the anticodon stem. All schematic representations of crystal structures were prepared with the program RIBBONS.[41]

14.4.2 Transglycosylation Using Two Aspartate Residues

Huang and co-workers solved the structure of the *Zymomonas mobilis* TGT bound to a substrate ASL at 2.9 Å resolution.[36] This enzyme introduces PreQ$_1$ (**12**, Scheme 14.6) into position 34 of tRNAs with the sequence UGU at positions 33–35. To trap the enzyme in a suicide complex, these authors replaced G34 of their substrate ASL with 9-deazaguanosine (**13**). The structure demonstrated that aspartate residue 280 forms a covalent adduct with the anomeric carbon of the ribose of residue 34. Soaking of crystals trapped in this state with PreQ$_1$ results in the transglycosylation reaction being carried out in the crystalline state. A second aspartate (D102 in the *Z. mobilis* TGT) functions as a general base to deprotonate the N9 of the incoming PreQ$_1$ so that it may in turn displace D280 from the covalent intermediate. Comparison of the two RNA-bound TGT structures indicates that residue D280 of the *Z. mobilis* enzyme is equivalent to D249 of the *P. horikoshii* ArcTGT, and that the two

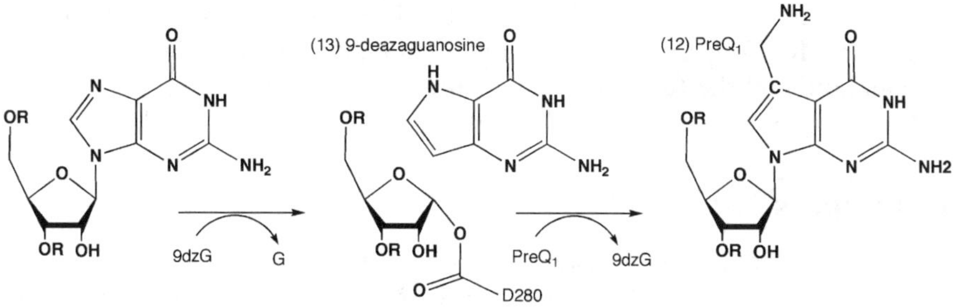

Scheme 14.6

enzymes employ the same catalytic mechanism. Because the bacterial enzyme modifies the ASL of tRNA rather than the D-loop, its substrate recognition strategy is similar to that of TadA.

14.5 Conclusions

Our survey of RNA recognition by enzymes responsible for post-transcriptional nucleobase modification is at present limited by the small number of co-crystal structures available. However, even the modest set of structures reviewed here demonstrates a wide variety of strategies that confer on these enzymes their exquisite sequence specificity. The ability of proteins to make sequence specific interactions with everted nucleobases is not surprising, and has been documented for RNA-binding proteins since the pioneering work of Steitz and coworkers on glutaminyl-tRNA sythetase[24] and Nagai and coworkers on the spliceosomal protein U1A.[37] However, the RluA and ArcTGT structures demonstrate the utilization by these proteins of the capacity of RNA to serve as its own internal template, or an internal guide sequence. This is also clearly shown by the structure of the *E. coli* 23S rRNA methyltransferase RumA bound to a fragment of the rRNA, which was solved by Stroud and co-workers.[38] RumA also appears to recognize the three-dimensional structure into which its substrate folds, analogous to what the Ψ synthase TruB does. RNA forms stable tertiary structures, and ArcTGT is an example of a protein exploiting this property to achieve sequence specificity. Perhaps the clearest demonstration of the interplay of RNA primary, secondary and tertiary structure and specific protein recognition is in the guide-RNA directed Ψ synthases.[27,39] The astonishing variety of post-transcriptional RNA modifications and their ubiquity across phylogeny implies that many more remarkable instances of the synergy between RNA and protein structure will be discovered in future studies.

Acknowledgements

The author is a Distinguished Young Scholar in Medical Research of the W.M. Keck Foundation. Work on RNA-modifying enzymes in the author's laboratory has been supported by grants from the National Institutes of Health (GM63576), the Leukemia and Lymphoma Society, the W.M. Keck Foundation, and the Rita Allen Foundation.

References

1. J.R. Warner, *Trends Biochem. Sci.*, 1999, **24**, 437–440.
2. V. Anantharaman, E.V. Koonin and L. Aravind, *Nucleic Acids Res.*, 2002, **30**, 1427–1464.
3. P. Cramer, *Curr. Opin. Struct. Biol.*, 2002, **12**, 89–97.
4. M. Ibba and D. Söll, *Annu. Rev. Biochem.*, 2000, **69**, 617–650.
5. M.J. Fedor and J.R. Williamson, *Nat. Rev. Mol. Cell Biol.*, 2005, **6**, 399–412.
6. J. Rozenski, P.F. Crain and J.A. McCloskey, *Nucleic Acids Res.*, 1999, **27**, 196–197.
7. H. Grosjean and R. Benne, *Modification and Editing of RNA*, American Society for Microbiology Press, Washington DC, 1998.
8. M. Olejniczak, T. Dale, R.P. Fahlman and O.C. Uhlenbeck, *Nat. Struct. Mol. Biol.*, 2005, **12**, 788–793.
9. V. de Crécy-Lagard, Finding missing tRNA modification genes: a comparative genomics goldmine, in *Practical Bioinformatics*, ed. J.M. Bujnicki, Springer-Verlag, New York, 2004, pp. 169–190.
10. W.F. Waas, V. de Crecy-Lagard and P. Schimmel, *J. Biol. Chem.*, 2005, **280**, 37616–37622.
11. J.S. Reader, D. Metzgar, P. Schimmel and V. de Crecy-Lagard, *J. Biol. Chem.*, 2004, **279**, 6280–6285.
12. A.C. Bishop, J. Xu, R.C. Johnson, P. Schimmel and V. de Crecy-Lagard, *J. Biol. Chem.*, 2002, **277**, 25090–25095.
13. E.M. Phizicky and E.J. Grayhack, *Crit. Rev. Biochem. Mol. Biol.*, 2006, **41**, 315–327.
14. F. Xing, S.L. Hiley, T.R. Hughes and E.M. Phizicky, *J. Biol. Chem.*, 2004, **279**, 17850–17860.
15. W. Gu, J.E. Jackman, A.J. Lohan, M.W. Gray and E.M. Phizicky, *Genes Dev.*, 2003, **17**, 2889–2901.
16. A. Alexandrov, E.J. Grayhack and E.M. Phizicky, *RNA*, 2005, **11**, 821–830.
17. M.L. Wilkinson, S.M. Crary, J.E. Jackman, E.J. Grayhack and E.M. Phizicky, *RNA*, 2007, **13**, 404–413.
18. Y. Ikeuchi, N. Shigi, J. Kato, A. Nishimura and T. Suzuki, *Mol. Cell*, 2006, **21**, 97–108.
19. S. Dunin-Horkawicz *et al.*, *Nucleic Acids Res.*, 2006, **34**, D145–149.
20. C. Hoang and A.R. Ferré-D'Amaré, *Cell*, 2001, **107**, 929–939.
21. S. Maas, A. Rich and K. Nishikura, *J. Biol. Chem.*, 2003, **278**, 1391–1394.
22. H.C. Losey, A.J. Ruthenburg and G.L. Verdine, *Nat. Struct. Mol. Biol.*, 2006, **13**, 153–159.

23. J.M. Ogle et al., Science, 2001, **292**, 897–902.
24. M.A. Rould, J.J. Perona and T.A. Steitz, Nature, 1991, **352**, 213–218.
25. S. Cusack, A. Yaremchuk, I. Krikliviy and M. Tukalo, Structure, 1998, **6**, 101–108.
26. R. Sankaranarayanan et al., Cell, 1999, **97**, 371–381.
27. T. Hamma and A.R. Ferré-D'Amaré, Chem. Biol., 2006, **13**, 1125–1135.
28. C. Hoang et al., Mol. Cell, 2006, **24**, 535–545.
29. L. Huang, M. Pookanjanatavip, X. Gu and D. V. Santi, Biochemistry, 1998, **37**, 344–351.
30. X. Gu, Y. Liu and D.V. Santi, Proc. Natl. Acad. Sci. U.S.A., 1999, **96**, 14270–14275.
31. C.J. Spedaliere and E.G. Mueller, RNA, 2004, **10**, 192–199.
32. R. Ishitani et al., Cell, 2003, **113**, 383–394.
33. L. Aravind and E.V. Koonin, J. Mol. Evol., 1999, **48**, 291–302.
34. A.R. Ferré-D'Amaré, Curr. Opin. Struct. Biol., 2003, **13**, 49–55.
35. T. Hamma, S.L. Reichow, G. Varani and A.R. Ferré-D'Amaré, Nat. Struct. Mol. Biol., 2005, **12**, 1101–1107.
36. W. Xie, X. Liu and R.H. Huang, Nat. Struct. Biol., 2003, **10**, 781–788.
37. C. Oubridge, N. Ito, P.R. Evans, C.-H. Teo and K. Nagai, Nature, 1994, **372**, 432–438.
38. T.T. Lee, S. Agarwalla and R.M. Stroud, Cell, 2005, **120**, 599–611.
39. S.L. Reichow, T. Hamma, A.R. Ferré-D'Amaré and G. Varani, Nucleic Acids Res., 2007, **35**, 1452–1464.
40. H. Shi and P.B. Moore, RNA, 2000, **6**, 1091–1105.
41. M. Carson, Methods Enzymol., 1997, **277**, 493–505.

Subject Index

A-form DNA structure, 3–6
α-helixes, B-form DNA recognition, 6, 9
A-minor interactions, 7, 153
"A-tracts", 72, 188, 190–1, 193, 206
Ac/Ds (Activator/Dissociator) transposing elements, 270
accessible surface area (ASA), 16–19
accessory sites
   serine recombinases, 324–6
   tyrosine recombinases, 313, 328
active-site clusters, 34
adaptability
   K-turn RNAs, 162
   Puf domain recognition strategies, 111–12
adenine
   deamination to inosine, 369
   RNA tertiary structures, 7, 153
adenylylimidodiphosphate (ADPNP or AMPPNP), 244, 253
affinity, 8, 14–15
   See also specificity
   anion effects, 28, 33
   cation effects, 34, 36
   cosolutes and, 36, 38
   DNA bending and, 80, 178, 195–8, 203–9
   enhancement, 53–4, 60–2
   retained water and, 26–7
   RNA and, 157, 159, 222, 226, 229–30, 368
   SSNA binding and, 113, 115, 117, 119

allosteric effects
   cyclicAMP binding, 201
   SSNA binding, 110–11, 117–20
   use by transcriptional regulators, 51
amide group hydrogen bonding, 98
amino-acid residues
   See also individual amino acids
   aromatic amino acids, 95, 157
   base-stacking interactions, 95, 157
   polar amino acids, SSNA and, 96
   side-chain intercalation, 184–5, 190, 204, 211–12, 340
AMPPNP, 244, 253
anion effects
   on protein-DNA interactions, 28, 32–4
   on protein-protein interactions, 30
   salt bridges, 93–4
annealing of DNA, 244, 309
annealing of RNA
   MRP1/MRP2 complexes, 165
   RCA proteins and, 222–7
anti-codon loops, 67, 95, 368
   loop eversion, 369–70
anti-codon stem loops (ASL), 370–5, 378–9
anti-topoisomerase agents, 238
antibiotic resistance, 271–3, 294, 296
antisense RNAs, 164, 224, 229
APE1 nuclease, 350–2, 355, 362
archaeosine transglycosylase ArcTGT, 376–9
architectural transcription factors, 54

Subject Index

arginine
  base stacking and, 95, 356
  histone-DNA binding, 193
  peptide-RNA reactions, 154
  salt bridges and, 93–4, 206
  serine recombinase role, 320
Argonaute protein, 157–9, 226
aromatic amino acids, 95, 157
asparagine as a base mimic, 95–6
aspartic acid
  pseudouridine synthase and, 374–5
  transglycosylation and, 378–9
AT-rich regions and bending, 188, 197, 202, 206
atomic force microscopy (AFM), 207, 223
ATPase domains, Type IIA topos, 250, 253, 255–6
autonomous transposons, 273

B-form DNA, 4–6
  α-helix major groove fitting, 9
  B to A transitions, 67
  twist angle, 185–6
β-hairpin, DNA recognition, 5
β-ribbon, DNA recognition, 53
Bacillus chlorismate mutase-like fold, 337, 361
bacteriophage integrases, 304–7, 311–13, 318, 324–6
  Cre recombinase in P1, 310
bacteriophages
  Gin recombinase, 323
  large serine recombinases, 326
  Mu transposase, 275, 284–8
  phage λ Xis protein, 194
bactoeriophage endonucleases *See* T4 endonuclease; T7 endonuclease
BamHI endonuclease, 18–24, 26–33
base-modifying enzymes, 11
base pair mimicry, 96–7
base-pair parameters, 69–70
base-pair positions, direct readout, 77
base pair steps (dinucleotide steps), 71, 186, 189–91
base-pairing, protein-induced, 374

base pairs
  folded RNA, 7
  helix conformation and, 4
  junction-resolving enzymes and, 138
  non-Watson-Crick, 10
  steric discrimination, 96
base stacking
  with aromatic amino-acid residues, 95, 157–8
  B-DNA polymorphisms and, 70–2
  disruption in by endonucleases, 138–9, 142
  disruption in kinked complexes, 85
  electrostatic and steric repulsion, 188
  exocyclic groups and, 188
  extrahelical bases, IntI action, 316
  helix rigidity and, 179–80
  non-helical regions, rRNA, 154
  propeller twists and, 186
base stacking energies, 71
base Y (wybutosine), 367–8
bases, chemically modified, in RNA, 367
basic helix-loop-helix (bHLH) recognition, 50–2
basic leucine zipper (bZIP) recognition, 50–2, 182
BBP (branch point-binding protein), 108, 114
bending *See* DNA bending
β-sheets
  RNA protein interactions, 151–64
  SSNA binding, 99, 108, 110
binding constant, $K_{A\text{-DNA}}$, 93, 117
box C/D RNAs, 154, 161–2
branch migration, four-way junctions, 132–4, 138
branch point-binding protein (BBP), 108, 114
branch site sequence recognition, 155, 163
branch sites, intron RNA, 95, 98, 107–8, 114–15
BRCT domain, 258
bulge-helix-bulge (BHB) motif, 169

C-terminal domains (CTD), 75, 256–7
C-terminal extensions (CTE), 57–9
cancer, 47, 53, 238, 274
CAP (catabolite activator protein)
  direct readout, 76–9
  DNA bending by, 201–3, 212
  mechanism of working, 74–6
  mutants, 85
  sequence recognition by, 73–85
CAP domains *See* winged helix
  domains
carboxylate-bearing amino acids, 31,
  33–5, 96–7, 107
catabolite activator protein *See* CAP
catenanes, 235, 237
cation–π interactions, 169
cations
  displacement, 30–2
  DNA crystallography and, 184
  enzyme activity and DNA binding,
  34–5
Cbf5 protein, 163–4
CBP2 splicing factor, 230
Cce1 enzyme, 135, 137–40, 144
Cdc13, 94, 102–3, 113
chaperone activity *See* RNA
  chaperones (RCA)
charge, nucleic acid backbones, 2
charge clusters, 30, 34, 37
circular DNA, 6, 244, 249, 253, 304
Class II transposons, 270
cleavage
  See also DNA cleavage mechanisms
  RNA susceptibility to, 3
cloning, 139, 310
cold-shock domains (CSD), 166, 227
cold-shock proteins, 222, 227–8
colicin E7, 347
combinatorial DNA interactions, 59–61
complex formation, 8
composite transposons, 271
computational methods, water
  release, 18–19, 25, 40
conformation changes *See* DNA
  geometry
conformers, four-way junctions, 132–3

conjugative transposons, 318
"contact free energy" values, 18
cooperativity
  combinatorial regulation, 48, 59–60, 62
  RNA binding, 160, 163, 168
  SSNA binding, 101–3, 108, 111,
  115–17
copy-in and copy-out transposition,
  277, 281, 294
cosolutes
  ion effects and, 36
  sensitivity to, and binding
  specificity, 37–9
  steric repulsion and, 20
  in water release studies, 18–26
Cre recombinases, 306, 309–13
CRP (cyclic AMP receptor protein)
  See CAP
crystallography
  adenosine deaminase, 371
  dinucleotide steps, 189–90
  DNA bending at high resolution,
  183–91
  invisible regions, 18
  junction-resolving enzymes, 135,
  140–1
  nuclease-DNA complexes, 352
  nucleosome DNA, 179
  oligonucleotide database, 183
  protein-RNA complexes, 151–2,
  226–7
  recombinase-DNA complexes,
  310–11
  retained water and, 26–7
  RNP crystal structure, 163
  sequence recognition studies, 67
  ternary CAP complexes, 75
  topoisomerases, 246, 249
  transposases, 276, 282, 291
cut-and-paste transposition, 277, 280–1,
  288, 292–3
cyclic AMP receptor protein (CRP)
  See CAP
cyclisation of DNA, 72, 79, 178–80,
  201, 204, 211
  See also circular DNA

Subject Index 385

Cys$_2$Cys$_2$ and Cys$_2$His$_2$, 52
cysteine, interactions with zinc, 52, 54, 136–7
Cyt-18, 228, 230
cytolethal toxins, 337, 351
cytosine deamination, 369

DDD and DDE transposases, 276–8, 281, 290–3
Debye screening, 30
decatenase activity, 243, 250, 252, 257
dehydration, protein-DNA complexation, 9
deoxyribose
  B-form DNA polymorphism, 69
  ribose and, 3–4
DHH phosphoesterases fold, 337, 357–8
Dicer, 157–9, 168
diffusion ("sliding"), 29
  facilitated diffusion, 36, 38
dinucleotide (base pair) steps, 71, 186, 189–91
direct readout, 68, 76–9, 177
divalent metal ions
  DNA bending and crystallography, 184, 191
  four-way DNA junctions and, 131–3
  nucleases and, 141, 334, 339, 341, 345, 362
  ribonuclease dependence on, 165
  topoisomerase-induced DNA cleavage and, 238, 240, 260
  transposase-induced DNA cleavage and, 276
DNA
  See also protein-DNA complexes; SSNAs
  B to A transitions, 67
  chemical differences from RNA, 3–4
  primary structure and numbering, 2
  site recognition principles, 8–10
DNA bending
  B-form polymorphism, 69–72
  base stacking and, 179–80
  bending resistance, 6
  compaction by Fis protein, 198–201
  dinucleotide steps and, 183–91
  electrostatic forces and, 180–3
  endonuclease specificity and, 138–9, 143–4
  forces controlling rigidity, 178–83
  helical parameters and, 184–5
  by HMG transcription regulators, 54
  indirect readout and, 68, 79–80
  kinking and, in CAP complexes, 75, 80–5
  proteins capable of, 191–212
  supercoiling and, 6
  by Zn$_2$Cys$_6$ binuclear clusters, 55
DNA bending proteins
  CAP protein, 201–3
  Fis protein, 197–201
  histones, 191–4
  HMGB family, 208–12
  HPV E2 protein, 194–7
  lambda Xis protein, 194, 195
  prokaryotic HU/IHF family, 203–8
DNA cleavage mechanisms
  cis cleavage, 310, 312
  DDE transposases, 277–9, 295
  nucleases, 334, 339, 346–7, 351, 359
  second strand cleavage, 280–1
  site-specific recombinases, 310
  topoisomerases, 237–8, 240, 246–8, 254–6
  trans cleavage, 283, 287, 310, 312
DNA deformability energy functions, 73
"DNA exit wedge", 211
DNA geometry
  See also A-form; B-form; DNA bending
  average parameters, B-DNA, 71
  bending and kinking in protein complexes, 75, 79–82, 204
  canonical A- and B-forms, 4–6
  common reference frame, 69–71
  conformational rearrangements, 10–11
  helical parameters controlling, 184–8
  inderect readout and, 68
  Z-form, 69

DNA hairpins, 10, 280, 283–4, 292–3, 295–8
DNA junctions
　distortion by resolving enzymes, 137–9
　proteins interacting with, 134
　structure and dynamics of four-way junctions, 129–34
DNA linking number (Lk), 237, 248, 251–2, 319
DNA nucleases
　See also ribonucleases
　Bacillus chlorismate mutase-like fold, 337, 361
　biological functions, 362
　DHH phosphoesterases fold, 337, 357–8
　DNase I-like fold, 337, 350–2
　GIY-YIG endonuclease fold, 337, 358–60
　grouping by fold type, 333–6
　His-Me finger endonuclease fold, 336, 345–8
　homing endonuclease-like fold, 336, 343–5
　metallo-dependent phosphatases fold, 337, 360–1
　phospholipase C/P1 nuclease fold, 337, 352–4
　phospholipase D/nuclease fold, 337, 354–5
　restriction endonuclease-like folds, 334–5, 337–41
　RNase H-like fold, 335–6, 341–3
　SAM domain/PIN domain-like fold, 336–7, 348–9
　TIM beta/alpha barrel fold, 337, 355–7
DNA replication
　nucleases and, 349
　ssDNA and, 101–3, 116, 120
　topoisomerases and, 235, 252, 258
　Y transposases and, 294, 296
DNA strand exchange, 307–9, 311, 315–17, 319–20, 323–6
DNA strand passage, 240, 243, 253, 255, 258, 260–1
DNA transposition *See* transposases

DNase I-like fold, 166, 337, 350–2
double helix *See* DNA geometry
double-stranded RNA
　protein interactions with, 150–1, 157–8
　RNase III specificity for, 167–8
　topoisomerase transport, 255
Drosha, 157, 168

eclectic proteins, 92, 98, 113
EcoRI endonuclease, 18–24, 26–30, 38–9, 67
EcoRV endonuclease, 18–24, 26–30, 38–9, 177
electron density maps, 67, 111, 114
electron microscopy, 116, 255, 287, 292
electrostatic interactions
　DNA bending and, 180–3
　four-way junction folding and, 131–2
　proteins and DNA, 28, 79–80
　RCA annealing and, 228
　SSNA-binding proteins, 93–4
electrostatic repulsion (strain), 27–8, 34–5
Elk-1 biding, 60
emergent properties, SSNA binding, 110
　allostery, 117–20
　cooperativity, 115–17
　recognition specificity, adaptability and degeneracy, 111–15
energetics of DNA deformation, 80–1
energetics of protein-DNA interactions, 20, 36–7
enhancers *See* accessory sites
enthalpy and protein binding, 16, 31, 93, 118
entropy
　RNA binding, 118, 151
　site specific interactions, 23, 32, 37
entropy exchange, 230
environmental parameters, 14, 38–9
enzyme systems using recombinases, 304
equilibrium association constant, Ki, 14, 21–4
estrogen receptor (ER), 57–8
ETS-domain transcription factors, 49
ETS family ternary complexes, 60–1

Subject Index 387

eukaryotes
  bZIP proteins restricted to, 51
  DNA transposon types, 273–4
  zinc-binding domains, 52
everted and inverted half-sites, 55–6
everted RNA nucleotides, 370, 378
evolution, 1, 135, 274, 338, 369
  SSNA-binding proteins and, 91–2, 101–3, 115, 119–20
extrahelical bases, 284, 316–18

"facilitated diffusion", 36, 38
FBF *See* Puf folding domains
FEN-1 enzymes, 336–7, 348–9
FinO protein, 229–30
Fis protein
  DNA bending by, 197–201
  recombinase enhancement and, 313–14, 325–6
Flp recombinases, 134, 305–6, 309–15, 318, 327–8
folding of nucleic acids
  See also RNA chaperones
  four-way DNA junctions, 131–2
  misfolding, 11, 221
  recognition strategies, 9–10
  RNA folding, 7
  stabilizing forces, 1–2
folding of proteins *See* protein folding motifs
fork head domains *See* winged helix-turn-helix
Fos-Jun complexes, 62, 182
four-way DNA junctions
  See also Holliday junctions
  branch migration, 132–4
  stacked X-structures, 131–2
  structure and dynamics, 129–34
four-way RNA junctions, 134
free energy
  affinity and specificity and, 14–15, 138
  anion interactivity and, 32–3
  binding free energy, 80, 93–4, 113
  "contact free energy" values, 18–20, 26
  DNA relaxation and, 248

FRET (Fluorescence Resonance Energy Transfer) spectroscopy
  DNA-protein complexes, 182, 197, 204, 314, 349
  efficiency, 133
  four-way junctions, 131
  RNA binding, 228–9
functional groups, major and minor grooves, 5

G-segment (Gate) binding, 255–6, 260–1
Gal4 protein, 55
γδ-resolvases, 321–4
Gar1 protein, 164
GATA-1 transcription factor, 51–2
GCN4 protein, 51, 182
gel electrophoresis
  protein-DNA complexes, 180, 182, 190, 198, 201, 204
  RNA chaperones, 225
genetic recombination, 129, 139
GHKL (Gyrase/Hsp90/histidine-Kinase/MutL)-family domain, 250, 253–4, 260–1
Gin recombinase, 323–4, 325
GIY-YIG endonuclease fold, 337, 358–60
glucocorticoid receptor (GR), 57–8
glutamine as a base mimic, 96
gRNA (guide RNA)
  annealing of protein-bound, 226–22
  MRP1 and MRP2 complexes, 94, 109, 111, 164
  RNP and, 164
grooves
  helix conformation and, 4
  recognition and functional groups in, 8–9, 154
  width changes and DNA bending, 186
guanine
  amino group and site recognition, 8, 154, 316–17
  base mimickry and, 97
  base stacking effects, 188
  replacement by TGT, 376

guanosine, 7-deaza, 376
guanosine, 9-deaza, 378
guanosine, 7-methyl, 367–8
guanosine- 7-methyltransferase, 368
"GyrA-box" motif, 257
gyrase, DNA, 249–50, 254–6
  reverse gyrase, 243–4

hairpin loops, RNA, 224
hairpin ribozyme, 134
hairpins, DNA, 10, 280, 283–4, 292–3, 295–8
half-sites
  CAP interaction with, 75–80
  DNA bending, 202–3
  transcription regulator biding, 54–9
  YR core half-sites, 310, 314, 318
hammerhead ribozyme catalysis, 222–3
hAT family transposases, 274–5, 290–3
Hbb protein, 204–5
heat capacity
  anion effects on, 30–1, 33
  specific binding and, 37
  trapped water and, 27
  water release effects on, 16
helical parameters, 69–70, 83, 184–8
"helical phasing", 72
helix-hairpin-helix (HhH) motif, 249–50, 339
helix-internal loop-helix structure, 161, 163
helix-turn-helix (HTH) recognition, 48–50, 67, 198, 201
  γδ-resolvases, 321
  MuA transposases, 285
  Tc1/*mariner* transposases, 289
  type IA topoisomerases, 240
helix-two-turn-helix (H2TH) folding, 259
Hermes transposase, 290–3
Hfq protein
  RCA activity, 225–6, 229
  RNA complex, 96
  Sm-folding, 104–6
high mobility group proteins *See* HMG
Hin recombinase, 323–4, 325–6

His-Cis box class endonucleases, 347, 358
His-Me finger endonuclease fold, 336, 345–8
histidine-hydrophobic residue-histidine (HUH) motif, 294, 296, 298
histone binding to DNA, 191–4
HJ *See* Holliday junctions
HMG box proteins, 134
HMG domain, 51, 54
HMGB family, 208–12
HNF-3 (hepatocyte nuclear factor 3), 49
hnRNP K-homology (KH) proteins
  eclectic nature, 92
  KH/QUA2 domain, 98
  RNA chaperone activity, 222
  SSNA binding, 98, 106–8
Holliday junctions (HJ), 10, 129, 134, 138
  cleavage by endonucleases, 347
  formation by tyrosine recombinases, 307–9, 311–12, 314–15, 328
  resolution by topoisomerase III, 243
  resolvases, 334, 337, 346, 361–2
homing endonuclease-like fold, 336, 343–5
homing endonucleases, 347, 358, 362
homologous recombination, 10, 129, 132, 270–1, 343, 358
"Hot-spots", telomere DNA recognition, 113
HPV E2, 194–8, 212
HU/IHF protein family, 203–8, 212
human papillomavirus E2 protein, 194–8, 212
hydration influences
  binding specificity and cosolute sensitivity, 37–8
  protein-DNA interactions, 15–17
hydration layers/shells, 17, 25
hydrogen bonding
  base pair differences, 72
  CAP indirect readout and, 79–80
  cation displacement by, 30
  complex formation, 8
  direct readout and, 68
  DNA nuclease activity, 345, 350–4

groove geometry and, 4–5
propeller twists and, 186, 188
RNA 2'-hydroxyl groups and, 3
sequence recognition studies, 67
solvent layers, 28
SSNA protein binding, 96–8
hydrolysis, susceptibility of RNA, 3
"hydrophobic effect", 16, 18, 49–50
hydrophobic interactions, 8, 95–6

I-CreI nuclease, 343–5
IAS (internal activating sequence), 285, 288
IHF *See* integration host factor
immunoglobulin fold recognition, 53
Imp3p and Imp4p, 227
indirect readout, 9, 68–73
  CAP and, 79–83
  conformational change and, 177
  direct readout contrasted, 68
  rRNA and, 154
  sequence specific endonucleases, 340
  universality of, 72–3
induced fit, 10–11
inhibitors, anti-topoisomerases, 238
inosine, 369
insertion sequences (IS), 271, 275, 282, 292
integrase enzyme family, 137, 316–18
  See also site-specific recombinases
integration host factor (IHF)
  as β-ribbon motif example, 53
  binding enthalpy, 33
  DNA bending by, 178, 190, 194, 203–8, 212
  HU/IHF protein family, 203–8, 212
  Mu genome and, 285, 288
  recombinases and, 313–15
integron integrases, 316–18
interaction types, SSNA-binding proteins, 92–8
interactive anions, 29–30, 32–4
intercalation
  amino-acid side-chains, 184–5, 190, 204, 211–12, 340
  water, 81–2

internal activating sequence (IAS), 285, 288
IntI enzyme, 316
introns
  branch site sequences, 95, 97–8, 107–8, 114
  pseudouridine and, 373
  splicing, 152, 169, 222–3, 225, 228–30, 367
invertases *See* site-specific recombinases
inverted terminal repeats (ITRs), 274–6
ions
  See also anion effects; cations
  effects on specific binding, 36–9
  small, effects on protein-DNA interactions, 28–32
IS (insertion sequences), 271, 275, 282, 292
ISHp608 transposon, 296–8

junction-resolving enzymes, 134–42, 144
  See also T4 endonuclease; T7 endonuclease
  Holliday junction resolvases, 334–7, 346, 361–2

K-turn (kink-turn) RNA motif, 154, 161–3
KH folding domain, 92, 99, 106–8
KH/QUA2 domain, 98, 108, 114, 155, 163
kink-turn (K-turn) RNA motif, 154, 161–3
kinks *See* DNA bending
Kirkwood–Buff theory of solutions, 23–4
"kissing loops", RNA, 224, 228
Klenow fragments, 278, 336, 343, 349

λ-exonuclease, 339
λ-integrase family, 305–7, 309, 311–16, 318, 324–6, 328
Lac and Pur repressor proteins, 212
L7Ac protein, 163
large serine recombinases (LSRs), 326–7

linking number (Lk) *See* DNA linking number
looping-condensation of DNA by Fis, 200
Lsm proteins, 103–4, 106
Lymphoid enhancer factor (LEF-1), 54
lysine and salt bridges, 93–4

"magnetic tweezers", 198–200
major groove DNA-binding proteins, 194–203
melting *See* strand displacement
Met repressor, 53
metallo-dependent phosphatases fold, 337, 360–1
metals ions
  See also divalent metal ions
  four-way DNA junctions and, 131–2
  ribonuclease dependence on, 165
methionine wedge, 210–11
Mg^{2+}
  See also divalent metal ions
  four-way junction folding and, 131–2
  nuclease cofactor, 343, 358–9, 361
Michael addition reactions, 374–5
microRNA (miRNA), 157–8, 168
minor groove DNA-binding proteins, 203–12
mitochondrial RNA binding protein *See* MRP
Mn(II) and Mn^{2+}, 339, 344, 348–52, 357–62
mobile DNA, 10, 270–3, 316
Mobile DNA II (book), 307
modular proteins, 92
molecular recognition *See* recognition strategies
molecular ruler strategy, 171
Mos1 transposase, 290
mRNA (messenger RNA)
  editing, 164
  intron removal, 169
  TRAP complexes, 94
MRP1 and MRP2, 94, 109–11, 164
  gRNA complex, 226–7
Mu transposase, 275, 284–8

NCp7 protein, 228
NCp (viral nucleocapsid protein), 222
nebularine, 370, 372
NFAT (nuclear factor of activated T cell) transcription factors, 62
Nhp6A protein, 209–11
nicked DNA, 145, 178–80, 235, 237–8, 247–8, 294
non-helical regions, rRNA, 153–4
noncanonical forms, 6
  recognition strategies, 10
  restriction endonucleases, 338
  RNA, 161, 163, 166
nonspecific binding
  electrostatics in, 93
  protein complexes, 9, 14–15, 29, 35–7
  RNA, 163, 166
Nop10 protein, 163
nuclear magnetic resonance (NMR) spectroscopy
  DNA bending and, 183, 190
  protein-RNA complexes, 151–2
nuclear receptors, 57–9
nucleases *See* DNA nucleases
nucleosides
  modified, in RNA, 4, 367
  steric discrimination, 96
nucleosomes, 177–9, 185–6, 189–94, 197
nucleotide analog interference modification (NAIM) studies, 161
nucleotide looping, 114
"Nucleotide Shuffling", TEBP-α/β, 113–14
nucleotides, modified, in RNA, 4, 367

OB-fold (oligonucleotide/oligosaccharide-binding), 98–103
  See also cold-shock domains
  TEBP related proteins, 113, 115–17
Okazaki fragments, 349
oligonucleotide/oligosaccharide-binding *See* OB-fold
open reading frames (ORFs), 271, 296, 367
osmolytes *See* cosolutes
"osmotic stress" analysis, 20, 24

Subject Index

P1 nuclease fold, 337, 352–4
p53 tumor suppressor, 53
papillomavirus E2 protein, 194–8, 212
PAZ domains, 157–9
PDB (protein database) codes
  CAP half-complex, 75, 79
  DNA nucleases, 335–8, 341–2, 344, 346, 348–51, 354, 355–61
  ribosomal proteins, 155–6, 158, 163, 167
  site-specific recombinases, 306, 311–12, 322
  topoisomerases, 239, 241–2, 247, 251, 257, 260
persistence length, 178
phage *See* bacteriophages
pharmaceuticals, 238, 252
phase variation, 323
phenylalanine wedge, 211–12
phosphates, cation binding, 30
phosphodiesters
  helix rigidity and, 179–80
  hydrolysis, 3, 141–2
  topoisomerase attacks by tyrosine, 235
  transposase cleavage mechanisms, 276
phospholipase C/P1 nuclease fold, 337, 352–4
phospholipase D/nuclease fold, 337, 354–5
phosphorus atom separation, A- and B-helixes, 6
phosphotyrosine and phosphoserine intermediates, 293–8, 304
PIWI domains, 158–9
poisons and inhibitors, 238
  See also toxins
polar amino acids, SSNA and, 96
polar surfaces, water release, 16
polymorphism
  homing endonuclease specificity and, 345
  K-turn RNA, 163
  RNA junctions, 134
  sequence-dependent, B-form DNA, 69–72
positive heterotypic cooperativity, 117

positive roll deformations (kinks), 81–4
post-transcriptional gene silencing (PTGS) *See* RNA interference
Pot1 (protection of telomere-1) protein, 96, 102–3, 116, 119
pre-mRNA splicing, 4
"presentation platforms", 165
proline intercalation, 204, 207
propeller twists, 69, 72, 186–90
protein-DNA complexes
  See also DNA bending
  anion effects, 33–4
  examples of DNA- bending proteins, 191–212
  histone binding, 191–4
  major groove binding proteins, 194–203
  minor groove binding proteins, 203–12
  single DNA molecule studies, 207–8, 211–12, 248, 328
  topoisomerases, 246
  transcription regulators, 47–63
  transposases, 274–81
protein folding motifs
  See also DNA nucleases
  classifying DNA nucleases, 333–4
  distinguishing between topoisomerases, 244
  junction-resolving enzymes, 140
  KH domain, 106–8
  OB-fold, 98–103
  Puf domain, 108
  RRM, 106
  Sm-fold, 103–6
  TOPRIM folds, 239–40
  TRAP domain, 108–9
  Whirly protein fold, 109–10
protein-induced base-pairing, 374
protein-protein interaction
  Fis dimers, 201
  MuA and MuB, 286
  salt concentration and, 29–30, 33
  SSNA binding, 117, 119
protein-RNA complexes, 150–2
  interactions in ribonucleases, 165–70

interactions in ribonucleoprotein
    particles, 152–65
  RCA and binding affinity, 228–31
  tertiary interactions, 154–5
proteins
  as allosteric effectors, 119–20
  classes involved in direct and
    indirect readout, 68
  classes involved in DNA bending,
    191–212
  cosolute accessibility, 26
  DNA distortion and, 18, 80, 85
  folding and SSNA binding, 98–110
  recognition strategies, 7–11
  with RNA chaperone activity, 222–30
protonation, binding-coupled, 31, 34–5, 37
protonation, leaving groups, 351, 355,
    358, 361
ψ–synthase
  pseudouridine synthase, R1uA, 371,
    372–6
  pseudouridine synthase, TruA, 375
  pseudouridine synthase, TruB, 369,
    374–5, 377, 379
PUF-8, 112
Puf (Pumilio/FBF homology) folding
    domains
  mRNA complex, 94, 97–8
  protein folding motifs, 108–9
  specific and adaptable recognition,
    112–13
purine-purine (RR) sequences, 187, 189–90
purine-pyrimidine (RY) sequences,
    71, 73, 190–1
purines See adenine; guanine
pyrimidine-purine (YR) sequences
  in base-pair steps, 72–3, 186, 189
  in DNA-protein complexes, 193,
    203, 206, 210
pyrimidine-pyrimidine (YY) sequences,
    189–90
pyrimidines
  See also cytosine; thymine; uracil
  distinguishing DNA and RNA, 3

queuosine, 368–9, 376

RCA See RNA chaperones
RecJ exonuclease, 357, 360–1
recognition elements, 48–54
recognition strategies, 7–11
  by CAP, 73–85
  DNA transposases, 274–81
  duplex DNA, 8–9
  history, 66–8
  junction-resolving enzymes, 137–9
  noncanonical DNA, 10
  ribonucleases, 165
  RNA, 369–70, 374, 376–8
  sequence specific endonucleases, 340
  single stranded nucleic acids, 9, 93–8
  specificity, adaptability and
    degeneracy, 111–15
  structural basis of, 47–63
  topoisomerases, 246, 254
recombinase binding elements
    (RBEs), 307–8, 319, 324, 326–7
recombinases See site-specific
    recombinases
RecQ-family helicases, 243
register slippage, 114
replication protein A See RPA
replicative transposition, 277, 279,
    281, 285
residence times, retained water, 26
resolvases
  Holliday junctions resolvases, 334–
    337, 346, 361–2
  telomere resolvases, 318
  Tn3 and γδ-resolvases, 321–4
restriction endonucleases, Type II,
    334, 338–9
restriction endonuclease-like folds,
    334–5, 337–41
restriction enzymes, 137, 140–2, 362
restrictocin, 168–70
retained water
  affinity and specificity and, 15, 26–7
  compaction and, 38
  kinked protein-DNA complexes
    and, 82
  thermodynamic effects of, 27–8
retinoid X receptor (RXR), 58–9

Subject Index 393

reverse gyrase, 243–4
ribonuclease H-like fold, 276, 288, 292, 335–6
ribonucleases
  protein-RNA interactions in, 165–70
  restrictocin, 168–70
  RNA splicing, 169–70
  RNase E, 165–6
  RNase II, 166–7
  RNase III, 167–8
  tRNase Z, 170
ribonucleoprotein particles
  functions, 152
  protein-RNA interactions in, 152–65
  small ribonucleoprotein particles, 160–1
ribose and deoxyribose, 3–4
ribosomes, 153–6, 370
ribothymidine, 368
RISC (RNA-induced silencing complex), 157–9, 226
RNA
  See also gRNA; mRNA; protein-RNA complexes; rRNA; tRNA
  annealing and strand displacement by RCA proteins, 222–3
  chemical differences from DNA, 3–4
  conformation of duplex, 4
  primary structure and numbering, 2
  secondary, tertiary and higher structures, 7, 150, 154, 160, 221, 376
  structural classes, 150
  susceptibility to cleavage, 3
RNA chaperones (RCA), 11, 91, 151
  binding affinity reciprocal relation, 228–9
  disordered proteins and, 230
  modes of action, 223–8
  proteins with RCA activity, 222–3
  in vivo and *in vitro* assays, 224–5
  website comparing, 222
RNA editing complexes, 164–5
RNA-induced silencing complex (RISC), 157–9, 226
RNA interference (RNAi), 156–9, 273

RNA modifying enzymes, 368–79
  Modomics database, 368
  scope of, 367–8
RNA polymerase holoenzyme (RNAP), 74
RNA splicing ribonucleases, 169–70
RNase H-like fold, 276, 288, 292, 335–6
RNP (ribonucleoprotein) motif *See* RRM
RNPs (ribonucleoprotein particles), 151–5, 161, 163–4
  hnRNP (heterogeneous nuclear), 92, 98, 106, 108, 222
  RNP1 and RNP2, 106, 227
  snRNPs (small nucleolar), 103, 108, 160–4
Ro60 protein, 222, 226
roll and tilt parameters, 184–5
roll angles of protein-DNA complexes, 193, 196, 202–3, 209–10
Rolling Circle Replication (RCR) initiators, 294
RPA (replication protein A), 95, 101–3, 116
RR and RY steps *See* purine-purine sequences; purine-pyrimidine sequences
RRM (RNA recognition motif), 106–7, 153, 156
rRNA methyltransferase RumA, 379
rRNA (ribosomal RNA), 154, 161, 227
RuvAB complex of *E. coli,* 134
RuvC, 135, 140

S1 domain, 166, 227
S1 ribosomal protein, 222, 227
S (serine) transposases, 276, 293–4
salt bridges, SSNA-binding proteins, 93–4
salt concentration
  binding specificity and, 38–9
  effects on binding, 28–32
  effects on four-way junctions, 132
SAM domain/PIN domain-like fold, 336–7, 348–9

SAP-1 binding, 60
sarcin/ricin loop (SRL), 167–8
second strand cleavage, transposase, 280
secondary structures
  in nucleic acids and proteins, 2
  RNA misfolding potential, 221
sequence recognition *See* recognition strategies
sequence specificity
  endonucleases, 340
  junction-resolving enzymes, 138
  modular and eclectic proteins, 92
serine (S) transposases, 276, 293–4
serine (SR) recombinase family, 304, 318–27
  Hin and Gin recombinases, 323–4
  large serine recombinases, 326–7
  regulation by accessory sites, 324–6
  subunit rotation model, 323
  Tn3 and γδ-resolvases, 321–3
serum response factor (SRF), 60–1
SF1 *See* splicing factor
Shapiro intermediates, 277
signal recognition particle (SRP), 152, 155, 159–60, 163
Sin recombinase, 305, 326
single-stranded nucleic acids *See* SSNAs
single-stranded RNA
  allostery, 118
  multiple cooperativity, 115–16
  protein interactions with, 150–1, 166
single-stranded transposon circles, 295
siRNA and miRNA, 157–9, 168, 226
site-specific recombinases.
  See also serine recombinase family; tyrosine recombinase family
  classification and functions, 303–5
  Mobile DNA II and, 307
  structures and PDB codes, 306
Sleeping Beauty transposon, 274, 290
"sliding", 29
Sm-folding motif, 103–6
Sm1 proteins, 105
small interference RNA (siRNA), 157

small molecules
  allosteric effectors, 118–19
  topoisomerase inihibitors, 238
  water release models, 16, 18, 20
small nuclear ribonucleoprotein (snRNP) biogenesis, 103–4
small nucleolar RNA (snoRNA), 227
small subunit processome (SSUP), 227
Sox2 protein, 209–10
specificity
  See also site-specific recombinases
  affinity distinguished from, 14–15
  design requirements for non-specific DNA, 15
  HMGB domains, 209
  RCA activity and, 229
  retained water and, 26–7
  water and ion effects on, 36–9
spliceosomes, 153, 373
splicing factor CBP2, 230
splicing factor SF1, 95, 98, 104, 114–15
splicing factor U1A, 156
SR *See* serine recombinase family
SSB (single stranded DNA-binding protein)
  anion effects, 39
  binding free energy, 91
  binding mechanism, 93–5, 100–3, 115–16, 118
  enthalpy effects, 32–3
SSNA-binding proteins
  cooperative binding, 115–17
  emergent properties, 110–20
  interaction types, 93–8
  modular and eclectic proteins, 92
  preferential affinity and annealing, 226
  protein folding motifs, 98–110
  RNA chaperones as, 223
SSNAs (single-stranded nucleic acids)
  occurrence, 91
  protein folding and function, 98–110
  recognition strategies, 9, 10
  ssRNA, 115–16, 118, 150–1, 166
stacked X-structures, 131–2
stacking interactions, SSNA binding, 95

Subject Index

step parameters, 69–70
steric complementarity, 107
steric effects
　See also allosteric effects
　constraining RNA to the A-form, 4
　DNA bending, Fis protein, 201
　DNA topo strand passage, 248
　minor groove binding, 352
steric exclusion, 20–1, 23, 30, 158
steric packing, SSNA interactions, 93, 95–6, 109
steric repulsion, 20, 23, 72, 188
StpA protein, 222, 225–6, 228–30
strand displacement, 222–3, 227–8
strand exchange, DNA, 307–9, 311, 315–17, 319–20, 323–6
strand passage, DNA, 240, 243, 253, 255, 258, 260–1
sugars and nucleaic acid geometry, 3–4, 69
suicide DNA, 246
supercoiled DNA, 6
　positive and negative supercoils illustrated, 235, 237
　serine recombinases, 319, 324, 327
　Type IA topoisomerases, 238, 240, 243
　Type IB topoisomerases, 244, 248–9
　Type IIA topoisomerases, 250, 252, 258
superfamily-2 (SF2) helicase, 243
superhelixes in nucleosomes, 191
"swivelases", 248–9
Sxl (sex-lethal protein), 106

T4 endonuclease VII, 139, 143
T-segment (Transport) binding, 255–6, 260–1
T7 endonuclease I, 137–44
T7 endonuclease VII, 137
T-stem-loop (TSL) RNA, 374, 377
target site duplications (TSD), 280
TATA-box binding protein (TBP), 26, 53, 190, 212
Tc1/*mariner* transposase family, 274–5, 288–90

TEBPs (telomere end-binding proteins)
　See also Cdc13; Pot1
　as an allosteric effector, 119
　cooperative binding and, 115
　cooperativity and anti-cooperativity, 116–17
　eclectic nature, 92
　electrostatics and, 94
　hydrogen boding and, 97
　hydrophobic interactions, 95–6, 98
　"nucleotide shuffling", 113–14
　OB-folding and, 101–3
telomere resolvases, 318
terminal inverted repeats (TIRs), 274–6
ternary DNA complexes, 59–61, 75
tertiary structures
　in nucleic acids, 2, 7
　resolvases, 322
　RNA, 150, 154, 160, 221, 228, 376
　topoisomerases, 235, 254
TFIIIA, 52
thermodynamic effects
　energetic components and specificity, 36–7
　environmental variable effects on binding, 14–15
　retained water, 27–8
thermodynamic nonideality factor, 20–2
thymidine, 96
　ribothymidine, 368
thymine
　CAP direct readout and, 77
　as a DNA characteristic, 2–3
　unpaired extrahelical, 284, 298
tilt parameter, 185
TIM beta/alpha barrel fold, 337, 355–7
Tn3 resolvases, 321–3
Tn3 transposase, 289
TN5 transposase, 273, 280, 282–4
Tn7 transposon, 273
Tn10 transposon, 271, 275, 280
TnpA transposase, 296–8
topoisomerases
　introduced, 234–8
　possible Type IC, 249
　small molecule agents, 238

subclasses, 235–6
topoisomerase III, 243
topoisomerase IV, 257–8
topoisomerase V, 249
topoisomerase VI, 259–60
Type IA, 235–6, 238–44
Type IB, 235–6, 244–9
Type IIA, 235–6, 249–59
Type IIB, 235–6, 259–61
"topological filters", 316, 324
TOPRIM (TOPoisomerase/PRIMase) folds, 239–40, 250–1, 253–6, 259–60
TOPS topology diagrams, 334
torsion angles, 2
torsional strain See supercoiled DNA
toxins, 167–8, 252, 347, 351
transcription regulators
  binding as dimers to half-sites, 54–9
  binding core DNA elements, 48–54
  binding motif combinations, 59–62
  classes of, 48
transglycosylation, 376–9
transposase binding site (TBS), 275–6, 290, 292
transposases
  classification and nomenclature, 271–4
  DDD and DDE transposases, 276–8, 290–3
  DNA recognition by, 274–81
  forming covalent intermediates, 293–8
  hAT family transposases, 290–3
  Mu transposase, 284–8
  P element transposase, 275–6
  serine (S) transposases, 276, 293–4
  Tc1/*mariner* transposases, 288–90
  TN5 transposase, 282–4
  tyrosine (Y) transposases, 276, 293–8
transposons
  Class I and Class II, 270
  conjugative transposons, 318
  ends of, 274–6
  ISHp608 transposon, 296–8
  silencing, 273
  simple and complex, 271
transpososomes, 281, 286–8, 290

TRAP (tryptophan and RNA-binding attenuation protein)
  allostery, 118
  folding domain, 108–10
  mRNA complex, 94
TREX exonucleases, 336, 341–2
tRNA adenosine deaminase TadA, 369–72
tRNA archaeosine transglycosylase ArcTGT, 376–9
tRNA guanyltransferase, 368
tRNA pseudouridine synthase R1uA, 372–6
tRNA (transfer RNA)
  anti-codon loops, 67, 95
  λ-form, 377–8
  maturation, 104
  RNA chaperones, 228–30
  RNA modification, 367–70, 372–4, 376–9
  RNA-protein interactions, 152, 164–5, 167–70
tRNase Z, 170
Trp repressor, 48–9, 51
tryptophan, HTH binding, 49
tryptophan and RNA-binding attenuation protein See TRAP
TTR motif, 205–6
twist angle, 185–6
Type IA topoisomerases and paralogues, 235–6, 238–44
Type IB topoisomerases, 235–6, 244–9
  tyrosine recombinase transfers and, 309
Type IC topoisomerases, 249
Type II restriction endonucleases, 334, 338–9
Type IIA topoisomerases and paralogues, 235–6, 249–59
Type IIB topoisomerases, 235–6, 259–61
tyrosine
  ETS family complexes, 60–1
  phosphodiester attacks by topoisomerase, 235
  phosphodiester attacks by transposases, 276

tyrosine recombinase (YR) family, 304, 307–18
  conjugative transposons, 318
  Cre and Flp recombinases, 309–13
  four-way junctions and, 129
  integron integrases, 316–18
  λ-integrase and XerCD recombinases, 313–16
  telomere resolvases, 318
  Type IB topoisomerases and, 244–5, 309
tyrosine (Y) transposases, 276, 293–8
tyrosyl DNA phosphodiesterase (Tdp1), 238

U2AF65 splicing factor, 115
unstacked X-structures, 138, 144
uracil, as characteristic of RNA, 3
uridine, 96, 369
  2-thio, 368–9
  dihydro, 367–8

van der Waals contacts *See* steric effects
V(D)J recombination, 274, 276, 280, 292
vibrational restrictions, 17
viral nucleocapsid proteins (NCp), 222
viral topoisomerases, 244–6, 256

water
  See also hydration; retained water
  adenosine deaminase activity, 370
  effects on specific binding, 23–4, 36–9
  intercalation, 81–2
  nucleophilic attack by, 340, 351, 354, 359
  phosphodiester hydrolysis, 141, 276
water-mediated protein-DNA interactions, 26–7, 49
water release
  computational/empirical methods, 18–19, 25, 40
  in protein-DNA association, 17–26
  surface polarity, enthalpy and heat capacity effects, 16

Watson-Crick base pairs
  deviations from, 67, 153, 221
  helix conformation and, 4–5, 7–10, 67
  helix rigidity and, 179–80, 186–7
  single stranded nucleic acids and, 96, 98
Werner exonuclease, 341
Whirly protein folding, 99, 108–10, 164
winged helix domains (WHDs), 194, 240, 254–5, 259
winged helix-turn-helix (wHTH) recognition, 48–50, 60, 74, 202
wybutosine, 367–8
wyosine, 368–9

X-ray crystallography *See* crystallography
X-structures *See* stacked X; unstacked X-structures
XerCD recombinases, 305–6, 311, 313–16
Xis protein (phage λ Xis), 194–5, 212, 313–15, 327
XPF nuclease, 335, 339

Y (tyrosine) transposases, 276, 293–4
  Y1 transposases, 296–8
  Y2 transposases, 294–6
Ydc2 enzyme, 136–8
YR *See* tyrosine recombinase family
YR and YY steps *See* pyrimidine-purine and pyrimidine-pyrimidine
YREK motifs, 282, 284, 290

Z-form DNA, 69
zinc-binding domains, 51–2, 57–9, 242, 292
zinc-bound water, 370
zinc finger motif, 51–2, 108, 228, 345
zinc ions
  cysteine stabilisation, 136–7
  DNA-nucleases and, 354–7
  in nuclear receptors, 57
zinc-knuckle motif, 108
Zn_2Cys_6 binuclear clusters, 54–6